Smart Nanomaterials Technology

Series Editors

Azamal Husen, Wolaita Sodo University, Wolaita, Ethiopia

Mohammad Jawaid, Department of Chemical and Petroleum Engineering, United Arab Emirates University, Al Ain, United Arab Emirates

Nanotechnology is a rapidly growing scientific field and has attracted a great interest over the last few years because of its abundant applications in different fields like biology, physics and chemistry. This science deals with the production of minute particles called nanomaterials having dimensions between 1 and 100 nm which may serve as building blocks for various physical and biological systems. On the other hand, there is the class of smart materials where the material that can stimuli by external factors and results a new kind of functional properties. The combination of these two classes forms a new class of smart nanomaterials, which produces unique functional material properties and a great opportunity to larger span of application. Smart nanomaterials have been employed by researchers to use it effectively in agricultural production, soil improvement, disease management, energy and environment, medical science, pharmaceuticals, engineering, food, animal husbandry and forestry sectors.

This book series in Smart Nanomaterials Technology aims to comprehensively cover topics in the fabrication, synthesis and application of these materials for applications in the following fields:

- Energy Systems—Renewable energy, energy storage (supercapacitors and electrochemical cells), hydrogen storage, photocatalytic water splitting for hydrogen production
- Biomedical—controlled release of drugs, treatment of various diseases, biosensors,
- Agricultural—agricultural production, soil improvement, disease management, animal feed, egg, milk and meat production/processing,
- Forestry—wood preservation, protection, disease management
- Environment—wastewater treatment, separation of hazardous contaminants from wastewater, indoor air filters.

Nadeesh M. Adassooriya · Aruna Manipura ·
Azamal Husen
Editors

Industrial Scale Production of Nanoparticles

Springer

Editors
Nadeesh M. Adassooriya
Department of Chemical and Process Engineering
University of Peradeniya
Peradeniya, Sri Lanka

Aruna Manipura
Department of Chemical and Process Engineering
University of Peradeniya
Kandy, Sri Lanka

Azamal Husen
Wolaita Sodo University
Wolaita, Ethiopia

Department of Biotechnology
Graphic Era (Deemed to be University)
Dehradun, Uttarakhand, India

ISSN 3004-8273 ISSN 3004-8281 (electronic)
Smart Nanomaterials Technology
ISBN 978-981-96-9175-3 ISBN 978-981-96-9176-0 (eBook)
https://doi.org/10.1007/978-981-96-9176-0

This Springer imprint is published by the registered company Springer Nature Singapore Pte Ltd.
The registered company address is: 152 Beach Road, #21-01/04 Gateway East, Singapore 189721, Singapore

Preface

This book provides a detailed review of scaled up production of nanoparticles. A recent estimation suggests that the global nanotechnology business is a Trillian dollar business with exponential rising of nanotechnology-based industries worldwide. In a such scenario nanoparticle production plays a vital role. Thus, this book describes the main top-down and bottom-up approaches used for nanoparticle synthesis, mechanism of nanoparticle formation, their key operating conditions, opportunities, and challenges associated with synthesizing processes.

There are 20 chapters which discuss about various aspects of nanomaterial synthesis techniques, applications and industrial scale production covering both top-down and bottom-up approaches. The first chapter "Industrial Scale-Up Production of Nanoparticles—Challenges and Opportunities" gives an overview about the book while chapters from "Nanocrystalisation for Scaled Up Manufacturing of Nanoparticles"–"Microbial Factories for Scaled-Up Production of Nanoparticles" are focusing on the synthesis techniques, chapters "Product Separation, Purification, Storage and Packaging of Nanoparticles"–"Health and Safety Aspects of Nanoparticle Production at Industrial Scale: An Overview" are concerned about product separation and purification, quality assurance, process economics, life cycle assessment and finally health and safety aspects of nanoparticle production at industrial scale. The second chapter "Nanocrystalisation for Scaled Up Manufacturing of Nanoparticles" discussed about nanocrystalisation for scaled up manufacturing of nanoparticles and challenges of batch vs. continuous crystallization. The third chapter "Spray Drying Techniques for Nanoparticle Production" presents spray drying techniques for nanoparticle production, applications and challenges and opportunities in spray drying as nanomaterial synthesis technique. The fourth chapter "Mechanical Milling for Scaled Up Manufacturing of Nanoparticles." focuses on mechanical milling for scaled up manufacturing of nanoparticles. The fifth chapter "Emulsion Techniques for Nanoparticle Production" presents emulsion techniques for nanoparticle production for liquid phase nanoparticle synthesis. The sixth chapter "Vapor-Phase Production of Nanomaterials" looked at vapor-phase production of nanomaterials focusing on techniques such as chemical vapor deposition (CVD), physical vapor deposition (PVD), and vapor-liquid-solid (VLS) methods for manufacturing various nanostructures. Microreactors for nanoparticles production are discussed in the seventh chapter "Microreactors for Nanoparticles Production". The eighth chapter "High Pressure Homogenization Techniques for the Production of Nanomaterials" presents high pressure homogenization techniques to produce nanomaterials whereas the ninth chapter "Extrusion for Nanomaterial Production" is dedicated for extrusion for nanomaterial production.

Further, the tenth chapter "Solgel and Solvothermal Methods for Nanoparticles Productions" presented solgel and solvothermal methods for nanoparticles productions. Supercritical fluid technologies for nanoparticle production are presented in the eleventh chapter "Supercritical Fluid Technologies for Nanoparticle Production" highlighting supercritical technologie's role in clean nanoparticle production. The twelfth chapter "Plasma and Laser Technology for Nanoparticles Production" discusses plasma and laser technology for nanoparticles production. The thirteenth chapter "Sonochemical Methods for Nanoparticle Production" presents an overview of sonochemical methods for energy efficient nanoparticle production. The fourteenth chapter "Nanolithographic Techniques for Scaled-Up Manufacturing of Nanostructures" presents nanolithographic techniques for scaled-up manufacturing of nanostructures. The focus of the fifteenth chapter "Microbial Factories for Scaled-Up Production of Nanoparticles" is microbial factories for scaled-up production of nanoparticles. The sixteenth chapter "Product Separation, Purification, Storage and Packaging of Nanoparticles" provides an overview of product separation, purification, storage and packaging of nanoparticles. Quality assurance in nanoparticles production is discussed in the seventeenth chapter "Quality Assurance in Nanoparticles Production". The eighteenth chapter "Process Economics of Nanomaterials Production—A Decision Analysis Framework for Assessing Technical, Business and Environmental Viability" presents a process economics of nanomaterials production with a decision analysis framework for assessing technical, business, and environmental viability. Life cycle assessment (LCA) of nanomaterials production is discussed in the nineteenth chapter "Life Cycle Assessment of Nanomaterials Production". The twentieth chapter "Health and Safety Aspects of Nanoparticle Production at Industrial Scale: An Overview" reviewed health and safety aspects of nanoparticle production at industrial scale.

Overall, this book provides an insight for practitioners and researchers during the selection of nanoparticle production technologies in industrial scale nanoparticle manufacturing for various applications. In addition, this book serves as a valuable resource for advanced students, faculty, scientists, and researchers. We are deeply grateful to all the contributors for accepting my invitation, sharing their expertise in specialized areas of research, and incorporating suggestions to enhance their contributions.

Peradeniya, Sri Lanka — Nadeesh M. Adassooriya
Peradeniya, Sri Lanka — Aruna Manipura
Wolaita, Ethiopia — Azamal Husen

Contents

About the Editors

Prof. Nadeesh M. Adassooriya is currently attached to the Department of Chemical and Process Engineering, Faculty of Engineering, University of Peradeniya, Sri Lanka. He holds a B.Sc. Honors University of Sri Jayewardenepura, M.Sc. in chemical and process engineering at University of Moratuwa, and a Ph.D. in engineering in the field of energy materials from University of Cambridge, UK. In 2022, he completed a research fellowship at Massachusetts Institute Technology (MIT), USA, in the field of nanocrystals. He has authored/co-authored over 100 research publications, which were presented at well-reputed journals and international/national conferences, and his publications have been cited over 2000 times with H-index of 22 to date. He holds two US patents to his name on nanofertlizers, and one of them holds the record of first nanotechnology-based US patent from Sri Lanka. He has contributed to over 30 book chapters in nanotechnology-related areas published by leading international publishers.

Dr. Aruna Manipura was a senior lecturer attached to the Department of Chemical and Process Engineering, Faculty of Engineering, University of Peradeniya, Sri Lanka, and currently he works as an independent chemical and environmental engineering consultant. He obtained B.Sc. Eng. in chemical engineering from University of Peradeniya, Sri Lanka, and M.Sc. from University of Moratuwa, Sri Lanka. He completed his Ph.D. in biotechnology from Rhodes University, South Africa, in 2007. He worked as a research associate for University of Manchester, UK, and University of Newcastle Upon Tyne, UK, until 2013. He has contributed to over 50 publications and has contributed to numerous industry projects as a chemical engineering consultant.

Prof. Azamal Husen is a distinguished scholar with over two decades of expertise in nanobiotechnology, plant physiology,environmental botany, medicinal plants, and clonal propagation. He has held senior academic roles, including Full Professor and Head of the Department of Biology at the University of Gondar, Ethiopia, and currently serves as a Foreign Delegate at Wolaita Sodo University, Ethiopia, and Adjunct Professor at Graphic Era University, India. He has also served as a Professor at Sankalchand Patel University and Visiting Faculty at leading Indian institutions. Prof. Husen has led multiple nationally and internationally funded research projects, contributed extensively to curriculum development, trained academic professionals worldwide, delivered various lecture and has been also honored by the University of Gondar for outstanding service. Named among the World's Top 2% Scientists by Stanford University, he is a prolific author, editor, and contributor to several international journals and book series.

Industrial Scale-Up Production of Nanoparticles—Challenges and Opportunities

Nadeesh M. Adassooriya, Aruna Manipura, and Azamal Husen

Abstract Since the "Gold rush" type enthusiasm among scientists and engineers around the world on nanoparticles have developed many potential applications due to significant performance improvements in many artefacts by changing the shape and the size of the matter. The global market for nanomaterials was valued at USD 12.42 billion in 2023 and USD 14.34 billion in 2024 with over 15% of compound annual growth rate (CAGR) of the sector. Staggering number of scientific publications produced each year by these communities is an unformidable task for any new comer to the field in grasping the knowledge created about the nanoparticles. Therefore, key approaches of nanoparticles synthesis and potential industrial production aspects may assist enthusiasts of nanoparticles to embark on their journey for further enhancement of the field and commercial exploitation of nanoparticle manufacturing industry. Thus, this book presents the main top-down and bottom-up approaches used for nanoparticle synthesis, mechanism of nanoparticle formation, their key operating conditions, opportunities, and challenges associated with synthesizing processes. It is expected to ignite the minds of budding nanomaterial scientists and engineers, business communities and tech entrepreneurs to develop next level of emerging nanoparticles by focusing saving energy, natural resources and reducing the impact on global environment via greener and safer processes at industrial scale.

Keywords Nanoparticles · Synthesis methods · Industrial scale production

N. M. Adassooriya (✉) · A. Manipura
Department of Chemical and Process Engineering, Faculty of Engineering, University of Peradeniya, Peradeniya, Sri Lanka
e-mail: nadeeshm@eng.pdn.ac.lk

A. Husen
Wolaita Sodo University, Wolaita, Ethiopia

Department of Biotechnology, Graphic Era (Deemed to be University), Dehradun, India

N. M. Adassooriya et al. (eds.), *Industrial Scale Production of Nanoparticles*, Smart Nanomaterials Technology, https://doi.org/10.1007/978-981-96-9176-0_1

1 Introduction

Nanoparticles play an important role in many industrial applications due to their unique properties arising from their sizes and shapes. Though the name nano came to exist in scientific literature in the twentieth century, nanomaterials have been used by humans since the fourth century AD, by the Roman [1]. John Utynam patented gold nanoparticle-based glass in 1449 and in the sixteenth century, Theophrastus von Hohenheim, a Swiss doctor better known as Paracelsus, used gold nanoparticles to treat patients suffering from various ailments [2]. The American physicist and Nobel Prize laureate Richard Feynman introduced the concept of nanotechnology in 1959 during annual meeting of the American Physical Society, a lecture entitled "There's Plenty of Room at the Bottom" at the California Institute of Technology [1]. Nano is referred to a Greek prefix meaning 'dwarf' or something very small and depicts one thousand millionth of a meter (10^{-9} m). Nanoscience is the study of structures and molecules on the scales of nanometers ranging between 1 and 100 nm, and the technology that utilizes it in practical applications such as devices etc. is called nanotechnology [1].

The global market for nanomaterials was valued at USD 12.42 billion in 2023 and USD 14.34 billion in 2024, assuming a steady growth rate [3]. These estimations are greatly varying due to unavailability of true global demand, type of nanomaterials, uncertainties associated with emerging applications, inability to access internal market information of various companies, and different estimation techniques used by different market research organizations. Phenomenal development of instrumentation and analytical equipment for materials characterization at comparatively lower cost has contributed our understanding of shape, size and functionality of various nanomaterials during the last 50 years. In the backdrop challenges related to climate change and responsible use of limited natural resources and growing global population have led many scientists and engineers to explore possibilities of use of nanomaterials in many sectors such as energy, food, water, health, and shelter as basic needs of all humans. Further, emerging and established manufacturing technologies (e.g. use of artificial intelligence, internet of things (IoT), process intensification, modular processing plants) have opened up many opportunities to combine these technologies and nanomaterials synthesis. Various formulations of nanomaterials in consumer products at lower costs and environmental footprint are not only a necessity, but also it helps to break the barriers of scale of economies that usually face by chemical manufacturing industry via increasing the demand for such nanomaterials.

2 Overall Objectives of the Book and the Target Audience

There are many excellent books written about nanomaterials synthesis and applications. Thus, this book is not intended to replace those, but to give an overview nanomaterials synthesis, applications and specially to highlight the opportunities and challenges associated with industrial scale. Our target audience is advanced undergraduate students, early graduate students, and anyone who is interested to learn about nanomaterial synthesis, applications and scale up challenges at industrial scale production.

- Overview of main techniques of nanoparticles synthesis,
- Working principles and mechanisms behind those methods,
- Various applications and selected case studies, and,
- Opportunities and challenges associated with those techniques for industrial scale nanoparticle production.

3 Organization of the Book

There are 20 chapters which discuss about various aspects of nanomaterial synthesis techniques, applications and industrial scale production covering both top-down and bottom-up approaches. The book addresses main techniques though it is not an exhaustive list, which is sufficient for someone who is now to the subject and develop the appetite for the field. While Chapters "Nanocrystalisation for Scaled Up Manufacturing of Nanoparticles" to "Microbial Factories for Scaled-Up Production of Nanoparticles" are focusing on the synthesis techniques, Chapters "Product Separation, Purification, Storage and Packaging of Nanoparticles" to "Health and Safety Aspects of Nanoparticle Production at Industrial Scale: An Overview" are focused on product separation and purification, quality assurance, process economics, life cycle assessment and finally health and safety aspects of nanoparticle production at industrial scale. The books can be read at any sequence based on the need to the reader as each chapter does not depend on the other. A short summary of each chapter is given in the following section.

Chapter "Nanocrystalisation for Scaled Up Manufacturing of Nanoparticles" discussed about nanocrystalisation for scaled up manufacturing of nanoparticles focusing applications and science behind crystallization, top-down and bottom approaches of nanocrystalisation and advantages and challenges of batch versus continuous crystallization. Chapter "Spray Drying Techniques for Nanoparticle Production" presents spray drying techniques for nanoparticle production and explains how spray drying deals with metastability related to nanomaterials, which, in most cases, is driven through their high surface-to-volume ratio into tendencies like aggregation or structural changes when the systems seek to reduce surface energy. Authors discuss number of applications and challenges and opportunities in spray drying as nanomaterial synthesis technique. Chapter "Mechanical Milling for Scaled Up Manufacturing of Nanoparticles" focuses on mechanical milling for scaled up manufacturing of nanoparticles. A detailed analysis of as a top-down approach, a case study of scaled up synthesis of nanoparticles and challenges for the scaled-up production on nanomaterial by mechanical milling are presented. Chapter "Emulsion Techniques for Nanoparticle Production" presents emulsion techniques for nanoparticle production. Emulsion technique is popular especially in liquid phase nanoparticle synthesis. Depending on the energy requirement, nanoemulsion techniques can be categorized under high-energy and low-energy techniques. Chapter "Vapor-Phase Production of Nanomaterials" looked at vapor-phase production of nanomaterials. Mainly, vapor-phase nanomaterial synthesis technique is applied to produce high-quality nanomaterials to achieve above benefits. For example, vapor-phase technique like chemical vapor deposition (CVD), physical vapor deposition (PVD), and vapor–liquid–solid (VLS) methods-provide versatile solutions for manufacturing various nanostructures to achieved the growing demand for nanomaterials. Microreactors for nanoparticles production is presented in Chapter "Microreactors for Nanoparticles Production". This technique provides precise control over reaction parameters, product quality, safety, and overall reaction efficiency. Microreactor-based production methods of various types of nanoparticles, including metallic, semiconductor, magnetic, organic, and hybrid, with varying sizes, shapes, crystalline structures, and surface chemistries are also discussed. Chapter "High Pressure Homogenization Techniques for the Production of Nanomaterials" presents high pressure homogenization techniques to produce nanomaterials. High pressure homogenization is a valuable top-down technique for both the small scale and the industrial production of nanoparticles and nanovesicles. A series of examples and discussions are brought together to suggested that structure- and surface-specific characteristics (e.g. molecular arrangement on the crystal facets, surface adhesion, existence of cleavage planes, morphology/shape, particle aspect ratio) are to be incorporated in future studies related to HPH and nanocrystals. Extrusion for nanomaterial production is presented in Chapter "Extrusion for Nanomaterial Production". Extrusion is one of the main nanomaterial production methods modified, scaled-down, and combined to fabricate nanomaterials in solid lipid nanoparticles, nanoemulsions, nanoparticles, liposomes, nanofibers, etc. Few classifications of nanomaterial extrusion were identified referring to literature. Among them, the hot-melt extrusion method can be identified as the classic version of extrusion. Chapter "Sol–Gel and Solvothermal Methods for Nanoparticles Productions" presented sol–gel and solvothermal methods for nanoparticles productions. Further, this chapter discusses on relative benefits depending on the desired physical and chemical characteristics of the material and the intended application, thus qualifying the two methods as valuable tools in nanomaterials fabrication. Supercritical fluid technologies for nanoparticle production is presented in Chapter "Supercritical Fluid Technologies for Nanoparticle Production" highlighting supercritical technologies role in clean nanoparticle production, exploring its application in drug delivery, catalysis, and electronic materials, and highlights ongoing research for optimizing process variables and scaling up operations. Chapter "Plasma and Laser Technology for Nanoparticles Production" discusses plasma and laser technology for nanoparticles production. In this chapter, principles of plasma synthesis techniques and laser-based nanoparticle synthesis techniques is presented along with overview of some key research and their contributions to the development of the field. Chapter "Sonochemical Methods for Nanoparticle Production" presents an overview of sonochemical methods for energy efficient nanoparticle production. These methods can be utilized to produce them through various means such as radical formation, sono-fragmentation, and nucleation. Most of these mechanisms are made possible due to a phenomenon known as acoustic cavitation, which is the nucleation, growth and implosion of bubbles. Scalability is of extreme importance due to being able to produce in a large scale, here authors discuss ways in which sonochemical methods can be introduced in to already existing systems to efficiently synthesize nanoparticles. Chapter "Nanolithographic Techniques for Scaled-Up Manufacturing of Nanostructures" presents nanolithographic techniques for scaled-up manufacturing of nanostructures. Nanolithography, a cornerstone of modern nanotechnology, has evolved significantly with techniques ranging from traditional photolithography to advanced methods like extreme Ultraviolet lithography (EUVL), electron beam lithography (EBL), ion beam lithography, and soft lithography. Emerging trends include the integration of artificial intelligence for process optimization in large-scale manufacturing, the development of 3D nanofabrication techniques for complex structures, and the exploration of quantum and biomimetic approaches for novel patterning methods. Microbial factories for scaled-up production of nanoparticles are discussed in Chapter "Microbial Factories for Scaled-Up Production of Nanoparticles". Adoption of novel green technologies by utilizing microorganisms as "microbial cell factories" use bacteria, fungi, and microalgae, for the intracellular/extracellular synthesis of nanoparticles. The challenges and limitations associated with the scaled-up production of biogenic nanoparticles, and potential solutions are also discussed, to distinguish the scope of future research enhancing the economic feasibility and sustainability.

Chapter "Product Separation, Purification, Storage and Packaging of Nanoparticles" provides an overview of product separation, purification, storage and packaging of nanoparticles. Separation of nanoparticles often involves techniques such as centrifugation, filtration, and magnetic separation, depending on the type of nanoparticle and the medium they are synthesized in. The goal is to isolate the nanoparticles from unwanted byproducts, residual solvents, or aggregation products that may have formed during synthesis. Purification aims to remove any contaminants or byproducts from the nanoparticle suspension. Common methods include dialysis, size-exclusion chromatography, and solvent exchange techniques. Storage of nanoparticles is another critical factor, as they are often prone to agglomeration, oxidation, or chemical degradation over time. Packaging considerations include ensuring that nanoparticles are protected from environmental factors while minimizing any health risks during handling. Quality assurance in nanoparticles production is discussed in Chapter "Quality Assurance in Nanoparticles Production". Process variables and post-processing steps determine the uniformity of the synthesized nanomaterials and the degree of reproducibility of the nanomaterial synthesis protocol. Quality assurance aims to minimize the variability of the targeted properties of synthesized nanoparticles characterized using various techniques. Chapter "Process Economics of Nanomaterials Production—A Decision Analysis Framework for Assessing Technical, Business and Environmental Viability" presents a process economics of nanomaterials production with a decision analysis framework for assessing technical, business, and environmental viability. The proposed framework consists of decision analysis for market and business viability, process chemistry R&D focusing nanomaterial synthesis, process design and equipment selection, and overall optimization of the whole manufacturing process. Then, a case study is presented by applying the proposed decision analysis framework for evaluating process economics of graphene oxide production at industrial scale. Life cycle assessment (LCA) of nanomaterials production is discussed in Chapter "Life Cycle Assessment of Nanomaterials Production". This chapter elaborates the current knowledge on life cycle assessment of nanomaterials and nano-enabled products. It is important to perform LCAs for identifying the product life cycle stages and operations with significant impacts on the environment and thereby find solutions to mitigate those impacts. Chapter "Health and Safety Aspects of Nanoparticle Production at Industrial Scale: An Overview" reviewed health and safety aspects of nanoparticle production at industrial scale. Understanding health and safety related risk perception on nanomaterials, prevailing industry standards, occupational health and safety risk management methodologies, assessment and quantification of risks, and regulatory frameworks and standards are also elaborated in this emerging sector.

4 Challenges and Opportunities in Nanoparticle Production in Industrial Scale

There are many challenges faced by nanoparticle production at industrials scale. Those can be listed as below.

- As stoichiometric reactions cannot be used to predict the yield of nanoparticle synthesis, optimization of nanoparticle formation conditions/mechanisms and scalability of such conditions at large scale equipment is necessary to transform nanomaterials synthesis from "an art to science."
- Predominantly literatures present the laboratory scale data which is insufficient for process scale up for industrial scale production.
- Industry-developed scale up data does not publicly available by commercial players, yet those information should have developed by such entities in-house.
- Due to smaller quantity demand, small scale production may be appropriate for high value low volume product formulation (e.g. API of pharmaceuticals, additives for bulk chemicals).
- Instead of process scaling using traditional chemical engineering approaches, emerging technologies (e.g. microfluidic systems) and principles of process intensification may help numbering up of nanoparticle synthesis to meet the varying demand.

Thus, we invite the readers to explore the opportunities and challenges associated with industrial scale nanoparticle production.

References

1. Bayda S, Adeel M, Tuccinardi T, Cordani M, Rizzolio F (2020) The history of nanoscience and nanotechnology: from chemical–physical applications to nanomedicine. Molecules 25:112. https://doi.org/10.3390/molecules25010112
2. Rafique M, Tahir MB, Rafique MS, Hamza M (2020) History and fundamentals of nanoscience and nanotechnology. In: Nanotechnology and photocatalysis for environmental applications. Elsevier Inc. https://doi.org/10.1016/B978-0-12-821192-2.00001-2
3. Grand View Research (2024) https://www.grandviewresearch.com/industry-analysis/nanotechnology-and-nanomaterials-market

Nanocrystalisation for Scaled Up Manufacturing of Nanoparticles

Roshan T. Bandara, Nadeesh M. Adassooriya, and Francesco Civati

Abstract Nanoparticles and nanocrystals have emerged as important areas in the development of diverse industries, including pharmaceuticals, agriculture, and electronics, due to their unique properties and wide-ranging applications. In this chapter, the crystallization process was discussed through the lens of classical nucleation theory and transitions to nanocrystallization with a focus on spatial and time confinement. Both top-down and bottom-up techniques can be used in the production of nanoparticles and nanocrystals; however, bottom-up methods such as sol–gel processing, aerosol techniques, and chemical vapor deposition often face limitations in achieving precise control over crystallinity. Techniques like spray drying, non-thermal plasma synthesis, and membrane-assisted crystallization address the challenges of those methods by enabling the production of high-quality nanocrystals with enhanced structural control. For large-scale production, process intensification strategies, such as reactive crystallization, ultra-sound assisted crystallization, and continuous crystallization aim to optimize energy consumption while enhancing product yield and consistency. The integration of in-situ monitoring systems, Process Analytical Tools (PATs), and microfluidic technology have made a rapid shift from traditional batch crystallization to efficient continuous systems by providing improved control over nucleation, growth, and crystallization kinetics, effectively bridging the gap between laboratory-scale to industrial-scale applications.

Keywords Nanoparticles · Nanocrystals · Nanocrystallization · Scaling up · Process Intensification

R. T. Bandara · N. M. Adassooriya
Department of Chemical and Process Engineering, Faculty of Engineering, University of Peradeniya, Peradeniya, Sri Lanka

F. Civati (✉)
Department of Chemical Engineering, Massachusetts Institute of Technology, Cambridge, USA
e-mail: Fre.civati@gmail.com

N. M. Adassooriya et al. (eds.), *Industrial Scale Production of Nanoparticles*, Smart Nanomaterials Technology, https://doi.org/10.1007/978-981-96-9176-0_2

1 Introduction

1.1 History of Nanoparticles and Nanocrystals

The history of nanotechnology spans centuries, with significant developments since the Industrial Revolution. It dates back to the ninth century, with early uses in color-changing glass and exceptionally strong steel swords [1]. As per history, nanocrystals have been used in glass, gold, and ceramic industries [46]. Richard Zsigmondy introduced the term "nanometer" for measuring particle sizes, laying the foundation for nanoparticle study. Richard Feynman, known as the father of modern nanotechnology, proposed manipulating matter at the atomic level in his 1959 lecture, which inspired future innovations. In the 1970s, Norio Taniguchi first used "nanotechnology" to describe processes at the nanometer scale [27]. After that nanotechnology was developed over the years and boosted after 2000 with government interventions.

In simple terms, nanocrystals are crystalline structures with dimensions on the nanometer scale, offering significant advantages over conventional nanoparticles primarily due to their unique structural characteristics. In the pharmaceutical industry, nanocrystals demonstrate enhanced dissolution rates, resulting in improved bioavailability. Additionally, they exhibit greater stability compared to amorphous nanoparticles, thereby minimizing the risk of recrystallization. Recrystallization, particularly in pharmaceutical applications, can lead to the transformation of the active ingredient into a polymorphic form with distinct physical and chemical properties. This transformation can significantly impact drug solubility, dissolution rate, and bioavailability. A transition to less a bioavailable form may reduce therapeutic efficacy, while a shift to a more bioavailable or reactive form could increase systematic exposure, potentially elevating the risk of adverse effects for the patient. Therefore, the enhanced stability of nanocrystals not only ensures consistent performance but also mitigates safety concerns associated with uncontrolled polymorphic changes.

These attributes, combined with their industry-specific properties, have made nanocrystals highly versatile, finding applications in electronics, photonics, catalysis, agriculture, pharmaceuticals, and drug delivery systems. Several examples are summarized in Table 1, showcasing their efficiency, durability, and superior performance in various domains.

Over the years, many studies have been conducted on the development of large-scale production of nanocrystals to fully exploit their potential. However, scaling up production poses numerous challenges, limiting their application in certain industries. This chapter discusses the principles of nanocrystallization, explores the various production methods, and highlights technological advancements that could propel further growth in the field of nanocrystallization.

Table 1 Applications of nanocrystals in different industries

Industry	Nanocrystal	Applications or benefits	Citations
Electronics and photonics	Trihalide Perovskite	Lighting, lasing, photon detection, and photovoltaics	He et al. [20]
	Silicon Nanocrystals	Photonic components and solar cells	Priolo et al. [56]
	Quantum dots (CdSe/CdS and PbS/CdS)	Solar cells, light-emitting diodes, and other optoelectronic devices	Panthani and Korgel [52], Priolo et al.[56]
Catalysts	Cuprous oxide (Cu_2O)	CO and propylene oxidation reactions	Chen et al. [8], Huang [26]
	$CoSe_2$ and $NiSe_2$	Water splitting	Kwak et al. [38]
	Ru nanocrystals	Oxygen evolution reactions	Zhao et al. [73]
Agriculture	Cellulose nanocrystals	Renewable and biocompatible agrochemicals	Vinzant et al. [67]
	Nanosensors	Detecting plant diseases, pollutants, and nutrient deficiency	Zhao et al. [74]
Pharmaceuticals	Caffeine	Improved tropical drug delivery	Zhai et al. [72]
	Dantrolene sodium	Short production time and small infusion	Keck and Müller [31]
	Fenofibrate	Low dosage requirement	Junghanns and Müller [30]
	Rapamycin	Improves solubility and therapeutic effect	Joshi et al. [29]

1.2 Applications of Nanoparticles and Nanocrystals

1.2.1 Electronics and Photonics

Due to the possibility of both amorphous and crystalline nature along with high surface-to-volume ratio and quantum effects, nanoparticles are functionalized accordingly for miniaturization and improved performance to use in electronic devices such as transistors, sensors, and memory devices [61, 65]. Moreover, their tunable optical and electronic properties allow for diverse functionalities in displays, high-resolution imaging systems, and sophisticated labeling techniques due to their ability to emit light at distinct wavelengths and sizes, enabling vibrant color displays and precise bioimaging [59].

In the semiconductor industry, the ordered structure of nanocrystals is particularly advantageous. Their structural regularity enables precise control over electron mobility, making them integral in the production of thin-film transistors, solar cells, and high-performance batteries [22, 25]. Thin films made from nanocrystals also improve charge carrier dynamics, enhancing device efficiency and longevity. In the field of photonics, the integration of nanocrystals into LEDs, lasers, and photodetectors has revolutionized light-based technologies. The crystalline nature of these nanoparticles enhances their light absorption and emission capabilities, leading to significant advancements in energy efficiency and color purity [37]. For instance, quantum dots, a type of semiconductor nanocrystals, are extensively used in high-efficiency displays and lighting due to their exceptional luminescence and tunability. Furthermore, nanophotonic devices leverage the unique optical properties of nanoparticles to manipulate light at the nanoscale, enabling breakthroughs in optical communication, sensing, and quantum computing.

By harnessing the extraordinary properties of nanoparticles, electronics, and photonics continue to evolve, paving the way for smaller, faster, and more energy-efficient technologies that have applications ranging from consumer electronics to advanced scientific instrumentation.

1.2.2 Catalysis and Chemical Synthesis

Nanostructured catalysts both nanoparticles and nanocrystals play a major role in chemical reactions. High surface-to-volume area leads to an increase in active sites of catalysts compared to conventional catalysts enhancing overall catalytic activity and easy tunability of the catalysts by changing size, shape, and composition making them further superior. Nano size makes them more reactive due to quantum effects that may alter electronic properties along with several physical and chemical property changes. The nanoscale size of these catalysts introduces quantum effects that alter their electronic properties, often resulting in unique reactivity and selectivity not observed in larger-scale materials. These quantum effects can also bring about changes in physical and chemical properties, such as improved conductivity, magnetism, or stability, further enhancing catalytic performance [63, 70] Nanocrystals, in particular, exhibit an ordered structure that can be finely tuned to favor specific reactions, making them ideal for selective catalytic processes. This feature not only improves the catalytic activity but also increases the lifetime and sustainability of the catalyst, reducing the need for frequent replacements and thus lowering operational costs in industrial processes [50]. The versatility of nanostructured catalysts makes them invaluable across a range of industries [7]. In water treatment, nanocatalysts are utilized for the degradation of pollutants, offering faster and more efficient purification processes [66]. In photocatalysis, their enhanced light absorption capabilities enable the decomposition of organic contaminants under UV or visible light, making them essential in environmental remediation [63]. Nanocatalysts are also key in material synthesis, facilitating reactions that require high precision and control, such as the production of fine chemicals and nanomaterials. In the biomedical field, they are being explored for drug delivery systems, where their size and surface properties can be engineered to target specific cells or tissues, improving therapeutic outcomes [69]. Furthermore, in sensor technologies, nanocatalysts contribute to the development of highly sensitive detection systems, where their ability to interact with minute amounts of chemical substances enables rapid and accurate sensing [58].

1.2.3 Paintings and Coatings

Nanostructured materials have revolutionized the painting and coatings industry. The incorporation of nanoparticles such as silica, alumina, and titanium dioxide has notably improved the mechanical properties of coatings, including increased hardness and resistance to scratches, which result in more durable and long-lasting surfaces [3]. In the automation industry, nanocoatings have been widely adopted in paints due to their ability to reduce scratches, enhance protection, and improve durability and aesthetic appeal [44, 60]. The topmost layer of these coatings acts as a protective barrier against sunlight and extreme environmental conditions, effectively preventing scratches and cracks [60]. Furthermore, when integrated with polymers, nanocoatings exhibit additional benefits such as improved corrosion resistance, abrasion resistance, and hardness [44].

Other modifications to coatings have been introduced to protect against UV radiation, microbes, wear, corrosion, thermal degradation, and moisture absorption. For instance, in the aerospace industry, the incorporation of nanoparticles into gradient coatings has significantly enhanced wear resistance. Fe_3Al/Al_2O_3 nanocoatings applied via laser cladding on 45 steel plates have shown a remarkable 14-fold improvement in wear resistance. Similarly, nanoscale SiO_2 composite coatings on steel plates have demonstrated superior corrosion resistance [24].

Moreover, titanium dioxide nanoparticles are the most frequently incorporated nanoparticles in self-cleaning coatings. Among the most innovative applications of nanostructured coatings are self-cleaning surfaces, Titanium dioxide (TiO_2) nanoparticles are frequently incorporated into these coatings due to their photocatalytic properties. Under UV light, TiO_2 nanoparticles activate and facilitate chemical reactions that break down organic dirt and contaminants, allowing surfaces to remain clean without the need for traditional cleaning methods. This technology is widely used in applications ranging from self-cleaning windows to outdoor facades and automotive glass [51].

In the construction industry, the integration of nanoparticles into coatings has led to chemically and physically enhanced properties. For example, nanocoatings are widely applied to steel components to improve weldability, corrosion resistance, and formability, thereby enhancing overall structural performance and longevity [43].

Overall, the integration of nanoparticles into coatings not only enhances mechanical properties but also adds multifunctionality, offering protection against a wide range of environmental factors and improving the longevity and maintenance of various surfaces.

1.2.4 Nano Agriculture

Nanotechnology can revolutionize agriculture by offering innovative solutions that improve crop yields while minimizing environmental and health risks. By integrating nanomaterials into agricultural practices—such as nano fertilizers, nano pesticides, soil quality enhancers (like nano zeolites and hydrogels), plant growth stimulants, and nanosensors for monitoring—nanoagriculture holds the potential to address several challenges facing conventional agriculture. These advancements not only promote food security but also reduce the ecological footprint of farming practices [16].

The increased surface area and targeted delivery of nanomaterials enable more efficient absorption of agrochemicals by plants, enhancing the effectiveness of fertilizers and pesticides and minimizing environmental impact [42]. Therefore, nanotechnology is increasingly being utilized to develop stable and biodegradable pesticides and fertilizers. The excessive use of conventional pesticides in agriculture has also led to significant environmental damage, including pesticide leaching into soil and water bodies, contamination, and the development of pest resistance [36]. To address these issues, biopolymeric nanoparticles have been introduced. These nanoparticles offer numerous advantages, including targeted delivery, reduced toxicity, improved stability, and biocompatibility. Additionally, biopolymeric nanoformulations enhance soil aeration and promote microbial growth, benefiting both plants and the environment [36].

However, despite these promising benefits, uncertainties remain regarding the long-term ecological impacts of nanoagriculture. The potential effects on beneficial insects, microorganisms, and soil health are not yet fully understood. Concerns have also been raised about the accumulation of nanoparticles in soil and their interactions with living organisms, which could lead to unintended consequences for biodiversity. To address these challenges, the concept of green synthesis approaches has been introduced, utilizing plants and microbes as cost-effective, efficient, and bio-compatible agents for the synthesis of nanoparticles in agricultural applications [9]. Nonetheless, addressing global issues requires ongoing research. Fraceto et al. [16] emphasize the need for hybrid carrier development under sustainable practices, scaling up lab production, conducting risk and life-cycle assessments of nanoagriculture, and updating regulatory frameworks as critical areas for further exploration.

1.2.5 Pharmaceuticals and Drug Delivery

The pharmaceutical industry continues to grow and expand its efforts in developing medicines to combat various diseases, thereby improving human health and increasing life expectancy. However, a significant challenge in drug development lies in the poor water solubility of many newly discovered drugs, leading to reduced dissolution rates and, consequently, poor bioavailability [62]. To address these limitations, the introduction of nanoparticles, particularly nanocrystals, has revolutionized the pharmaceutical landscape [28]. These advancements have been especially impactful for poorly soluble Class II and IV drugs under the Biopharmaceutics Classification System (BCS), enhancing therapeutic potential and achieving cost efficiencies [21].

Nanocrystals offer distinct advantages due to their unique properties, such as enhanced surface area, which improves the bioavailability of poorly soluble drugs [39]. Unlike traditional nanoparticles that require polymeric carriers or excipients, nanocrystals minimize the need for additional components, enabling high drug-loading capacity and simplifying the formulation process. This approach reduces the reliance on supplementary chemicals beyond the Active Pharmaceutical Ingredient (API), making nanocrystals a preferred choice for pharmaceutical applications [17]. Examples of nanocrystal-based drugs include Rapamune® (Sirolimus) for immunosuppression, Emend® (Aprepitant) for nausea and vomiting, and Tricor® (Fenofibrate) for hypercholesterolemia, all of which leverage nanocrystal technology to improve their bioavailability [19, 32, 48].

The small size of nanocrystals also enhances saturation solubility by increasing their curvature, leading to higher dissolution pressure and better solubility [40]. Furthermore, their increased surface area results in faster dissolution and more efficient absorption within the body, thereby improving overall bioavailability. For instance, Megace® ES (Megestrol acetate) has improved patient compliance through enhanced bioavailability, while Cesamet® (Nabilone) effectively addresses chemotherapy-induced nausea and vomiting with its nanocrystal formulation [41].

Beyond solubility and dissolution, nanocrystals provide versatility in drug delivery systems, enabling oral, parenteral, pulmonary, ocular, and dermal administration. This flexibility broadens therapeutic applications and facilitates personalized treatment approaches [40]. Additionally, the reduced reliance on complex carriers simplifies the manufacturing process, enabling cost-efficient production and easier regulatory compliance, as evidenced by several nanocrystal-based drugs achieving FDA approval.

2 Crystallization

Crystallization refers to the process by which atoms, molecules, or ions organize themselves into a highly ordered, repetitive structure, forming a crystal lattice. Although crystallization can occur in the solid, liquid, or gaseous phases, it is most commonly discussed in the context of solid crystal formation from a liquid or solution. Despite the diversity of phases, all crystal formation processes are governed by the same underlying crystallization theories. While the steps involved in crystal formation are typically described as distinct stages, these stages are, in reality, interconnected and continuous, often occurring simultaneously within the system [4].

2.1 Supersaturation

The initiation step of crystallization is the achievement of supersaturation, as the system must be in a non-equilibrium state for a new phase to form spontaneously [14]. In this stage, solutes—whether atoms, molecules, or ions—are present in a solvent, and the maximum amount of solute that the liquid can hold at equilibrium is exceeded, resulting in a supersaturated, non-equilibrium state. Supersaturation can be achieved through various methods, including cooling or heating the solution based on the solubility-temperature relationship, solvent evaporation, pH adjustment, or altering the solvent composition through mixing.

The difference in chemical potential between the solutes in supersaturated and saturated conditions drives the nucleation and subsequent growth of crystals [4]. This driving force per molecule can be expressed as:

$$\Delta\mu = k_B T \ln(C/C_s) \tag{1}$$

where C is the concentration of solutes before crystallization, C_s is the equilibrium concentration (solubility), k_B Is the Boltzmann constant, and T is the absolute temperature.

Although various methods have been proposed to define supersaturation, the complexity of factors influencing it makes it difficult to establish a universal equation. In this context, the supersaturation ratio $\beta = C/C_s$ is selected in defining supersaturation [4].

2.2 Nucleation

Once supersaturation reaches the required level, nucleation begins, which is a critical step in determining the size and structure of the resulting crystals. During nucleation, solutes aggregate to form clusters. If the concentration of these solutes exceeds their solubility in the medium, nucleation begins. A nucleus can be defined as the smallest amount of a new phase that can exist independently [14]. Nucleation occurs via two mechanisms: homogeneous and heterogeneous nucleation. In homogeneous nucleation, solute particles come together within the bulk solution to form nuclei, typically under conditions of very high supersaturation.

Heterogeneous nucleation, on the other hand, occurs on foreign surfaces or impurities and is more likely when supersaturation is lower. One of the key impurities in nucleation processes is the solvent itself. When a nucleus or solute adsorbs an impurity, the interfacial energy is reduced, resulting in an increase in the nucleation rate, which is characteristic of heterogeneous nucleation.

According to Classical Nucleation Theory (CNT), both types of nucleation require overcoming an energy barrier—the activation free energy of nucleation. This activation energy consists of two components: surface-free energy and volume-free energy.

For a spherical nucleus, the total free energy ΔG is expressed as:

$$G = \Delta G_V + \Delta G_s \quad (2)$$

1. Surface Free Energy (ΔG_s): This energy is associated with the creation of a new solid–liquid interface, which resists nucleus formation because additional energy is required to form the surface. As a result, the change in free energy due to the formation of a nucleus is initially positive and thus energetically unfavorable.

$$\Delta G_s = 4\pi r^2 \gamma \quad (3)$$

Here, γ represents the interfacial free energy between the nucleus and the solution.

2. Volume Free Energy (ΔG_V): This term, also known as bulk free energy, favors nucleation as it is negative, reflecting the thermodynamic stability of the solid phase over the liquid. As particles form a crystalline structure, the system's free energy decreases since the solid state is more energetically stable than the liquid.

$$\Delta G_V = \frac{4\pi r^3}{3\Omega} k_B T \ln(\beta) \quad (4)$$

where and Ω is the volume of a molecule within the crystal.

Since surface free energy is proportional to r^2 (where r is the nucleus radius) and volume-free energy is proportional to r^3 the system tends to stabilize as the nucleus grows. Initially, nuclei exist in a metastable state, where they are prone to shrinking or dissolving. However, once they reach a critical radius, the total free energy reaches a maximum, and the nucleus becomes stable, allowing crystal growth as volume-free energy begins to dominate.

The maximum free energy reached during nucleation is the critical energy threshold that must be overcome for nucleation to commence. This energy required to form a stable nucleus can be expressed as:

$$\Delta G^* = \frac{16\pi \Omega^2 \gamma^3}{3(k_B T \ln(\beta))^2} \quad (5)$$

The radius of the nucleus at this point is known as the critical radius (r^*).

$$r^* = \frac{2\Omega\gamma}{k_B T \ln(\beta)} \quad (6)$$

For nuclei with a radius smaller than the critical radius ($r < r^*$), the system will tend to shrink or dissolve the nucleus due to the dominance of surface-free energy. Conversely, when the nucleus exceeds the critical radius ($r > r^*$), it becomes energetically favorable for the nucleus to grow spontaneously, as volume-free energy begins to dominate, reducing the system's overall free energy.

Fig. 1 Free energy diagram

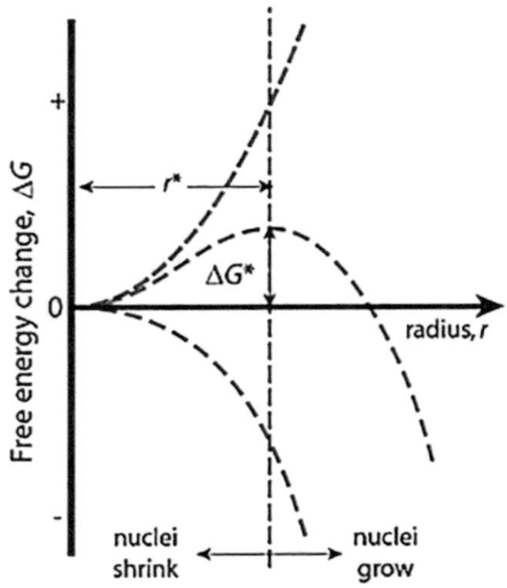

Alternatively, it can also be described as one-third of the energy needed to form a stable surface at the critical radius:

$$\Delta G^* = \frac{1}{3}\left(4\pi r^{*2}\gamma\right) \tag{7}$$

This relationship between critical radius and the activation energy for nucleation suggests that by modifying conditions such as temperature, increasing supersaturation, or reducing surface free energy, the energy barrier for nucleation can be minimized (Fig. 1).

The rate at which nuclei form in a unit time per unit volume, known as the nucleation rate, can be expressed under CNT as:

$$J = K_0 e^{\frac{-\Delta G^*}{k_B T}} \tag{8}$$

where K_0 is the kinetic coefficient.

The nucleation rate is primarily influenced by the solubility of the solute, the interfacial free energy, and the level of supersaturation. Solutes with higher solubility and lower interfacial surface energy can nucleate at lower supersaturation levels compared to sparingly soluble solutes, which require higher levels of supersaturation to initiate nucleation.

This dynamic leads to a situation in which the solution remains in a metastable state with a low nucleation rate until it reaches a critical supersaturation, at which point an exponential increase in nucleation occurs. Controlling supersaturation is essential because, at lower supersaturation, nucleation may take considerable time, whereas, at higher supersaturation, rapid nucleation may result in poor crystallinity or the formation of amorphous phases (Fig. 2).

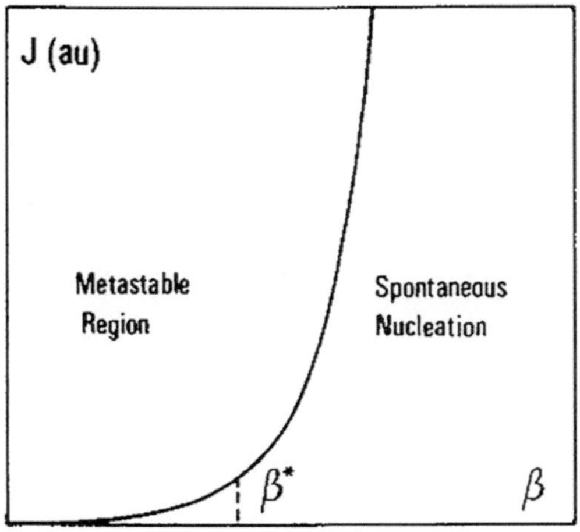

Fig. 2 Nucleation rate versus supersaturation

2.3 *Crystal Growth*

After nucleation, crystal growth begins as solute particles continue to attach to the nucleus. This process is influenced by both internal and external factors. Solute particles, whether in solution or gas, can be transported to the nucleus either by diffusion or desolvation. In diffusion, the solutes move with the medium and adhere to the nucleus. In desolvation, solute particles, initially surrounded by solvent molecules, disperse or remove these solvent molecules to integrate into the crystal lattice.

Hu et al. divided the growth process into two categories [23]:

1. Diffusion-controlled growth occurs when the monomer concentration drops below the minimum level required for nucleation. In this case, nucleation stops and growth becomes the dominant process.
2. Surface-controlled growth happens when monomer diffusion from the bulk solution to the crystal surface occurs rapidly, allowing for faster crystal growth.

Once solutes reach the growing nucleus, they organize into a repeating, ordered three-dimensional arrangement, forming a lattice structure. This regular, periodic organization releases energy, further driving the crystal growth process. The spatial arrangement of these units varies depending on the crystal system (e.g., cubic, tetragonal, hexagonal), which determines the distinct shape of the crystals.

Crystal growth mechanisms can be described by several theories. Under the Periodic Bond Chain (PBC) theory, growth involves the continuous formation of strong bonds, and crystals are classified into three types of faces based on how these bonds are organized:

1. F faces: PBC in at least two directions (flat). These faces have a low number of growth sites and grow through two-dimensional nucleation or spiral growth mechanisms.

2. S faces: PBC in one direction (stepped). Growth on these faces is governed by a one-dimensional nucleation mechanism.
3. K faces: No PBC in any direction, meaning that all sites on this face are growth sites.

Two-dimensional nucleation growth is the dominant mechanism when crystals are defect-free. Once the nucleus radius surpasses the critical radius, it becomes stable and creates kink positions where growth continues. Growth can occur either through the mononuclear model (where a single nucleus drives growth) or the polynuclear model (where multiple nuclei are present).

When defects like screw dislocations are present, the spiral growth mechanism becomes dominant. Impurities can also influence the growth process by altering the energetics of the growth sites.

2.4 Phase Transitions and Ostwald Ripening

In closed systems, as crystallization progresses, supersaturation gradually decreases, which in turn reduces nucleation and growth rates. This leads to phase transitions. When multiple phases coexist in a solution, the phase with the highest solubility (or lowest supersaturation) crystallizes first. As growth continues, the less soluble phase becomes supersaturated and crystallizes. The first phase then becomes unstable and dissolves, allowing the growth of the more stable phase. If the initial phase remains without dissolving, it will exist in a metastable state [4].

Ostwald Ripening can occur when the system contains crystals of varying sizes (due to differences in growth time and rate). This process is governed by volume diffusion, where smaller crystals dissolve and larger ones grow, increasing the overall mean crystal size. This typically happens as supersaturation decreases, with smaller crystals disappearing as the system approaches equilibrium [4].

2.5 Limitations of Classical Nucleation Theory

Classical nucleation theory, while a useful model, is based on several simplifying assumptions that deviate from real-world nucleation and crystal growth processes [14]:

- Clusters are modeled as spherical droplets with uniform density and sharp interfaces.
- The curvature or size dependence and temperature dependence of surface tension are ignored.
- Clusters are assumed to be at rest without translational, vibrational, or rotational motion.
- The nucleation rate is assumed to be time-independent.

- Clusters are considered incompressible.

While CNT provides a reasonable approximation for single-component nucleation, its applicability to systems with mixed solutes is oversimplified. To address these discrepancies, two-step nucleation theory is often employed, which considers more complex interactions and processes in actual systems.

2.6 Influencing Factors

The degree of supersaturation plays a critical role in controlling nucleation and growth rates. When supersaturation levels are high, bulk nucleation occurs, increasing the volume of free energy and reducing the critical energy requirement for nucleation [49]. However, excessively high supersaturation can lead to broad crystal size distributions and inconsistent product quality.

The metastable zone width (MSZW) is the range between the solubility curve and the nucleation point where crystal growth occurs without immediate nucleation. Operating within this zone allows for controlled growth while minimizing uncontrolled nucleation events [49]. Higher temperatures tend to narrow the MSZW, reducing the energy barrier for nucleation and increasing the solubility of solutes, which accelerates growth.

For optimal crystal size distribution (CSD), it is important to operate near the metastable zone limit [47]. Deviating too far from this range can either result in excessive nucleation, leading to lower purity and longer filtration times, or slow growth, resulting in extended process times.

3 Nanocrystallization

Confinement of time and space in crystallization is the core principle behind nanocrystallization by controlling Ostwald ripening. As the initial step, rapid supersaturation needs to be provided which acts as the time confinement, where antisolvent precipitation comes into play. This state leads to a critical solute concentration necessary for instantaneous nucleation, which reduces the concentration of solute way below critical concentration and stops nucleation. Then, the nuclei formed start growing. For uniform nanocrystal production, the nuclei growth rate must be minimal [45].

The second confinement for nanocrystallization: spatial confinement, achieved via emulsions, micelles, or pores, provides space to form nanoscale reactors which limits the size of the crystal by controlling both nucleation and growth [45]. This confinement further helps in controlling polymorphism and stabilizing metastable polymorphs Dwyer et al. [13]. However, a challenge with spatial confinement is

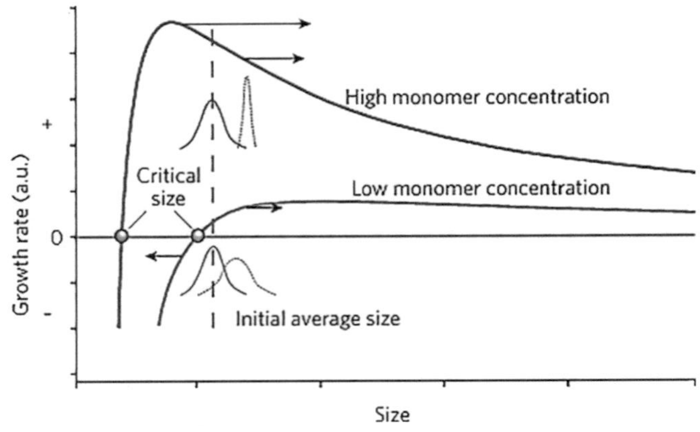

Fig. 3 Growth rate variation with monomer concentration [71]

ensuring that crystals form within the porous matrix rather than on the matrix surface [13, 18].

Liquid antisolvent precipitation is a straightforward method. An antisolvent (in which the solute is insoluble but the initial solvent miscible) is introduced to the solution made of solute and initial solvent, causing rapid supersaturation and then precipitation. This reduces the time available for Ostwald ripening, yielding fine particles. The limitation of this method is the need for post-processing to remove the antisolvent. Therefore, as a solution to this, supercritical antisolvent precipitation can be used, which utilizes a supercritical fluid as the antisolvent, which can be removed easily after nanocrystal production by pressurizing and converting to gas. In addition, supercritical antisolvent precipitation also produces smaller particles than conventional antisolvent precipitation due to rapid diffusion [57]. To address imbalances between nucleation and crystal growth in this method, stabilizers are used [12].

In addition, the growth mechanism is believed to shape the final morphology of the nanocrystal by initially determining the morphology of the seed. Another significant factor in this process is the ratio between the stabilizer and the precursor, which plays a huge role in controlling particle size. Adjusting this ratio can either increase or decrease the final nanoparticle size [11]. Moreover, the concentration of active monomers in the system also has a significant impact on particle size; if it falls below the desired level, nucleation decreases, leading to fewer nuclei and a tendency to form larger nanoparticles Fig. 3 [71].

3.1 *Nanoparticle and Nanocrystal Production Methods*

While nanoparticle and nanocrystal production methods share similarities, distinct differences exist, primarily as some techniques for nanoparticles do not ensure a crystalline structure. Generally, these production methods are classified as either top-down—where larger particles are reduced to the nanoscale—or bottom-up, in which particles form from atomic or molecular precursors. Each approach requires precise control over factors governing nanocrystallization to produce nanocrystals with the desired size, shape, and distribution.

3.1.1 Bottom-Up Techniques—Nanoparticle

Sol–Gel Processing

Sol–gel processing is simply transforming a solution into a gel, which is then further processed to yield nanoparticles. This low-temperature, cost-effective, highly efficient technique offers precise control over product composition, making it widely used for continuous industrial nanoparticle production [5].

Initially, a homogeneous solution is prepared by mixing a precursor with a solvent, often water or alcohol. Common precursors for metal oxide nanomaterials include Tetraethoxysilane (TEOS), Tetramethoxysilane (TMOS), and Titanium tetraisopropoxide [5]. After homogenization, hydrolysis occurs, where water reacts with the precursor to form a colloidal mixture. If Brownian motion prevents the settling of particles, the colloidal mixture remains without settling, improving final product quality. Chemical or physical interactions then transform these colloidal particles into a gel, which, once dried and ground, produces nanoparticles.

Aerosol Processes

Aerosol processes offer advantages over wet and solid-phase chemical methods, such as no liquid by-products, simple particle collection, fewer steps, and high purity [55]. However, high temperatures in these processes may lead to particle agglomeration, necessitating precise control [6].

Flame spray pyrolysis and plasma are the most common nanoparticle synthesis methods under the aerosol process. As the initial step raw material preparation is done, where the most critical aspect of precursor preparation is done. Solid, liquid, or gas precursors can be used with modifications to standard reactors like flame, hot-wall, plasma, or microwave reactors which operate at high temperatures to vaporize the precursor, followed by cooling to control particle growth and crystallinity. This process either produces soft agglomerates or hard agglomerates which need to be de-agglomerated to obtain better particle size in the final product [6].

Chemical Vapor Deposition

Though chemical vapor deposition is not widely used for nanoparticles due to the need for vacuum and high temperatures, it offers high control over structure, morphology, and purity [54]. Precursor is introduced at high pressure with a carrier gas. Then, the precursor vapor adsorbs onto a substrate and forms an intermediate product, which decomposes under heat, initiating nucleation and growth. Chamber pressure, carrier gas purity, flow rate of precursor, substrate temperature, and deposition time of particles are key parameters for defining the nanoparticle quality [54].

All these methods have not resulted in producing nanocrystals due to no proper control of the crystallinity of the final product. The following top-down and bottom-up techniques are currently used to produce nanocrystals, which control both crystallization and nanocrystallization governing factors.

3.1.2 Bottom-Up Techniques—Nanocrystals

Spray Drying

Spray drying is widely used in industries for powder production, and adapted for nanoscale material production. Instead of conventional nozzles, nanospray dryers use a vibrating mesh technology-based nebulizer to atomize particles at very high pressure. The spray mesh, which moves in a vertical motion directs the droplets into the chamber. Then, rapid supersaturation and solvent evaporation take place inside the chamber when atomized particles with solvent and high surface area are met with hot dry air. This led to an increase in droplet concentration and droplet–droplet interactions and formation of the nanocrystals. Thus, created nanocrystals are collected using electrostatically charged collectors to ensure high collection efficiency without damage to the final product [64].

The nebulizer must be properly controlled by adjusting mesh size, flow rate, pressure, etc. as they influence droplet–droplet interactions and nanocrystal formation. Controlling feed air temperature, is the mostly controlled critical process parameter (CCP), as higher temperature leads to quick solvent evaporation and then a high potential for agglomeration and hence large particle size [64].

Non-thermal Plasma Synthesis

The nonthermal plasma method synthesizes high-purity nanocrystals, which require high temperatures. The nonequilibrium environment maintained under this method helps in reaching temperature exceeding gas temperature, reducing agglomeration, and enabling nanocrystal doping [33]. With no liquid phase involved, nanocrystals produced via this method have high purity. The initial negative charge on nanoparticles minimizes agglomeration, yielding uniform size distributions and small nanocrystals. Nucleation occurs via supersaturated vapor condensation or sequential reactions, where growth halts as particles become negatively charged [33]. After that point surface growth begins by reacting with the remaining precursor in the solution.

3.1.3 Top-Down Techniques—Nanocrystals

Dry and Wet Milling

Milling is a common material production process. Different types of milling (e.g., dry, wet) impart energy that fractures particles through compression, shear, and impact. After a certain milling time, particle size reduction reaches a plateau, with attrition as the main comminution mechanism [53]. Process parameters like speed, bead concentration, and temperature should be set based on the desired nanocrystal size, balancing between speed and duration for efficiency [40]. Nanocrystals can be obtained by either a combination of low milling speeds and long milling times or high milling speeds and short milling times [53]. However, a long time may lead to degradation, contamination, and increased cost. Dry milling is simple and cost-effective but may lead to amorphous structures, while wet milling reduces agglomeration due to the liquid medium. However, due to the usage of a liquid, there is space to reduce agglomeration by using the liquid, but need to remove the liquid to get pure crystals, even though the liquid addition is very minimal. During the dry milling after critical pressure within the milling chamber, the lattice structure may tend to break and form an amorphous phase, which is energetically stable compared to a crystalline structure. However, this is not present in wet milling, as the liquid present will reduce the glass transmission temperature [53].

In milling, both the phenomena of size reduction due to collisions and growth due to collisions are possible, where controlling milling parameters is very important. After the particle size reaches some level, it won't further reduce particle size but may lead to an increase in particle size [53].

High-Pressure Homogenization (HPH)

In HPH, two principles are used: microfluidization and piston-gap homogenization for nanocrystal size reduction [29]. In microfluidization, suspension circulates at high velocity through flow paths, but it's costly and can yield micron-sized particles [31].

Due to limitations in the milling and HPH-micro fluidization techniques, piston-gap homogenization was introduced. Suspension is sent through a high-pressure gap. Due to the reduction of cross-sectional area, the dynamic pressure increases at the gap, and suspension may tend to boil and form gas bubbles once the flow passes the gap it again reaches normal pressure conditions and bubbles get imploded. The number of cycles through the gap, velocity, and pressure drop determines the final size of the final product [15]. Under this method, the agglomeration of nanocrystals can be reduced by modifying the initial suspension with surfactants.

3.2 *Nanoparticle Agglomeration*

After nanoparticle production, agglomeration is a significant challenge, especially for metallic nanoparticles, as it leads to clustering and loss of nanoscale properties. As explained previously introducing stabilizers during production helps control particle size and shape. It also helps in reducing post-production agglomeration [71]. Stabilized nanoparticles have also remained stable under standard conditions, including aqueous environments, without requiring special storage or handling as mentioned by Das et al. [12].

4 Process Intensification in Crystallization

Process intensification focuses on reducing energy consumption while improving product yield and consistency. In crystallization, recent developments such as reactive crystallization, membrane crystallization, and ultrasound-assisted crystallization are being explored to enhance nucleation and growth efficiency [2].

Membrane-assisted crystallization promotes heterogeneous nucleation by using membranes as foreign bodies, which induce nucleation on their surface. This method is both environmentally friendly and energy-efficient, providing an alternative to traditional methods of crystal formation. By carefully controlling the supersaturation rate and optimizing the membrane surface area, this technique has shown promise in reducing energy costs while enhancing crystal quality.

Another technique gaining popularity is continuous crystallization, which allows for better control over supersaturation and the ability to separate product crystals from the medium in a single process. Unlike batch crystallization, which can suffer from scale-up challenges and inconsistent product quality, continuous systems such as MSMPR (Mixed Suspension Mixed Product Removal) crystallizers provide greater process control [10].

4.1 Technological Advances in Crystallization: Monitoring and Automation

Real-time monitoring of nucleation and crystal growth is a crucial aspect of controlling the size, shape, and composition of nanocrystals, helping to optimize synthesis techniques. Effective in-situ monitoring systems must allow rapid, short-interval observations, endure extreme conditions (high temperature, pressure), and avoid disrupting the reaction process [34]. Electron microscopy is a common monitoring method, though it's limited mainly to metallic nanoparticles in liquid media. For semiconductor nanocrystals, UV–vis absorption techniques, capable of withstanding high temperatures and capturing microsecond interval data, are preferred for real-time monitoring [34]. X-ray diffraction also serves in growth monitoring, while in-situ fluorescent microscopy has been explored to study crystallization kinetics [68].

These technologies enable researchers to identify governing factors like viscosity, surface tension, and local solute concentration [2]. Computational modeling and simulations have also contributed to a better understanding of the numerous variables that govern nucleation and growth. Though complex, these models provide new insights into how crystals form and grow under various conditions, which can then be applied to optimize industrial processes [2].

Furthermore, Process Analytical Tools (PATs) such as FBRM (Focused Beam Reflectance Measurement) and Mid-IR (Mid-Infrared Spectroscopy) have become integral in both batch and continuous crystallization systems. These tools offer real-time feedback on crystal growth and enable model-based predictive control (MPC) systems to adjust process parameters dynamically, ensuring better consistency and product quality [10].

Microfluidic technology is another promising tool in crystallization, offering precise control over nucleation and growth in continuous processes. By confining individual crystals and leveraging the high surface-to-volume ratios of microfluidic devices, this technology offers better heat and mass transfer and reduced particle size distribution. It has been applied to nanocrystal production, showing improved performance in reaction control and mixing [35].

4.2 Batch Versus Continuous Crystallization: Advantages and Challenges

Batch crystallization has traditionally been the dominant method in industry. However, it suffers from several drawbacks, including high manufacturing costs, scale-up issues, and inconsistent product quality [2]. Continuous crystallization processes, by contrast, provide more time and cost-efficiency while addressing many of these challenges.

Continuous systems such as MSMPR and plug flow crystallizers (PFC) offer better control over nucleation and supersaturation, which leads to more consistent CSD. For example, laminar-flow tubular crystallizers (LFTC) and coiled flow inverters (CFI) are notable designs that have been used in the pharmaceutical industry for continuous crystallization. The transition to continuous systems has been used by the Novartis-MIT Center for Continuous Manufacturing, which has optimized process design and control using mathematical models [2].

The challenge lies in accurately transferring kinetic data from batch to continuous processes, as nucleation and growth mechanisms behave differently in each system. Nevertheless, the adoption of online monitoring tools and automated control systems in continuous crystallization offers promising potential for increased efficiency and product consistency [35].

5 Conclusion

In conclusion, the integration of advanced production techniques and process intensification strategies has greatly enhanced the quality, efficiency, and scalability of nanocrystal production. Methods such as spray drying, non-thermal plasma synthesis, and membrane-assisted crystallization provide superior control over particle size, distribution, and crystallinity. Process intensification strategies, including continuous crystallization and ultrasound-assisted techniques, offer significant advantages over traditional batch processes by improving control over supersaturation, nucleation, and crystal growth. Real-time monitoring systems, including in-situ electron microscopy, UV–vis absorption, and PATs, play a vital role in optimizing crystallization conditions, ensuring consistency, and predicting product outcomes. While challenges persist in transferring batch kinetic data to continuous processes and fine-tuning process parameters, the integration of these technologies continues to enable the production of high-quality nanocrystals.

References

1. Amen R, Mukhtar A, Saqib S, Ullah S, Al-Sehemi AG, Mehdi SEH, Babar M, Bustam MA (2021) History and development of nanomaterials. In: Nanomaterials: synthesis, characterization, hazards and safety. https://doi.org/10.1016/B978-0-12-823823-3.00008-2
2. Aspillaga L, Jan Bautista D, Daluz SN, Hernandez K, Renta JA, Lopez ECR (2023) Nucleation and crystal growth: recent advances and future trends. Eng Proc 56(1). https://doi.org/10.3390/ASEC2023-15281
3. Barna E, Bommer B, Kürsteiner J, Vital A, Trzebiatowski OV, Koch W, Schmid B, Graule T (2005) Innovative, scratch proof nanocomposites for clear coatings. Composites Part A: Appl Sci Manuf 36(4). https://doi.org/10.1016/j.compositesa.2004.10.014
4. Boistelle R, Astier JP (1988) Crystallization mechanisms in solution. J Cryst Growth 90(1–3). https://doi.org/10.1016/0022-0248(88)90294-1
5. Bokov D, Turki Jalil A, Chupradit S, Suksatan W, Javed Ansari M, Shewael IH, Valiev GH, Kianfar E (2021) Nanomaterial by sol-gel method: synthesis and application. Adv Mater Sci Eng (2021). https://doi.org/10.1155/2021/5102014
6. Buesser B, Pratsinis SE (2012) Design of nanomaterial synthesis by aerosol processes. Annu Rev Chem Biomol Eng 3. https://doi.org/10.1146/annurev-chembioeng-062011-080930
7. Chaturvedi S, Dave PN, Shah NK (2012) Applications of nano-catalyst in new era. J Saudi Chem Soc 16(3). https://doi.org/10.1016/j.jscs.2011.01.015
8. Chen S, Xiong F, Huang W (2019) Surface chemistry and catalysis of oxide model catalysts from single crystals to nanocrystals. Surface Sci Rep 74(4). https://doi.org/10.1016/j.surfrep.2019.100471
9. Chowdappa P, Gowda S (2013) Nanotechnology in crop protection : status and scope. Pest Manag Hortic Ecosyst 19(2)
10. Chung SH, Ma DL, Braatz RD (1999) Optimal seeding in batch crystallization. Can J Chem Eng 77(3). https://doi.org/10.1002/cjce.5450770322
11. Dang TMD, Le TTT, Fribourg-Blanc E, Dang MC (2012) Influence of surfactant on the preparation of silver nanoparticles by polyol method. Adv Natural Sci Nanosci Nanotechnol 3(3). https://doi.org/10.1088/2043-6262/3/3/035004

12. Das S, Bandyopadhyay K, Ghosh MM (2021) Effect of stabilizer concentration on the size of silver nanoparticles synthesized through chemical route. Inorg Chem Commun 123.https://doi.org/10.1016/j.inoche.2020.108319
13. Dwyer LM, Michaelis VK, O'Mahony M, Griffin RG, Myerson AS (2015) Confined crystallization of fenofibrate in nanoporous silica. CrystEngComm 17(41). https://doi.org/10.1039/c5ce01148e
14. Erdemir D, Lee AY, Myerson AS (2009) Nucleation of crystals from solution: classical and two-step models. Accounts Chem Res 42(5). https://doi.org/10.1021/ar800217x
15. Fernández-Ronco MP, Kluge J, Mazzotti M (2013) High pressure homogenization as a novel approach for the preparation of co-crystals. Crystal Growth Des 13(5). https://doi.org/10.1021/cg400098c
16. Fraceto LF, Grillo R, de Medeiros GA, Scognamiglio V, Rea G, Bartolucci C (2016) Nanotechnology in agriculture: which innovation potential does it have? Front Environ Sci 4(MAR). https://doi.org/10.3389/fenvs.2016.00020
17. Gigliobianco MR, Casadidio C, Censi R, Di Martino P (2018) Nanocrystals of poorly soluble drugs: Drug bioavailability and physicochemical stability. Pharmaceutics 10(3). https://doi.org/10.3390/pharmaceutics10030134
18. Ha JM, Wolf JH, Hillmyer MA, Ward MD (2004) Polymorph selectivity under nanoscopic confinement. J Am Chem Soc 126(11). https://doi.org/10.1021/ja049724r
19. Haeri A, Osouli M, Bayat F, Alavi S, Dadashzadeh S (2018) Nanomedicine approaches for sirolimus delivery: a review of pharmaceutical properties and preclinical studies. Artif Cells Nanomedicine Biotechnol 46(sup1). https://doi.org/10.1080/21691401.2017.1408123
20. He X, Qiu Y, Yang S (2017) Fully-inorganic trihalide perovskite nanocrystals: a new research frontier of optoelectronic materials. Adv Mater 29(32). https://doi.org/10.1002/adma.201700775
21. Heng D, Lee SH, Ng WK, Tan RB (2011) The nano spray dryer B-90. Expert Opin Drug Deliv 8(7). https://doi.org/10.1517/17425247.2011.588206
22. Hossain N, Mobarak MH, Mimona MA, Islam MA, Hossain A, Zohura FT, Chowdhury MA (2023) Advances and significances of nanoparticles in semiconductor applications—a review. Results Eng 19. https://doi.org/10.1016/j.rineng.2023.101347
23. Hu H, Singh M, Wan X, Tang J, Chu CW, Li G (2020) Nucleation and crystal growth control for scalable solution-processed organic-inorganic hybrid perovskite solar cells. J Mater Chem A 8(4). https://doi.org/10.1039/c9ta11245f
24. Hu J (2024) Nanomaterials in aerospace: revolutionizing flight and exploration through nanoscale advancements. Appl Comput Eng 60(1):1–5. https://doi.org/10.54254/2755-2721/60/20240821
25. Hu MZ, Zhu T (2015) Semiconductor nanocrystal quantum dot synthesis approaches towards large-scale industrial production for energy applications. Nanoscale Res Lett 10(1). https://doi.org/10.1186/s11671-015-1166-y
26. Huang W (2016) Oxide nanocrystal model catalysts. Acc Chem Res 49(3). https://doi.org/10.1021/acs.accounts.5b00537
27. Hulla JE, Sahu SC, Hayes AW (2015) Nanotechnology: history and future. Hum Exp Toxicol 34(12). https://doi.org/10.1177/0960327115603588
28. Jarvis M, Krishnan V, Mitragotri S (2019) Nanocrystals: a perspective on translational research and clinical studies. Bioeng Trans Med 4(1). https://doi.org/10.1002/btm2.10122
29. Joshi K, Chandra A, Jain K, Talegaonkar S (2019) Nanocrystalization: an emerging technology to enhance the bioavailability of poorly soluble drugs. Pharm Nanotechnol 7(4). https://doi.org/10.2174/2211738507666190405182524
30. Junghanns JUAH, Müller RH (2008) Nanocrystal technology, drug delivery and clinical applications. Int J Nanomedicine 3(3)
31. Keck CM, Müller RH (2006) Drug nanocrystals of poorly soluble drugs produced by high pressure homogenisation. Eur J Pharm Biopharm 62(1). https://doi.org/10.1016/j.ejpb.2005.05.009

32. Kevadiya BD, Chen L, Zhang L, Thomas MB, Davé RN (2019) Fenofibrate nanocrystal composite microparticles for intestine-specific oral drug delivery system. Pharmaceuticals 12(3). https://doi.org/10.3390/ph12030109
33. Kortshagen UR, Sankaran RM, Pereira RN, Girshick SL, Wu JJ, Aydil ES (2016) Nonthermal plasma synthesis of nanocrystals: fundamental principles, materials, and applications. Chem Rev 116(18). https://doi.org/10.1021/acs.chemrev.6b00039
34. Kovalenko MV, Manna L, Cabot A, Hens Z, Talapin DV, Kagan CR, Klimov VI, Rogach AL, Reiss P, Milliron DJ, Guyot-Sionnnest P, Konstantatos G, Parak WJ, Hyeon T, Korgel BA, Murray CB, Heiss W (2015) Prospects of nanoscience with nanocrystals. ACS Nano 9(2). https://doi.org/10.1021/nn506223h
35. Kulkarni SA, Myerson AS (2017) Methods for nano-crystals preparation. In: NATO science for peace and security series a: chemistry and biology, vol PartF1. https://doi.org/10.1007/978-94-024-1117-1_16
36. Kumar R, Kumar N, Rajput VD, Mandzhieva S, Minkina T, Saharan BS, Kumar D, Sadh PK, Duhan JS (2022) Advances in biopolymeric nanopesticides: a new eco-friendly/eco-protective perspective in precision agriculture. Nanomaterials 12(22). https://doi.org/10.3390/nano12223964
37. Kumar Sahu M (2019) Semiconductor nanoparticles theory and applications. Int J Appl Eng Res 14(2)
38. Kwak IH, Im HS, Jang DM, Kim YW, Park K, Lim YR, Cha EH, Park J (2016) $CoSe_2$ and $NiSe_2$ nanocrystals as superior bifunctional catalysts for electrochemical and photoelectrochemical water splitting. ACS Appl Mater Interfaces 8(8). https://doi.org/10.1021/acsami.5b12093
39. Lee BK, Yun YH, Park K (2015) Smart nanoparticles for drug delivery: boundaries and opportunities. Chem Eng Sci 125. https://doi.org/10.1016/j.ces.2014.06.042
40. M Malamatari KMG Taylor S Malamataris D Douroumis K Kachrimanis 2018 Pharmaceutical nanocrystals: production by wet milling and applications Drug Discovery Today 23 3 534 547 https://doi.org/10.1016/J.DRUDIS.2018.01.016
41. Megestrol acetate NCD oral suspension—par pharmaceutical: megestrol acetate NanoCrystal® dispersion oral suspension, PAR 100.2, PAR-100.2 (2007). In Drugs in R and D, vol 8, no 4. https://doi.org/10.2165/00126839-200708040-00005
42. Mgadi K, Ndaba B, Roopnarain A, Rama H, Adeleke R (2024) Nanoparticle applications in agriculture: overview and response of plant-associated microorganisms. Front Microbiol 15. https://doi.org/10.3389/fmicb.2024.1354440
43. Mohajerani A, Burnett L, Smith JV, Kurmus H, Milas J, Arulrajah A, Horpibulsuk S, Kadir AA (2019) Nanoparticles in construction materials and other applications, and implications of nanoparticle use. Materials (19). https://doi.org/10.3390/ma12193052
44. Mohseni M, Ramezanzadeh B, Yari H, Moazzami M (2012) The role of nanotechnology in automotive industries. New Adv Veh Technol Automot Eng. https://doi.org/10.5772/49939
45. Monnier V, Sanz N, Botzung-Appert E, Bacia M, Ibanez A (2006) Confined nucleation and growth of organic nanocrystals in sol-gel matrices. J Mater Chem 16(15). https://doi.org/10.1039/b509833p
46. Montanarella F, Kovalenko MV (2022) Three millennia of nanocrystals. ACS Nano 16(4). https://doi.org/10.1021/acsnano.1c11159
47. Nagy ZK, Aamir E (2012) Systematic design of supersaturation controlled crystallization processes for shaping the crystal size distribution using an analytical estimator. Chem Eng Sci 84.https://doi.org/10.1016/j.ces.2012.08.048
48. Olver I, Shelukar S, Thompson KC (2007) Nanomedicines in the treatment of emesis during chemotherapy: focus on aprepitant. Int J Nanomed 2(1). https://doi.org/10.2147/nano.2007.2.1.13
49. Ouda A, Bajón-Fernández Y, McAdam E (2024) Methods to modify supersaturation rate in membrane distillation crystallisation: control of nucleation and crystal growth kinetics (including scaling). J Membr Sci 691.https://doi.org/10.1016/j.memsci.2023.122249
50. Özkar S (2024) Increasing the catalytic efficiency of rhodium(0) nanoparticles in hydrolytic dehydrogenation of ammonia borane. Int J Hydrog Energy 54.https://doi.org/10.1016/j.ijhydene.2023.03.322

51. Padmanabhan NT, John H (2020) Titanium dioxide based self-cleaning smart surfaces: a short review. J Environ Chem Eng 8(5). https://doi.org/10.1016/j.jece.2020.104211
52. Panthani MG, Korgel BA (2012) Nanocrystals for electronics. Annu Rev Chem Biomol Eng 3. https://doi.org/10.1146/annurev-chembioeng-062011-081040
53. Peltonen L, Hirvonen J (2010) Pharmaceutical nanocrystals by nanomilling: critical process parameters, particle fracturing and stabilization methods. J Pharm Pharmacol 62(11). https://doi.org/10.1111/j.2042-7158.2010.01022.x
54. Piszczek P, Radtke A (2018) Silver nanoparticles fabricated using chemical vapor deposition and atomic layer deposition techniques: properties, applications and perspectives: review. In: Noble and precious metals—properties, nanoscale effects and applications. https://doi.org/10.5772/intechopen.71571
55. Pratsinis SE (2010) Aerosol-based technologies in nanoscale manufacturing: from functional materials to devices through core chemical engineering. AIChE J 56(12). https://doi.org/10.1002/aic.12478
56. Priolo F, Gregorkiewicz T, Galli M, Krauss TF (2014) Silicon nanostructures for photonics and photovoltaics. Nat Nanotechnol 9(1). https://doi.org/10.1038/nnano.2013.271
57. Reverchon E (1999) Supercritical antisolvent precipitation of micro- and nano-particles. J Supercrit Fluids 15(1). https://doi.org/10.1016/S0896-8446(98)00129-6
58. Roy A, Elzaki A, Tirth V, Kajoak S, Osman H, Algahtani A, Islam S, Faizo NL, Khandaker MU, Islam MN, Emran TB, Bilal M (2021) Biological synthesis of nanocatalysts and their applications. Catalysts 11(12). https://doi.org/10.3390/catal11121494
59. Sahi S, Magill S, Ma L, Xie J, Chen W, Jones B, Nygren D (2018) Wavelength-shifting properties of luminescence nanoparticles for high energy particle detection and specific physics process observation. Sci Rep 8(1). https://doi.org/10.1038/s41598-018-28741-y
60. Shafique M, Luo X (2019) Nanotechnology in transportation vehicles: an overview of its applications, environmental, health and safety concerns. Materials 12(15). https://doi.org/10.3390/ma12152493
61. Shipway AN, Willner I (2001) Nanoparticles as structural and functional units in surface-confined architectures. Chem Commun 1(20). https://doi.org/10.1039/b105164b
62. Sivanathan G, Rajagopal S, Mahadevaswamy G, Angamuthu G, Dhandapani NV (2024) Pharmaceutical nanocrystals: an extensive overview. Int J Appl Pharm 16(6):1–9. https://doi.org/10.22159/ijap.2024v16i6.52257
63. Stewart S, Wei Q, Sun Y (2021) Surface chemistry of quantum-sized metal nanoparticles under light illumination. Chem Sci 12(4). https://doi.org/10.1039/d0sc04651e
64. Strojewski D, Krupa A (2022) Spray drying and nano spray drying as manufacturing methods of drug-loaded polymeric particles. Polimery w Medycynie 52(2). https://doi.org/10.17219/pim/152230
65. Trindade T, O'Brien P, Pickett NL (2001) Nanocrystalline semiconductors: synthesis, properties, and perspectives. Chem Mater 13(11). https://doi.org/10.1021/cm000843p
66. Vaiano V, Sannino D, Sacco O (2020) The use of nanocatalysts (and nanoparticles) for water and wastewater treatment by means of advanced oxidation processes. In: Nanotechnology in the beverage industry: fundamentals and applications. https://doi.org/10.1016/B978-0-12-819941-1.00009-2
67. Vinzant K, Clouse DE, Amit SK, Ivanov IN, Davis VA, Khodakovskaya MV (2024) Cellulose nanocrystals are a renewable and biocompatible nanocarrier of agrochemicals directly to plant cells. Adv Sustain Syst 8(5). https://doi.org/10.1002/adsu.202300511
68. Wu X, Omagari S, Gao J, Vacha M (2024) In Situ monitoring of nanocrystal formation and ion migration in lead halide perovskite metal–organic framework composites. Adv Opt Mater 12(3). https://doi.org/10.1002/adom.202301479
69. Yang B, Chen Y, Shi J (2019) Nanocatalytic medicine. Adv Mater 31(39). https://doi.org/10.1002/adma.201901778
70. Yang F, Deng D, Pan X, Fu Q, Bao X (2015) Understanding nano effects in catalysis. Nat Sci Rev 2(2). https://doi.org/10.1093/nsr/nwv024

71. Yin Y, Alivisatos AP (2005) Colloidal nanocrystal synthesis and the organic-inorganic interface. Nature 437(7059). https://doi.org/10.1038/nature04165
72. Zhai X, Lademann J, Keck CM, Müller RH (2014) Nanocrystals of medium soluble actives—novel concept for improved dermal delivery and production strategy. Int J Pharm 470(1–2). https://doi.org/10.1016/j.ijpharm.2014.04.060
73. Zhao M, Chen Z, Lyu Z, Hood ZD, Xie M, Vara M, Chi M, Xia Y (2019) Ru Octahedral nanocrystals with a face-centered cubic structure, {111} facets, thermal stability up to 400 °C, and enhanced catalytic activity. J Am Chem Soc 141(17). https://doi.org/10.1021/jacs.9b01640
74. Zhao W, Zhang J, Kong F, Ye T (2023) Application of perovskite nanocrystals as fluorescent probes in the detection of agriculture- and food-related hazardous substances. Polymers 15(13). https://doi.org/10.3390/polym15132873

Spray Drying Techniques for Nanoparticle Production

P. G. S. Rahul Piyadigama, Yasun Y. Kannangara, Induni W. Siriwardane, and Chanaka Sandaruwan

Abstract This report deals with metastability related to nanomaterials, which, in most cases, is driven through their high surface-to-volume ratio into tendencies like aggregation or structural changes when the systems seek to reduce surface energy. Such instability poses a great challenge in preserving nanomaterials, especially in the case of transitioning from synthesis in aqueous or solvent mediums to dry forms that are necessary for practical applications. However, conventional drying methods, such as oven and vacuum drying, may present certain undesirable effects that include material aggregation, re-stacking, and sintering, which could deteriorate the structural integrity and functionality of the nanomaterials. Against this backdrop, spray drying has emerged as a more economical alternative to freeze-drying, which, though effective, is prohibitively expensive for large-scale applications. This paper also presented the general application of the spray drying technique in preserving all kinds of nanomaterials. In particular, pharmaceuticals, nutraceuticals, and fertilizers have been highlighted to take advantage of this technology. Some classifications of utilized materials and their properties are also summarized. Additionally, a few tables containing studies conducted by research groups regarding nanomaterials such as pharmaceuticals, nutraceuticals, nanofertilizers, cosmeceuticals, and other industrial applications. This comparative study shows the economic and functional benefits of spray drying for the preservation of nanomaterials.

P. G. S. Rahul Piyadigama · I. W. Siriwardane (✉)
Center for Advanced Materials and Smart Manufacturing (CAMSM), Faculty of Computing and Technology, University of Kelaniya, Kelaniya, Sri Lanka
e-mail: induni@kln.ac.lk

Y. Y. Kannangara · C. Sandaruwan (✉)
Sri Lanka Institute of Nanotechnology, Homagama, Sri Lanka
e-mail: supurachchi.chanaka@ku.ac.ae

C. Sandaruwan
Department of Aerospace Engineering, Khalifa University, Abu Dhabi, United Arab Emirates

Research & Innovation Center in 2D Materials, Khalifa University, Abu Dhabi, United Arab Emirates

N. M. Adassooriya et al. (eds.), *Industrial Scale Production of Nanoparticles*, Smart Nanomaterials Technology, https://doi.org/10.1007/978-981-96-9176-0_3

Keywords Nanomaterials preservation · Spray drying · Nanoparticle stability · Nano fertilizer · Pharmaceuticals

1 Introduction

Nanomaterials have been chosen to be the miracle material in numerous applications due to the unique features they exhibit. Conventionally defined as material in the size range of 1–100 nm, nanomaterials display high reactivity and also very unique chemical, and physical properties compared to their bulk state counterparts. The origin of such distinct properties can be attributed to the surface science of these nanomaterials. Nanotechnology is the stronghold of innovation in modern science and industry, finding much appeal in pharmaceuticals, nutraceuticals, and even agriculture [31, 46, 103]. This is possible due to the manipulation of material at a nanometre scale, which allows for special control of its chemical, physical, and biological properties [37, 45, 63, 119]. The unique properties, such as high surface area [101], enhanced solubility [6, 109], improved stability, and controlled release capability [52], have extended the applications of nanomaterials in almost every field. From the various effective technologies for processing nanomaterials, spray drying can change liquid dispersions, emulsions, or solutions into dry powder forms with preserved or improved functional characteristics [117]. The scalability, reasonable costs, and flexibility of spray drying have been vastly utilized across many industries [33, 121]. It allows for the control of the particle size and morphology with high precision, a crucial factor in the fabrication of nanomaterials with specific properties for desired applications [67]. This chapter deals with the type, properties, applications, advantages, and disadvantages of different nanomaterials produced through different spray drying methods used across many industries (Figs. 1, 2, 3, 4, 5 and 6).

Fig. 1 Material properties manipulation

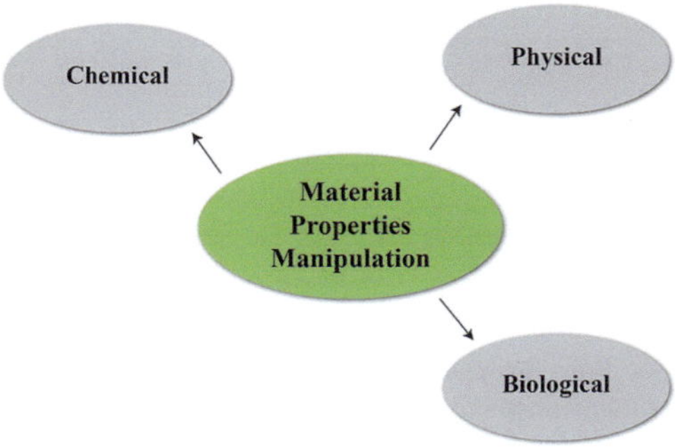

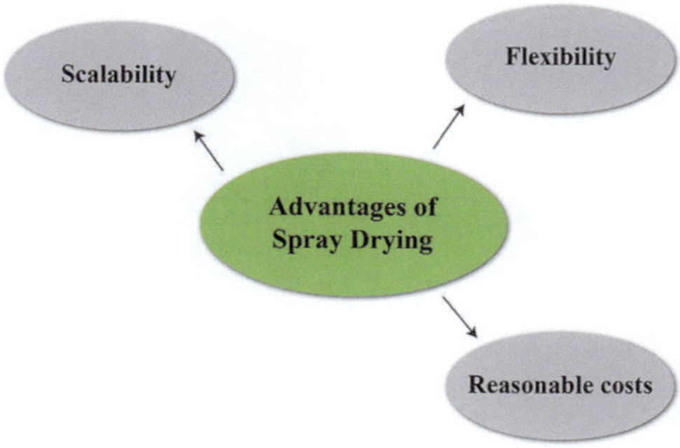

Fig. 2 Advantages of spray drying

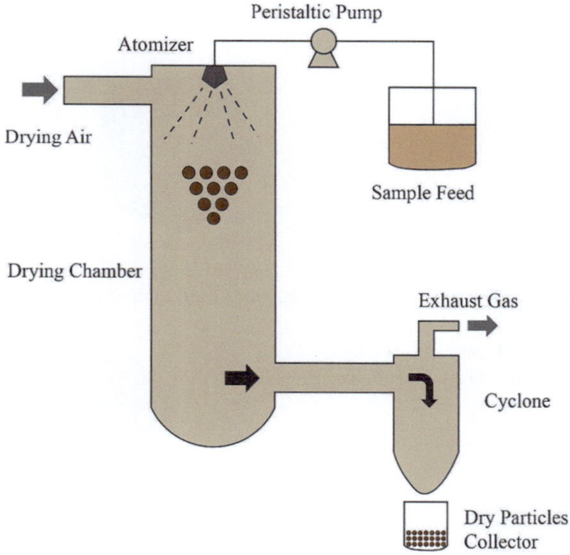

Fig. 3 Schematic diagram of a spray drying process

2 Stability of Nanoparticles

As the materials are becoming smaller in size, the surface area to volume ratio of the material is also increasing exponentially. In a material with more surfaces exposed to the outer environment, the availability of surface atoms is higher. Surface atoms are coordinatively unsaturated (having dangling bonds) compared to the atoms in the bulk material, hence are becoming more reactive. In other words, the surface energy of the nanoscale material is higher compared to that of interior atoms. Therefore, their attempts to reduce the surface energy to achieve better thermodynamic stability result in the very high reactivity of nanomaterials. On the other hand, the same phenomenon imposes challenges in controlling the nanoparticle size, as agglomeration and aggregation of nanomaterial happen in minimizing the surface energy of the synthesized nanomaterials. The nanomaterials are considered as metastable materials that are in

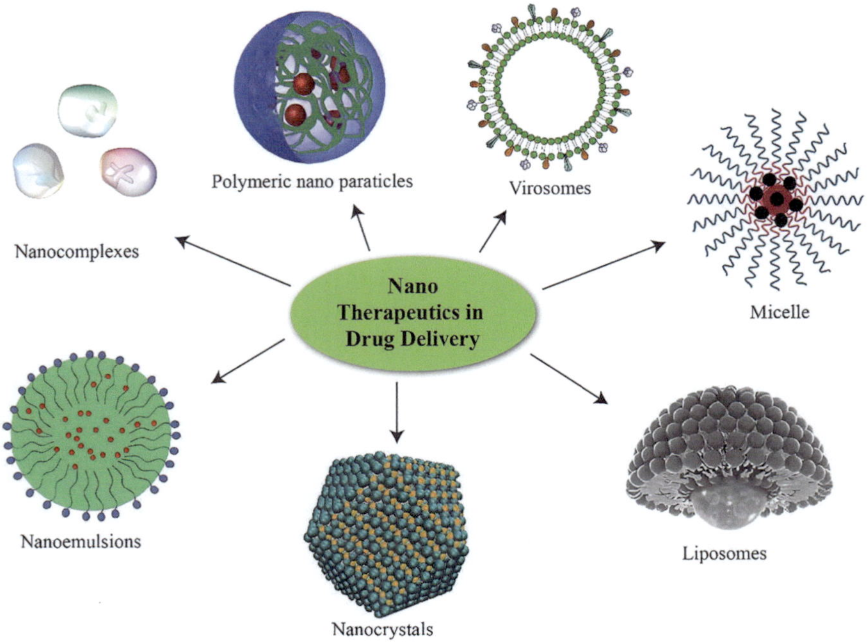

Fig. 4 The types of nano-therapeutics used in drug delivery

Fig. 5 Health-promoting activities of nutraceuticals

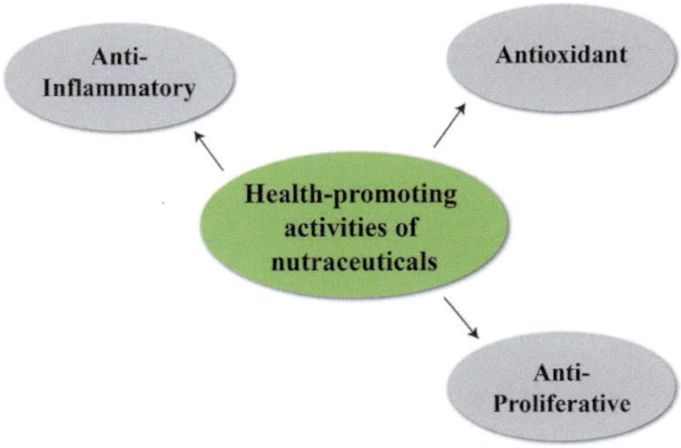

Fig. 6 Properties of nano-fertilizers

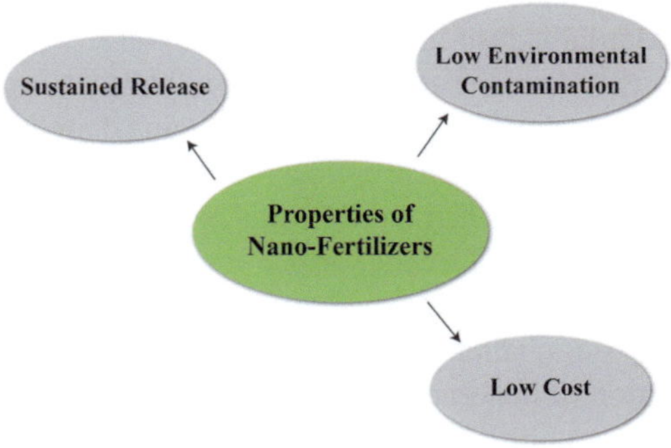

continuous conversion into larger particles, but in an absolutely slow process, which lowers their surface energy.

Several physical and chemical stabilization methods undergo in minimizing the surface energy of nanomaterials [44]. The physical stabilization methods include isotropic minimization by forming spherical particles, anisotropic stabilization by forming multi-faceted particles, geometric agglomeration by Ostwald ripening of particles in gas and liquid phases, physisorption of other atoms and molecules to form a monolayer of adsorbates, surface restructuring, and relaxation methods, sintering of material to achieve grain boundary stabilization, and maximization of packing efficiency of atoms by forming law-index crystal planes to maximize the nearest neighbor atom. The chemical stabilization of nanoparticles includes the use of surfactants to make hydrophobic-hydrophilic interactions, steric stabilization by ligand interactions, electrostatic stabilization by formic electrical double layers, chemisorption methods by forming weak covalent bonds and chemical termination methods where strong covalent bonds are formed.

3 Challenges in Nanomaterial Synthesis and Drying

With the evolution of nanotechnology, nanomaterial synthesis methods have been developed in multiple avenues. Among these diverse techniques, wet chemical synthesis methods such as hydrothermal or solvothermal methods remain the most popular technique for both lab-scale and large-scale syntheses of nanomaterial. In these wet chemical methods, most often, the surfactants are used as the size-controlling mechanism while other parameters such as temperature, pH, electrolyte concentration etc. are also carefully controlled during the synthesis. However, in nanomaterial synthesis, the stability of should be ensured not only in terms of its size, but also in several other aspects. The term "nanoparticle stability" is widely used to describe the preservation of a particular nanostructure property ranging from aggregation, composition, crystallinity, shape, size, and surface chemistry. Phan and Haes [91] Once the nanomaterials are synthesized in liquid media, the removal of the liquid phase while preserving the stability remains a challenge.

Water molecules, which are 0.35 nm in size, have interactions of the solid nanoparticle surface, thereby maintaining a continuous surface of water around the particle surface to form a water meniscus in such samples. Pakowski [85] Hence, the interaction forces between molecules of gas, molecules of liquid, and nanoparticles which are in different magnitudes have to be carefully handled during the drying process of a thin layer of suspension of nanoparticles in liquid. The surfaces that have attained an equilibrium to minimize their surface energy tend to achieve different surface configurations, upon removal of water molecules from the surfaces, which leads to the phenomena like aggregation, and sintering of nanoparticles during drying processes.

4 Nano-spray Drying

Spray drying is a widely used technique which transform solution, suspension, emulsion, liposome, elixir, and slurry or even paste into a dry solid powder. A spray drying techniques consist of four main steps, namely atomization, drying, collection, and waste filtration. Wong and John [122] First, during atomization, when the liquid feed is passing through a nozzle. Next, the dispersed liquid during atomization is then passed through an air stream with the aid of a compressed gas to dry. The drying zone consists of an air intake system, heater, drying chamber, and aspirator. During the drying process, air will be filtered and heated to the required temperature and moved to contact with the sprayed droplets to elicit heat and mass transfer for drying, thereby removing the solvent layers around the particles. After that, the dried particles exit the drying chamber and are collected and separated from the gas stream in the collection chamber. Usually, the collection of dried particles equips mechanisms such as a cyclone separator and/or less commonly an electrostatic precipitator, textile filter, and scrubber. Wong and John [122] The spraying of liquid feed is accompanied by the formation of millions of micrometre-sized droplets. Having a large specific surface area, these droplets experience a very short drying time due to highly effective heat and mass transfer between the droplets and hot air. Finally, the unintended particles and wet air are introduced into the waste filtration zone where particle leakage into the immediate environment and recycling of the air would take place.

Spray drying is a well-mastered technique with applications in various laboratory and industrial scales. However, the nanometric spray drying requires two more additional steps. Pakowski [85] First, the reduction of the initial spray size should be done using special techniques like ultrasonic nebulization or electrostatic spraying. Ultrasonic nebulization produces droplets ca. 1.5 μm in size, while electrostatic spraying allows the production of even smaller droplets. Pakowski [85] Second, the initial concentration of the solution droplets should be very small as to contain the solvent as the major content in the droplet. It is reported that to obtain 100 nm particles from 5 μm droplets of solution in water the initial concentration must be as low as 0.000008% [47].

The spray drying of nanomaterials should be done with care, since inhaled aerosols may end up in the lungs and easily penetrate the vacuole. It is commonly accepted that at this scale of dispersion, nanoparticles can have adverse effects, even if they are composed of nontoxic material.

5 Spray-Dried Nanomaterials for Pharmaceuticals

The pharmaceutical industry extensively employs nanomaterials for drug delivery, controlled release, and improved bioavailability [60, 69, 76]. Spray drying is a commonly used method to convert these nanomaterials into stable, easy-to-handle dry powder forms for various medical applications [15]. When focusing on the utilization of spray-dried nanomaterials for pharmaceuticals a few types of nanoparticles can be highlighted. In the typical pulmonary drug delivery by using dried powder formation for disease situations such as asthma, cystic fibrosis, chronic obstructive pulmonary disorder (COPD), pulmonary arterial hypertension, and lung cancers various problematic instances have occurred. Deposition of particles in the airway dividing, and deep part of the lungs by impaction and sedimentation have been reported [67]. Several studies have proved pulmonary particle deposition by using airway geometries, measurements of airway flow, and effects of drug inhalation and deposition patterns and drug inhalation time [23, 56, 70]. Particles in the nano range smaller than 50 nm and particle sizes of 1–5 μm are seen in depositions, but particles larger than that are inhaled into the lungs, where there is a significant chance of expiration before deposition [42, 95]. With these consequences, the requirement of drugs in the nano range with low or no deposition and also, with high efficiency in treating the disease arises. When addressing the above issues and innovating nano forms for inhalation drugs, a few other challenges would have to be faced. Constant particle size in both the synthesis state and dry state, preserving the active ingredients (biotherapeutics) throughout the whole synthesis procedure, prevention of aggregation of nanoparticles, and well dispersion of nano-sized drug particles in lung fluid are some of them [79]. Different types of nanotherapeutics available in the drug delivery systems, nanoemulsions, nano complexes, liposomes, polymeric micelles, virosomes, polymeric nanoparticles, and nanocrystals are popular in the field [68]. Apart from these pulmonary drug delivery systems, researchers are interested in the usage of nanomaterials in pharmaceutical applications such as antibiotics delivery, gene delivery, cell tracking, cancer treatments, and delivery of therapeutic molecules as well [26, 27, 81]. Drug formulations have included nanomaterials as active ingredients (nanocrystals), drug transporters (liposomes), excipients (drug-metal complexes), or complexes/ conjugates (drug-protein) [27]. For all of the above-discussed nanomaterials used in pharmaceuticals and also to address common issues improve the delivery and provide accelerated healing various types of synthesis methods have been used and considering the pros and cons of each technique researchers have been recommended for a few types of such techniques. When discussing such methodologies spray drying is one technique used to produce dry nanomaterial-based pharmaceuticals.

Recently, Munir et al. have developed inhalable peptide-DNA nanoparticles for pulmonary delivery by spray drying where they expected spherical nanoparticles to be achieved after the spray drying procedure, and by the scanning electron microscopic (SEM) images the initial expectation was fulfilled. Spherical-shaped inhalable peptide-DNA particles show smooth surfaces in the SEM images [41]. In this nanomaterial-based pharmaceutical synthesis, Healy et al. introduced the nanoporous microparticles spray-dried to manufacture excipient-free nanoporous particles [78]. Another development of microparticles has been recorded where lactose-leucine microparticles for pulmonary delivery by spray drying targeting for antibacterial actions [112]. Amr Maged et al. have successfully developed spray-dried rosuvastatin nanoparticles for promoting hair growth and it has been tested on albino rats which shows increased hair growth. For this, they used a nanospray drier and methyl-β-cyclodextrin as a carrier for obtaining expected nanoparticles and after spray drying, it was further dried by Nitrogen gas [66]. The application of spray drying covers a vast area, even in pharmaceutical nanoparticle production it goes into numerous regions.

It is proved by the applications and studies of different types of drug nanoparticles and the way that they are delivered. One such example can be highlighted as follows. Yigong Guo et al. have produced nanoparticles of insulin to be orally delivered by using spray drying and freeze drying. For the spray drying process, a Buchi mini spray dryer B-290 (BÜCHI, Flawil, Switzerland) was used and obtained insulin nanoparticles which showed superior reconstitution ability and higher loading content. They have assessed the impact of spray drying to produce these insulin-loaded chitosan nanoparticles. In the end, that study showcased insulin nanoparticles can be dehydrated without the aid of cryoprotectants by utilizing the technique of spray drying, which directly affects the lower additive need and higher loading capacity which leads to lower operating costs concerning traditional freeze-drying techniques in the synthesis procedure [39]. Rutin is one of the common flavonoids that are available in fruits and vegetables which possesses Vaso protective, anti-inflammatory, antiviral, anticarcinogenic, and antioxidant properties. However, the limited oral bioavailability of rutin has an impact on its biological actions and to address this circumstance a group of researchers has conducted a study of loading rutin to bovine serum albumin nanoparticles by spray drying which resulted in producing spherical particles with an average particle size of 316 nm [87]. As spray drying is a multifunctional technique in nanoparticle production as well as a preservation method the studies have gone deep up to the brain drug delivery as well. Alaa H. Salama et al. have worked on developing a process for nano spray drying to dry gabapentin-loaded nanoparticles for the treatment of PTZ-induced seizures. The initial nanoparticle matrix was formed by using chitosan and D-α-Tocopherol polyethylene glycol 1000 succinate (TPGS) [125] a non-ionic surfactant [22] as a stabilizer that ultimately promotes the formation of smooth and good spherical shaped nanoparticles [9, 90]. At the end of their study gabapentin-loaded nanoparticles with an average size of 107 ± 13 nm were obtained by spray drying [90].

These studies have been conducted to synthesize nanoparticle-based pharmaceuticals, for numerous applications using the spray-drying technique, demonstrating the suitability and appropriateness of spray-drying for them. A few more such nanoparticle-based spray-dried pharmaceuticals are summarized in Table 1.

6 Spray-Dried Nanomaterials for Food and Nutraceuticals

Pharmaceutical forms that contain food phytochemicals as active principles are called nutraceuticals. These nutraceuticals are dietary supplements that provide a concentrated form of an assumed bioactive ingredient from food in a non-food matrix. They are used at higher dosages than what might be gained from typical foods to improve health [30, 123]. Nutraceuticals offer several advantages to humans. From them, a few can be listed as the capacity to treat metabolic diseases, such as obesity and

Table 1 Nanoparticle-based spray-dried pharmaceuticals

#	Type of nanomaterial	Application	Description	Reference
1	Vancomycin coated magnetic nanoparticles	Localized lung delivery	Vancomycin absorbed on the core of magnetic nanoparticles (MN) by spray drying. MN diameter of14.7 ± 5.9 nm	Abdelaziz et al. [1]
2	Acetazolamide from ammonia solutions	Glaucoma management	Addresses the weak diuretic activity of Acetazolamide by spray drying by spherical particles	Di Martino et al. [25]
3	Nanocrystalline microparticles of chlorothiazide	Producing additive-free chlorothiazide nanocrystalline microparticles	Making poorly compactable chlorothiazide into direct compactable powder by spray drying	Paluch et al. [86]
4	Amorphous celecoxib and celecoxib-PVP-meglumine ternary mixture	Making powder suitable for direct compaction for tablets	They found spray dried powder have great tensile strength compared to crystalline celecoxib	Joshi et al. [5, 50]
5	Naproxen and naproxen sodium with HPMC	Higher mechanical strength	Naproxen's crystallinity decreased and its dihydrate content increased as a result of co-processing	Al-Zoubi et al. [5]
6	Tenofovir alafenamide-chitosan nanoparticles loaded oleo gels	Long-acting injectable depot system of anti-HIV drug	Sesame oil and ethyl cellulose were used to create oleo gels filled with polymeric nanoparticles for sustained release for the first time	Naryan et al. [80]
7	Silica nanoparticles-sodium alginate	Biohybrid drug carrier of doxorubicin	Improved functionality of Dox-biohybrid by spray drying	Mishra et al. [75]
8	Cefaclor monohydrate containing nanoparticles	Extended antibacterial effect	Sustained release when taken orally, cefaclor monohydrate nanoparticles	Öztürk et al. [84]

(continued)

Table 1 (continued)

#	Type of nanomaterial	Application	Description	Reference
9	Simvastatin Nanoparticles	Pulmonary Arterial Hypertension	Dry powder formation of simvastatin for inhalation with particle size below 100 nm	Zendehdel et al. [124]
10	Cellulose nanocrystals and metal oxide-based nanoparticles	Antimicrobial and antibiofilm activity against Gram-negative and Gram-positive bacteria	Matrix compositions (ratios) in controlling the toxicity and antibacterial qualities of metal oxide biomaterials	Baldelli et al. [12]

heart disease, etc., the capacity to treat bone conditions, including osteoporosis and arthritis etc., the ability to handle long-term inflammatory conditions, acting as immunity boosters, and preventing cancer and diabetes [97]. In nutraceutical synthesis, substances such as vitamins like vitamin A, vitamin B, vitamin C, vitamin D, and vitamin E, and minerals like Calcium, Magnesium, Ferrous, Zinc, and Manganese are focused [92]. Also, incorporating plant extracts into the nutraceutical is popular and it showcases the positive effect of them in the usage. Potentially health-promoting activities, anti-inflammatory, antioxidant, and anti-proliferative properties are highlighted properties among them [111]. The importance of encapsulating such plant extracts into nutraceuticals is demonstrated by several studies. Such research shows that anthocyanins and berry extracts can improve memory and safeguard brain function by lowering oxidative ischemia damage and improving cognitive performance [64, 115, 120].

When focusing on the involvement of nanotechnology in nutraceuticals, active ingredients are encapsulated according to the food grade or Generally Recognized as Safe (GRAS) [40]. Those ingredients are encapsulated in nanocarriers such as liposomes, nanoemulsions, nanodispersions, and nanoparticles [104]. For the encapsulation procedure of an active ingredient to a carrier to produce a nutraceutical, different methods are used. Solvent evaporation [62], electrohydrodynamic techniques electrospinning and electrospraying [21, 82], complex coacervation [63, 99], fluidized bed coating [65], ionic gelation [59], melt extrusion [93, 107], spray chilling [83], freeze-drying [4], and spray drying [10] are some of the common techniques. In this chapter, the utilization of spray drying to produce nanomaterials-based nutraceuticals is discussed.

When going through the studies conducted regarding spray drying, a group of researchers Iara et al. has successfully studied the possibility of loading rosemary (*Rosmarinus officinalis*) into well-organized lipid structures to enhance its physical–chemical characteristics. Rosemary is a herb that has an inherent antioxidative activity and is usually used as a spice to improve taste and preserve foods. In there, extracts of rosemary with stearic and oleic acids, as well as a variable mixture of gum Arabic or whey protein concentrate as drying carriers and poloxymer 188 as a surfactant in the aqueous phase. Then, the mixture was spray-dried to obtain the powered version [13]. Miranda et al. have investigated encapsulating a functionalized component in nano- and microparticles of alginate-pectin that are packed full of different natural active ingredients and successfully produced them [73]. Another study done by Fei-Ping et al. reports the encapsulation of curcumin nano complexes with soy protein isolate by spray drying. In there, mainly encapsulation, bioactivity, assessing bioaccessibility, water dispersion, and encapsulation are conducted. In their study, they were able to produce a higher concentration of curcumin-loaded soy proteins [19]. Rocío et al. have worked on encapsulating folic acid, and formulations were spray-dried to obtain electrospraying and the final result of the capsules looks identical to each other. The folic acid was spry-dried with food hydrocolloids. The microscopic images show spherical submicron-shaped particles [89]. Amr et al. spray-dried *Nigella sativa* (black cumin) extract, extracted from hexane to apply on foods and nutraceuticals. This black cumin oil has several advantages over the other types of plant seed oils. The rich content of omega-6 and omega-9 fatty acids, oleic and linoleic showcases the nutritional value which has protective properties of liver and nerve cells [3, 43, 116]. In the study, they investigated the handling and protection of the available oil in the extracts in the powder production procedure by spray drying [29].

The discussed instances show the wide usage of the spray drying technique in the synthesis of different foods and nutraceuticals, which demonstrates the advantageous situations and benefits of using spray drying over other methods. Table 2. Summarises a few more applications and studies that have been conducted to synthesis and drying of nanomaterial-based nutraceuticals.

7 Spray-Dried Nanomaterials for Fertilizers and Agricultural Products

With the current requirements and demand for sustainable agriculture with higher harvests and minimum environmental damage, researchers are interested in discovering new types of highly efficient fertilizers and other agricultural requirements. The problem facing agriculture is to produce food using sustainable methods for a population that is constantly expanding. According to UN predictions, there will be 10.9 billion people on the planet by the year 2100 [28]. This means the food production also should increase accordingly. Fertilizers boost output since they provide vital nutrients for plant development [54]. However, because of the lixiviation and volatilization processes, which cause 30–50% of fertilizer loss, high ion mobility in the soil can contaminate rivers, lakes, and aquifers [36, 49, 102, 113]. To address the consequences of conventional fertilizer with properties such as slow release, reduced amount to apply, and considerable manufacturing cost are developed. The materials known as enhanced efficiency fertilizers mix fertilizer with components of interest to create a finished product with beneficial qualities [20, 113].

Table 2 Nanoparticle-based spray-dried nutraceuticals

#	Type of nanomaterial	Application	Description	Reference
1	Carbohydrate matrix	Antioxidant property	Catechins encapsulated nanoparticles carbohydrate matrix to improve bioavailability	Ferreira et al. [32]
2	Grapefruit nutraceutical powder	The possible health impacts of bioactive substances	Grapefruit (*Citrus paradisi*) powder used in spray drying as a nutraceutical	Freddy et al. [38]
3	Encapsulation of Bioactive Peptides	Alpha-calcitonin gene-related peptide	Synthesis using alginate as the carrier and water as a solvent to get Vasodilator	Carmen et al. [58]
4	Nettle Leaf Extract Powder	Encapsulation of polyphenols	Ground leaf powder extracts and spray dried to encapsulation	Ena et al. [16]
5	Venlafaxine with Chitosan microparticles	Reducing side effects of depression treatment	Venlafaxine drug-controlled release by spray drying with chitosan	Inmaculada et al. [7]
6	β-carotene encapsulated Pickering emulsion with chitosan and seaweed polyphenol	Preventing destroying of β-carotene in the drying process	Antioxidant, anti-inflammatory, anti-diabetic, anti-viral, hepatoprotective and neuroprotective activities encapsulation	Meng et al. [71]
7	Ascorbic acid-loaded gliadin nanoparticles	Enhancing stability of antioxidants	Encapsulation in a matrix of vegetal proteins composed of gliadin, the prolamin derived from wheat kernels	Voci et al. [118]
8	Chlorophyll with maltodextrin and whey protein isolate	Healthy green pigment	Chlorophyll was extracted from a seaweed and rotary evaporated then spray dried	Ledari et al. [61]

When it focuses on the involvement of nanotechnology in agriculture, it is mainly to enhance properties. Nanotechnology might present a chance to enhance fertilizers. However, it could also raise issues. Nanomaterial's larger surface area can result in greater reactivity and quicker dissolution [17]. With these advantages, the utilization of nano-material-based fertilizers as well as other agricultural requirements, have been innovated. When it is focusing on nano fertilizers different types of new formations have been discovered with enhanced properties. A group of researchers Silva et al. have synthesized urea-hydroxyapatite nanohybrids for the slow releasing of required nutrients to plants. They have successfully developed an innovative, readily scalable, energy-efficient, environmentally friendly, one-step bottom-up mechanochemical method for creating Urea-HA nanohybrids in situ, which results in a slow-release, functional nano fertilizer [24]. Another study conducted by Kottegoda et al. has developed modified layered nanohybrid structures for the slow release of urea using layered double hydroxides and montmorillonite [55]. Those novel fertilizers have been designed to magnify their performances and ultimately improve the harvest, reduce the cost, and save the environment.

In this part of the chapter, spray drying is used to synthesize nano-materials-based fertilizers and other agriculture-based products. When focusing on such works, Chella Divyah et al. planned to produce a micronutrient nanoparticles fertilizer mixture for banana cultivation by spray drying methodology [18] Another study shows that Alberto et al. tried to enhance water solubility of low water-soluble foliar fertilization by using spray drying. Scanning electron microscopic images of this study show non-porous, irregular, and rough surfaces of the particles [2]. Miranda-Villagómez et al. studied about releasing mechanism of phosphorous and potassium by producing biopolymeric particles loaded with phosphorus and potassium and determined a controlled release [74]. Juan-Martínez et al. conducted a study to synthesize calcium-loaded gelatin nanoparticles to be used as a plant fertilizer. In there, type-A gelatin, Calcium, and citric acid were used together for the spray drying procedure, and they investigated the releasing mechanism of calcium using an atomic absorption spectrometer (PinAAcle 900, PerkinElmer, USA) according to the method that has been followed by another group of scientists (previous study) who were investigated phosphorous and potassium release [74]. They have determined that Calcium is slowly released with the time. Furthermore, by FTIR analysis, they have proved the incorporation of Calcium in the synthesized particles [51]

Discussed examples show that spray drying of nano fertilizers and other agricultural products is studied for a lower coverage compared to other types such as pharmaceuticals and nutraceuticals. Apart from the studies that are explained above, a few other works are summarized in Table 3.

Table 3 Nanoparticle-based spray-dried fertilizers and agricultural products

#	Types of nanomaterial/ material	Application	Description	Reference
1	Chitosan, chitosan/ montmorillonite clay	Slow released fertilizer	Greater swelling values (approximately 500%) and a slower fertilizer release rate (less than 70% in the initial measurements)	França et al. [35]
2	Mineral fertilizer by polysulfide	Slow released fertilizer	NPK 6–20-30 fertilizer, polysulfide (PSF) coating, and N, N dimethylformamide	Tomaszewska et al. [114]
3	Chitosan, and nanocellulose, NPK fertilizer solution	Slow released fertilizer	The CS microparticles caused a decrease in the initial release rate of NPK	Messa et al. [72]
4	Cellulose and cellulose nanofibrils (CNFs)	Improve the biodegradability of the coating	Cellulose and cellulose nanofibrils to encapsulate nitrogen- and potassium-based nutrients	Franca et al. [34]

8 Spray-Dried Nanomaterials for Cosmeceuticals and Other Industries

The spray drying technique has extended for numerous applications in different scales such as macroscopic, micron scale, and nanoscale. Other than nano-level applications in pharmaceuticals, nutraceuticals, and fertilizers, there are some other applications also. However, those are not much extended and popular among researchers. One such field is cosmeceuticals. Currently, people in the world are looking for products and substances that provide personal care for them. It is not specific to women but also applies to men for the improvement of existing appearance. For this purpose, cosmetics are used. Cosmetics are materials that are applied to enhance appearance, improve attractiveness, and other requirements such as washing and beautifying [8]. Cosmeceuticals are usually developed by incorporating some active ingredients and allowed to be applied in normal conditions, but due to their drug-like active ingredients, they influence the skin's biological functions [105]. The incorporation of nanoformulations from nanotechnology into these cosmeceuticals and efficient delivery of the incorporated ingredients is achieved [106]. It has reported the incorporation of liposomes in moisturizing creams nearly four decades ago [11]. Since then, numerous types of cosmeceuticals based on nanomaterials have been developed for different applications, such as shampoos, conditioners, lip-care products, nail polishing items, and many more [53, 88, 96, 110]. Focusing on spray drying and synthesizing these cosmeceuticals, a few studies have been reported. Su et al. have successfully synthesized polymethyl methacrylate and boron nitride powder by spray drying. Even though the particle size is on a micro-scale, with the particle size of

30.5 μm and spherical shape. One of the expected properties of these particles was the ability of oil absorption, which it has shown 259 g/100 g due to their higher number of pores within the composite particles [108]. Morland et al. have investigated the photocatalytic activity of TiO_2 nanoparticles that are encapsulated by chitosan for utilization in sun-blocking applications by using spray drying. Through their study, they have determined that the photoactivity of TiO_2-encapsulated chitosan nanoparticles has been reduced by a considerable amount concerning the original version of TiO_2 nanoparticles. Using crystal violet, an organic dye, as the target for degradation and subjecting it to UV radiation, the photocatalytic properties were evaluated [77]. Another study done by Jafari et al. assessed the activity of emulsifying in emulsifiers by d-limonene nanoparticles encapsulation by spray drying. In that study, they used tween 20 as the emulsifying agent and found that it greatly affected in reducing the emulsion size to below 200 nm [48].

Apart from the utilization of spray drying to synthesize nanomaterials-based cosmeceuticals, there are also other types of industries and activities where nanomaterials-based products are manufactured by spray drying. Some of them are summarized in Table 4.

Table 4 Miscellaneous nanoparticle-based spray-dried cosmeceuticals and other industrial materials

#	Types of nanomaterial/material	Application/Type	Description	Reference
1	Cyclosporin A and dexamethasone with poly(d,l-lactide-co-glycolide) (PLGA)	Improving water solubility Improve the particle size and the obtained yield	Cyclosporin A and dexamethasone encapsulation in poly(d,l-lactide-co-glycolide)	Schafroth et al. [100]
2	Cardamom oleoresin using Arabic, maltodextrin, and modified starch	Food related	Encapsulation of cardamom oleoresin	Krishnan et al. [57]
3	Hydroalcoholic with maltodextrin	Food related	Synergistic antioxidant mushroom extracts microencapsulated	Ribeiro et al. [94]
4	Cinnamon infusions with maltodextrin	Food related	Microencapsulation of cinnamon infusions via spray drying	Santiago-Adame et al. [98]
5	Probiotic fruit juice powders	Food nutritional related Maintain viability of probiotics	Incorporation of probiotic cultures to fruit juices by spray drying to improve functional properties	Barbosa et al. [14]

9 Conclusion

It is a continuous challenge to preserve the structural integrity of nanomaterials without functionality compromise since, naturally, they will have high metastability due to the huge surface-to-volume ratio. The inherent instability makes nanoparticles prone to aggregation and other structural changes, especially under drying processes. Oven drying and vacuum drying raise very critical issues of material aggregation, re-stacking, and sintering that usually limit the practical usability of the dried nanomaterials. While these have been some of the challenges, spray drying has become a viable alternative that meets most of these without affecting nanomaterials' structural and functional properties. Compared to freeze drying which though very effective, is highly expensive spray drying can be more effectively scaled up and economically viable, especially for industrial applications. This has evidenced its versatility, from preserving pharmaceuticals, nutraceuticals, and even fertilizers to setting it as one of the key methods in the preservation of nanomaterials. The further direction of study should be at optimizing spray-drying parameters regarding specific properties that might differ with nanomaterial type. Providing a comprehensive set of datasets shows the way to create a solid solution for preserving nanomaterials. The result of such efforts will be the better optimization of spray drying techniques, which ensures wider applications in various sectors and opens up new avenues for the application of nanomaterials both in R&D and industry.

References

1. Abdelaziz MM, Hefnawy A, Anter A, Abdellatif MM, Khalil MAF, Khalil IA (2022) Respirable spray dried vancomycin coated magnetic nanoparticles for localized lung delivery. Int J Pharm 611:121318
2. Alberto L, da Luz MS, dos Santos KG, Okura MH (2022) Enhanced solubility of foliar fertilizer via spray dryer: process analysis and productivity optimization by response surface methodology. Ciência e Agrotecnologia 46:e002422
3. Al-Okbi SY, Mohamed DA, Hamed TE, Edris AE (2013) Potential protective effect of Nigella sativa crude oils towards fatty liver in rats. Eur J Lipid Sci Technol 115(7):774–782
4. Alonso S (2016) Novel preservation techniques for microbial cultures. Nov Food Ferment Technol 7–33
5. Al-Zoubi N, Odeh F, Partheniadis I, Gharaibeh S, Nikolakakis I (2021) Spray drying of naproxen and naproxen sodium for improved tableting and dissolution–physicochemical characterization and compression performance. Pharm Dev Technol 26(2):193–208
6. Andrievski RA (2014) Review of thermal stability of nanomaterials. J Mater Sci 49:1449–1460
7. Aranaz I, Paños I, Peniche C, Heras Á, Acosta N (2017) Chitosan spray-dried microparticles for controlled delivery of venlafaxine hydrochloride. Molecules 22(11):1980
8. Arora N, Agarwal S, Murthy RSR (2012) Latest technology advances in cosmaceuticals. Int J Pharm Sci Drug Res 4(3):168–182
9. Arpagaus C (2011) Nano spray dryer B-90: literature review and applications. Büchi Inf Bull 63(8)
10. Arpagaus C (2019) Production of food bioactive-loaded nanoparticles by nano spray drying. In: Nanoencapsulation of food ingredients by specialized equipment. Elsevier, pp 151–211

11. Ashtiani HRA, Bishe P, Lashgari N-A, Nilforoushzadeh MA, Zare S (2016) Liposomes in cosmetics. J Skin Stem Cell 3(3)
12. Baldelli A, Etayash H, Oguzlu H, Mandal R, Jiang F, Hancock REW, Pratap-Singh A (2022) Antimicrobial properties of spray-dried cellulose nanocrystals and metal oxide-based nanoparticles-in-microspheres. Chem Eng J Adv 10:100273
13. Baldim I, Souza CRF, Durazzo A, Lucarini M, Santini A, Souto EB, Oliveira WP (2020) Spray-dried structured lipid carriers for the loading of rosmarinus officinalis: new nutraceutical and food preservative. Foods 9(8):1110
14. Barbosa J, Teixeira P (2017) Development of probiotic fruit juice powders by spray-drying: a review. Food Rev Intl 33(4):335–358. https://doi.org/10.1080/87559129.2016.1175016
15. Baumann JM, Adam MS, Wood JD (2021) Engineering advances in spray drying for pharmaceuticals. Annu Rev Chem Biomol Eng 12(1):217–240
16. Cegledi E, Garofulić IE, Zorić Z, Roje M, Dragović-Uzelac V (2022) Effect of spray drying encapsulation on nettle leaf extract powder properties, polyphenols and their bioavailability. Foods 11(18):2852
17. Chahal AS, Madgulkar AR, Kshirsagar SJ, Bhalekar MR, Dikpati A, Gawli P (2012) Amorphous nanoparticles for solubility enhancement. J Adv Pharm Sci 2:167–178
18. Chella Divyah D, Ramya Devi D, Vedha Hari BN, Winnifer Jose W, Isabell Rupa R, Jeya Baskaran KJ, Ramajayam D (2024) Efficiency of micronutrient fertilizer mixture as nanoformulation for improving banana cultivation. In: Responsible production and consumption: proceedings of the international conference on responsible consumption and production: agriculture sustainability and food security, p 280
19. Chen F-P, Liu L-L, Tang C-H (2020) Spray-drying microencapsulation of curcumin nanocomplexes with soy protein isolate: Encapsulation, water dispersion, bioaccessibility and bioactivities of curcumin. Food Hydrocolloids 105:105821
20. Chen M, Schievano A, Bosco S, Montero-Castaño A, Tamburini G, Pérez-Soba M, Makowski D (2023) Evidence map of the benefits of enhanced-efficiency fertilisers for the environment, nutrient use efficiency, soil fertility, and crop production. Environ Res Lett 18(4):043005
21. Coelho SC, Estevinho BN, Rocha F (2021) Encapsulation in food industry with emerging electrohydrodynamic techniques: electrospinning and electrospraying—a review. Food Chem 339:127850
22. Cortés H, Hernández-Parra H, Bernal-Chávez SA, Del Prado-Audelo ML, Caballero-Florán IH, Borbolla-Jiménez FV, González-Torres M, Magaña JJ, Leyva-Gómez G (2021) Non-ionic surfactants for stabilization of polymeric nanoparticles for biomedical uses. Materials 14(12):3197
23. Darquenne C (2012) Aerosol deposition in health and disease. J Aerosol Med Pulm Drug Deliv 25(3):140–147
24. de Silva M, Sandaruwan C, Hernandez FCR, Sahin O, Ashokkumar M, Ajayan PM, Karunaratne V, Amaratunga GAJ, Kottegoda N (2022) A greener mechanochemical approach to the synthesis of urea-hydroxyapatite nanohybrids for slow release of plant nutrients
25. Di Martino P, Scoppa M, Joiris E, Palmieri GF, Andres C, Pourcelot Y, Martelli S (2001) The spray drying of acetazolamide as method to modify crystal properties and to improve compression behaviour. Int J Pharm 213(1–2):209–221
26. Di paolo A (2004) Liposomal anticancer therapy: pharmacokinetic and clinical aspects. J Chemother 16(sup4):90–93
27. D'Mello SR, Cruz CN, Chen M-L, Kapoor M, Lee SL, Tyner KM (2017) The evolving landscape of drug products containing nanomaterials in the United States. Nat Nanotechnol 12(6):523–529
28. Dorling D (2021) World population prospects at the UN: our numbers are not our problem? In: The struggle for social sustainability. Policy Press, pp 129–154
29. Edris AE, Kalemba D, Adamiec J, Piątkowski M (2016) Microencapsulation of Nigella sativa oleoresin by spray drying for food and nutraceutical applications. Food Chem 204:326–333
30. Espín JC, García-Conesa MT, Tomás-Barberán FA (2007) Nutraceuticals: facts and fiction. Phytochemistry 68(22–24):2986–3008

31. Fallah Z, Zare EN, Ghomi M, Ahmadijokani F, Amini M, Tajbakhsh M, Arjmand M, Sharma G, Ali H, Ahmad A (2021) Toxicity and remediation of pharmaceuticals and pesticides using metal oxides and carbon nanomaterials. Chemosphere 275:130055
32. Ferreira I, Rocha S, Coelho M (2007) Encapsulation of antioxidants by spray-drying. Chem Eng Trans 11(9):713–717
33. Filková I, Mujumdar AS (2020) Industrial spray drying systems. In: Handbook of industrial drying. CRC Press, pp 263–307
34. Franca D, de Barros JRS, Faez R (2021) Spray-dried cellulose nanofibrils microparticles as a vehicle for enhanced efficiency fertilizers. Cellulose 28:1571–1585
35. França D, Medina ÂF, Messa LL, Souza CF, Faez R (2018) Chitosan spray-dried microcapsule and microsphere as fertilizer host for swellable—controlled release materials. Carbohyd Polym 196:47–55
36. García-Garizábal I, Causapé J, Abrahão R (2012) Nitrate contamination and its relationship with flood irrigation management. J Hydrol 442:15–22
37. Golovin YI (2021) Nanoindentation and mechanical properties of materials at submicro-and nanoscale levels: recent results and achievements. Phys Solid State 63:1–41
38. González F, García-Martínez E, del Mar Camacho M, Martínez-Navarrete N, Sarmento B, Fernandes I, Freitas V, Rodrigues F, Oliveira B (2019) Insights into the development of grapefruit nutraceutical powder by spray drying: Physical characterization, chemical composition and 3D intestinal permeability. J Sci Food Agric 99(10):4686–4694
39. Guo Y, Baldelli A, Singh A, Fathordoobady F, Kitts D, Pratap-Singh A (2022) Production of high loading insulin nanoparticles suitable for oral delivery by spray drying and freeze drying techniques. Sci Rep 12(1):9949
40. Hallagan JB, Hall RL, Drake J (2020) The GRAS provision-the FEMA GRAS program and the safety and regulation of flavors in the United States. Food Chem Toxicol 138:111236
41. Healy AM, McDonald BF, Tajber L, Corrigan OI (2008) Characterisation of excipient-free nanoporous microparticles (NPMPs) of bendroflumethiazide. Eur J Pharm Biopharm 69(3):1182–1186
42. Heyder J (2004) Deposition of inhaled particles in the human respiratory tract and consequences for regional targeting in respiratory drug delivery. Proc Am Thorac Soc 1(4):315–320
43. Hobbenaghi R, Javanbakht J, Sadeghzadeh SH, Kheradmand D, Abdi FS, Jaberi MH, Mohammadiyan MR, Khadivar F, Mollaei Y (2014) Neuroprotective effects of Nigella sativa extract on cell death in hippocampal neurons following experimental global cerebral ischemia-reperfusion injury in rats. J Neurol Sci 337(1–2):74–79
44. Hornyak GL, Tibbals HF, Dutta J, Moore JJ (2008) Introduction to nanoscience and nanotechnology. CRC Press
45. Hwang M, Yeom B (2020) Fabrication of chiral materials in nano-and microscale. Chem Mater 33(3):807–817
46. Ioannou A, Gohari G, Papaphilippou P, Panahirad S, Akbari A, Dadpour MR, Krasia-Christoforou T, Fotopoulos V (2020) Advanced nanomaterials in agriculture under a changing climate: the way to the future? Environ Exp Bot 176:104048
47. Iskandar F, Gradon L, Okuyama K (2003) Control of the morphology of nanostructured particles prepared by the spray drying of a nanoparticle sol. J Colloid Interface Sci 265(2):296–303. https://doi.org/10.1016/S0021-9797(03)00519-8
48. Jafari SM, He Y, Bhandari B (2007) Encapsulation of nanoparticles of d-limonene by spray drying: role of emulsifiers and emulsifying techniques. Drying Technol 25(6):1069–1079
49. Jarosiewicz A, Tomaszewska M (2003) Controlled-release NPK fertilizer encapsulated by polymeric membranes. J Agric Food Chem 51(2):413–417
50. Joshi AB, Patel S, Kaushal AM, Bansal AK (2010) Compaction studies of alternate solid forms of celecoxib. Adv Powder Technol 21(4):452–460
51. Juan-Martínez ES, Sandoval-Villa M, Trejo-Téllez LI, Jiménez-Flores Y, Aguilar-Méndez MÁ (2020) Calcium-loaded gelatin nanoparticles as a source of agricultural fertilizer. Ingeniería Agrícola y Biosistemas 12(1):69–77

52. Kamaly N, Yameen B, Wu J, Farokhzad OC (2016) Degradable controlled-release polymers and polymeric nanoparticles: mechanisms of controlling drug release. Chem Rev 116(4):2602–2663
53. Kaul S, Gulati N, Verma D, Mukherjee S, Nagaich U (2018) Role of nanotechnology in cosmeceuticals: a review of recent advances. J Pharm 2018(1):3420204
54. Koli P, Bhardwaj NR, Mahawer SK (2019) Agrochemicals: harmful and beneficial effects of climate changing scenarios. In: Climate change and agricultural ecosystems. Elsevier, pp 65–94
55. Kottegoda N, Sandaruwan C, Perera P, Madusanka N, Karunaratne V (2014) Modified layered nanohybrid structures for the slow release of urea. Nanosci Nanotechnol-Asia 4(2):94–102
56. Kourmatzis A, Cheng S, Chan H-K (2018) Airway geometry, airway flow, and particle measurement methods: implications on pulmonary drug delivery. Expert Opin Drug Deliv 15(3):271–282
57. Krishnan S, Bhosale R, Singhal RS (2005) Microencapsulation of cardamom oleoresin: Evaluation of blends of gum arabic, maltodextrin and a modified starch as wall materials. Carbohydrate Polym 61(1), 95–102. https://doi.org/10.1016/j.carbpol.2005.02.020
58. Kumar A, Belhaj M, DiPette DJ, Potts JD (2021) A novel alginate-based delivery system for the prevention and treatment of pressure-overload induced heart failure. Front Pharmacol 11:602952
59. Kurozawa LE, Hubinger MD (2017) Hydrophilic food compounds encapsulation by ionic gelation. Curr Opin Food Sci 15:50–55
60. Lai H, Liu S, Yan J, Xing F, Xiao P (2020) Facile fabrication of biobased hydrogel from natural resources: l-cysteine, itaconic anhydride, and chitosan. ACS Sustain Chem Eng 8(12):4941–4947
61. Ledari SA, Milani JM, Shahidi S-A, Golkar A (2024) Comparative analysis of freeze drying and spray drying methods for encapsulation of chlorophyll with maltodextrin and whey protein isolate. Food Chemistry: X 21:101156
62. Li B, Li X, Liu Y, Zhang D, Lin J, Mu W, Liu F (2018) Easily tunable membrane thickness of microcapsules by using a coordination assembly on the liquid-liquid interface. Front Chem 6:387
63. Li J, Alfares A, Zheng Y (2022b) Optical manipulation and assembly of micro/nanoscale objects on solid substrates. Iscience 25(4)
64. Li P, Feng D, Yang D, Li X, Sun J, Wang G, Tian L, Jiang X, Bai W (2021) Protective effects of anthocyanins on neurodegenerative diseases. Trends Food Sci Technol 117:205–217
65. Lopes DB, Speranza P, Macedo GA (2016) A new approach for flavor and aroma encapsulation. In: Novel approaches of nanotechnology in food. Elsevier, pp 623–661
66. Maged A, Mahmoud AA, Salah S, Abd-Elmonsif NM, Ghorab MM (2020) Spray-dried rosuvastatin nanoparticles for promoting hair growth. AAPS PharmSciTech 21:1–12
67. Malamatari M, Charisi A, Malamataris S, Kachrimanis K, Nikolakakis I (2020) Spray drying for the preparation of nanoparticle-based drug formulations as dry powders for inhalation. Processes 8(7):788
68. Malamatari M, Taylor KMG, Malamataris S, Douroumis D, Kachrimanis K (2018) Pharmaceutical nanocrystals: production by wet milling and applications. Drug Discovery Today 23(3):534–547
69. Marco-Dufort B, Willi J, Vielba-Gomez F, Gatti F, Tibbitt MW (2020) Environment controls biomolecule release from dynamic covalent hydrogels. Biomacromol 22(1):146–157
70. Martin AR, Moore CP, Finlay WH (2018) Models of deposition, pharmacokinetics, and intersubject variability in respiratory drug delivery. Expert Opin Drug Deliv 15(12):1175–1188
71. Meng W, Sun H, Mu T, Garcia-Vaquero M (2024) Spray-drying and rehydration on β-carotene encapsulated Pickering emulsion with chitosan and seaweed polyphenol. Int J Biol Macromol 268:131654
72. Messa LL, Faez R (2020) Spray-dried chitosan/nanocellulose microparticles: synergistic effects for the sustained release of NPK fertilizer. Cellulose 27(17):10077–10093

73. Miranda-Linares V, Quintanar-Guerrero D, Del Real A, Zambrano-Zaragoza ML (2020) Spray-drying method for the encapsulation of a functionalized ingredient in alginate-pectin nano-and microparticles loaded with distinct natural actives: stability and antioxidant effect. Food Hydrocolloids 101:105560
74. Miranda-Villagómez E, Aguilar-Méndez MÁ, Gómez-Merino FC, Ronquillo de Jesús E, Sandoval-Villa M, Sánchez-García P, Trejo-Téllez LI (2019) Synthesis of biopolymeric particles loaded with phosphorus and potassium: characterisation and release tests. IET Nanobiotechnol 13(5):493–497
75. Mishra A, Pandey VK, Shankar BS, Melo JS (2021) Spray drying as an efficient route for synthesis of silica nanoparticles-sodium alginate biohybrid drug carrier of doxorubicin. Colloids Surf, B 197:111445
76. Mitchell MJ, Billingsley MM, Haley RM, Wechsler ME, Peppas NA, Langer R (2021) Engineering precision nanoparticles for drug delivery. Nat Rev Drug Discovery 20(2):101–124
77. Morlando A, Sencadas V, Cardillo D, Konstantinov K (2018) Suppression of the photocatalytic activity of TiO_2 nanoparticles encapsulated by chitosan through a spray-drying method with potential for use in sunblocking applications. Powder Technol 329:252–259
78. Munir M, Kett VL, Dunne NJ, McCarthy HO (2022) Development of a spray-dried formulation of peptide-DNA nanoparticles into a dry powder for pulmonary delivery using factorial design. Pharm Res 39(6):1215–1232
79. Muralidharan P, Malapit M, Mallory E, Hayes Jr D, Mansour HM (2015) Inhalable nanoparticulate powders for respiratory delivery. Nanomedicine: Nanotechnol, Biol Med 11(5):1189–1199
80. Narayanan VHB, Lewandowski A, Durai R, Gonciarz W, Wawrzyniak P, Brzezinski M (2022) Spray-dried tenofovir alafenamide-chitosan nanoparticles loaded oleogels as a long-acting injectable depot system of anti-HIV drug. Int J Biol Macromol 222:473–486
81. Nguyen VP, Le Trung H, Nguyen TH, Hoang D, Tran TH (2021) Synthesis of biogenic silver nanoparticles with eco-friendly processes using Ganoderma lucidum extract and evaluation of their Theranostic applications. J Nanomater 2021(1):6135920
82. Noruzi M (2016) Electrospun nanofibres in agriculture and the food industry: a review. J Sci Food Agric 96(14):4663–4678
83. Okuro PK, de Matos Junior FE, Favaro-Trindade CS (2013) Technological challenges for spray chilling encapsulation of functional food ingredients. Food Technol Biotechnol 51(2):171
84. Öztürk A, Aygül A (2020) Design of cefaclor monohydrate containing nanoparticles with extended antibacterial effect by nano-spray dryer: a nanoenglobing study. J Res Pharm 24(1)
85. Pakowski Z (2007) Modern methods of drying nanomaterials. Transp Porous Media 66(1):19–27. https://doi.org/10.1007/s11242-006-9019-x
86. Paluch KJ, Tajber L, Adamczyk B, Corrigan OI, Healy AM (2012) A novel approach to crystallisation of nanodispersible microparticles by spray drying for improved tabletability. Int J Pharm 436(1–2):873–876
87. Pedrozo RC, Antônio E, Khalil NM, Mainardes RM (2020) Bovine serum albumin-based nanoparticles containing the flavonoid rutin produced by nano spray drying. Braz J Pharm Sci 56:e17692
88. Pereira L, Dias N, Carvalho J, Fernandes S, Santos C, Lima N (2014) Synthesis, characterization and antifungal activity of chemically and fungal-produced silver nanoparticles against Trichophyton rubrum. J Appl Microbiol 117(6):1601–1613
89. Pérez-Masiá R, López-Nicolás R, Periago MJ, Ros G, Lagaron JM, López-Rubio A (2015) Encapsulation of folic acid in food hydrocolloids through nanospray drying and electrospraying for nutraceutical applications. Food Chem 168:124–133
90. Petroff OAC, Rothman DL, Behar KL, Hyder F, Mattson RH (1999) Effects of valproate and other antiepileptic drugs on brain glutamate, glutamine, and GABA in patients with refractory complex partial seizures. Seizure 8(2):120–127
91. Phan HT, Haes AJ (2019) What does nanoparticle stability mean? The J Phys Chem C 123(27):16495–16507

92. Rajat S, Manisha S, Robin S, Sunil K (2012) Nutraceuticals: a review. Int Res J Pharm 3(4):95–99
93. Ren Y, Mei L, Zhou L, Guo G (2019) Recent perspectives in hot melt extrusion-based polymeric formulations for drug delivery: applications and innovations. AAPS PharmSciTech 20(3):92
94. Ribeiro A, Ruphuy G, Lopes JC, Dias MM, Barros L, Barreiro F, Ferreira ICFR (2015) Spray-drying microencapsulation of synergistic antioxidant mushroom extracts and their use as functional food ingredients. Food Chem 188:612–618. https://doi.org/10.1016/j.foodchem.2015.05.061
95. Rogueda PGA, Traini D (2007) The nanoscale in pulmonary delivery. Part 1: deposition, fate, toxicology and effects. Expert Opin Drug Deliv 4(6)595–606
96. Rosen J, Landriscina A, Friedman AJ (2015) Nanotechnology-based cosmetics for hair care. Cosmetics 2(3):211–224
97. Sachdeva V, Roy A, Bharadvaja N (2020) Current prospects of nutraceuticals: a review. Curr Pharm Biotechnol 21(10):884–896
98. Santiago-Adame R, Medina-Torres L, Gallegos-Infante JA, Calderas F, González-Laredo RF, Rocha-Guzmán NE, Ochoa-Martínez LA, Bernad-Bernad MJ (2015) Spray drying-microencapsulation of cinnamon infusions (Cinnamomum zeylanicum) with maltodextrin. LWT Food Sci Technol 64(2):571–577. https://doi.org/10.1016/j.lwt.2015.06.020
99. Santos DT, Meireles MAA (2010) Carotenoid pigments encapsulation: fundamentals, techniques and recent trends. Open Chem Eng J 4(1)
100. Schafroth N, Arpagaus C, Jadhav UY, Makne S, Douroumis D (2012) Nano and microparticle engineering of water insoluble drugs using a novel spray-drying process. Colloids Surf B: Biointerfaces 90:8–15. https://doi.org/10.1016/j.colsurfb.2011.09.038
101. Shaji A, Zachariah AK (2017) Surface area analysis of nanomaterials. In: Thermal and rheological measurement techniques for nanomaterials characterization. Elsevier, pp 197–231
102. Shalev N, Burg A, Gavrieli I, Lazar B (2015) Nitrate contamination sources in aquifers underlying cultivated fields in an arid region—the Arava Valley, Israel. Appl Geochem 63:322–332
103. Singh AR, Desu PK, Nakkala RK, Kondi V, Devi S, Alam MS, Hamid H, Athawale RB, Kesharwani P (2022) Nanotechnology-based approaches applied to nutraceuticals. Drug Deliv Transl Res 1–15
104. Singh M, Singh N, Chandrasekaran B, Deb PK (2020) Nanomaterials in nutraceuticals applications. Integr Nanomedicine New Ther 405–435
105. Singhal M, Khanna S, Nasa A (2011) Cosmeceuticals for the skin: An overview. Asian J Pharm Clin Res 4(2):1–6
106. Souto EB, Fernandes AR, Martins-Gomes C, Coutinho TE, Durazzo A, Lucarini M, Souto SB, Silva AM, Santini A (2020) Nanomaterials for skin delivery of cosmeceuticals and pharmaceuticals. Appl Sci 10(5):1594
107. Srinarong P, de Waard H, Frijlink HW, Hinrichs WLJ (2011) Improved dissolution behavior of lipophilic drugs by solid dispersions: the production process as starting point for formulation considerations. Expert Opin Drug Deliv 8(9):1121–1140
108. Su C-Y, Wang J-C, Chen C-Y, Chu K, Lin C-K (2019) Spherical composite powder by coupling polymethyl methacrylate and boron nitride via spray drying for cosmetic application. Materials 12(5):706
109. Sun X, Ning H, Shi Y, Yu G, Jia Z, Han M, Zhang Y (2023) Study of CO_2 solubility enhancement by nanomaterials in carbonated water: Implications for enhanced oil recovery and CO_2 storage. J Clean Prod 396:136562
110. Sundari PT, Anushree H (2017) Novel delivery systems: current trend in cosmetic industry. Eur J Pharm Med Res 4(8):617–627
111. Taroncher M, Vila-Donat P, Tolosa J, Ruiz MJ, Rodriguez-Carrasco Y (2021) Biological activity and toxicity of plant nutraceuticals: an overview. Curr Opin Food Sci 42:113–118
112. Thiyagarajan D, Huck B, Nothdurft B, Koch M, Rudolph D, Rutschmann M, Feldmann C, Hozsa C, Furch M, Besecke KFW (2021) Spray-dried lactose-leucine microparticles

for pulmonary delivery of antimycobacterial nanopharmaceuticals. Drug Deliv Transl Res 11(4):1766–1778
113. Timilsena YP, Adhikari R, Casey P, Muster T, Gill H, Adhikari B (2015) Enhanced efficiency fertilisers: a review of formulation and nutrient release patterns. J Sci Food Agric 95(6):1131–1142
114. Tomaszewska M, Jarosiewicz A (2006) Encapsulation of mineral fertilizer by polysulfone using a spraying method. Desalination 198(1–3):346–352
115. Tran PHL, Tran TTD (2021) Blueberry supplementation in neuronal health and protective technologies for efficient delivery of blueberry anthocyanins. Biomolecules 11(1):102
116. Tulukcu E (2011) A comparative study on fatty acid composition of black cumin obtained from different regions of Turkey, Iran and Syria. Afr J Agric Res 6(4):892–895
117. Vega C, Roos YH (2006) Invited review: spray-dried dairy and dairy-like emulsions—compositional considerations. J Dairy Sci 89(2):383–401
118. Voci S, Gagliardi A, Fresta M, Cosco D (2022) Ascorbic acid-loaded gliadin nanoparticles as a novel nutraceutical formulation. Food Res Int 161:111869
119. Wang L, Liu H, Zhuang J, Wang D (2022) Small-scale big science: From nano-to atomically dispersed catalytic materials. Small Sci 2(11):2200036
120. Wen H, Cui H, Tian H, Zhang X, Ma L, Ramassamy C, Li J (2020) Isolation of neuroprotective anthocyanins from black chokeberry (Aronia melanocarpa) against amyloid-β-induced cognitive impairment. Foods 10(1):63
121. Weng Y, Li Y, Chen X, Song H, Zhao C-X (2024) Encapsulation of enzymes in food industry using spray drying: recent advances and process scale-ups. Crit Rev Food Sci Nutr 64(22):7941–7958
122. Wong TW, John P (2016) Advances in spray drying technology for nanoparticle formation. In: Aliofkhazraei M (Ed) Handbook of nanoparticles. Springer International Publishing, pp 329–346. https://doi.org/10.1007/978-3-319-15338-4_18
123. Zeisel S (1999). Regulation of "Nutraceuticals." Science 285:1853–1855
124. Zendehdel Baher S, Yaqoubi S, Asare-Addo K, Hamishehkar H, Nokhodchi A (2022) Dry powder formulation of simvastatin nanoparticles for potential application in pulmonary arterial hypertension. Pharmaceutics 14(5):895
125. Zheng Y, Zhao C, Chen B, Teng H, Ai C, Chen L (2023) D-α-tocopherol polyethylene glycol 1000 succinate-based microemulsion delivery system: stability enhancement of physicochemical properties of luteolin. Food Chem 426:136587

Mechanical Milling for Scaled Up Manufacturing of Nanoparticles

Diptajyoti Gogoi, Nabadeep Kalita, and Ranjit Thakuria

Abstract Nanoparticles are materials with average particles size less than 100 nm. Due to their smaller particle size compared to macro-sized materials, they have larger surface area that enhances various physical and chemical properties such as mechanical strength, stability, reactivity, solubility, dissolution rate, bioavailability etc. As a result nanomaterials have wide applications in various fields ranging from chemical, pharmaceutical, agro, as well as paint industry. Here in this chapter, we have discussed about scaled up manufacturing of nanoparticles prepared using mechanical milling, their advantages as well as challenges encountered during their preparation.

Keywords Nanoparticles · Mechanochemistry · Cocrystal · LAG · API

1 Introduction

The word 'Nano' was originated from the Greek word 'Nanos', which signifies dwarf or minute. Nanoparticles are extremely small particles with a size ranging from 1 to 100 nm, having a unique chemical and physical properties compared to their larger-sized counterparts [1]. People are interested to use these materials in a variety of fields due to their enhanced reactivity, increased mechanical strength, larger surface area and better chemical stability. During the last two decades, there has been a tremendous development in the fields of nanoscience and technology, ushering in a new era across a variety of fields such as medicine, energy, information technology, homeland security, food safety, environmental science etc. [2]. However, setting up a large-scale nanoparticle production method still presents significant difficulties and a very few works on the production of nanoparticles at the industrial scale has been reported [3]. To synthesize nanoparticles for commercial usage and clinical trials, companies need production lines that are easy to maintain, quickly implementable and adjustable in terms of formulation type and sterility requirements. The process of moving a synthetic method from a laboratory scale to a larger scale is intricate and,

D. Gogoi · N. Kalita · R. Thakuria (✉)
Department of Chemistry, Gauhati University, Guwahati 781014, Assam, India
e-mail: ranjit.thakuria@gauhati.ac.in; ranjit.thakuria@gmail.com

N. M. Adassooriya et al. (eds.), *Industrial Scale Production of Nanoparticles*, Smart Nanomaterials Technology, https://doi.org/10.1007/978-981-96-9176-0_4

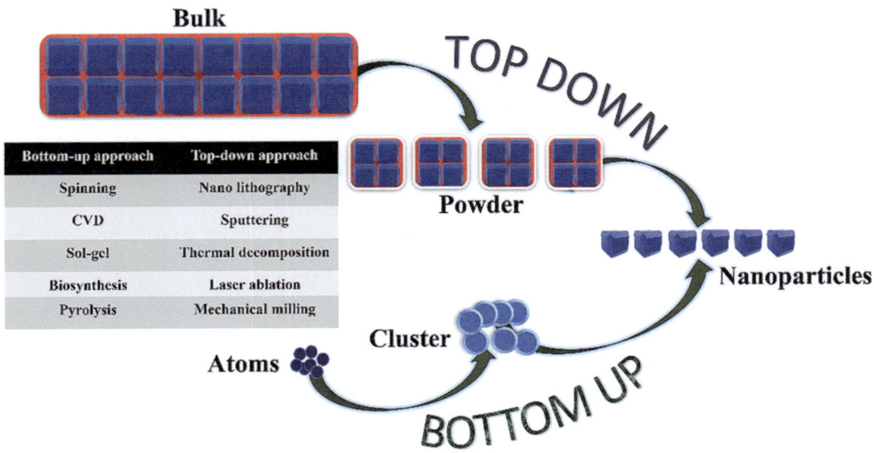

Fig. 1 Schematic diagram for different approaches of nanoparticle synthesis

if not done properly, can result in variable product quality [4]. It is very difficult to synthesize nanomaterials by a simple, cheaper synthetic method with a high yield. For the commercial manufacturing of nanomaterials, numerous top-down and bottom-up approaches have been established so far (Fig. 1).

The bottom-up method of nanofabrication makes use of chemical or physical forces operating at the nanoscale in order to integrate building blocks into functional structures. In this mode, materials are synthesized from the atomic to the cluster to the nanoparticle level. Sol–Gel [5], spinning [6], chemical vapor deposition (CVD) [7], pyrolysis [8] and biosynthesis [9] are some common examples of bottom-up approach. In contrast, a substance is broken into its constituent atomic parts in case of top-down method. This method is well suited for structures with long range order and linkages at the macroscopic level. Mechanical milling [10], nanolithography [11], laser ablation [12], sputtering [13] and thermal decomposition [14] are some common examples of top-down approach.

The adaptation of current technology depends on the quality, target nano-product specifications, desired physico-chemical properties and sterility criteria. Three criteria viz. type of nanoparticles, type of approach used (bottom-up vs. top-down) and regulatory requirements; are taken into consideration when the pharmaceutical industry chooses its large scale nanoparticle production techniques [15]. Studies on the large-scale production of nanosuspensions using top-down methods are scarce. High pressure homogenization (HPH) and wet media milling are the most commonly used techniques for this purpose [16]. It is well established that due to the application of extreme shear forces in HPH, bead milling is a more resilient process compared to HPH. High energy ball milling has been used extensively among all top-down approaches for the production of different nanomaterials, nanograins, nanoalloys, nanocomposites and nanoquasicrystalline materials [17].

2 Mechanical Milling

A high energy mill is used in mechanical milling to combine an appropriate milling medium with a powder charge, which is usually a combination of elements. Reduced particle size and improved particle blending in new phases are the main goals of milling. Additionally, it has been reported that the effects of high energy ball milling method can occasionally results in a local high pressure of several GPa [18] and temperature above 1000 °C [19]. Because of this nature, the ball milling has been applied as a mechanochemical synthetic method to fabricate distinct nanostructures with novel chemical properties (Fig. 2) [20].

Different ball milling techniques that impact the powder charge can be utilized to synthesize nanomaterials [21]. The energy that is transferred from the balls to the powder during milling determines the kinetics of mechanical milling [10]. The extent of energy transfer is associated with numerous factors, including type of the mill, the powder used to power the milling chamber, the frequency at which the mill is operating, size and distribution of balls, wet or dry milling, temperature and milling time [22].

The process of dry grinding or neat grinding involves combining and grinding solid materials without the use of liquid medium to produce nanoparticles. The starting material is exposed to mechanical forces produced by the milling media [23]. This method has the benefits of not requiring the use of any type of solvents, making the process simpler because the separation is easier and lowering the possibility of any solvent related issue. The unpredictable environmental conditions of dry milling are one of the major drawbacks. The production of nanoparticle can vary depending on a number of factors including humidity, temperature and oxygen exposure [24]. Due to the absence of liquid media, the mixture is not homogeneous during the dry milling process, resulting in a distribution of particles of varying size [25].

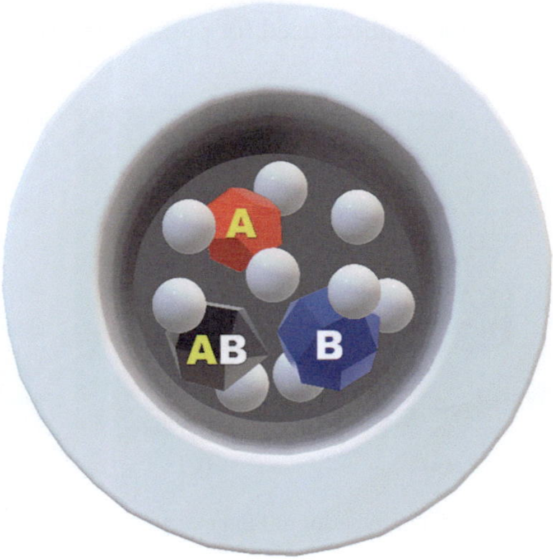

Fig. 2 Schematic representation for the mechanical milling process

In contrast, wet milling or liquid assisted grinding (LAG) is performed in presence of a liquid media. In case of nanococrystal, a medium-typically a solvent or a stabilizer is used to distribute the coformer and the API in the micrometer range. LAG causes collisions between the coformer and API particles and the milling jar walls. LAG results a narrow size distribution of particles due to homogeneous nature of the mixture [26]. The separation of the liquid or solvent used during grinding from the components that have been milled is one of the major drawbacks of liquid assisted grinding. Additionally, when the liquid stuck in the crystal lattice, solvates may appear.

2.1 Types of Milling Machines

The motion of the balls inside mills has been modeled using a variety of techniques, such as analytical techniques or discrete element approaches [27]. Some of the milling machines are shown in Fig. 3.

(a) **Planetary Milling Machines**

For decades, planetary ball mills have been utilized for reducing particle size on laboratory as well as pilot scales. However, their application has been extended to include mechanochemical methods in recent times [28]. Due to the high stress intensities attained from high accelerations, planetary ball mills are appropriate for wet and dry, fine and ultra-fine grinding of particles down to the nanometer range. These types of mills are typically used for developing new materials and releasing minerals and other materials for analysis. Planetary ball mills are especially helpful tool for mechanochemical methods because of the extremely effective material mixing that occurs in the grinding chamber [19]. Use of planetary ball milling instead of traditional chemical or processing methods has the advantage of reducing the amount of reagents without need for purification, speeding up reaction time and increasing

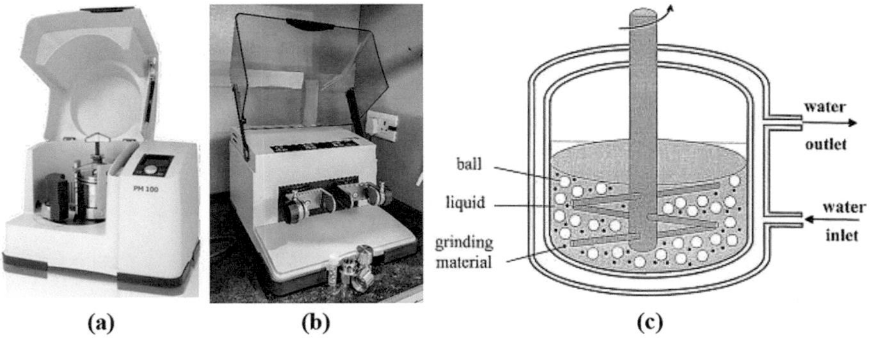

Fig. 3 Different types of milling machines: (**a**) Planetary milling machines, (**b**) Vibratory milling machine and (**c**) Attrition mill

yield. Typically, planetary ball mills have two or four pots and the pots rotate around their respective axes and are joined to a disk that revolves around a shared central axis. To ensure an effective grinding performance, the milling balls inside the pots have a substantial impact energy due to the high rotational speed of the disk and the pots [29].

(b) **Vibratory Milling Machine**

Vibratory milling is an effective milling technique with a wide range applications in multiple sectors like food processing, pharmaceuticals, ceramics, material science etc. This mechanical technique moves the grinding media and breaks down the materials using vibrational energy [30]. This method produces high quality particles with a restricted size distribution. Particle size reduction results from the high energy interaction between the substance and the media caused by the vibratory motion. The desired particle size distribution can be attained by varying the frequency and amplitude of vibration. Compared to other milling methods, vibratory milling has a number of advantages including versatility, reduced contamination and improved product quality. However, higher equipment cost, higher maintenance cost and limited scalability are some of the disadvantages of this technique [31].

(c) **Attrition Mill**

A type of milling apparatus used to grind materials into fine particles is an attrition mill, sometimes referred to as a stirred ball mill. Its distinctive operation, involving an agitator element and a high-speed rotating shaft, is what makes it special. The material is added to the milling chamber together with the grinding media, which is usually composed of steel or ceramics. Through a combination of impact, compression and shear forces, the grinding media impacts and grinds the materials into small particles as the shaft and agitator parts rotate at high speeds. This type of mill is an adaptable device that may be used for both wet and dry grinding. Compared to other conventional grinding mills, the attrition mill has a number of advantages including high grinding efficiency, fine particle size distribution, versatility and low maintenance which make it an ideal choice for a wide range of industries [32, 33].

The selection of different milling machines for the synthesis of nanoparticles is governed by certain critical factors such as particle size requirements, scalability and energy efficiency. Planetary ball milling are highly effective for producing ultra-fine, high-purity nanoparticles, but their industrial applicability is constrained by significant energy consumption and limited scalability [18]. Vibratory mills, though characterized by a simpler design and moderate scalability, are less effective in achieving uniform and finely controlled particle sizes [19]. In contrast, attrition mills are particularly well-suited for large scale industrial applications, offering enhanced energy efficiency, scalability and precise control over particle morphology and size distribution, making them a preferred option for mass production [20].

All the mentioned mills usually operate under atmospheric pressure and ambient temperature. But, there are some reports that indicates the use of ball milling at cryogenic temperatures (below 123 K) for the synthesis of nanoparticles in large scale is also very effective [34, 35]. Moreover, for materials that are thermally sensitive, the implementation of cooling systems may be necessary to control the temperature. Residence time in planetary mill is influenced by the properties of the material being processed, desired particle size as well as operational parameters such as feed rate of the material being milled and volume of the mill [36]. Yield of nanoparticles produced is influenced by factors such as milling duration, material properties and energy input. However, attrition mill provides the highest efficiency among the three milling methods influenced by variables such milling time, material properties and operational conditions [37].

Quantifying the precise energy consumption for the production of unit mass of NPs across different milling machines is challenging task due to the influence of variables such as material properties, desired particle size and operational parameters. However, vibratory milling is more energy-efficient than planetary milling for producing unit mass of nanoparticles. Primary advantages of attrition mills include their high energy efficiency, rapid and effective fine grinding capabilities, precise temperature control and simple operational method [37]. Energy efficiency can be improved by optimizing milling parameters such as rotational speed and ball-to-material ratio, as well as by utilizing larger milling chambers to enable higher volume processing [28].

3 Advantages of Mechanical Milling for Scaled Up Nanoparticle Production

Nanoparticles have garnered immense attention across various fields, such as material science, medicine, electronics etc. because of their unique properties. But despite the increase of demand for nanoparticles with tailored properties and their pivotal role in industrial applications, scalable and efficient production of nanoparticles remains a significant challenge. Various bottom up and top-down methods have been employed for the production of nanoparticles and among them mechanical milling has emerged as a robust, cost-effective and scalable technique. Mechanical milling helps in the reduction of particle size by the application of mechanical energy. It involves repeated fracturing, welding and of materials transforming bulk materials into smaller particles. And varying the parameters used such as ball to powder ratio, milling time, frequency etc. the method can be customized to generate nanoparticles with precise size distributions and structural characteristics. The high energy ball milling is the most extensively utilized approach for synthesizing nanomaterials including nanocomposite, nanograin, nanoalloys etc. [17]. Unlike traditional chemical methods which involve very expensive reagents and complex procedures, mechanical milling is highly regarded for its cost efficiency and scalability. It utilizes relatively simple equipment, such as ball mills, planetary mills etc. which are readily accessible. And this enables a smooth transition from laboratory-scale production to industrial-scale manufacturing with minimal adjustments to the process or equipment. In addition, its ability to produce nanoparticles with tailored properties boosts cost effectiveness, as manufacturers can obtain the exact materials needed without investing in multiple expensive synthetic processes. Another key reason for its cost effectiveness is energy efficiency. Although mechanical milling demands a significant input of mechanical energy yet it reduces energy cost by avoiding high temperature or complex chemical reactions. Moreover, versatility in producing nanoparticles from a wide range of materials is one of the significant advantages of mechanical milling. This process enables the production of nanoparticles with diverse morphology, sizes and compositions with consistency, making them highly versatile for a wide range of applications. Mechanochemical milling also has the ability to induce phase transformation and alloying reactions which are difficult to achieve through other methods. Moreover, mechanical milling is considered as an environmentally benign process, in contrast to methods involving chemical synthesis that may involve toxic solvents, hazardous precursors, or complicated waste disposal systems. It eliminates the need of solvent or toxic chemicals, minimizing the environmental impact of nanoparticle production. Again, mechanical milling maintains atom economy and produces very minimal waste, since nearly all input materials are transformed into nanoparticles. Furthermore, mechanical milling can recycle waste materials, supporting circular economy practices. Mechanical milling offers significant advantages for large scale production while also yielding nanoparticles with enhanced and tunable properties by controlling the milling parameters, such as duration, speed and milling atmosphere. All these advantages make it a versatile and efficient approach for material synthesis.

4 Case Studies of Scaled Up Synthesis of Nanoparticles by Mechanical Milling

Due to cost effectiveness, versatility in material processing, eco-friendly approach, enhanced physical and chemical properties resulted from mechanical milling, more and more people are interested to use this technique for the scaled-up synthesis of nanoparticles nowadays. However, only a few works have been reported so far. Arbain et al. reported [38] the synthesis of Fe_2O_3 nanoparticle by using a planetary mill with the variation in rotational speed and milling time. They also investigated the size reduction and mechanochemical effect by X-ray diffraction line broadening, specific surface area, particle size and morphology analysis. They obtained a minimum particle size of 76.6 nm at 600 rpm for 10 h. Xing et al. [39] demonstrated a high energy ball milling method for the production of nitrogen-doped carbon nanoparticles on a massive scale. These nanoparticles can be can be applied as electro-catalyst for oxygen reduction reactions following structural refinement with controlled annealing. Lam et al. reported [40] the synthesis of ultrafine Si nanoparticles by ball milling of graphite and SiO_2 powders. Epur et al. also reported [41] the scaled-up synthesis of Si nanoparticles with high specific surface area. Si nanoparticles were obtained by the mechano-chemical reduction reaction of magnesium silicide and silicon monoxide using a high energy ball milling method followed by acid leaching. In a recent report, Nugrahani and Auli [25] utilized top-down approach like wet milling and neat grinding, while the bottom-up technique was performed through the globule inversion phase and fast evaporation assisted microwaving in order to prepare nano-diclofenac-proline-cocrystal. The synthesized nanococrystals were characterized using various microscopic as well as spectroscopic techniques such as dynamic light scattering, binocular microscope, scanning electron microscopy, transmission electron microscopy, differential scanning calorimetry, powder X-ray diffractometry, FT-IR spectrophotometry and investigated their diffusion and dissolution profile. From their study it was observed that the fast evaporation assisted microwaving provided the best nano-cocrystal with a mean diameter of 598.2 ± 63.2 nm and a polydispersity index of 0.278 ± 0.062. In another report by Santos et al. [42] investigated in situ synthesis of pharmaceutical nanococrystal of (S)-naproxen with nicotinamide in 2:1 ratio through grinding assisted by minimal amounts of solutions of non-ionic surfactants or by PEG 6000, as stabilizers. This one-step synthesis of nanococrystals through surfactant- assisted grinding proved to be a greener and efficient way to obtain nanoparticles with improved oral bioavailability. Xiao et al. [43] developed a mechanochemical method based on synergistic supramolecular self-assembly in order to prepare a nanococrystal material composed of the herbicide clopyralid and coformer phenazine. The synthesized nanococrystal enhances pesticide performance, exhibiting a marked improvement in stability, with reductions in hygroscopicity and volatility by approximately 38%. This nanococrystal also showed improved crop safety and reduction in genotoxicity compared to conventional herbicide formulations. Moores and coworkers [44] together synthesized a flame-retardant cellulose nanocrystals by phosphorylation of cellulose nanocrystals poly(ethylene glycol), poly(vinyl alcohol), poly(vinyl chloride) and lignin with phosphorus pentoxide through mechanochemical processes.

5 Challenges for the Scaled-Up Production on Nanomaterial by Mechanical Milling

Mechanochemical method has emerged as one of the most versatile and cost-effective methods for synthesizing nanomaterials. Despite its advantages, such as scalability and ability to process a wide variety of materials, several challenges arise while moving from laboratory-scale production to industrial-scale manufacturing. The challenges include the possibility of contamination, particle agglomeration, formation of nanomaterial with non-uniform shapes [45] and equipment limitations. Contamination is one of the primary challenges that is observed in scaling up mechanical milling for nanomaterial production [4]. The contamination may be introduced by the milling media or from the atmosphere. For instance, using steel balls in a metallic milling jar may introduce iron or chromium as a contaminant into the milled product. Similarly, milling in presence of air may result oxidation or chemical reaction with the material because of the introduction of oxygen, nitrogen, or water vapour [46]. Cross-contamination may also happen during switching between different materials in the same milling equipment [47]. If traces of previously milled product left in the equipment, that may also lead to unintended reaction or impurities. Such contaminations can be mitigated by using milling balls and lining materials that are inert with the target materials. Similarly, milling under inert conditions may reduce the risk of oxidation or chemical contamination. Therefore, to prevent cross-contamination one must ensure thorough cleaning between operations and maintaining the equipment regularly.

Nanoparticles have high surface area as well as high surface energy. Because of which they have a strong tendency to agglomerate and this is a significant issue in the production of nanoparticles via mechanical milling. Because of this agglomeration the properties of the final product changes which is not good for large scale production. This agglomeration can be prevented by the addition of surfactant or dispersing agents during wet milling which create a steric or electrostatic barrier around the particles preventing them to agglomerate. Again, optimization of the milling parameters can also help to minimize agglomeration [48].

The scaling up production of nanomaterials often requires larger and more powerful equipment. However, increasing the size of the milling chamber and the number of balls can decrease the overall efficiency of the process. Also, larger operating systems require more power to achieve the same impact energy as in the laboratory-scale equipment leading to higher operating costs. Such drawbacks can be tackled by using some modular milling system where instead of increasing the size of a single unit, a set of milling units are added which improve process control and provide more flexibility in production capacity.

6 Conclusion

Mechanical milling, which combines cost effectiveness, adaptability and efficiency, has become a key technique for the scalable production of nanomaterials. By using mechanical energy to break down bulk materials into nano-sized particles, this technology makes it possible to synthesize a variety of materials with distinct properties. Additionally, the particle size and shape can be tuned by adjusting variables like milling time, speed and media type, which makes it appropriate for a range of industries like pharmaceuticals, material science, catalysis etc. Furthermore, mechanical milling is beneficial for large-scale manufacturing because of its scalability, which can be readily modified to satisfy industrial demands, and its low equipment requirements. However, to optimize the product quality and performance, issues including heat generation, contamination, and agglomeration risk need to be resolved. Considering all these points, mechanical milling is a viable method for producing nanoparticles in large quantities, opening the door for developments in nanotechnology and its uses. To ensure that mechanical milling continues to be a major player in the manufacture of nanoparticles, research and developments in the areas of process optimization and material characterization will make significant improvements to its capabilities.

Funding Sources This work was funded by Science and Engineering Research Board under the Teachers Associateship for Research Excellence (TARE) grant (Project No. TAR/2021/000251).

References

1. Hasan S (2015) A review on nanoparticles: their synthesis and types. Res J Recent Sci 2277:2502
2. Khan Y, Sadia H, Ali Shah SZ, Khan MN, Shah AA, Ullah N, Ullah MF, Bibi H, Bafakeeh OT, Khedher NB (2022) Classification, synthetic, and characterization approaches to nanoparticles, and their applications in various fields of nanotechnology: a review. Catalysts 12:1386
3. Shegokar R, Müller RH (2010) Nanocrystals: industrially feasible multifunctional formulation technology for poorly soluble actives. Int J Pharm 399:129–139
4. Shegokar R, Nakach M (2020) Large-scale manufacturing of nanoparticles—an industrial outlook. In: Drug delivery aspects. Elsevier, pp 57–77
5. Verma R, Gangwar J, Srivastava AK (2017) Multiphase TiO_2 nanostructures: a review of efficient synthesis, growth mechanism, probing capabilities, and applications in bio-safety and health. RSC Adv 7:44199–44224
6. Hassellöv M, Readman JW, Ranville JF, Tiede K (2008) Nanoparticle analysis and characterization methodologies in environmental risk assessment of engineered nanoparticles. Ecotoxicology 17:344–361
7. Dion CD, Tavares JR (2013) Photo-initiated chemical vapor deposition as a scalable particle functionalization technology (a practical review). Powder Technol 239:484–491
8. Akerman ME (2005) Targeting and inhibiting tumor angiogenesis. University of California, San Diego
9. Grasso G, Zane D, Dragone R (2019) Microbial nanotechnology: challenges and prospects for green biocatalytic synthesis of nanoscale materials for sensoristic and biomedical applications. Nanomaterials 10:11
10. Tjong S, Chen H (2004) Nanocrystalline materials and coatings. Mater Sci Eng R Rep 45:1–88
11. Pimpin A, Srituravanich W (2012) Review on micro- and nanolithography techniques and their applications. Eng J 16:37–56
12. Amendola V, Meneghetti M (2009) Laser ablation synthesis in solution and size manipulation of noble metal nanoparticles. Phys Chem Chem Phys 11:3805–3821
13. Sharotri N, Sharma D (2022) Approaches for nanomaterial lab scale synthesis and manufacturing. In: Nanomaterials in manufacturing processes. CRC Press, pp 163–188
14. Palacios-Hernández T, Hirata-Flores GA, Contreras-López OE, Mendoza-Sánchez ME, Valeriano-Arreola I, González-Vergara E, Méndez-Rojas MA (2012) Synthesis of Cu and Co metal oxide nanoparticles from thermal decomposition of tartrate complexes. Inorg Chim Acta 392:277–282
15. Ranjita S (2013) Nanosuspensions: a new approach for organ and cellular targeting in infectious diseases. J Pharm Investig 43:1–26
16. Leung DH, Lamberto DJ, Liu L, Kwong E, Nelson T, Rhodes T, Bak A (2014) A new and improved method for the preparation of drug nanosuspension formulations using acoustic mixing technology. Int J Pharm 473:10–19
17. Yadav TP, Yadav RM, Singh DP (2012) Mechanical milling: a top down approach for the synthesis of nanomaterials and nanocomposites. Nanosci Nanotechnol 2:22–48
18. Koch CC (2006) Nanostructured materials: processing, properties and applications. William Andrew
19. Suryanarayana C (2001) Mechanical alloying and milling. Prog Mater Sci 46:1–184

20. DeCastro CL, Mitchell BS (2002) Nanoparticles from mechanical attrition. In: Synthesis, functionalization, and surface treatment of nanoparticles, p 5
21. Koch CC, Whittenberger J (1996) Mechanical milling/alloying of intermetallics. Intermetallics 4:339–355
22. Benjamin JS (1990) Mechanical alloying—a perspective. Met Powder Rep 45:122–127
23. Braga D, Maini L, Grepioni F (2013) Mechanochemical preparation of co-crystals. Chem Soc Rev 42:7638–7648
24. Rasenack N, Müller BW (2004) Micron-size drug particles: common and novel micronization techniques. Pharm Dev Technol 9:1–13
25. Nugrahani I, Auli WN (2020) Diclofenac-proline nano-co-crystal development, characterization, in vitro dissolution and diffusion study. Heliyon 6
26. Junyaprasert VB, Morakul B (2015) Nanocrystals for enhancement of oral bioavailability of poorly water-soluble drugs. Asian J Pharm Sci 10:13–23
27. Mishra B, Murty C (2001) On the determination of contact parameters for realistic DEM simulations of ball mills. Powder Technol 115:290–297
28. Burmeister CF, Kwade A (2013) Process engineering with planetary ball mills. Chem Soc Rev 42:7660–7667
29. Mio H, Kano J, Saito F, Kaneko K (2004) Optimum revolution and rotational directions and their speeds in planetary ball milling. Int J Miner Process 74:S85–S92
30. Stolle A, Szuppa T, Leonhardt SE, Ondruschka B (2011) Ball milling in organic synthesis: solutions and challenges. Chem Soc Rev 40:2317–2329
31. Tang J, Qiao J, Liu Z, Zhou X, Yu G, Zhao J (2018) Mechanism characteristic analysis and soft measuring method review for ball mill load based on mechanical vibration and acoustic signals in the grinding process. Miner Eng 128:294–311
32. Koch C (1993) The synthesis and structure of nanocrystalline materials produced by mechanical attrition: a review. Nanostruct Mater 2:109–129
33. Kim HJ, Lee S, Kim J, Mitchell RJ, Lee JH (2013) Environmentally friendly pretreatment of plant biomass by planetary and attrition milling. Bioresour Technol 144:50–56
34. Kumar N, Biswas K (2019) Cryomilling: an environment friendly approach of preparation large quantity ultra refined pure aluminium nanoparticles. J Mater Res Technol 8:63–74
35. Tiwary CS, Verma A, Kashyp S, Biswas K, Chattopadhyay K (2013) Preparation of freestanding Zn nanocrystallites by combined milling at cryogenic and room temperatures. Metall Mater Trans A 44:1917–1924
36. Cho HC, Kim KH, Lee H, Kim DJ (2011) Study of residence time distribution and mill hold-up for a continuous centrifugal mill with various G/D ratios in a dry-grinding environment. Miner Eng 24:77–81
37. Shinohara K, Golman B, Uchiyama T, Otani M (1999) Fine-grinding characteristics of hard materials by attrition mill. Powder Technol 103:292–296
38. Arbain R, Othman M, Palaniandy S (2011) Preparation of iron oxide nanoparticles by mechanical milling. Miner Eng 24:1–9
39. Xing T, Sunarso J, Yang W, Yin Y, Glushenkov AM, Li LH, Howlett PC, Chen Y (2013) Ball milling: a green mechanochemical approach for synthesis of nitrogen doped carbon nanoparticles. Nanoscale 5:7970–7976
40. Lam C, Zhang Y, Tang Y, Lee C, Bello I, Lee S (2000) Large-scale synthesis of ultrafine Si nanoparticles by ball milling. J Cryst Growth 220:466–470
41. Epur R, Minardi L, Datta MK, Chung SJ, Kumta PN (2013) A simple facile approach to large scale synthesis of high specific surface area silicon nanoparticles. J Solid State Chem 208:93–98
42. Santos JA, Baptista JA, Santos IC, Maria TM, Canotilho J, Castro RA, Eusébio MES (2022) Pharmaceutical nanococrystal synthesis: a novel grinding approach. CrystEngComm 24:947–961
43. Xiao Y, Wu C, Liu Y, Zhou L, Wu S, Yin Q (2024) Biocompatible nano-cocrystal engineering for targeted herbicide delivery: enhancing efficacy through stimuli-responsive release and reduced environmental losses. ACS Appl Mater Interfaces 16:51283–51300

44. Fiss BG, Hatherly L, Stein RS, Friščić T, Moores A (2019) Mechanochemical phosphorylation of polymers and synthesis of flame-retardant cellulose nanocrystals. ACS Sustain Chem Eng 7:7951–7959
45. Piras CC, Fernández-Prieto S, De Borggraeve WM (2019) Ball milling: a green technology for the preparation and functionalisation of nanocellulose derivatives. Nanoscale Adv 1:937–947
46. Tao J, Shih AJ, Ni J (2008) Experimental study of the dry and near-dry electrical discharge milling processes
47. Thompson G, Bankston DC (1970) Sample contamination from grinding and sieving determined by emission spectrometry. Appl Spectrosc 24:210–219
48. Bhakay A, Merwade M, Bilgili E, Dave RN (2011) Novel aspects of wet milling for the production of microsuspensions and nanosuspensions of poorly water-soluble drugs. Drug Dev Ind Pharm 37:963–976

Emulsion Techniques for Nanoparticle Production

H. A. D. B. Amarasiri and Lasantha R. Sendanayake

Abstract Nanoparticles are synthesized using various physical and chemical methods. Emulsion technique is popular especially in liquid phase nanoparticle synthesis. Depending on the energy requirement, nanoemulsion techniques can be categorized under high-energy and low-energy techniques. Ultrasonication, homogenization, and mechanical stirring are some of the high-energy techniques which can be fit for large scale industrial manufacturing compared to low-energy techniques such as phase inversion and self-emulsification. However, energetically cost-effective low-energy methods need attention to enhance safety and thereby apply in major industries like the pharmaceutical industry.

Keywords Nanoemulsions · Nanoparticles · High energy techniques · Low energy techniques

1 Introduction

2 Fundamentals of Emulsions

An emulsion is a mixture of two or more immiscible liquids that adopt a macroscopic homogeneity due to a specific transformation process but are still microscopically heterogeneous [11]. It is important to know the fundamentals of emulsions and their unique characteristics compared to the other liquids. Francesco Selmi studied a special liquid mixture where it seems all the components fully dissolved to make one perfect solution. However, careful investigation revealed the components only *appeared to be completely dissolved*. Therefore, he named it as "pseudosolution" [65]. Later, Thomas Grahm invented the term "colloid" which means "glue"

H. A. D. B. Amarasiri (✉) · L. R. Sendanayake
Department of Chemistry and Biochemistry, College of Arts and Sciences, Ohio University, Athens, USA
e-mail: da438421@ohio.edu

N. M. Adassooriya et al. (eds.), *Industrial Scale Production of Nanoparticles*, Smart Nanomaterials Technology, https://doi.org/10.1007/978-981-96-9176-0_5

in Greek to describe pseudosolutions [65]. Colloids consist of tiny droplets, particles, or bubbles in one phase (dispersed phase) and spread in another phase which is named as the dispersion medium [11, 88]. Both of these phases or one phase can be in a liquid, gas, solid, or supercritical state. The diffusion rate of the colloidal matter within the dispersion medium is low which evidences their relatively large size [11, 65, 89]. At least one dimension lies in the range of 1–1000 nm which overlaps in the dimensions of nanoparticles [92]. Due to this tiny size, they show different physicochemical properties compared to their macroscopic material [88]. The dispersed substance state and dispersion medium state determine the type of colloid. If both phases are solids, they are named as sols. Whereas foams are made by dispersing gas particles in liquid/solid and aerosols have liquid/solid particles in the gaseous dispersion medium. When both phases are liquid state, it is called an emulsion. The basic characteristics and properties of emulsions can be discussed under colloids.

3 Characteristics of the Dispersed Phase

Generally, atoms and molecules are extremely small compared to their macroscopic large particles. Interestingly, colloid size is in between these two extremes [89]. Thomas Grahm defined a range of particle sizes to categorize liquids. According to that, solutions are in atomic/molecular dispersion (< 1 nm), emulsions are in colloidal dispersion (1–1000 nm) and suspensions are in coarse dispersion (> 1000 nm) [65]. This tiny particle size of colloids gives rise to a higher surface area than their volume which results in neither float nor sink. Blood, milk, rubber latex, and egg whites are good examples of natural colloids [67].

In these colloids, the preparation method determines the shape of the dispersed phase and the affinity towards the dispersion medium. The diversity of common shapes includes spheres, globular, short/long rods, prolate/oblate ellipsoids, flakes, thread, loosely coiled threads, and branched threads that can be observed under microscopes [59]. The colloidal systems generally have a high specific surface area (A_{sp}) causing surface forces to dominate over body forces. The specific surface area is defined as surface area per unit volume or unit weight/mass [59, 67]. High surface area property facilitates the adsorption action (ex: catalyst), separates particles, and causes different colors compared to parent material. Furthermore, due to the small size of the colloidal particles, a considerable number of atoms/ions can be located on the surface/interface [80]. The core of the particles is electronically neutral but ionization of functional groups (ex:–COOH) on the surface and/or adsorption of ions (ex: H_3O^+ in pure water) in the dispersion medium possess a charge on their surface (Fig. 1) [6].

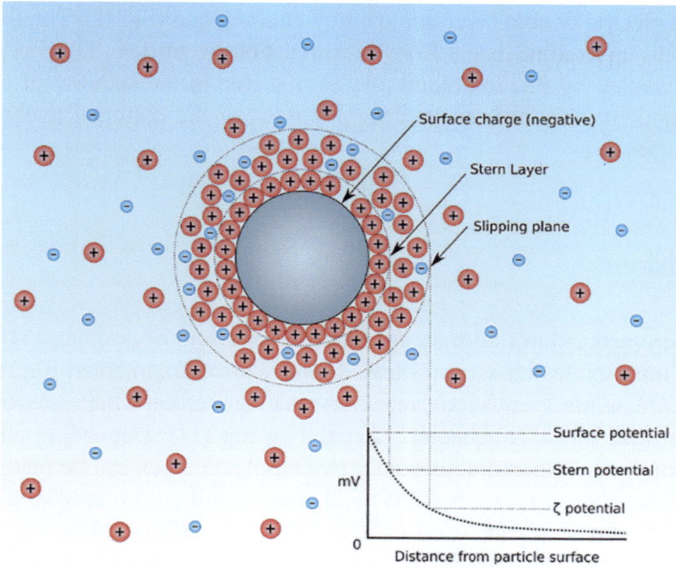

Fig. 1 Particles possess charge on their surfaces [6]

4 Emulsion Properties

The colloidal/emulsion properties can be divided into three categories; kinetic, optical, and electric [14, 80, 89]. The motion of the dispersed phase compared to the dispersion medium is discussed under kinetic properties and it helps to determine/predict the stability [105]. Colloidal particles are in continuous random movements as a result of collision among them. This is also known as Brownian motion which work against gravitational forces [23, 97]. However, increased particle size lowers the velocity which ultimately gives a chance for gravitational force to be dominant. This random/Brownian motion can be modified by altering the viscosity of the dispersion medium. Furthermore, random movement is the main reason for the dispersed phase to diffuse throughout the dispersion medium [105]. Other properties (colligative properties) depend on the concentration of the dispersed phase, such as the freezing point, boiling point, vapor pressure, and osmotic pressure [28]. Molecular weight determination using osmotic pressure is one of the examples for the usage of colligative properties [28].

The primary optical property of the colloids is the Faraday-Tyndall effect, where the dispersed phase scatters the light resulting in a cone-shaped light beam (Tyndall cone) when passing through the colloidal solution [31]. The surface charge of the dispersed phase is the main topic to discuss under the electric properties. Electrostatic adsorption of ions onto the surface of dispersed particles creates a primary adsorption/stern layer (Fig. 1) which repulse each particle when they come to close proximity [56]. On the other hand, they attract oppositely charged matter into the vicinity forming an electric double layer and thereby an ion cloud (Fig. 1). The formed ion cloud spreads up to some extent from the surface of the particle. In general electric properties created by this ion cloud play a vital role in the stability of emulsions and its magnitude depends on the concentrations of the colloidal matter and the electrolyte [56, 67].

5 Emulsions

Emulsion droplets exhibit all the conventional behaviors of colloids [54]. Simply, when two immiscible liquids are sheared, one will be fragmented to result in an emulsion [72]. In most emulsions, one phase is hydrocarbon which is referred to as oil and the other phase is aqueous referred as water [11]. Depending on the type of dispersion medium/continuous phase, two emulsion types can be readily distinguished. Those are oil-in-water (O/W) and water-in-oil (W/O) emulsions. The oil droplets are dispersed in water in oil-in-water (O/W) emulsions and the water droplets are dispersed in oil in water-in-oil (W/O) emulsions. There are two other types that are comparatively complex in structure compared to the previously mentioned basic forms. They are oil-in-water-in-oil (O/W/O) and water-in-oil-in-water (W/O/W) which are also termed double emulsions. As the name implies, oil droplets are dispersed in water and again the resulting phase is dispersed in another oil phase in O/W/O emulsion [11, 46, 88]. There can be even more complex emulsions such as O/W/O/W [11]. Though emulsions are stable for a considerable period such as weeks, emulsifying agents/stabilizers are added to enhance their thermodynamic stability for months/years. An emulsifying agent/surfactant also makes the formulation process more convenient by reducing the interfacial tension, charging the surface, and building a protective layer around dispersed droplets [92]. Furthermore, some emulsions are stabilized by using particles (ex: solid/colloidal) and are known as Pickering emulsions which are characterized by long-term stability [34]. In the last few decades, Pickering emulsions were developed to apply in various fields using solid nanoparticles such as nanotubes, clay, silica, and various other nanoparticles such as magnetic particles [48]. In 2018, Sun et al. developed Janus hydroxyapatite nanoparticles coated with bovine hemoglobin to stabilize Pickering emulsions [10]. Interestingly emulsion techniques are becoming popular to prepare nanoparticles among the currently available physical and chemical methods. This chapter aims to describe emulsion techniques used for nanoparticle production. There are several techniques to prepare nanoemulsions and the preparation methods can be classified based on the energy requirement. High-pressure homogenization, microfluidization, and ultrasonication can be accounted for in high-energy methods, whereas phase inversion and self-emulsification in low energy methods.

6 Nanoemulsion Preparation

6.1 High Energy Methods for Nanoemulsion Preparation

A high mechanical energy is used to disintegrate large-dispersed emulsion droplets into tiny nanosized droplets. High-pressure homogenizers, microfluidizers, and ultrasonicators are the main and common mechanical devices used at laboratory and industrial levels [26, 63]. This high energy level produces nanoemulsions with high kinetic energy and enables greater control over particle size, rheology, uniformity, and also color [27, 71].

6.1.1 Ultrasonication

This method involves the usage of high-energy ultrasonic waves to provide cavitation energy and thereby break macrosize dispersed droplets into nanosize droplets (Fig. 2). Depending on the application, we can change wavelengths (20 kHz to 1 GHz) emitted by the ultrasonicator probe to control the energy, surfactant, and operation time to ultimately control particle size [53]. Acoustic cavitation provides the main shear force in this process to form and grow microbubbles. Later microbubbles are broken down as a result of acoustic wave pressure fluctuation [45]. This phenomenon causes intense turbulence to produce nanosized droplets [15, 55]. Even in the absence of emulsifying agents, ultrasonic waves provide enough energy to overcome the energy barrier to build new interfaces [20, 42]. Compared to the other high-energy approaches, ultrasonication is better in operation and cleaning especially in the field of medicine and food industry [55, 63]. Nanoemulsion production using ultrasonication demonstrated great stability, especially in the food industry [24, 85, 98]. However, this technique is not highly compatible with heat-sensitive nanoemulsion preparation and has a low mixing efficiency.

6.1.2 Microfluidization

This is a micro-level mixing technology where a microfluidizer is utilized to produce nanoemulsions (Fig. 2). The premixed dispersed phase and the dispersion medium are forced to pass under high pressure (500–20,000 psi) through microchannels that allow a microsize level mixing to a reaction chamber [82, 101]. Two streams of microemulsions will strike each other with a high velocity and this collision produces shearing and cavitation forces to make the desired submicrolevel emulsion [42, 60, 61, 63]. According to the reported studies, this technique produces fine nanoparticles compared to the homogenizer technique [78, 104]. In 2003, Pinnamaneni et al. reported that microfluidization requires less amount of surfactant to produce stable nanoemulsions [79]. Furthermore, food-grade nanoemulsions can be produced with uniform size distribution and enhanced stability [25, 100, 108]. This technique is

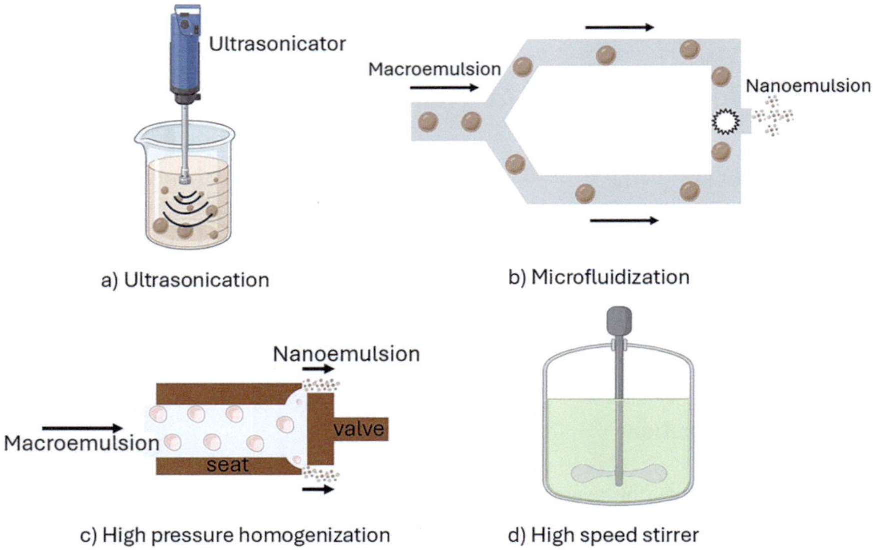

Fig. 2 High energy methods for nanoemulsion preparation

not fit to produce large-scale concentrated nanoemulsion as well as for prolonged operations due to droplet re-coalescence.

Jet disperser method

The jet disperser method is kind of similar to microfluidization, but the collision of macroemulsion is performed by two or more jets opposing each other. The diameter of the jet disperser bore is typically 0.3–0.5 mm. The macroemulsion droplets start to break mainly from the laminar elongation flow before the fed macroemulsion comes out from the bores. The nozzle acts as the microfluidizer under high pressure (300–400 MPa) and the jet disperser itself does not contain any moving part [44].

6.1.3 High-Pressure Homogenization

This technique provides high energy to produce small droplets and is widely used in stable nanoemulsion preparation on both laboratory scale and industrial scales. High pressure (500–5000 psi) is the intensive disruptive force that makes extremely small nanoparticles/droplets (< 1 nm) in most cases from the macroemulsion feed into the machine (Fig. 2) [84]. Several forces, namely cavitation, hydraulic shear, and turbulence work together in the process of macrodroplet breakdown [19, 90]. The particle size can be determined by several factors such as the type of the high-pressure homogenizer, sample composition, experimental parameters and conditions (ex: pressure and temperature), and operation duration [83]. Homogenization is inversely proportional to the particle/droplet size and it is recommended to use small-molecule emulsifying agents [4, 5, 41, 42]. This approach is popular in the pharmaceutical, food, and biotechnology research and industries due to its ability to operate even under low temperatures [35, 107]. Therefore thermolabile nanoemulsions can also be formulated. Despite those advantages, high-pressure homogenization consumes a lot of energy and is not suitable for biologics such as proteins and DNA.

6.1.4 High-Speed Stirrer

Many studies have reported the usage of high-speed stirring to produce nanoemulsions (Fig. 2) [58, 73]. In 2015, Scholz and Keck used a rotor–stator system (rotor diameter was 40 mm with 0.1 mm gap and slit size of 0.5 mm) to produce a nanoemulsion. The used MICCRA D27 stirrer which could go up to 36,000 rpm stir speed with the capacity of 45 $Lmin^{-1}$. The resulting nanoemulsion after 5 min operation time had a droplet size of ~ 135 nm and was physically stable for three months. They also utilized a high pressure homogenizer to compare the particle size and found it produced fine nanoparticles compared to used MICCRA D27 stirrer [87]. This process is cost-effective and efficient to use in large industrial-scale nanoparticle production.

6.2 Low Energy Methods for Nanoemulsion Preparation

As the category name implies, low external energy requirement makes these approaches more energy efficient. Interestingly, internal chemical energy is the main energy entity used for nanoparticle production [94]. Phase inversion emulsification and self-emulsification are the two main subclasses and these techniques are not much encouraged especially in the food industry as they require relatively a high number of surfactants [52].

6.2.1 Phase Inversion Emulsification

Spontaneous curvature/arrangement into spherical shapes of emulsifying agents causes a phase inversion during the emulsifying process (Fig. 3). The process is governed by experimental/operational parameters such as temperature and composition of raw materials [95]. There are two types of phase inversion methods; (i) transitional phase inversion (TPI) and (ii) catastrophic phase inversion (CPI).

Transitional phase inversion is a result of the change in the affinity of the surfactants based on the provided operational conditions such as temperature and composition [37, 94, 95]. The catastrophic phase inversion is a result of the continuous addition of dispersed phases until they get together to form a lamellar/bicontinuous

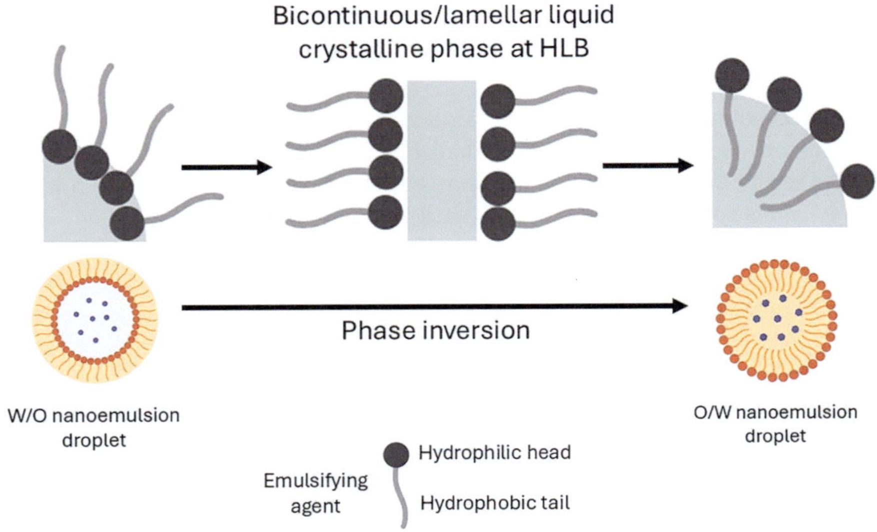

Fig. 3 Low energy methods for nanoemulsion preparation- phase inversion

structural phase [37]. The sudden change in emulsion behavior resulting from continuous addition is the reason for using the phrase "catastrophic". Furthermore, it is a main requirement to feed the emulsifying agent along with the dispersed phase to occur phase inversion with a high coalescence rate [2]. The main difference in TPI compared to CPI is the change in emulsifying agent affinity and/or curvature whereas CPI does not. TPI again can be subdivided into two, phase inversion temperature (PIT and phase inversion composition (PIC. Temperature change is the main driving factor for PIT to inverse the surfactant curve around the dispersed phase. Generally, oil, water, and non-ionic surfactants are mixed at ambient temperature to form O/W emulsions under PIT. After that, the temperature is gradually increased causing dehydration of polyolefin elastomer (POE groups in non-ionic surfactants such as polyethoxylated surfactants making them more lipophilic [68, 94]. In the beginning, the surfactant had an affinity towards the water phase, but with this chemical change its affinity is changed (inverse) towards the oil phase. In other words, the initial emulsion is an O/W emulsion, but at the end, it is a W/O emulsion which causes it to inverse the phase through a bi-continuous/lamellar structure or intermediate liquid crystalline. Rapid temperature change (that is either quick cooling or heating) can cause an efficient phase inversion as well as kinetically stable nanoemulsions [94]. At the hydrophile-lipophile balanced (HLB) temperature, the surfactant has nearly the same affinity for both oil and water, showing no curvature [38–40]. Generally, the PIT technique is suitable for thermostable surfactants and materials.

Phase inversion composition changes the components of the emulsion system at a constant temperature compared to PIT [91]. One component/phase is added first into the reactor/container and then the other component/phase. The surfactant can be added either into the water phase or oil phase depending on its affinity. Any type of surfactant can be employed based on the application. The POE groups of the supplemented surfactant (ex: nonionic surfactants) start to hydrate following the addition of the water phase into the oil phase resulting in a change in the affinity from the oil phase to the water phase. Just as in PIT, there is a critical point where the surfactant starts to show an equal affinity for both water and oil phases at HLB causing no significant curvature. During this transition, a bi-continuous/lamellar structure is formed. Further addition of water positions the hydrophilic portion of the surfactant towards the water and surrounds the oil phase as droplets (phase transition) [94, 99]. Salt addition and pH adjustment of the emulsion system also help to produce nanoemulsification under PIC [62, 66, 91].

Emulsion inversion point (EIP) is categorized under CPI where phase inversion occurs as a result of the change in the dispersed phase fraction rather than any chemical change in the surfactant/s [2, 18, 37, 66, 91, 95]. For instance, let's consider a gradual water addition to a surfactant-water mixture. Once started, it is a W/O emulsion, but the addition of water will exceed the amount of oil at one point making it an O/W emulsion. At the phase inversion point, a bi-continuous/lamellar structure is formed as usual. According to Bancroft's rule, the emulsifying agent should always be with the dispersion medium to have a stable normal emulsion [77]. However, in the EIP approach (in the above example), the emulsifying agent was with the dispersed phase from the beginning [2, 18, 91]. Therefore, we can presume that the EIP procedure contradicts Bancroft's rule to produce unstable emulsions. The size of the droplets can be modified by varying the second phase addition rate and stir speed.

6.2.2 The Self-nanoemulsification Method

This approach does not change the curvature of the emulsifying/stabilizing agent/s. Nanoemulsions are a result of rapid diffusion (a passive process) of the dispersed phase within the dispersion medium and therefore the method is also known as the spontaneous emulsification method [93, 94, 96]. This is the fundamental phenomenon for self-nano emulsifying drug delivery systems where contain a large number of emulsifying agents or co-emulsifying agents with a low oil content [1]. These drug particles added isotropic mixtures can be diluted in an aqueous in vivo environment with the aid of a light agitation (ex: peristalsis movement) to form fine and clear O/W nanoemulsions [8, 51]. As a drug delivery system, it is a promising approach to deliver hydrophobic drugs to increase bioavailability as well as for bioactive food components [3, 75, 76].

6.2.3 D Phase Emulsification

D-phase emulsification is another low-energy technique to prepare homogeneous nanoemulsions. Sagitani et al. proposed this method in 1983, where O/W nanoemulsions can be prepared by diluting an oil-in surfactant gel emulsion (O/D gel emulsion).

Oil droplets are surrounded by the surfactant/D phase [30]. Compared to the other methods, D-phase emulsification is an easy approach to form nanoemulsions and it is not mandatory to control the temperature of the process as well as HLB. Generally, the nonionic surfactant is mixed with a polyol to make the surfactant solution and then the oil is added dropwise to make an O/D emulsion. Then the nanoemulsion can be obtained by diluting the O/D emulsion/gel in water. Researchers presume that the nanoemulsion formation results from a special dynamic surfactant arrangement that is spontaneous [49]. In 2019 Zhang et al. prepared nanoemulsions using this technique and studied their stability. According to the results, the stability of the prepared nanoemulsions where droplet size was ~ 150 nm was mainly decided by the amount of surfactant used [110].

6.3 *Emulsion Polymerization*

The most common industrial polymer nanoparticle production techniques are heterophase polymerization processes and among them, emulsion polymerization is the most popular approach [43, 64]. Briefly, the hydrophobic monomers are polymerized and then emulsified in an aqueous medium (O/W emulsion) by mechanical agitation [17]. For the other type of emulsion, i.e. W/O, hydrophilic monomers such as methacrylic acid are dispersed in an oil phase. In this approach, radical initiators need to be added to the reaction mixture to start polymerization, and initiator selection depends on the type of emulsion [17]. As usual, surfactants need to be added to the system to stabilize the emulsion and meantime to prevent coagulation of emulsion droplets [12]. Other than them, buffers, acids, bases, chain transfer agents, anti-aging agents, and biocides can be there [10]. This polymerization mechanism in an emulsion environment results in sub-micrometer particles with diameters typically 10–1000 nm [9, 69]. If we consider the number of molecules, typically a dispersed droplet is composed of 1–10,000 macromolecules and each polymer molecule has 100 to 10^6 monomers [10].

The emulsion polymerization mechanism occurs in three stages; (I) particle formation, (II) particle growth, and (III) completion [10, 22]. The initial stage/particle formation is also known as particle nucleation. Once the initiator is added to the mixture, formed free radicals start the polymerization in the continuous phase and diffuse into the monomer-swollen micelles. These micelles provide a venue for the initiators and monomers to meet and carry out polymerization and therefore act as a nucleus. Following the particle nucleation process, the particle growth stage starts where simultaneous polymerization and monomer diffusion into micelles take place at a constant rate. With time, monomer quantity gradually decreases and ends the particle growth stage and brings the whole system into the final stage, the completion stage. Here, the polymerization continues as long as monomers are there and come to the end of the whole process.

There are three types of emulsion polymerization processes used at the industrial scale; batch, semi-continuous, and continuous. This classification is based on behavior of the stages II and III [86]. In batch process, all the ingredients are placed in the reactor and mechanical agitation is applied. Once the initiator is applied the polymerization reaction starts. The temperature of the system should be controlled as needed to optimize the operation. Semi-continuous type is common in both industry and laboratory scales as it provides great flexibility in operation. Here, the reactants are added continuously into the reactor and after the operational period, the products are taken out [102, 106]. In the continuous process, reactants are continuously fed to the reactor from one end to maintain a continuous rate and the products are collected at the other end of the reactor [16, 47, 86]. This is a tubular system and not much satisfactory in the industrial section as they need to maintain a lengthy tubular system.

When this technique produces emulsions with nanosize droplets they are named nanoemulsions/miniemulsions. A number of research have reported the utilization of emulsion polymerization techniques to synthesize nanoparticles. In 2016, Zhao et al. developed pH and thermosensitive core–shell nanoparticles using the emulsion polymerization technique which was used to study photoluminescence as a cancer diagnostic reagent [111]. They displayed different colors upon being transferred into cancer cells to distinguish them from healthy cells. In one study emulsifier/stabilizer-free emulsion polymerization was used to develop magnetic core–shell nanoparticles for magnetic resonance imaging (MRI) [109]. Here the shell of the nanoparticle (110–170 nm) was polymerized with gadolinium and europium complexes to coat the core composed of sodium undecylenate-modified iron(III) oxide and oleic acid.

6.4 Emulsion-Solvent Evaporation Method

Emulsion-solvent evaporation technique is used for nanoencapsulation or nanoparticle production [33]. Typically, a polymer (dispersed phase) is emulsified in an aqueous solution (dispersion medium) that is supplemented with surfactants/emulsifying agents. Once the emulsion preparation process is complete, the solvent is gradually evaporated by using continuous stirring, vacuum, or high temperature [33]. Particle size can be adjusted by changing the operational/experimental parameters such as evaporation temperature and stir speed [43]. During this method, solvents such as chloroform or dichloromethane with low boiling points are used to prepare the polymer solution. At the end of the evaporation, the prepared solution will be emulsified in an aqueous medium. This method is not limited to nonpolar solvents, even polar solvents can be employed. For instance, in 2013 Bohlender et al. developed a series of nanoparticles using a polar solvent; hexafluoroisopropanol (HFIP) [13]. In 2017, Guo et al. synthesized Poly(lactic-co-glycolic) acid nanoparticles which encapsulated hypocrellin A (an anticancer agent) by single-emulsion solvent-evaporation [29]. The developed nanoparticles had a 7% drug-loading content and 57.5% drug encapsulation efficiency. However, the single-emulsion solvent-evaporation technique has some drawbacks such as poor encapsulation efficiency for hydrophobic molecules. The double-emulsion solvent evaporation is a good solution in this case as it is suitable for both hydrophobic and hydrophilic materials [36]. In 2010, Liu et al. reported remarkably enhanced drug encapsulation efficiency (>80%) for daunorubicin in poly(d,l-lactic-co-glycolic acid) and poly(d,l-lactic acid) using a modified double-emulsion solvent-evaporation [57]. All the nanoparticles produced in the aforementioned studies ranged from 200 to 300 nm [36, 57].

7 Scale Up Challenges and Solutions

Generally, new products are developed at laboratory scale with the expectation to manufacture in large scale at industry. For example, a new drug dosage form is developed at the research and development section of a pharmaceutical company in small scale. Once it is successful, the next step is to move into large scale manufacturing. In a similar manner, nanoemulsions can be developed at laboratory scale and eventually shift to large scale manufacturing. However, it is not always simple and there will be a number of challenges such as maintaining the consistent particle size, stability, controlling heat generation [103]. Particle size maintenance is the key challenge with respect to nanoemulsion. However, application of strategies to rectify issues completely depend on the type of nanoemulsion and the intended field as explained previously. When it comes to large scale production, due to various shear forces, and mixing dynamics in the production vessel, particle size can be scattered [7]. Formulation design optimization by addition of surfactants to balance HLB, co-emulsifier addition to further stabilize fine tuned particle size as well as addition of polymers (stabilizing agents) to prevent aggregation are some of the potential solutions to maintain a good particle size distribution as well as to improve final product stability and lower phase separation resulting from creaming, coalescence, and flocculation [70, 81]. High-shear homogenization methods such as high pressure homogenizers can produce excess heat which ultimately affect thermolabile compounds/products [32]. This excess heat can be managed by introducing cooling systems [21]. Another crucial challenge is to ensure consistency of the product within batches [7]. This can be addressed by strict adherence to operational parameters and implementing a good, continuous quality control and quality assurance process [32, 103]. The cost of preparation plays a major role in determining the market price of the final product. Scaling up requires especial equipment, surfactants, and energy which highly impact the overall procedure and the value of the product [32, 103]. Judicial selection of the equipment based on the desire product and volume contribute greatly to cut off unnecessary costs including the cost for energy [103]. Safety is a top priority when it comes to the pharmaceutical and food industry nanoparticle production. We can minimize organic solvent usage by employing new solvent free methods such as SolEmuls technology to improve safety of nanoparticles [50, 74]. Practically, it is not easy to fulfill all the requirements when scaling up production. It is a matter of balancing factors to achieve optimal state.

8 Conclusion

A number of nanoemulsion preparation methods/techniques can be found in the literature. Often it is hard to distinguish some preparation methods from each other, however, careful investigation allows the identification of the significant differences between them. In this chapter, we used energy consumption as the foundation to discuss nanoemulsion preparation. Nanoemulsion techniques are useful and popular in the preparation of nanoparticles both on the industrial and laboratory scale. Cosmetics, pharmaceutical, and food industries are predominantly involved in the use of nanoemulsions as an approach for formulating nanoparticles. Frequently, the prepared nanoparticles are kept in the emulsion/liquid form due to stability concerns. The selection of emulsion-based nanoparticle production depends on several factors such as the nature of the material (ex: protein), its thermolability, and pH sensitivity. Low-energy nanoemulsion preparation methods are cost-friendly, but they require extreme chemical conditions and high emulsifier amounts (occasionally artificial emulsifying agents) which can imply safety issues in the pharmaceutical, cosmetics, and food industries. Furthermore, these industries require large-scale preparations where many high-energy nanoemulsion preparation techniques can fit. In our point of view, high-energy techniques are suitable for large scale nanoparticle production whereas low-energy nanoemulsion techniques require further improvements to be compatible with large scale production.

References

1. Agrawal S, Giri TK, Tripathi DK, Alexander A (2012) A review on novel therapeutic strategies for the enhancement of solubility for hydrophobic drugs through lipid and surfactant based self micro emulsifying drug delivery system: a novel approach. Am J Drug Discov Dev 2(4):143–183https://doi.org/10.3923/ajdd.2012.143.183
2. Armanet L, Hunkeler D (2007) Phase inversion of polyacrylamide-based inverse-emulsions: influence of inverting-surfactant type and concentration. J Appl Polym Sci 103(6):3567–3584. https://doi.org/10.1002/app.25062
3. Meena AK, Sharma K, Mullangi R, Kandaswamy M, Rajagopal S (2012) Formulation development of an albendazole self-emulsifying drug delivery system (SEDDS) with enhanced systemic exposure/Razvoj samoemulzifirajućeg sustava za isporuku albendazola (SEDDS) s pojačanom sistemskom apsorpcijom. Acta Pharm 62(4):563–580. https://doi.org/10.2478/v10007-012-0031-0
4. Azeem A, Rizwan M, Ahmad FJ, Iqbal Z, Khar RK, Aqil M, Talegaonkar S (2009) Nanoemulsion components screening and selection: a technical note. AAPS PharmSciTech 10(1):69–76. https://doi.org/10.1208/s12249-008-9178-x
5. Azeem A, Rizwan M, Ahmad F, Khar R, Iqbal Z, Talegaonkar S (2009) Components screening and influence of surfactant and cosurfactant on nanoemulsion formation. Curr Nanosci 5(2):220–226. https://doi.org/10.2174/157341309788185505
6. Bajpai P (2018) Papermaking chemistry. In: Biermann's handbook of pulp and paper. Elsevier, pp 207–236. https://doi.org/10.1016/B978-0-12-814238-7.00010-6
7. Bamisaye A, Adegoke KA, Alli YA, Bamidele MO, Idowu MA, Ogunjinmi OE (2023) Recent advances in nanoemulsion for sustainable development of farm-to-fork systems. J Clean Prod 429:139226. https://doi.org/10.1016/j.jclepro.2023.139226

8. Bandyopadhyay S, Katare O, Singh B (2014) Development of optimized supersaturable self-nanoemulsifying systems of ezetimibe: effect of polymers and efflux transporters. Expert Opin Drug Deliv 11(4):479–492. https://doi.org/10.1517/17425247.2014.877885
9. Baumann H (1967) Hochpolymere, physikalisch bindende Leime (synthetische Produkte). In: Baumann H (ed) Leime und Kontaktkleber. Springer Berlin Heidelberg, pp 161–209. https://doi.org/10.1007/978-3-642-92929-8_14
10. Berber H (2013) Emulsion polymerization: effects of polymerization variables on the properties of vinyl acetate based emulsion polymers. In Ylmaz F (Ed) Polymer science. InTech. https://doi.org/10.5772/51498
11. Bibette J, Calderon FL, Poulin P (1999) Emulsions: basic principles. Rep Prog Phys 62(6):969–1033. https://doi.org/10.1088/0034-4885/62/6/203
12. Bjerk TR, Severino P, Jain S, Marques C, Silva AM, Pashirova T, Souto EB (2021) Biosurfactants: properties and applications in drug delivery, biotechnology and ecotoxicology. Bioengineering 8(8):115. https://doi.org/10.3390/bioengineering8080115
13. Bohlender C, Landfester K, Crespy D, Schiller A (2013) Unconventional Non-aqueous emulsions for the encapsulation of a phototriggerable NO-donor complex in polymer nanoparticles. Part Part Syst Charact 30(2):138–142. https://doi.org/10.1002/ppsc.201200067
14. Cai Z, Li Z, Ravaine S, He M, Song Y, Yin Y, Zheng H, Teng J, Zhang A (2021) From colloidal particles to photonic crystals: advances in self-assembly and their emerging applications. Chem Soc Rev 50(10):5898–5951. https://doi.org/10.1039/D0CS00706D
15. Canselier JP, Delmas H, Wilhelm AM, Abismaïl B (2002) Ultrasound emulsification—an overview. J Dispersion Sci Technol 23(1–3):333–349. https://doi.org/10.1080/01932690208984209
16. Chen J, Zhong W, Tang Y, Wu Z, Li Y, Yi P, Jiang J (2015) Amphiphilic BODIPY-based photoswitchable fluorescent polymeric nanoparticles for rewritable patterning and dual-color cell imaging. Macromolecules 48(11):3500–3508. https://doi.org/10.1021/acs.macromol.5b00667
17. Chern CS (2006) Emulsion polymerization mechanisms and kinetics. Prog Polym Sci 31(5):443–486. https://doi.org/10.1016/j.progpolymsci.2006.02.001
18. Fernandez P, André V, Rieger J, Kühnle A (2004) Nano-emulsion formation by emulsion phase inversion. Colloids Surf, A 251(1–3):53–58. https://doi.org/10.1016/j.colsurfa.2004.09.029
19. Floury J, Desrumaux A, Lardières J (2000) Effect of high-pressure homogenization on droplet size distributions and rheological properties of model oil-in-water emulsions. Innov Food Sci Emerg Technol 1(2):127–134. https://doi.org/10.1016/S1466-8564(00)00012-6
20. Gaikwad SG, Pandit AB (2008) Ultrasound emulsification: Effect of ultrasonic and physicochemical properties on dispersed phase volume and droplet size. Ultrason Sonochem 15(4):554–563. https://doi.org/10.1016/j.ultsonch.2007.06.011
21. Ganta S, Talekar M, Singh A, Coleman TP, Amiji MM (2014) Nanoemulsions in translational research—opportunities and challenges in targeted cancer therapy. AAPS PharmSciTech 15(3):694–708. https://doi.org/10.1208/s12249-014-0088-9
22. Gardon JL (1968) Emulsion polymerization. I. Recalculation and extension of the Smith-Ewart theory. J Polym Sci Part A-1: Polym Chem 6(3):623–641. https://doi.org/10.1002/pol.1968.150060318
23. Ghosh A, Bandyopadhyay M, Dattagupta S, Gupta S (2023) Independent-oscillator model and the quantum Langevin equation for an oscillator: a review. https://doi.org/10.48550/ARXIV.2306.02665
24. Ghosh V, Mukherjee A, Chandrasekaran N (2013) Ultrasonic emulsification of food-grade nanoemulsion formulation and evaluation of its bactericidal activity. Ultrason Sonochem 20(1):338–344. https://doi.org/10.1016/j.ultsonch.2012.08.010
25. Goh P, Ng M, Choo Y, Amru N, Chuah C (2015) Production of nanoemulsions from palm-based tocotrienol rich fraction by microfluidization. Molecules 20(11):19936–19946. https://doi.org/10.3390/molecules201119666

26. Gonçalves A, Nikmaram N, Roohinejad S, Estevinho BN, Rocha F, Greiner R, McClements DJ (2018) Production, properties, and applications of solid self-emulsifying delivery systems (S-SEDS) in the food and pharmaceutical industries. Colloids Surf, A 538:108–126. https://doi.org/10.1016/j.colsurfa.2017.10.076
27. Graves S, Meleson K, Wilking J, Lin MY, Mason TG (2005) Structure of concentrated nanoemulsions. J Chem Phys 122(13):134703. https://doi.org/10.1063/1.1874952
28. Gun'ko VM, Turov VV (2024) Colligative properties of various liquid blends vs. Temperature under confined space effects in pores of different adsorbents. Himia, Fizika Ta Tehnologia Poverhni 15(1):3–34. https://doi.org/10.15407/hftp15.01.003
29. Guo L-Y, Yan S-Z, Li Q, Xu Q, Lin X, Qi S-S, Yu S-Q, Chen S-L (2017) Poly(lactic-co-glycolic) acid nanoparticles improve oral bioavailability of hypocrellin A in rat. RSC Adv 7(67):42073–42082. https://doi.org/10.1039/C7RA04748G
30. Sagitani H, Hattori T, Nabeta K, Nagai M (1983) Formation of O/W emulsions having fine and uniform droplets by the surfactant (D) phase emulsification method. Nippon Kagaku Kaishi 1399–1404
31. Hamans RF, Parente M, Garcia-Etxarri A, Baldi A (2022) Optical properties of colloidal silver nanowires. J Phys Chem C 126(20):8703–8709. https://doi.org/10.1021/acs.jpcc.2c01251
32. Herdiana Y, Wathoni N, Shamsuddin S, Muchtaridi M (2022) Scale-up polymeric-based nanoparticles drug delivery systems: development and challenges. OpenNano 7:100048. https://doi.org/10.1016/j.onano.2022.100048
33. Hoa LTM, Chi NT, Triet NM, Nhan LNT, Chien DM (2009) Preparation of drug nanoparticles by emulsion evaporation method. J Phys: Conf Ser 187:012047. https://doi.org/10.1088/1742-6596/187/1/012047
34. Hohenstein WP, Mark H (1946) Polymerization of olefins and diolefins in suspension and emulsion. Part II. J Polym Sci 1(6):549–580. https://doi.org/10.1002/pol.1946.120010611
35. Hsieh C-W, Li P-H, Lu I-C, Wang T-H (2012) Preparing glabridin-in-water nanoemulsions by high pressure homogenization with response surface methodology. J Oleo Sci 61(9):483–489. https://doi.org/10.5650/jos.61.483
36. Iqbal M, Zafar N, Fessi H, Elaissari A (2015) Double emulsion solvent evaporation techniques used for drug encapsulation. Int J Pharm 496(2):173–190. https://doi.org/10.1016/j.ijpharm.2015.10.057
37. Ishak KA, Annuar MSM (2016) Phase inversion of medium-chain-length poly-3-hydroxyalkanoates (mcl-PHA)-incorporated nanoemulsion: effects of mcl-PHA molecular weight and amount on its mechanism. Colloid Polym Sci 294(12):1969–1981. https://doi.org/10.1007/s00396-016-3957-9
38. Izquierdo P, Esquena J, Tadros TF, Dederen JC, Feng J, Garcia-Celma MJ, Azemar N, Solans C (2004) Phase behavior and nano-emulsion formation by the phase inversion temperature method. Langmuir 20(16):6594–6598. https://doi.org/10.1021/la049566h
39. Izquierdo P, Esquena J, Tadros ThF, Dederen C, Garcia MJ, Azemar N, Solans C (2002) Formation and stability of nano-emulsions prepared using the phase inversion temperature method. Langmuir 18(1):26–30. https://doi.org/10.1021/la010808c
40. Izquierdo P, Feng J, Esquena J, Tadros TF, Dederen JC, Garcia MJ, Azemar N, Solans C (2005) The influence of surfactant mixing ratio on nano-emulsion formation by the pit method. J Colloid Interface Sci 285(1):388–394. https://doi.org/10.1016/j.jcis.2004.10.047
41. Jafari SM, Assadpoor E, He Y, Bhandari B (2008) Re-coalescence of emulsion droplets during high-energy emulsification. Food Hydrocolloids 22(7):1191–1202. https://doi.org/10.1016/j.foodhyd.2007.09.006
42. Jafari SM, He Y, Bhandari B (2007) Production of sub-micron emulsions by ultrasound and microfluidization techniques. J Food Eng 82(4):478–488. https://doi.org/10.1016/j.jfoodeng.2007.03.007
43. Jamekhorshid A, Sadrameli SM, Farid M (2014) A review of microencapsulation methods of phase change materials (PCMs) as a thermal energy storage (TES) medium. Renew Sustain Energy Rev 31:531–542. https://doi.org/10.1016/j.rser.2013.12.033

44. Jasmina H, Džana O, Alisa E, Edina V, Ognjenka R (2017) Preparation of nanoemulsions by high-energy and lowenergy emulsification methods. In: Badnjevic A (ed) CMBEBIH 2017, vol 62. Springer Singapore, pp 317–322. https://doi.org/10.1007/978-981-10-4166-2_48
45. Jayasooriya SD, Bhandari BR, Torley P, D'Arcy BR (2004) Effect of high power ultrasound waves on properties of meat: a review. Int J Food Prop 7(2):301–319. https://doi.org/10.1081/JFP-120030039
46. Jenjob R, Phakkeeree T, Seidi F, Theerasilp M, Crespy D (2019) Emulsion techniques for the production of pharmacological nanoparticles. Macromol Biosci 19(6):1900063. https://doi.org/10.1002/mabi.201900063
47. Jenjob R, Seidi F, Crespy D (2018) Recent advances in polymerizations in dispersed media. Adv Coll Interface Sci 260:24–31. https://doi.org/10.1016/j.cis.2018.08.002
48. Joris GG (1945) Synthetic rubber: *synthetic rubber from alcohol.* A survey based on the Russian literature by Anselm Talalay and Michel Magat. 298 pp. New York, N.Y.: Interscience Publishers. 1945. $5.00. Science 101(2632):587–587. https://doi.org/10.1126/science.101.2632.587.a
49. Kaizu K, Alexandridis P (2016) Effect of surfactant phase behavior on emulsification. J Colloid Interface Sci 466:138–149. https://doi.org/10.1016/j.jcis.2015.10.016
50. Kaur G, Panigrahi C, Agarwal S, Khuntia A, Sahoo M (2024) Recent trends and advancements in nanoemulsions: production methods, functional properties, applications in food sector, safety and toxicological effects. Food Phys 1:100024. https://doi.org/10.1016/j.foodp.2024.100024
51. Khan AW, Kotta S, Ansari SH, Sharma RK, Ali J (2015) Self-nanoemulsifying drug delivery system (SNEDDS) of the poorly water-soluble grapefruit flavonoid Naringenin: design, characterization, in vitro and in vivo evaluation. Drug Delivery 22(4):552–561. https://doi.org/10.3109/10717544.2013.878003
52. Komaiko JS, McClements DJ (2016) Formation of food-grade nanoemulsions using low-energy preparation methods: a review of available methods. Compr Rev Food Sci Food Saf 15(2):331–352. https://doi.org/10.1111/1541-4337.12189
53. Landfester K, Eisenblätter J, Rothe R (2004) Preparation of polymerizable miniemulsions by ultrasonication. J Coat Technol Res 1(1):65–68. https://doi.org/10.1007/s11998-004-0026-y
54. Leal-Calderon F, Poulin P (1999) Progress in understanding emulsion metastability and surface forces. Curr Opin Colloid Interface Sci 4(3):223–230. https://doi.org/10.1016/S1359-0294(99)00038-2
55. Leong TSH, Wooster TJ, Kentish SE, Ashokkumar M (2009) Minimising oil droplet size using ultrasonic emulsification. Ultrason Sonochem 16(6):721–727. https://doi.org/10.1016/j.ultsonch.2009.02.008
56. Li D, Zhao Y, Wang X, Tang H, Wu N, Wu F, Yu D, Elfalleh W (2020) Effects of (+)-catechin on a rice bran protein oil-in-water emulsion: droplet size, zeta-potential, emulsifying properties, and rheological behavior. Food Hydrocolloids 98:105306. https://doi.org/10.1016/j.foodhyd.2019.105306
57. Liu J, Qiu Z, Wang S, Zhou L, Zhang S (2010) A modified double-emulsion method for the preparation of daunorubicin-loaded polymeric nanoparticle with enhanced in vitro anti-tumor activity. Biomed Mater 5(6):065002. https://doi.org/10.1088/1748-6041/5/6/065002
58. Lochana EAL, Kawmudhi PAS (n.d.) general formulation of nanoemulsions using high-energy and low-energy principal methods: a brief review. 4
59. Lotito V, Zambelli T (2022) Playing with sizes and shapes of colloidal particles via dry etching methods. Adv Coll Interface Sci 299:102538. https://doi.org/10.1016/j.cis.2021.102538
60. Lovelyn C, Attama AA (2011) Current state of nanoemulsions in drug delivery. J Biomaterials Nanobiotechnology 02(05):626–639. https://doi.org/10.4236/jbnb.2011.225075
61. Maa Y-F, Hsu CC (1999) Performance of sonication and microfluidization for liquid-liquid emulsification. Pharm Dev Technol 4(2):233–240. https://doi.org/10.1081/PDT-100101357
62. Maestro A, Solè I, González C, Solans C, Gutiérrez JM (2008) Influence of the phase behavior on the properties of ionic nanoemulsions prepared by the phase inversion composition method. J Colloid Interface Sci 327(2):433–439. https://doi.org/10.1016/j.jcis.2008.07.059

63. Mahdi Jafari S, He Y, Bhandari B (2006) Nano-emulsion production by sonication and microfluidization—a comparison. Int J Food Prop 9(3):475–485. https://doi.org/10.1080/10942910600596464
64. Mamaghani MY, Pishvaei M, Kaffashi B (2011) Synthesis of latex based antibacterial acrylate polymer/nanosilver via in situ miniemulsion polymerization. Macromol Res 19(3):243–249. https://doi.org/10.1007/s13233-011-0307-0
65. Martin A, Bustamante P, Chun AHC (1993) Physical pharmacy: physical chemical principles in the pharmaceutical sciences (4th ed). Waverly
66. McClements DJ, Rao J (2011) Food-grade nanoemulsions: formulation, fabrication, properties, performance, biological fate, and potential toxicity. Crit Rev Food Sci Nutr 51(4):285–330. https://doi.org/10.1080/10408398.2011.559558
67. Mitra S, Khandelwal P (2009) Are all colloids same? How to select the right colloid? Indian J Anaesth 53(5):592–607
68. Moreira De Morais J, David Henrique Dos Santos O, Delicato T, Azzini Gonçalves R, Alves Da Rocha-Filho P (2006) Physicochemical characterization of canola oil/water nano-emulsions obtained by determination of required HLB number and emulsion phase inversion methods. J Dispers Sci Technol 27(1):109–115.https://doi.org/10.1081/DIS-200066829
69. Morton M (1981) History of synthetic rubber. J Macromol Sci: Part-Chem 15(7):1289–1302. https://doi.org/10.1080/00222338108056786
70. Mushtaq A, Mohd Wani S, Malik AR, Gull A, Ramniwas S, Ahmad Nayik G, Ercisli S, Alina Marc R, Ullah R, Bari A (2023) Recent insights into nanoemulsions: their preparation, properties and applications. Food Chemistry: X 18:100684. https://doi.org/10.1016/j.fochx.2023.100684
71. Neslihan Gursoy R, Benita S (2004) Self-emulsifying drug delivery systems (SEDDS) for improved oral delivery of lipophilic drugs. Biomed Pharmacother 58(3):173–182. https://doi.org/10.1016/j.biopha.2004.02.001
72. Niknam SM, Escudero I, Benito JM (2020) Formulation and preparation of water-in-oil-in-water emulsions loaded with a phenolic-rich inner aqueous phase by application of high energy emulsification methods. Foods 9(10):1411. https://doi.org/10.3390/foods9101411
73. Nor Bainun I, Alias NH, Syed-Hassan SSA (2015) Nanoemulsion: formation, characterization, properties and applications—a review. Adv Mater Res 1113:147–152. https://doi.org/10.4028/www.scientific.net/AMR.1113.147
74. Paliwal R, Babu RJ, Palakurthi S (2014) Nanomedicine scale-up technologies: feasibilities and challenges. AAPS PharmSciTech 15(6):1527–1534. https://doi.org/10.1208/s12249-014-0177-9
75. Patel AR, Vavia PR (2007) Preparation and in vivo evaluation of SMEDDS (self-microemulsifying drug delivery system) containing fenofibrate. AAPS J 9(3):E344–E352. https://doi.org/10.1208/aapsj0903041
76. Patel G, Shelat P, Lalwani A (2016) Statistical modeling, optimization and characterization of solid self-nanoemulsifying drug delivery system of lopinavir using design of experiment. Drug Delivery 23(8):3027–3042. https://doi.org/10.3109/10717544.2016.1141260
77. Perazzo A, Preziosi V, Guido S (2015) Phase inversion emulsification: Current understanding and applications. Adv Coll Interface Sci 222:581–599. https://doi.org/10.1016/j.cis.2015.01.001
78. Perrier-Cornet JM, Marie P, Gervais P (2005) Comparison of emulsification efficiency of protein-stabilized oil-in-water emulsions using jet, high pressure and colloid mill homogenization. J Food Eng 66(2):211–217. https://doi.org/10.1016/j.jfoodeng.2004.03.008
79. Pinnamaneni S, Das NG, Das SK (2003) Comparison of oil-in-water emulsions manufactured by microfluidization and homogenization. Harmazie 58:554–558
80. Polarz S (2011) Shape matters: anisotropy of the morphology of inorganic colloidal particles—synthesis and function. Adv Func Mater 21(17):3214–3230. https://doi.org/10.1002/adfm.201101205
81. Preeti, Sambhakar S, Malik R, Bhatia S, Al Harrasi A, Rani C, Saharan R, Kumar S, Geeta, Sehrawat R (2023) Nanoemulsion: an emerging novel technology for improving the bioavailability of drugs. Scientifica 1–25.https://doi.org/10.1155/2023/6640103

82. Qadir A, Faiyazuddin MD, Talib Hussain MD, Alshammari TM, Shakeel F (2016) Critical steps and energetics involved in a successful development of a stable nanoemulsion. J Mol Liq 214:7–18. https://doi.org/10.1016/j.molliq.2015.11.050
83. Qian C, McClements DJ (2011) Formation of nanoemulsions stabilized by model food-grade emulsifiers using high-pressure homogenization: factors affecting particle size. Food Hydrocolloids 25(5):1000–1008. https://doi.org/10.1016/j.foodhyd.2010.09.017
84. Rai VK, Mishra N, Yadav KS, Yadav NP (2018) Nanoemulsion as pharmaceutical carrier for dermal and transdermal drug delivery: Formulation development, stability issues, basic considerations and applications. J Control Release 270:203–225. https://doi.org/10.1016/j.jconrel.2017.11.049
85. Salvia-Trujillo L, Rojas-Graü MA, Soliva-Fortuny R, Martín-Belloso O (2014) Impact of microfluidization or ultrasound processing on the antimicrobial activity against *Escherichia coli* of lemongrass oil-loaded nanoemulsions. Food Control 37:292–297. https://doi.org/10.1016/j.foodcont.2013.09.015
86. Schaeffel D, Staff RH, Butt H-J, Landfester K, Crespy D, Koynov K (2012) Fluorescence correlation spectroscopy directly monitors coalescence during nanoparticle preparation. Nano Lett 12(11):6012–6017. https://doi.org/10.1021/nl303581q
87. Scholz P, Keck CM (2015) Nanoemulsions produced by rotor–stator high speed stirring. Int J Pharm 482(1–2):110–117. https://doi.org/10.1016/j.ijpharm.2014.12.040
88. Schramm LL (2005) Emulsions, foams, and suspensions: fundamentals and applications (1st ed). Wiley. https://doi.org/10.1002/3527606750
89. Schramm LL, Schramm LL (2001) Dictionary of colloid and interface science (2nd ed). Wiley
90. Schultz S, Wagner G, Urban K, Ulrich J (2004) High-Pressure Homogenization as a Process for Emulsion Formation. Chem Eng Technol 27(4):361–368. https://doi.org/10.1002/ceat.200406111
91. Sokolov YV (2014) Nanoemulsion formation by low-energy methods: a review. Vìsnik Farmacìï 0(3(79)):16–19. https://doi.org/10.24959/nphj.14.1981
92. Solans C, Izquierdo P, Nolla J, Azemar N, Garcia-Celma MJ (2005) Nano-emulsions. Curr Opin Colloid Interface Sci 10(3–4):102–110. https://doi.org/10.1016/j.cocis.2005.06.004
93. Solans C, Morales D, Homs M (2016) Spontaneous emulsification. Curr Opin Colloid Interface Sci 22:88–93. https://doi.org/10.1016/j.cocis.2016.03.002
94. Solans C, Solé I (2012) Nano-emulsions: Formation by low-energy methods. Curr Opin Colloid Interface Sci 17(5):246–254. https://doi.org/10.1016/j.cocis.2012.07.003
95. Solè I, Pey CM, Maestro A, González C, Porras M, Solans C, Gutiérrez JM (2010) Nano-emulsions prepared by the phase inversion composition method: preparation variables and scale up. J Colloid Interface Sci 344(2):417–423. https://doi.org/10.1016/j.jcis.2009.11.046
96. Solè I, Solans C, Maestro A, González C, Gutiérrez JM (2012) Study of nano-emulsion formation by dilution of microemulsions. J Colloid Interface Sci 376(1):133–139. https://doi.org/10.1016/j.jcis.2012.02.063
97. Stojkoski V, Sandev T, Basnarkov L, Kocarev L, Metzler R (2020) Generalised geometric brownian motion: theory and applications to option pricing. Entropy 22(12):1432. https://doi.org/10.3390/e22121432
98. Tiwari SB, Shenoy DB, Amiji MM (2006) Nanoemulsion formulations for improved oral delivery of poorly soluble drugs. NSTI-Nanotech 1:475–478
99. Vandamme T, Anton V (2010) Low energy nanoemulsification to design veterinary controlled drug delivery devices. Int J Nanomedicine, 867.https://doi.org/10.2147/IJN.S13273
100. Villalobos-Castillejos F, Granillo-Guerrero VG, Leyva-Daniel DE, Alamilla-Beltrán L, Gutiérrez-López GF, Monroy-Villagrana A et al (2018) Fabrication of nanoemulsions by microfluidization. In: Jafari SM, McClements DJ (eds). Nanoemulsions: formulation, applications, and characterization. Academic Press
101. Whitesides GM (2006) The origins and the future of microfluidics. Nature 442(7101):368–373. https://doi.org/10.1038/nature05058
102. Willmann JK, Van Bruggen N, Dinkelborg LM, Gambhir SS (2008) Molecular imaging in drug development. Nat Rev Drug Discovery 7(7):591–607. https://doi.org/10.1038/nrd2290

103. Wilson RJ, Li Y, Yang G, Zhao C-X (2022) Nanoemulsions for drug delivery. Particuology 64:85–97. https://doi.org/10.1016/j.partic.2021.05.009
104. Wooster TJ, Golding M, Sanguansri P (2008) Impact of oil type on nanoemulsion formation and ostwald ripening stability. Langmuir 24(22):12758–12765. https://doi.org/10.1021/la801685v
105. Xu Z, Sun Y, Niu Z, Xu Y, Wei X, Chen X, Pan D, Wu W (2020) Kinetic determination of sedimentation for GMZ bentonite colloids in aqueous solution: effect of pH, temperature and electrolyte concentration. Appl Clay Sci 184:105393. https://doi.org/10.1016/j.clay.2019.105393
106. Yu M, Zhang P, Krishnan BP, Wang H, Gao Y, Chen S, Zeng R, Cui J, Chen J (2018) From a molecular toolbox to a toolbox for photoswitchable fluorescent polymeric nanoparticles. Adv Func Mater 28(46):1804759. https://doi.org/10.1002/adfm.201804759
107. Yuan Y, Gao Y, Zhao J, Mao L (2008) Characterization and stability evaluation of β-carotene nanoemulsions prepared by high pressure homogenization under various emulsifying conditions. Food Res Int 41(1):61–68. https://doi.org/10.1016/j.foodres.2007.09.006
108. Zhang J, Peppard TL, Reineccius GA (2015) Preparation and characterization of nanoemulsions stabilized by food biopolymers using microfluidization. Flavour Fragr J 30(4):288–294. https://doi.org/10.1002/ffj.3244
109. Zhang L, Liang S, Liu R, Yuan T, Zhang S, Xu Z, Xu H (2016) Facile preparation of multifunctional uniform magnetic microspheres for T1–T2 dual modal magnetic resonance and optical imaging. Colloids Surf, B 144:344–354. https://doi.org/10.1016/j.colsurfb.2016.04.014
110. Zhang W, Qin Y, Gao Z, Ou W, Zhu H, Zhang Q (2019) Phase behavior and stability of nano-emulsions prepared by D phase emulsification method. J Mol Liq 285:424–429. https://doi.org/10.1016/j.molliq.2019.04.112
111. Zhao Y, Shi C, Yang X, Shen B, Sun Y, Chen Y, Xu X, Sun H, Yu K, Yang B, Lin Q (2016) PH- and temperature-sensitive hydrogel nanoparticles with dual photoluminescence for bioprobes. ACS Nano 10(6):5856–5863. https://doi.org/10.1021/acsnano.6b00770

Vapor-Phase Production of Nanomaterials

Shalika Parakatawella, A. P. G. M. V. Samaraweera, and Thisari Maleesha Gunathilaka

Abstract Over the past few decades, nanotechnology has played an important role in the industrial sector because of the unique properties of nanomaterials, such as scalability, cost-effectiveness, high product quality and consistency, versatility, and environmental friendliness. The vapor-phase nanomaterial synthesis technique is mainly applied to produce high-quality nanomaterials to achieve the above benefits. For example, vapor-phase techniques like chemical vapor deposition (CVD), physical vapor deposition (PVD), and vapor–liquid-solid (VLS) methods provide versatile solutions for manufacturing various nanostructures to achieve the growing demand for nanomaterials. CVD facilitates reactions such as oxidation and decomposition to create high-purity thin films, while PVD provides sublimation and sputtering to deposit materials precisely onto substrates. The VLS method is specifically used for synthesizing nanowires. Furthermore, optimizing the structural, compositional, optical, and thermal properties is crucial for enhancing the performance of these nanomaterials. The characterization techniques allow researchers and industry professionals to fine-tune material qualities essential for advanced applications. Vapor-phase synthesized nanomaterials are largely used in electronics, energy, environmental remediation, and healthcare industries by driving innovation and meeting important social demands. As vapor-phase methods advance, they will continue shaping the future of nanotechnology, enabling the development of tailored nanomaterials for a wide range of industrial applications.

Keywords Nanomaterials · Nanotechnology · Vapor-phase synthesis · Deposition · Advantages · Industrial applications

S. Parakatawella (✉) · T. M. Gunathilaka
Graduate School of Science and Technology, Hamamatsu Campus, Shizuoka University, Hamamatsu, Shizuoka, Japan
e-mail: parakatawella.21@shizuoka.ac.jp

A. P. G. M. V. Samaraweera
Postgraduate Institute of Science, Peradeniya, Sri Lanka

N. M. Adassooriya et al. (eds.), *Industrial Scale Production of Nanoparticles*, Smart Nanomaterials Technology, https://doi.org/10.1007/978-981-96-9176-0_6

1 Introduction to Nanotechnology

In the modern age, nanotechnology is considered one of the most powerful scientific breakthroughs and has influenced several fields such as material science, electronics, medicine, environmental science, and energy industries. Nanotechnology, defined as the engineering and manipulation of matter on a nanoscale (1–100 nm), enables the creation of materials with unique chemical, biological, and/or physical properties. Nanoscale materials show unique characteristics compared to their bulk counterparts owing to a variety of characteristics such as quantum confinement effects, increased surface area-to-volume ratio, and modified interatomic interactions.

Applied nanotechnology spans many disciplines but integration of nanomaterials into commercial products will only be achievable through detailed knowledge of their synthesis- one that ensures precise control of size, morphology, composition, and surface properties. Many synthesis approaches have been developed that can be broadly classified into "top-down" or "bottom-up" strategies. The top-down approaches are those methods involving the mechanical or chemical reduction of bulk materials to the nanoscale, such as lithography, etching, ball milling, etc. Even though they usually allow for a high level of control of the final shape, these techniques can also introduce defects and contaminants that can endanger the designed material properties [29].

On the other hand, the bottom-up approach concentrates on building nanostructures from an atomic or molecular level, which provides enhanced control over the material's inherent properties. Typical bottom-up methods encompass sol–gel processes, hydrothermal synthesis, and chemical vapor deposition (CVD) [24]. These techniques present unique benefits and challenges, adding to the different kinds of nanomaterial synthesis respectively. Bottom-up approaches can enable the creation of materials with more consistent properties and fewer defects, which are essential for numerous applications.

Among the various synthesis methods, vapor-phase techniques have gained significant interest because they can produce high-purity nanomaterials while allowing for precise control over factors like crystallinity, composition, and morphology. These methods encompass CVD, Physical Vapor Deposition (PVD), and Molecular Beam Epitaxy (MBE). These techniques are particularly popular in the semiconductor industry for depositing thin films, nanowires, and other nanostructures with atomic-scale accuracy [51]. The significance of vapor-phase techniques is found not just in their capacity to produce materials of outstanding quality, but also in their flexibility and suitability for various applications [55].

In materials of vapor phase deposition, gaseous materials/substances are brought into a chemical reaction chamber, where the gaseous substances either chemically react or thermally decompose on a heated substrate to enable controlled material growth. Factors like temperature, pressure, and even the concentration of the precursors can be adjusted as needed to achieve the required material properties. Such capabilities as CVD provide the opportunity to deposit various materials, including metals, oxides, and nitrides, making it crucial for the manufacturing of electronic and photonic devices. With these methods, films of high purity and uniformity are formed in the regulated environment, which is important for application in quantum wells, heterostructures, and superlattices in advanced electronics.

Additionally, processes in the vapor-phase allow for multilayer applications, which is important when applying multilayer structures. By virtue of its inherent character, Atomic Layer Deposition (ALD), a modified version of CVD, exemplifies this advantage as one can deposit materials layer by layer, with each layer being an atomic layer. This technique ensures the remarkable consistency of films in morphological parameters such as thickness, enabling the application of coatings with uniform properties to intricate shapes. Thus, ALD has become one of the most popular processes in the semiconductor industry for fabricating gate dielectrics and other important layers in modern transistors [59]. ALD offers the necessary precision required for the production and design of advanced electronic components with complex material qualities.

The ongoing development of nanotechnology and the growing need for materials with specific properties require constant improvement and advancement of vapor-phase synthesis methods. These techniques are being modified to solve problems in different areas such as energy conversion, catalysis, and biomedicine. For example, when creating new catalysts for energy purposes, vapor-phase methods are commonly used to precisely control the shape and preparation of the catalyst, which is crucial for enhancing its performance [67]. In biomedicine, vapor-phase techniques are being used to create functionalized nanostructures for drug delivery and imaging, improving the effectiveness of treatments [39]. These techniques are flexible and play a crucial role in advancing nanotechnology in various industries.

2 Synthesis Techniques

Vapor phase synthesis techniques are essential for making nanomaterials. They have special benefits, such as controlling the makeup, structure, and shape of materials at the nanoscale. These methods work by turning material substances into vapor and then cooling these vapors to form solid nanomaterials in a controlled environment.

2.1 Chemical Vapor Deposition (CVD)

CVD is a commonly employed technique for synthesizing high-purity thin films and nanostructures. The main characteristics of CVD involve chemical reactions between substance gases on a substrate, producing high-purity materials. Also, this technique is used to deposit materials that are difficult to evaporate and allow for growing a crystalline layer on a substrate with a specific orientation. Thus, the CVD technique is used to produce electronic devices, sensors, and protective coatings.

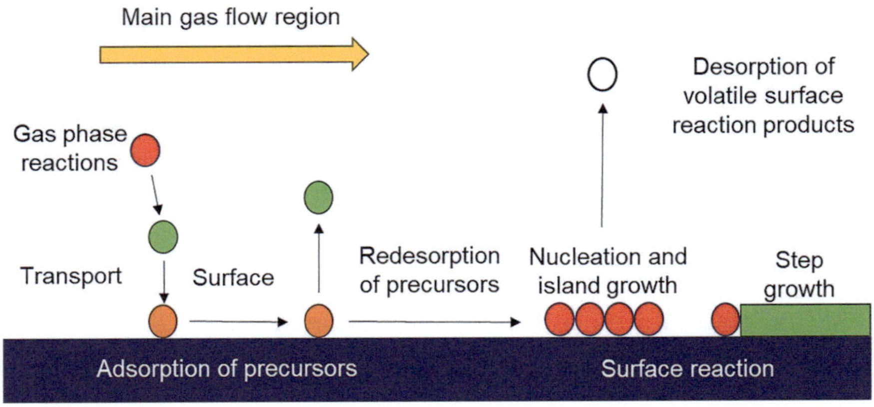

Fig. 1 Transport and reaction process of the reaction process

In the CVD process, reactant gases are transported to the substrate. Then, diffuse through the boundary layer, and adsorb onto the surface. After surface migration, the gases react and incorporate into the growing layer, forming by-products. Next, these by-products desorb, diffuse into the bulk gas, and are removed [64]. The schematic presentation of the CVD process is shown in Fig. 1 [31].

For an effective CVD process, precursors must be volatile (gaseous), reactions should be thermodynamically favorable (with a negative Gibbs Free Energy), and byproducts should be volatile [10].

The probability of a CVD reaction can be assessed by calculating the Gibbs free energy ΔG_r at specific temperature and pressure conditions. To determine ΔG_r, one must first find the free energy of formation ΔG_f for the reactants and products using Eq. (1) at temperature T, followed by Eq. (2). Here, ΔH_f^0 and S^0 Denote the standard enthalpy of formation and entropy at 298 K, respectively, while C_p Is the heat capacity. A negative ΔG_r indicates the reaction can proceed, whereas a positive value suggests it will not occur. When multiple thermodynamically feasible reactions exist, the one with the most negative ΔG_r is favored, as it results in the most stable products [10].

$$\Delta G_f(T) = \Delta H_f^0(298) + \int_{298}^{T} C_p dT - TS^0(298) - \int_{298}^{T} \left(\frac{C_p}{T}\right) dT \quad (1)$$

$$\Delta G_r = \Delta G_r(products) - \Delta G_r(reactants) \quad (2)$$

2.1.1 Types of CVD Reactions According to the Literature [10, 44]

Pyrolysis/Thermal Decomposition

Thermal decomposition of gaseous substances on hot substrates

Ex:

$$SiH_4(s) \rightarrow Si(s) + 2H_2(g) \text{ at } 600\text{–}1150\ ^\circ C$$
$$Fe(CO)_3(s) \rightarrow Fe(s) + 5CO(g) \text{ at } 370\ ^\circ C$$

Reduction

Using hydrogen gas to reduce carbonyl halides, halides, and oxyhalides

Ex:

$$SiCl_4(s) + 2H_2(g) \rightarrow Si(s) + 4HCl(g) \text{ at } 900\text{–}1200\ ^\circ C$$
$$WF_6(s) + 3H_2(g) \rightarrow W(s) + 6HF(g) \text{ at } 550\text{–}800\ ^\circ C$$

Oxidation

Produce oxides using oxygen gas.

Ex:

$$SiH_4(s) + 2O_2(g) \rightarrow SiO_2(s) + 2H_2O(g) \text{ at } 350\ ^\circ C$$
$$TiCl_4(s) + O_2(g) \rightarrow TiO_2(s) + 2Cl_2(g) \text{ at } 350\ ^\circ C$$

Compound Formation

Formation of carbide, nitride, and boride films

Ex:

$$TiCl_4(s) + BCl_3(s) + 5H_2(g) \rightarrow TiB_2(s) + 10HCl(g) \text{ at } 1100\ ^\circ C$$
$$Ga(CH_3)_3(s) + AsH_3(s) \rightarrow GaAs(s) + 3CH_4(g) \text{ at } 1100\ ^\circ C$$

Disproportionation

Deposition of two valence state solid metals of a metal exists at different stable temperatures.

Ex:

$$2GeI_2(s) \rightleftharpoons Ge(s) + GeI_4(s) \quad \text{at } 300\text{–}600\ ^\circ C$$

Reversible Transfer

Temperature-dependent deposition or etching

Ex:

$$As_4(s) + As_2(s) + 6GaCl(s) + 3H_2(g) \rightleftharpoons 6GaAs(s) + 6HCl(g)(750\ ^\circ C \rightleftharpoons 850\ ^\circ C)$$

Hydrolysis

The metal halides combine with hydrolyzing compounds near the substrate surface to prevent premature reactions upon contact.

Ex:

$$AlCl_3(s) + 3CO_2(g) + 3H_2(g) \rightleftharpoons Al_2O_3(s) + 6HCl(g) + 3CO(g)(800\ ^\circ C \rightleftharpoons 500\ ^\circ C)$$

Additionally, two types of nucleation occur in CVD [10]:

- Homogeneous: Nuclei form in vapor before deposition
- Heterogeneous: Nuclei form on the substrate surface

Gas transport in CVD is crucial for uniform deposition. Key considerations include laminar vs. turbulent flow, boundary layer formation, and diffusion through the stagnant layer. The deposition rate (DR) in a CVD process is determined by using pressure (P), temperature (T), activation energy (E_a), and diffusion coefficient (D). The deposition rate is influenced by the following equation [31]:

$$DR = \frac{N_A.P.D}{R.T}.e^{-\frac{E_a}{RT}} \tag{3}$$

where N_A is the Avogadro number, R is the universal gas constant, T is the temperature, E_a is the activation energy, P is the pressure and D is the diffusion coefficient. To experimentally determine the activation energy (E_a) the Arrhenius equation is applied:

$$DR = A.e^{-\frac{E_a}{RT}} \tag{4}$$

where A is the pre-exponential factor. By plotting (DR) against 1/T, the slope of the resulting line, $-E_a/R$, can be used to calculate the activation energy.

Further, key steps in determining E_a involve,

1. Conduct deposition experiments at different temperatures, keeping other parameters like pressure, flow rate of gas, and reactant gas concentration constant
2. Record the (DR) and corresponding T
3. Use the Arrhenius plot to calculate E_a from the slope of the line, which follows the equation $-E_a/R$.

This approach provides critical insight into how deposition rates are affected by various process parameters in the CVD technique for better control and optimization of material deposition.

2.2 *Physical Vapor Deposition (PVD)*

The PVD technique consists of thermal evaporation and sputtering. This involves vaporizing material from a solid source and condensing it onto a substrate in a vacuum environment forming a high pure and uniform thin film or nanostructure. The primary reactions are sublimation and sputtering. PVD has a wide range of applications in optoelectronics, surface protection, and ceramic industries.

Sublimation Reaction

Solid materials transition directly into the gas phase in the sublimation process. For example, CaTe sublimates as follows:

$$2CdTe(s) \leftrightarrows 2Cd(g) + Te_2(g)$$

The vaporized materials afterward condense on a substrate to form a thin film [4].

Sputtering Reaction

In the sputtering, high-energy particles like ions and electrons bombard into a target material. Therefore, atoms eject that can be deposited on a substrate forming a thin film on the substrate. Magnetic sputtering is one of the most valuable deposition methods for complex alloys and non-metallic compounds because it offers excellent composition control [34].

2.3 *Vapor-Liquid-Solid (VLS) Growth*

The VLS growth mechanism is usually used for synthesizing one-dimensional nanostructures or nanowires. In this process, a liquid catalyst absorbs vaporized material until it becomes supersaturated, resulting in the nucleation of solid nanowires.

The VLS mechanism can be described, as vaporized material condensing in a liquid catalyst droplet, leading to the formation of solid nanowires as the droplet becomes supersaturated [23]. The schematic representation of VLS growth is represented in Fig. 2. This method is particularly advantageous for producing semiconductor nanowires. Such as Si, ZnO, Ge, GaN, and GaAs which are essential for applications in optoelectronics [9].

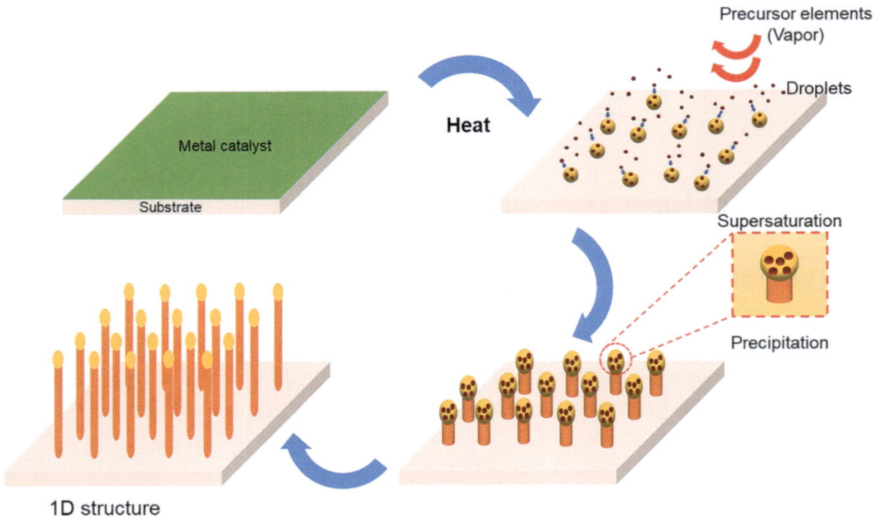

Fig. 2 Growth of 1D structures by VLS mechanism [9]

2.4 Atomic Layer Deposition (ALD)

ALD is a specific form of CVD technique that allows the deposition of materials with atomic-level precision. The ALD process involves alternating exposure of the substrate to different gas substances while controlling the growth of nanomaterials with an exact thickness and composition in each cycle. ALD involves a consecutive surface reaction, and each reaction step ensures that only a single atomic layer is deposited. This process results in a uniform and control of the film thickness [38, 67].

Because of atomic-scale precision, ALD is particularly applied in the semiconductor industry, where it is employed for the deposition of dielectric layers, metal films, and other materials that require accurate thickness control and excellent uniform deposition. On the other hand, ALD has been used in energy storage systems, catalysis, and advanced coatings, where uniform films are crucial [12, 18].

2.5 Molecular Beam Epitaxy (MBE)

MBE is a highly controlled PVD technique utilized for the epitaxial growth of crystalline materials with atomic precision. In MBE, molecular or atomic beams generated from heated compound or element sources are directed onto a heated substrate to form a thin film or nanostructure in an ultra-high vacuum environment. This technical process allows for precise control of both composition and thickness, allowing to growth of high-quality, defect-free materials on the substrate [46].

MBE deposition technique is used to fabricate semiconductor devices, quantum wells, superlattices, and heterostructures because atomic-scale control is essential for optimizing electronic and optical properties. Further, MBE is commonly used to develop new materials with tailored electronic, optical, and magnetic properties for various industries [28].

2.6 Spray Pyrolysis

The spray pyrolysis technique is useful for synthesizing nanomaterials through the atomization and thermal decomposition of precursor solutions. In this process, First, the precursor solution is atomized into fine droplets. Then the droplets are carried by a gas stream into a high-temperature zone, where they undergo rapid pyrolysis and form nanoparticles or thin films, with precise size, morphology, and composition [16, 33].

Spray pyrolysis is widely used in the fabrication of oxide nanoparticles, thin films, and coatings and this technique is suitable for industrial applications such as catalysis, energy storage devices, solar cells, and functional coatings [35].

2.7 Pulsed Laser Deposition (PLD)

The PVD technique that uses high-energy laser pulses to remove material from a solid target is known as PLD. This process resulting a plasma cloud that consist of vaporized atoms or ions [37]. These materials are then transported to a nearby substrate, where the materials condense and form a thin film or nanostructure [21].

PLD is mostly used for depositing complex oxide films, superconductors, and other materials that are challenging to grow with conventional methods. This technique's ability to preserve the stoichiometry of multi-component materials makes it valuable for the research and development of thin films with unique or tailored properties for electronic, optical, and superconducting applications [63].

Table 1 provides a concise summary of the pros and cons, along with the scale-up challenges and potential opportunities, associated with various synthesis techniques. Figure 3 illustrates a decision tree designed to facilitate the selection of a suitable synthesis technique based on specific requirements, ensuring clarity and ease of understanding.

Table 1 Summery of vapor-phase nanomaterial synthesis technique

Synthesis technique	Pros.	Cons.	Scaling-Up challenges	Opportunities	References
CVD	High-quality, uniform films	High energy consumption	Uniformity over large areas	Roll-to-roll processing for continuous production	[5, 49]
	Versatile for various materials (e.g., metals, semiconductors, oxides)	Expensive precursor materials	Cost of high-purity precursors and reactors	Integration with additive manufacturing	
	Scalable with batch processing	Potential toxic byproducts			
PVD	High purity and controlled thickness	Limited to line-of-sight deposition	Achieving uniform deposition on complex geometries	Modular systems for larger substrates	[52]
	Applicable for hard coatings and thin films	Requires vacuum systems, increasing costs	High capital costs for equipment	Hybrid methods (e.g., PVD-CVD combinations)	
	Low chemical waste				
VLS	Effective for nanowire synthesis	Requires a catalyst, which may contaminate the product	Catalyst control for large-scale uniformity	Use of template-assisted growth for scalability	[23]
	Controlled growth directions and aspect ratios	Limited to specific material systems	Managing high temperature requirements	Application in optoelectronics and sensors	
ALD	Atomic-scale precision and uniformity	Slow deposition rates	Increasing deposition speed without compromising precision	Spatial ALD for faster processing	[52]

(continued)

Table 1 (continued)

Synthesis technique	Pros.	Cons.	Scaling-Up challenges	Opportunities	References
	Excellent for conformal coatings on 3D structures	Expensive precursors and equipment	High costs for batch processing	Application in flexible electronics and nanostructures	
MBE	Atomic-scale precision and control over thickness and composition	Extremely expensive due to UHV systems and materials	High cost of UHV systems and precursors	Integration with other techniques (e.g., CVD) for hybrid systems	[51, 54]
	Ultra-high vacuum (UHV) ensures high purity	Low deposition rates	Limited throughput and substrate size	Advancements in automation to reduce operational costs	
		Limited to thin films and small substrates	Difficulty in scaling for industrial production due to slow deposition rates	Use in high-value, niche applications like quantum computing and photonics	
Spray pyrolysis	Simple and cost-effective	Limited control over particle size and distribution	Maintaining uniformity in large-scale systems	Integration with continuous flow systems	[16, 33]
	Scalable for nanoparticle production	Requires post-processing for high purity	Managing waste and emissions	Application in catalysis and battery materials	
PLD	Versatile for a wide range of materials (metals, oxides, polymers)	Poor uniformity over large areas	Non-uniformity on large substrates due to the point source nature of laser ablation	Development of advanced laser systems for better control and uniformity	[63]

(continued)

Table 1 (continued)

Synthesis technique	Pros.	Cons.	Scaling-Up challenges	Opportunities	References
	Simple setup compared to MBE	Particle formation (droplets) can reduce film quality	Particle formation (droplets) during ablation can degrade film quality	Use in additive manufacturing and prototyping for high-value applications	
	Capable of stoichiometric transfer of complex materials	Limited scalability for industrial applications	Limited scalability for industrial-scale manufacturing	Application in emerging fields like energy storage and multifunctional coatings	

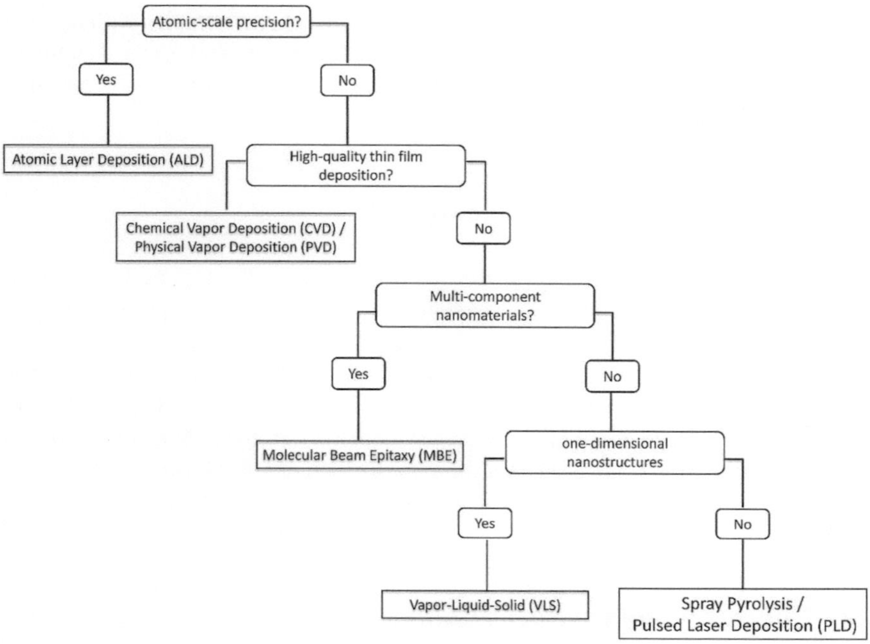

Fig. 3 Decision tree for selecting synthesis techniques based on precision, film quality, material composition, and structure

3 Characterization Techniques of Vapor-Phase Synthesized Nanomaterials

The progress of nanotechnology and the synthesis of nanomaterials through vapor-phase methods necessitate thorough characterization to understand their structural, optical, electrical, and chemical properties. Characterization techniques are essential for determining the shape, size, composition, and distribution of nanomaterials, which in turn influence their functionality and potential applications in both fundamental research and industrial applications.

3.1 Structural Characterization

3.1.1 Scanning Electron Microscopy (SEM)

This technique is used for examining the surface morphology and microstructure of nanomaterials. SEM provides high-resolution images by scanning a focused electron beam across the sample surface, producing secondary electrons that form an image and are particularly useful for determining particle size, shape, and distribution [2].

Further, the main advantage of SEM is its ability to analyze samples with minimal preparation, making it suitable for a variety of vapor-phase synthesized nanomaterials, including thin films and nanoparticles. The information obtained from SEM can be used for optimizing synthesis processes.

3.1.2 Transmission Electron Microscopy (TEM)

TEM Provides a more detailed view of nanostructures at the atomic level. In this technique, a beam of electrons is transmitted through a thin sample to capture a detailed image of internal structures and phases [14]. TEM is especially valuable for examining the crystalline structures and any defects within nanomaterials created through vapor-phase synthesis. Further, Selected Area Electron Diffraction (SAED) to provide information about the crystallinity and phase of the nanomaterials. High-resolution TEM (HRTEM) allows for the visualization of lattice fringes, enabling to analyze of crystal orientation and atomic arrangement in detail [8].

3.2 Compositional Characterization

3.2.1 Energy Dispersive X-Ray Spectroscopy (EDX)

EDX is often coupled with SEM or TEM to provide an elemental composition of nanomaterials. The fundamental of EDX is detecting the characteristic X-rays emitted from a sample when it is excited by an electron beam [62]. This technique allows for qualitative and quantitative elemental analysis to confirm the composition of a nanomaterial. Further, EDX is particularly useful for assessing the distribution of elements in heterogeneous nanostructures, such as core–shell nanoparticles or doped materials [47].

3.2.2 X-Ray Photoelectron Spectroscopy (XPS)

XPS is a surface-sensitive technique that provides information about the chemical state and electronic environment of elements in a material [13]. X-rays are used to irradiate the sample to emit photoemission of core electrons. Then, the free electrons are analyzed to determine the elemental composition and oxidation states. XPS is particularly valuable for vapor-phase synthesized nanomaterials because it can elucidate significantly impacts catalytic and electronic properties. Thus, this technique is crucial for understanding how surface modifications and functionalization can enhance the performance of nanomaterials in various applications [26].

3.3 *Optical Characterization*

3.3.1 Ultraviolet–Visible (UV–Vis) Spectroscopy

UV–Vis Spectroscopy is a powerful tool for investigating the optical properties of vapor-phase synthesized nanomaterials. This technique provides information about electronic transitions and band gap energy of a substance by measuring the absorbance and transmittance of light in the UV–Vis range [3]. UV–Vis spectroscopy is particularly useful for characterizing semiconductor nanomaterials because it can be used to determine the effect of shape and size on optical properties. Also, this technique can be used for real-time process optimization during the synthesis of nanostructures [40].

3.3.2 Photoluminescence (PL) Spectroscopy

PL spectroscopy is used to study the emission of light from a material after it has absorbed photons. This technique is valuable for assessing the optical quality and electronic structure of nanomaterials [60]. PL measurements can provide insights into defects, impurities, and the electronic states of vapor-phase synthesized nanostructures. It can reveal information about excitonic behavior and energy levels in semiconductor nanomaterials. The wavelength and intensity of the emitted light can be correlated with the size and morphology of the nanomaterials. Thus, this is a useful tool for optimizing synthesis conditions [32].

3.3.3 Cathodoluminescence Spectroscopy (CL)

This characterization technique is crucial for investigating the optical properties of vapor-phase synthesized nanomaterials. It reveals insights into electronic states and defects that are essential for assessing material quality by utilizing a focused electron beam to excite the sample. In CL, A high-energy electron beam interacts with the material to emit characteristic light. This analysis is performed using an SEM equipped with a CL detector to generate spectra that provide information about the material's electronic structure [11]. Thus, CL's ability to identify and characterize defects significantly influences the optical and electronic properties of nanomaterials [65].

Additionally, CL provides optical information that helps assess material homogeneity by identifying variations in crystallinity or composition. This technique provides essential data for evaluating their applicability in determining the bandgap energy of semiconductor nanomaterials [27]. Also, CL helps to ensure high-quality standards in the industry by enabling adjustments to synthesis parameters because this technique provides real-time information about the semiconductor production process.

3.4 *Thermal and Mechanical Characterization*

3.4.1 Thermogravimetric Analysis (TGA)

TGA is employed to assess the thermal stability and composition of vapor-phase synthesized nanomaterials. This technique provides information about decomposition temperatures, moisture content, and thermal stability by measuring the change in weight of a sample as it is heated. TGA is particularly useful for evaluating the stability of nanostructured materials under various temperature conditions. Thus, it is crucial for applications in energy storage and catalysis [7]. The results from TGA can guide the selection of materials and processing conditions for specific industrial applications.

3.4.2 Nanoindentation

The nanoindentation technique is used to determine the mechanical properties of materials at the nanoscale. It involves the use of a sharp indenter to penetrate the surface of the material for the determination of hardness and elastic modulus [20]. This technique is essential for vapor-phase synthesized nanomaterials because its mechanical properties can significantly affect their performance in applications like coatings, structural components, and nanocomposites [22].

4 Commercial Advantages of Vapor-Phase Synthesis Techniques

In the fast-changing field of material science and nanotechnology, vapor-phase synthesis methods have become essential for the industrial-scale production of high-quality nanomaterials. These techniques offer distinctive benefits that attract industries aiming to create advanced materials with specific properties for various uses. This chapter explores the commercial benefits of vapor-phase synthesis, focusing on factors like scalability, cost efficiency, product excellence, and versatility.

4.1 *Scalability*

One of the main commercial advantages of vapor-phase synthesis techniques is their scalability. CVD and ALD techniques are examples that show scalability. These techniques can be adapted to produce materials at both laboratory and industrial scales to smooth the transition from research to commercial production. For example, CVD processes can be implemented in large-scale reactors that accommodate the continuous flow of precursors and facilitate the mass production of thin films or nanostructures without compromising quality or consistency [17].

Further, scalability is critical in electronics industries because the demand for high-quality materials is continuously growing in these industries. The semiconductor industry depends on vapor-phase techniques to produce nanomaterials that meet the exact requirements of modern devices, such as transistors, solar cells, and sensors on the nanoscale. As consumer electronics become more sophisticated and miniaturized, the ability to scale up vapor-phase synthesis methods is essential for meeting market demands without significant delays [51].

4.2 Cost-Effectiveness

Cost considerations are essential in any commercial product, and vapor-phase synthesis techniques offer significant cost advantages over traditional synthesis methods. The precision and control afforded by these techniques often lead to higher yields of usable materials, minimizing waste and reducing production costs. Moreover, the ability to produce high-purity materials directly from gaseous precursors eliminates both time-consuming and expensive post-processing [51].

Furthermore, many vapor-phase synthesis processes can be performed at relatively low temperatures and pressures, which can translate to lower energy consumption compared to other synthesis methods. For example, ALD operates at low deposition rates but provides high-quality films. Thus, it is economically viable for the production of dielectric layers and other critical components in semiconductor fabrication [52]. The reduction in energy costs, coupled with increased efficiency in material use, positions vapor-phase techniques as a cost-effective solution for industries aiming to optimize their production processes.

4.3 Product Quality and Consistency

The quality of materials produced through vapor-phase synthesis techniques is one of their most compelling advantages. These methods allow for precise control over deposition parameters including pressure, temperature, and precursor concentration, and directly influence the morphology, crystallinity, and chemical composition of the resulting nanomaterials [51]. The high degree of control inherent in vapor-phase techniques is essential for high-performance applications, resulting in materials with uniform properties [25].

In the semiconductor industry, producing thin films with consistent thickness and composition is critical for device performance. Variability in material properties can lead to device failure or reduce the device's performance, making quality control a top priority. Particularly CVD and MBE vapor-phase techniques have been instrumental in providing the reproducibility and reliability required in this sector [51, 54]. This consistency not only enhances the performance of end products but also builds trust with manufacturers and customers, further driving demand for vapor-phase-synthesized materials.

4.4 Versatility

Another significant commercial advantage of vapor-phase synthesis techniques is their versatility. These methods can be used to produce a wide range of nanomaterials, such as metals, metal oxides, and semiconductors, making them suitable for various kinds of applications across multiple industries. For example, these techniques can be employed to improve wear resistance, corrosion resistance, or optical performance which are essential in fields such as aerospace, automotive, and renewable energy [51, 55].

Moreover, the adaptability of vapor-phase techniques allows for the deposition of various dopants and additives into the materials being synthesized. Thus, enabling the customization of properties to meet specific application requirements. This flexibility is very important in rapidly changing markets where the demand for novel materials is constantly evolving. Thus, these techniques are solutions for competitiveness and advancements in technology by providing manufacturers with the ability to innovate and customize materials efficiently [15].

4.5 Environmental Considerations

In recent years, a consideration has been made about the importance of sustainability and environmental responsibility in manufacturing processes. Because the traditional synthesis methods often involve the use of toxic solvents and produce hazardous byproducts, which can pose risks to human health and the environment vapor-phase synthesis techniques can contribute positively in this matter by minimizing waste and using precursors more efficiently. Because of this, these methods usually operate in closed systems with lower environmental impact [30]. Furthermore, many vapor-phase synthesis processes have low energy requirements, as reactions can be performed at ambient or low temperatures. Thus, vapor-phase synthesis techniques are well-positioned to meet both economic and environmental goals in the industrial sector [56].

5 Industrial Applications of Vapor-Phase Synthesized Nanomaterials

Over the past years, the field of nanotechnology has developed significantly due to the unique properties of nanomaterials, particularly those synthesized through vapor-phase methods. Because vapor-phase methods allow for precise control, the characteristics of nanomaterials-such as size, shape, and composition. Thus, the industrial applications of these nanomaterials expand multiple sectors, including electronics, energy, catalysis, environmental remediation, and healthcare.

5.1 Electronics Industry

Advanced vapor-phase synthesis techniques such as CVD and MBE are primarily used to fabricate high-performance materials for a broad range of applications in the electronics industry.

5.1.1 Semiconductor Devices

In semiconductor fabrication, the need for high-purity materials with specific electronic properties is dominant. CVD processes are commonly employed to deposit silicon, silicon nitride, and silicon carbide films, which serve as essential components in transistors and integrated circuits. The exceptional uniformity and conformity of CVD films enable the production of complex microelectronic devices, critical for enhancing device performance [5]. Moreover, the advent of two-dimensional materials like graphene, synthesized via CVD, has opened new avenues for flexible and high-speed electronic devices. Graphene's unique electronic properties, including high carrier mobility and mechanical strength, make it an ideal candidate for next-generation transistors and sensors [49, 68].

5.1.2 Display Technologies

Nanomaterials synthesized through vapor-phase techniques also play a significant role in display technologies. For example, organic light-emitting diodes (OLEDs) have benefited from the development of nanostructured organic materials that enhance light emission and efficiency. CVD is employed to produce thin films of organic semiconductors, which are essential for high-performance OLED displays used in smartphones and televisions [6].

5.2 Energy Sector

The energy sector is witnessing a rapid transformation, driven by the need for sustainable energy solutions. Vapor-phase synthesized nanomaterials have become critical in the development of advanced energy storage and conversion systems.

5.2.1 Batteries and Supercapacitors

Batteries are used in a wide range of applications, including portable electronics and electric vehicles. Vapor-phase synthesized nanomaterials play a crucial role in improving the performance of these batteries. For example, silicon nanostructures, particularly silicon nanowires, synthesized through VLS growth, exhibit high specific capacity and significantly enhance battery performance compared to traditional graphite anodes [48]. Furthermore, lithium-ion batteries and supercapacitors have benefited from the use of nanostructured carbon materials synthesized through CVD. Because the CVD technique provides increased surface area and conductivity to improve energy and power densities [58].

5.2.2 Photovoltaic Cells

In solar energy applications, vapor-phase techniques are important for the production of thin-film photovoltaic cells. Such as cadmium telluride (CdTe) and copper indium gallium selenide (CIGS) are common materials deposited using CVD. These films are efficient solar cells with lower production costs compared to traditional silicon solar cells. The achieved precise film thickness and composition are crucial for optimizing the performance of these solar cells [1].

5.3 Catalysis

Vapor-phase synthesized nanomaterials in catalysis provide significant advantages in terms of activity, selectivity, and stability. These nanoparticles are widely used in heterogeneous catalysis, where their high surface area and tunable properties lead to enhanced catalytic activity. Metal nanoparticles such as platinum, palladium, and gold, are synthesized through CVD or ALD and applied in various catalytic processes, including hydrogenation, oxidation, and catalytic converters in automotive applications [43, 50]. The reaction efficiencies and selectivity are improved because the ability to control the size and morphology of these nanoparticles allows for the optimization of catalytic properties, which are crucial for industrial-scale chemical manufacturing [57].

5.3.1 Environmental Catalysis

Environmental catalysis that is produced from vapor-phase techniques plays a crucial role. For example, TiO_2 nanoparticles are effective photocatalysts for the degradation of organic pollutants in wastewater treatment. Because of their high reactivity and surface area facilitate the breakdown of hazardous organic compounds under UV light [36]. Furthermore, these nanomaterials can be employed in air purification systems, where their photocatalytic properties help convert harmful gases into less toxic byproducts, significantly improving air quality [42].

5.4 Environmental Remediation

The environmental sector has gradually adopted vapor-phase synthesized nanomaterials for remediation applications. Because of their unique properties to remove the pollutants and contaminants in water and soil.

5.4.1 Water Treatment

Zero-valent iron and metal oxide nanoparticles are synthesized through vapor-phase methods. They have demonstrated excellent capabilities in treating contaminated water. For example, zero-valent iron nanoparticles show effective heavy metal ions and chlorinated solvent adsorption in groundwater. This provides a sustainable and efficient solution to remediate groundwater [19]. Additionally, vapor-phase synthesized titanium dioxide nanoparticles are used in photocatalytic water treatment systems, because of effectively degrade organic pollutants under UV light [45].

5.4.2 Soil Remediation

Vapor-phase synthesized nanomaterials are utilized to remediate heavy metal contamination in soil. Studies have shown that magnetite (Fe_3O_4) and silica nanoparticles, can effectively immobilize heavy metals in contaminated soils, preventing their leaching into groundwater. This approach offers a promising strategy for restoring contaminated lands to a safe and usable state [53].

5.5 Healthcare and Biomedical Applications

Nanomaterials are employed in drug delivery, imaging, and diagnostics for various healthcare and biomedical applications.

5.5.1 Drug Delivery Systems

Nanoparticles synthesized through vapor-phase methods, such as silica and lipid nanoparticles, are increasingly being explored for drug delivery applications. Because of their tunable size and surface properties allow for the encapsulation of therapeutic agents, enhancing drug solubility and bioavailability. For instance, silica nanoparticles functionalized with targeting ligands can selectively deliver drugs to cancer cells, improving treatment efficacy while minimizing side effects. This targeted drug delivery approach is particularly important in the diagnosis treatment, and management of cancer, where maximizing drug concentration at the tumor site can lead to better patient outcomes [6, 66].

5.5.2 Imaging and Diagnostics

Vapor-phase synthesized gold nanoparticles are prominent in biomedical imaging and diagnostics because of their unique optical properties. These nanoparticles can be used as contrast agents in imaging techniques such as computed tomography (CT) and magnetic resonance imaging (MRI) to provide high image resolution. Also, these gold nanoparticles can be employed in biosensing applications for the detection of biomolecules, enabling early disease diagnosis [41, 61].

6 Conclusion

Vapor-phase synthesis methods provide numerous commercial benefits that make them essential for producing high-quality nanomaterials. Their scalability, cost efficiency, product excellence, versatility, and compatibility with environmental standards position these techniques as a top choice across various industries. As the demand for advanced materials rises, vapor-phase synthesis is set to play a vital role in overcoming these challenges, fostering innovation, and influencing the future of nanotechnology. By harnessing these methods, manufacturers can boost their competitive edge and contribute to the creation of advanced technologies that address significant societal needs.

The processes involved in the vapor-phase synthesis of nanomaterials differ based on the technique used. CVD encompasses various reactions, such as thermal reduction, oxidation, decomposition, formation, and disproportionation, all essential for generating high-purity thin films and nanostructures. PVD employs sublimation and sputtering to vaporize material from a solid source onto a substrate. The VLS method is specifically utilized for synthesizing nanowires, where vapor interacts with a liquid catalyst to create solid nanowires. A thorough understanding of these mechanisms is vital for refining synthesis processes and advancing the field of nanotechnology.

Characterizing nanomaterials produced through vapor-phase synthesis is critical for grasping their properties and optimizing their performance across various applications. The methods outlined in this chapter, including structural, compositional, optical, and thermal characterization techniques, equip researchers and industry professionals with a robust toolkit. As nanotechnology progresses, these characterization approaches will remain essential for enhancing our knowledge and application of nanomaterials.

The industrial applications of nanomaterials synthesized through vapor-phase techniques are extensive and diverse, showcasing their versatility and importance in advancing technology across multiple sectors. These materials enhance electronic device performance, improve energy efficiency, facilitate environmental remediation, and transform healthcare, thereby reshaping industries and driving innovation. As the demand for advanced materials continues to escalate, vapor-phase synthesis methods will be crucial in providing high-quality, customized nanomaterials that meet the specific requirements of various industrial applications.

7 Future Directions

The future of vapor-phase synthesis for nanomaterials is characterized by outstanding progress in precision, sustainability, scalability, and the integration of emerging technologies like artificial intelligence and machine learning. Techniques such as ALD and MBE are at the forefront, enabling atomic-level control over material characteristics, with anticipated enhancements, in defect engineering and growth dynamics. The implementation of real-time monitoring methods, including spectroscopy and ellipsometry, will further improve the reproducibility and accuracy of nanomaterials. Additionally, the capability to create complex heterostructures, like Van Der Waals structures, will expand opportunities in areas such as electronics and catalysis.

Sustainability is becoming increasingly important, with future developments in vapor-phase synthesis likely to focus on green chemistry practices that aim to minimize energy use and harmful by-products. Techniques like low-temperature synthesis, the use of safer chemical precursors, and strategies aligned with the circular economy are vital for reducing environmental impacts. Furthermore, incorporating renewable energy sources, such as solar and wind, into vapor-phase reactors will contribute to a more sustainable process. While scalability presents challenges, advancements in reactor design and high-throughput systems are expected to address the rising demand for large-scale, high-quality nanomaterial production without sacrificing precision or material integrity.

Artificial intelligence (AI) and machine learning are set to revolutionize vapor-phase synthesis by optimizing process parameters in real-time and expediting the identification of new materials. AI can enhance the synthesis process by forecasting results, analyzing data, and automating intricate deposition techniques. As the field evolves, vapor-phase synthesis will be crucial in generating materials for next-generation applications, including quantum technologies, 2D nanomaterials, energy storage, and biomedicine. These advancements will not only improve established industries but also pave the way for new possibilities in nanotechnology, solidifying vapor-phase synthesis as an essential method for future scientific and technological innovations.

References

1. Abzieher T, Moore DT, Roß M, Albrecht S, Silvia J, Tan H, Jeangros Q et al (2024) Vapor phase deposition of perovskite photovoltaics: short track to commercialization? Energy Environ Sci 17(5):1645–1663
2. Akhtar K, Khan SA, Khan SB, Asiri AM (2018) Scanning electron microscopy: principle and applications in nanomaterials characterization. In: Handbook of materials characterization, pp 113–145
3. Alkallas FH, Salem A, Al-Ahmadi AN, Mwafy EA, Elsharkawy WB, Trabelsi ABG, Motawea MM et al (2024) Structural, optical, and dielectric characterization of NiO nanoparticles and their integration into PVA/SA blends for enhanced nanocomposite properties. J Electron Mater 53(10):6209–6220
4. Amin N, Rahman KS (2017) Close-spaced sublimation (CSS): a low-cost, high-yield deposition system for cadmium telluride (CdTe) thin film solar cells. Mod Technol Thin-Film Syst Coat
5. Baek SY, Park J, Koh T, Kim D, Woo J, Jung J, Park SJ et al (2024) Achievement of green and sustainable CVD through process, equipment, and systematic optimization in semiconductor fabrication. Int J Precis Eng Manuf-Green Technol 11(4):1295–1316
6. Baig N, Kammakakam I, Falath W (2021) Nanomaterials: a review of synthesis methods, properties, recent progress, and challenges. Mater Adv 2(6):1821–1871
7. Bannov AG, Popov MV, Kurmashov PB (2020) Thermal analysis of carbon nanomaterials: advantages and problems of interpretation. J Therm Anal Calorim 142(1):349–370
8. Bijelić L, Ruiz-Zepeda F, Hodnik N (2024) The role of high-resolution transmission electron microscopy and aberration-corrected scanning transmission electron microscopy in unraveling the structure-property relationships of Pt-based fuel cells electrocatalysts. Inorg Chem Front 11(2):323–341
9. Choi HJ (2012) Vapor-liquid-solid growth of semiconductor nanowires. NanoSci Technol 58:1–36
10. Choy K (2003) Chemical vapor deposition of coatings. Prog Mater Sci 48(2):57–170
11. Coenen T, Haegel NM (2017) Cathodoluminescence for the 21st century: learning more from light. Appl Phys Rev 4(3)
12. Coll M, Napari M (2019) Atomic layer deposition of functional multicomponent oxides. APL Materials 7(11)
13. Cui Y, Liao Y, Sun Y, Wang W, Wu J, Dai W, Huang T (2024) Advanced XPS-based techniques in the characterization of catalytic materials: a mini-review. Catalysts 14(9):595
14. Dai S, Gao W, Zhang S, Graham GW, Pan X (2017) Transmission electron microscopy with atomic resolution under atmospheric pressures. MRS Commun 7(4):798–812
15. Dhiman G, Kumar S, Kumar R, Brajpuriya R (2024) An improved CVD design for graphene growth and transfer improvements. J Electron Mater 53(10):5916–5925
16. Dimitriou C, Psathas P, Solakidou M, Deligiannakis Y (2023) Advanced flame spray pyrolysis (FSP) technologies for engineering multifunctional nanostructures and nanodevices. Nanomater 13(23):3006
17. Doll GL, Mensah BA, Mohseni H, Scharf TW (2010) Chemical vapor deposition and atomic layer deposition of coatings for mechanical applications. J Therm Spray Technol 19:510–516
18. Foroughi-Abari A, Cadien K (2012) Atomic layer deposition for nanotechnology. Nanofabrication 143–161

19. Galdames A, Ruiz-Rubio L, Orueta M, Sánchez-Arzalluz M, Vilas-Vilela JL (2020) Zero-valent iron nanoparticles for soil and groundwater remediation. Int J Environ Res Public Health 17(16):5817
20. Golovin YuI (2021) Nanoindentation and mechanical properties of materials at submicro- and nanoscale levels: recent results and achievements. Phys Solid State 63(1):1–41
21. Haider AJ, Alawsi T, Haider MJ, Taha BA, Marhoon HA (2022) A comprehensive review on pulsed laser deposition technique to effective nanostructure production: trends and challenges. Opt Quantum Electron 54(8):488
22. Herbert EG, Hackney SA, Thole V, Dudney NJ, Phani PS (2018) Nanoindentation of high-purity vapor deposited lithium films: a mechanistic rationalization of diffusion-mediated flow. J Mater Res 33(10):1347–1360
23. Huang MH, Wu Y, Feick H, Tran N, Weber E, Yang P (2001) Catalytic growth of zinc oxide nanowires by vapor transport. Adv Mater 13(2):113–116
24. Indiarto R, Indriana LPA, Andoyo R, Subroto E, Nurhadi B (2022) Bottom-up nanoparticle synthesis: a review of techniques, polyphenol-based core materials, and their properties. Eur Food Res Technol 248(1):1–24
25. Isaac NA, Pikaar I, Biskos G (2022) Metal oxide semiconducting nanomaterials for air quality gas sensors: operating principles, performance, and synthesis techniques. Microchim Acta 189(5):196
26. Isaacs MA, Davies-Jones J, Davies PR, Guan S, Lee R, Morgan DJ, Palgrave R (2021) Advanced XPS characterization: XPS-based multi-technique analyses for a comprehensive understanding of functional materials. Mater Chem Front 5(22):7931–7963
27. Jimenez J, Tomm JW (2016) Cathodoluminescence, spectroscopic analysis of optoelectronic semiconductors, vol 202. Springer Series in Optical Science, pp 213–263
28. Joyce BA (1988) Molecular beam epitaxy: recent trends and future developments. J Phys Chem Solids 49(3):237–242
29. Khan T, Singh B, Manikandan M (2023) Synthesis, characteristics, and applications of nanomaterials, smart nanomater. Technology 11–26
30. Kumari N, Kumar S, Chauhan P, Kaur GA, Kainthla I, Shandilya M (2024) Environmentally sustainable techniques for rGO synthesis: focus on spun calcination and clean technology advances. J Inorg Organomet Polym Mater
31. Łach Ł, Svyetlichnyy D (2024) Recent progress in heat and mass transfer modeling for chemical vapor deposition processes. Energies, Multidisciplinary Digital Publishing Institute (MDPI)
32. Lee H, Bin KY, Ryu JW, Kim S, Bae J, Koo Y, Jang D et al (2023) Recent progress of exciton transport in two-dimensional semiconductors. Nano Convergence 10(1):57
33. Leng J, Wang Z, Wang J, Wu HH, Yan G, Li X, Guo H et al (2019) Advances in nanostructures fabricated via spray pyrolysis and their applications in energy storage and conversion. Chem Soc Rev 48(11):3015–3072
34. Li J, Ren GK, Chen J, Chen X, Wu W, Liu Y, Chen X et al (2022) Facilitating complex thin film deposition by using magnetron sputtering: a review. JOM 74(8):3069–3081
35. Malik O, La Hidalga-Wade FJD, Amador RR (2017) Spray pyrolysis processing for optoelectronic applications. Pyrolysis
36. Mao T, Zha J, Hu Y, Chen Q, Zhang J, Luo X (2024) Research progress of TiO_2 modification and photodegradation of organic pollutants. Inorganics 12(7):178
37. Masood KB, Kumar P, Malik MA, Singh J (2021) A comprehensive tutorial on the pulsed laser deposition technique and developments in the fabrication of low-dimensional systems and nanostructures. Emerg Mater 4(3):737–754
38. Meng X, Wang X, Geng D, Ozgit-Akgun C, Schneider N, Elam JW (2017) Atomic layer deposition for nanomaterial synthesis and functionalization in energy technology. Mater Horiz 4(2):133–154
39. Mohammad SN (2020) The VLS mechanism, synthesis of nanomaterials, vol 307, pp 69–99
40. Mourdikoudis S, Pallares RM, Thanh NTK (2018) Characterization techniques for nanoparticles: comparison and complementarity upon studying nanoparticle properties. Nanoscale 10(27):12871–12934

41. Nguyen LTN, Van Vu H, Le TT (2024) Size-controlled synthesis of gold nanoparticles and related molecular imaging contrast for computed tomography. J Nanoparticle Res 26(6):113
42. Noman MT, Ashraf MA, Ali A (2019) Synthesis and applications of nano-TiO_2: a review. Environ Sci Pollut Res 26(4):3262–3291
43. Pathak R, Punetha VD, Bhatt S, Punetha M (2024) A review on copper-based nanoparticles as a catalyst: synthesis and applications in coupling reactions. J Mater Sci 59(15):6169–6205
44. Pierson HO (1999) Principles, technology, and applications. In: Handbook chemical vapor deposition. Noyes Publications
45. Plessl K, Russ A, Vollprecht D (2023) Application and development of zero-valent iron (ZVI) for groundwater and wastewater treatment. Int J Environ Sci Technol 20(6):6913–6928
46. Ptak AJ (2015) Principles of molecular beam epitaxy. In: Handbook of crystal growth, pp 161–192
47. Rades S, Hodoroaba VD, Salge T, Wirth T, Lobera MP, Labrador RH, Natte K et al (2014) High-resolution imaging with SEM/T-SEM, EDX, and SAM as a combined methodical approach for morphological and elemental analyses of single-engineered nanoparticles. RSC Adv 4(91):49577–49587
48. Ran F, Yang X, Shao L (2018) Recent progress in carbon-based nanoarchitectures for advanced supercapacitors. Adv Compos Hybrid Mater 1(1):32–55
49. Saeed M, Alshammari Y, Majeed SA, Al-Nasrallah E (2020) Chemical vapour deposition of graphene-synthesis, characterisation, and applications: a review. Molecules 25(17):3856
50. Sápi A, Rajkumar T, Kiss J, Kukovecz Á, Kónya Z, Somorjai GA (2021) Metallic nanoparticles in heterogeneous catalysis. Catal Lett 151(8):2153–2175
51. Sharafinia S, Rashidi A, Babaei B (2024). Vapor phase production of carbon nanostructures. In: Handbook of functionalized carbon nanostructures: from synthesis methods to applications, pp 651–677
52. Shen J, Roozeboom F, Mameli A (2023) Atmospheric-pressure plasma-enhanced spatial atomic layer deposition of silicon nitride at low temperature. At. Layer Depos
53. Singh A, Chaudhary S, Dehiya BS (2021) Fast removal of heavy metals from water and soil samples using magnetic Fe_3O_4 nanoparticles. Environ Sci Pollut Res 28(4):3942–3952
54. Su SS, Chang I (2018) Review of production routes of nanomaterials. In: Commercialization of nanotechnologies–a case study approach, pp 15–29
55. Subramanian A, Tiwale N, Nam CY (2019) Review of recent advances in applications of vapor-phase material infiltration based on atomic layer deposition. JOM 71(1):185–196
56. Wagare DS, Shirsath SE, Shaikh M, Netankar P (2021) Sustainable solvents in chemical synthesis: a review. Environ Chem Lett 19(4):3263–3282
57. Wan D, Li W, Wang G, Wei X (2016) Size-controllable synthesis of Fe_3O_4 nanoparticles through oxidation–precipitation method as heterogeneous Fenton catalyst. J Mater Res 31(17):2608–2616
58. Wang Y, Zhang L, Hou H, Xu W, Duan G, He S, Liu K et al (2021) Recent progress in carbon-based materials for supercapacitor electrodes: a review. J Mater Sci 56(1):173–200
59. Weimer AW (2019) Particle atomic layer deposition. J Nanopart Res 21(1):9
60. Wu S, Cheng L, Wang Q (2017) Excitonic effects and related properties in semiconductor nanostructures: roles of size and dimensionality. Mater Res Express 4(8):085017
61. Xi D, Dong S, Meng X, Lu Q, Meng L, Ye J (2012) Gold nanoparticles as computerized tomography (CT) contrast agents. RSC Adv 2(33):12515
62. Yang F, Han P, Yao N, Cheng G, Chen S, Luo W (2020) Inter-regulated d-band centers of the Ni_3B/Ni heterostructure for boosting hydrogen electrooxidation in alkaline media. Chem Sci 11(44):12118–12123
63. Yang Z, Hao J (2016) Progress in pulsed laser-deposited two-dimensional layered materials for device applications. J Mater Chem C 4(38):8859–8878
64. Yongdong X, Yan X-T (2010) Introduction to chemical vapour deposition. Integr Eng Des Adv Mater 1:1–2
65. Yoshikawa M (2023) Overview of Cathodoluminescence (CL) spectroscopy. Adv Opt Spectrosc Tech Semicond 3–45

66. Yusuf A, Almotairy ARZ, Henidi H, Alshehri OY, Aldughaim MS (2023) Nanoparticles as drug delivery systems: a review of the implication of nanoparticles' physicochemical properties on responses in biological systems. Polymers 15(7):1596
67. Zhang J, Li Y, Cao K, Chen R (2022) Advances in atomic layer deposition. Nanomanuf Metrol 5(3):191–208
68. Zhang X, Lai J, Gray T (2023) Recent progress in low-temperature CVD growth of 2D materials. Oxf Open Mater Sci 3(1)

Microreactors for Nanoparticles Production

Teshi Fernando and Hiran Chathuranga

Abstract Recent advances in microreactor technologies have enabled innovative approaches for synthesizing nanoparticles. These techniques provide precise control over reaction parameters, product quality, safety, and overall reaction efficiency. This chapter introduces the recent developments in nanoparticles synthesis using microreactors, including gas–liquid segmented, continuous-flow, and droplet-based systems. Microreactor-based production methods of various types of nanoparticles, including metallic, semiconductor, magnetic, organic, and hybrid, with varying sizes, shapes, crystalline structures, and surface chemistries are also discussed. Additionally, this chapter introduces the challenges and limitations of microreactor-based nanoparticles synthesis methods.

Keyword Microreactors · Nanoparticles · Nanoparticles production · Advanced manufacturing

1 Introduction

Miniaturized reaction systems or microreactors are small-scale reaction systems with characteristic dimensions typically ranging from micrometers to a few millimeters. In 1959, Richard Feynman introduced the concept of microreactors to the world [51]. However, the development of microreactors began during the oil crisis in the 1970s. This development was a part of process intensification efforts aimed at reducing energy consumption and improving process efficiency [35]. Spinning disk reactors were one of the key innovations in this period, pioneered by Colin Ramshaw. This technology became a foundational element in the design of micro-scale reaction systems. In the earlier stages, microreactor technology was mainly utilized in analytical applications, but in the present, it has expanded to areas like chemical synthesis, pharmaceuticals, and biotechnology [20, 37]. In these applications, microreactors

T. Fernando · H. Chathuranga (✉)
Department of Chemical and Process Engineering, Faculty of Engineering, University of Moratuwa, Moratuwa, Sri Lanka
e-mail: hiranc@uom.lk

N. M. Adassooriya et al. (eds.), *Industrial Scale Production of Nanoparticles*, Smart Nanomaterials Technology, https://doi.org/10.1007/978-981-96-9176-0_7

act as reaction condition controllers and help to increase the reaction efficiency and product quality. Also, microreactors use less hazardous materials, contributing to safer and more sustainable chemical processes [45].

Generally, microreactors operate on the principle of conducting chemical reactions within miniature channels, typically in the micrometer range. These microchannels enable different flow conditions within the reactor, allowing for precise control of reaction conditions like reactants mixing, heat and mass transfer, and residence time, leading to more uniform and reproducible chemical reactions. Furthermore, microreactors can be designed to support a variety of flow regimes like single-phase continuous or multi-phase segmented flow systems. Single-phase flow provides homogeneity and easier control, which makes it ideal in situations where stable and predictable reaction conditions are required [22]. On the other hand, multi-phase flow systems such as droplet-based or gas–liquid segmented reactors use passive or active mixing, which increases the mass transfer and reduces the variations in residence time [57]. These systems can successfully handle complex reactions that involve immiscible phases. This is achieved by optimizing mixing and reaction efficiency through unique flow patterns like slug or annular flows [15]. In addition, microreactors have a high surface-to-volume ratio, which increases heat and mass transfer, allowing quick temperature control and effective reaction management. This efficiency helps to manage exothermic reactions and avoid thermal runaway, which is a common issue in large-scale reactors. Many microreactors include integrated heat exchangers for precise temperature control. This design of microreactors improves the reaction performance and enhances safety. Additionally, effective mass transfer in microreactors ensures the even distribution of reactants and products. This uniform distribution is crucial for the production of high-quality nanoparticles [51].

Microreactors have been able to revolutionize several fields. In the pharmaceutical industry, microreactors support the synthesis of active pharmaceutical ingredients (APIs) with higher purity and yields. In biotechnology, it provides means for the development of new biochemical processes and the synthesis of bioactive molecules at a smaller scale, which is critical for rapid prototyping and screening [37]. Moreover, microreactors have demonstrated their value in environmental applications through the effective treatment of wastes, degradation of pollutants, and detection of environmental elements [10, 17]. Also, in tissue engineering, microreactors offer controlled microenvironments for cell cultures and tissue growth. This technology helps to develop in vitro tissues and organs for possible implantation [45]. Additionally, microreactors improve high-throughput screening (HTS) and drug testing by efficiently handling small volumes of samples and providing high sensitivity [14].

One of the most important uses of microreactors is their applications in the field of nanoparticle production. Nanoparticles are nano-scale (< 100 nm) particles with unique chemical, physical, and/or optical properties. Microreactors can be used to fabricate these nanoparticles with great precision. Also, numerous types of nanoparticles, including metallic, semiconductor, organic, hybrid, composite, and quantum dots can be easily synthesized using microreactors. The process of nanoparticles production using microreactors involves several steps, such as nucleation, crystal growth, and agglomeration. Moreover, to achieve the desired properties of nanoparticles, it is essential to carefully control factors like flow rates, residence time, reactant concentrations, and temperature [17, 51]. Therefore, the applications of microreactors in nanoparticle production are discussed in this chapter.

2 Design and Fabrication of Microreactors

Microreactors can be grouped into three categories; continuous-flow microreactors, gas–liquid segmented microreactors, and droplet-based microreactors. Continuous flow microreactors are more productive, and they increase the uniformity of reaction solutions, resulting in uniform products [12, 65]. Also, continuous flow in these reactors saves energy by avoiding the frequent heating and cooling cycles. Continuous-flow microreactors are further classified into three categories; capillary tubes microreactors, coaxial-flow microreactors, and 2-D micro-mixing microreactors. Capillary tube microreactors are a scaled-down version of regular tube reactors that contain micron-sized flow cross-sectional areas. As illustrated in Fig. 1a, Yang and colleagues developed a Silica (SiO_2) particles synthesising method using a PTFE helical capillary microreactor. SiO_2 particles are formed by hydrolyzing and condensing the precursor of silica, tetraethyl orthosilicate (TEOS) (Fig. 1b) [61]. Meanwhile, the coaxial flow microreactors solve problems such as high poly-dispersity and clogging in single-tube systems. For instance, Takagi et al. demonstrated that these reactors can be utilized to produce titania nanoparticles with a narrow size range by using tetraisopropoxide (TTIP) in the inner tube and isopropanol/water mixture in the outer tube [52]. 2-D micromixing-based reactors are unique due to their rapid and tunable mixing characteristics. Moreover, these reactors generally contain Y-shaped micro-channels, enabling precise control over nanoparticle formation [2].

Gas–liquid segmented microfluidic reactors facilitate the easy separation of the gas phase from the liquid phase. These segmented reactors use gas bubbles to improve mixing and produce segmented liquid sludge which can narrow the residence time distribution [16]. Moreover, gas bubbles help to separate small liquid reaction zones. This phenomenon reduces axial dispersion and improves the reaction consistency. For example, Nakamura and co-workers experimentally proved that the particle size distribution of CdSe can be narrowed down by using nitrogen bubbles to segment the reaction solution [36]. In another study, Khan et al. demonstrated that gas–liquid microfluidic reactors can produce a narrow size distribution of silica particles, which is challenging in traditional laminar flow reactors [26]. In addition, Jiang et al. developed a novel high throughput segmented flow microreactor for the continuous synthesis of Fe_3O_4 nanoparticles (Fig. 1c) [23]. Furthermore, they reported that this microfluidic device can be used to produce uniform 6 nm Fe_3O_4 nanoparticles with narrow size distribution.

Droplet-based microreactors use droplets in reactions to prevent clogging or contamination in microchannels [37]. This microfluidic droplet technique improves

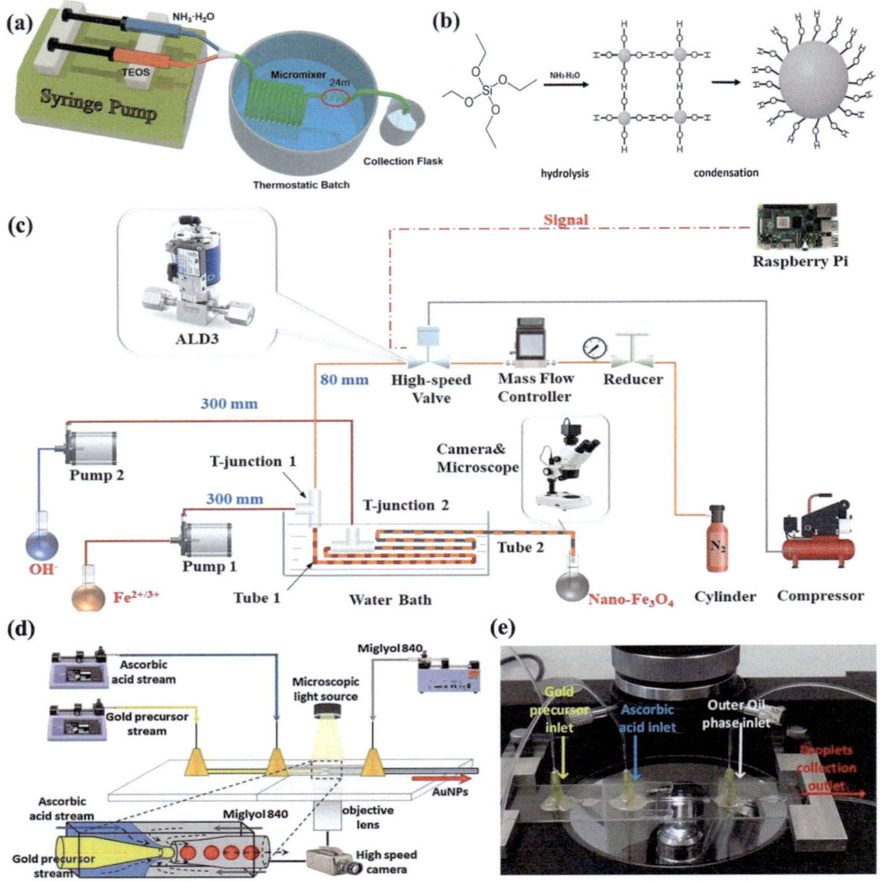

Fig. 1 **a** Schematic diagram of the capillary microreactor and **b** reaction mechanism of the silica particles production (Adapted with permission from [61]. Copyright (2020) Elsevier). **c** The experimental setup of the segmented flow microreactor used for the synthesis of nano-Fe_3O_4 (Adapted with permission from [23]. Copyright (2023) Elsevier). **d** Schematic diagram and **e** the photograph of the droplet-based microfluidic reactor utilized for gold nanoparticle production (Adapted with permission from [6]. Copyright (2019) Elsevier)

the mixing efficiency as well as provides control over reagent concentration. Droplet-based microreactors are further classified into three categories; single-emulsion reactors, double-emulsion reactors, and coalescence reactors. In the single-emulsion-based method, the reaction can occur in microfluidic channels and reactants are incorporated into droplets or plugs [48]. There are several microfluidic techniques to create double or multiple emulsions. To produce a double emulsion, some studies have used pneumatically controlled moving walls [64]. As shown in Fig. 1d, e, Bandulasena et al. designed a droplet-based microreactor to fabricate gold nanoparticles using tetrachloroauric acid trihydrate (gold precursor), polyvinylpyrrolidone

Table 1 Overview of microreactor types for nanoparticle synthesis

Microreactor type	Applications	Design dimensions	Throughput	Key references
Continuous-flow	Chemical synthesis, pharmaceuticals	Micron-sized cross-sections	High	[12, 65]
Gas–liquid segmented	Nanoparticle synthesis, polymerization	Micron-sized channels	Medium	[16, 26]
Droplet-based	Nanoparticle production, drug synthesis	Sub-micron droplets	Low to medium	[6, 64]
Coalescence-based	Nanoparticles synthesis	Micron-sized droplet fusion	Medium	[32, 53]

(coupling agent), L-ascorbic acid (reducing agent), and Miglyol 840 (carrier fluid). They found that the size of resultant gold nanoparticles reduces with decreasing droplet size and increases with increasing velocity difference between two reactant streams [6]. Meanwhile, coalescence-based reactors use two droplet reagents instead of one droplet to start a reaction [53]. These reactors precisely control droplet fusion even at high flow rates by using advanced microfluidics [32]. For example, circular expansion chambers can achieve high fusion efficiency, while double T-junction reactors improve the product quality. This approach is effective in producing nanoparticles such as $BaSO_4$ that are 15 nm in size [64]. In addition, Table 1 summarizes key microreactor types and their applications in nano particle production.

2.1 *Decision Framework for Selecting Microreactors in Nanoparticle Synthesis*

Selecting the appropriate type of microreactor for a specific application is crucial and depends on factors such as the reaction type, production scale, throughput, and control over particle size. Therefore, Table 2 outlines the key factors to consider when choosing a microreactor.

After selecting the most appropriate microreactor, scalability becomes an important consideration. There are two ways to increase the production capacity. (1) Scale-up approach: in this method, reactor size is increasing while maintaining operating conditions. For example, in continuous-flow reactors scaling up means increasing reactor dimensions or adding flow channels while preserving flow rates and mixing. (2) Numbering-up approach: this method uses multiple reactors in parallel or series to boost the throughput. Continuous-flow reactors and droplet-based reactors achieve this by integrating multiple units to handle higher production volumes efficiently.

Table 2 Microreactor selection guide

Criteria	Options	Recommended microreactor type
Reaction type	Single-phase	Continuous-flow microreactors
	Multi-phase	Gas–liquid segmented or droplet-based microreactors
Scale of production	Small-scale	Droplet-based microreactors
	Large-scale	Continuous-flow microreactors
Throughput	High throughput	Continuous-flow microreactors
	Low-to-moderate throughput	Droplet-based or segmented flow reactors
Nanoparticle size control	High control	Droplet-based microreactors
	Moderate control	Continuous-flow microreactors

2.2 Materials Used to Fabricate Microreactors

Different materials are used to fabricate microreactors, including silicon, polymers, glass, metals, and ceramics. Microreactors can be produced as micro-capillaries or chips [44]. Micro-capillary reactors are fabricated using the suitable tubing of the chosen material and the desired length. Chip-based reactors are fabricated using materials like silicon or plastic [40, 51]. Silicon is the most common material used to make microreactors owing to its affordability, availability, and processability. Polymers are also used due to their affordability, tunable properties, and the ease of using different fabrication techniques. Glass microreactors offer advantages like the ability to visualize reaction and flow, support for electroosmotic flow, and the ability to withstand high operating pressure. Using metal as the microreactor fabrication material also offers several advantages, such as durability and not requiring a clean room for fabrication [52].

2.3 Fabrication Techniques of Microreactors

Fabrication of microreactors includes numerous methods like micromachining, soft lithography, injection moulding, wet etching, photolithography, nanoimprinting, hot embossing, and electroforming [7, 44]. In micromachining, microreactors can be fabricated through decomposition, evaporation, melting, photo-ablation, or a combination of these methods [52]. In addition, micro electro-discharge machining (EDM) technology can also be used to machine micro-channels on metal surfaces. As illustrated in Fig. 2a, firstly, an electric field is created between the work-piece and the electrode by applying a gap voltage. Secondly, the gap between the work-piece and the electrode is adjusted to generate an electric spark. Thirdly, a plasma channel is formed, vaporizing and melting a small area of the work-piece material. Finally, a flushing system is used to remove debris from the machining area, avoiding short circuits or irregular sparks [21].

Meanwhile, the LIGA technique, a German acronym for Lithography, Galvanoformung, and Abformung (lithography, electrodeposition, and moulding), has also become popular for the fabrication of microreactors from materials like plastics, metals, and polymers [13, 52]. The LIGA technique consists of three main process

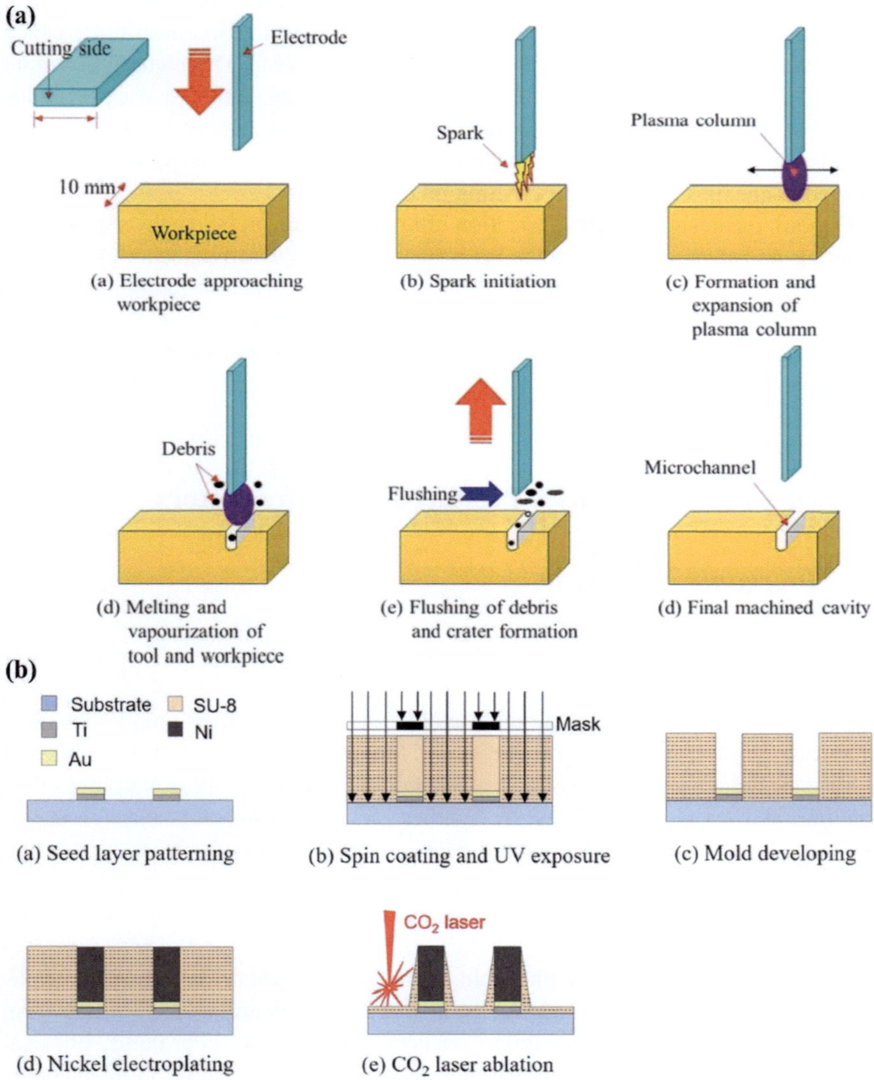

Fig. 2 Process steps of the **a** EDM (Adapted with permission from [21]. Copyright (2022) Springer Nature) and **b** LIGA techniques (Adapted with permission from [42]. Copyright (2023) Royal Society of Chemistry)

steps. Firstly, the lithography technique is used to transfer a pattern from a mask onto a photosensitive resin on a substrate. Secondly, using the electroplating method, metal is deposited into the patterned areas to form microstructures. Finally, the remaining resin is removed, and the metal structures are used as moulds to replicate micro-components (moulding) (Fig. 2b) [42]. Processes like casting, injection moulding, or embossing may additionally follow the above steps [13, 52]. Etching is also used to fabricate microreactors, which include removing material from a substrate through wet or dry methods. Wet etching is a simple method that uses basic lab equipment and chemicals. However, this method is limited by aspect ratio constraints, which often lead to semi-circular features due to isotropic removal. Dry etching is a more expensive process, but it allows more complex geometries and higher aspect ratios through anisotropic etching methods like plasma etching. Generally, this process includes steps like applying a photosensitive mask, exposing it to light, processing to remove unexposed areas, and creating precise patterns.

Overall, the designing and fabrication of microreactors include a diverse array of materials and advanced techniques. Each material and fabrication technique offers a wide range of advantages and disadvantages. Therefore, when designing and fabricating a microreactor, the materials and fabrication methods should be selected by considering factors like chemical compatibility, mechanical strength, fabrication precision, cost, scalability, etc.

3 Types of Nanoparticles Produced in Microreactors

3.1 Inorganic Nanoparticles

3.1.1 Metal/Metal Oxide Nanoparticles

Metal nanoparticles differ from bulk metal solids owing to their unique physical, chemical, thermal, and optical properties [8]. Metal nanoparticles have been utilized in various fields, including energy, electronics, bio-sensing, biotechnology, bio-imaging, and textiles [37]. They are synthesized in microreactors through stages like nucleation, formation of crystal nuclei, crystal growth through clustering, and particle agglomeration [17]. In these stages, factors like flow rates, residence time, reactant concentration, and reaction temperature are adjusted to control the size and shape of the nanoparticles [52].

Noble metal nanoparticles like gold, silver, and platinum have various applications across diverse fields, especially in medical applications due to their size, shape, and optical (surface plasmon resonance) properties [37]. For purely physical applications, the shape of the nanoparticles is essential. For example, Gold nanoparticles are especially recognized for their optical properties, with rod-shaped particles exhibiting better optical effects [64]. Silver nanoparticles have special optical, antimicrobial, and electrochemical properties. Gold and silver nanoparticles with sizes between 4 and 7 nm can be produced in microreactors by reducing metal salts with borohydride [56]. Meanwhile, Yao and colleagues developed a novel bamboo-based microreactor with polydopamine (PDA) modified bamboo micro-channels to Synthesize metallic nanoparticles like Ag, Pd, and Au (Fig. 3a). The resultant Pd-PDA nanoparticles exhibited excellent catalytic performance when used as a catalyst for the hydrogenation of nitro-aromatics [62].

Metal oxide nanoparticles are also a type of inorganic nanoparticle that is composed of metal atoms bonded to oxygen. Some of the commonly produced metal oxide nanoparticles are Titanium Dioxide (TiO_2), Zinc Oxide (ZnO), Cerium Oxide (CeO_2), Aluminum Oxide (Al_2O_3), etc. For instance, Cottam et al. reported that Y-shaped microfluidic chips can be used to produce TiO_2 nano-rods at the temperature of 90 °C [11]. Another method for the production of TiO_2 is using metal alkoxides and emulsion droplets in microreactors where water diffusion triggers hydrolysis [64]. Additionally, Yao et al. demonstrated that dual pipe microreactor design allows precise control over particle size by avoiding precipitation [63].

3.1.2 Semiconductor Nanoparticles

Semiconductor nanoparticles are also a special category of nanoparticles. These nanoparticles have various uses in areas like chemical catalysis, electronics, and solar energy conversion due to their size and structure-dependent properties [18, 34]. For the fabrication of these nanoparticles, continuous flow microreactors especially capillary or tubes are widely used owing to their precise temperature control capability [36, 64]. For instance, Nakamura et al. have synthesized high-quality CdSe nanoparticles using fused silica micro-capillaries [36], while Winterton et al. reported the production of high-quality semiconductor nanocrystals using a droplet-based microfluidic reactor with a two-phase flow, which effectively prevented the broadening of particle size distribution [59]. Similarly, Luan et al. produced highly luminescent CdSe/ZnS core–shell nanocrystals using microfluidic reaction technology [33].

3.1.3 Magnetic Nanoparticles

Magnetic nanoparticles are used in various applications like controlled drug delivery, bio-sensing, diagnosis, biomedical imaging, data storage devices and sample labelling [5, 9]. Magnetic nanoparticles have unique and important properties like magnetic response, super-paramagnetism, magnetic memory effect, chemical stability, and biocompatibility. Cobalt nanoparticles are an example of magnetic nanoparticles with numerous unique properties. For instance, by using a Y-type mixer and microchannel chips, Song et al. have successfully developed Cobalt nanoparticles with structures like face-centered cubic and hexagonal closed packed [49]. Other types of magnetic nanoparticles like iron oxide nanoparticles have also been produced using microfluidic techniques to increase the quality of nano-medical applications [37]. In another study, Saqib et al. managed to synthesize magnetic

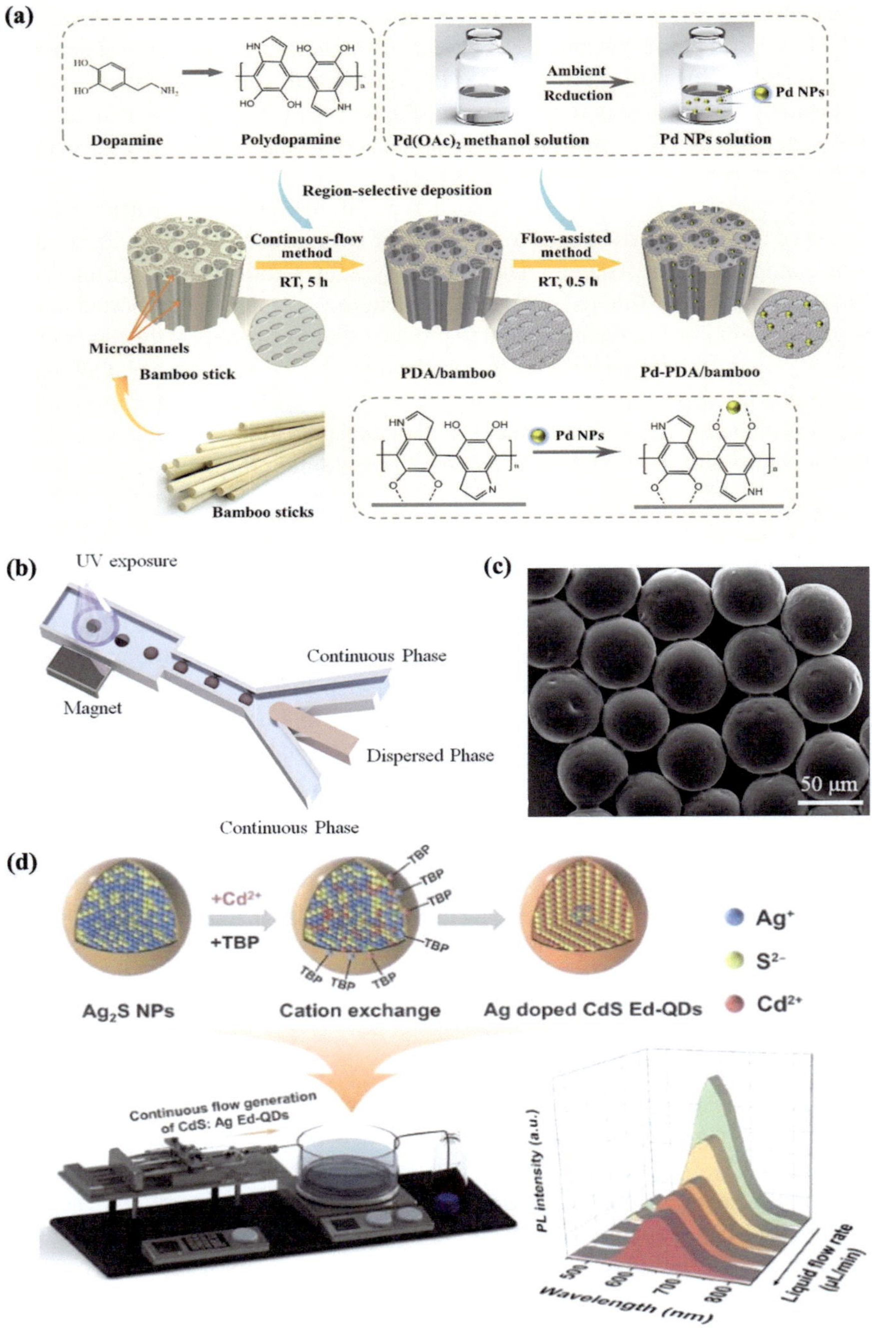
(a)
Dopamine
Polydopamine
Pd(OAc)₂ methanol solution
Ambient
Reduction
Pd NPs
Pd NPs solution
Region-selective deposition
Continuous-flow method
RT, 5 h
Flow-assisted method
RT, 0.5 h
Microchannels
Bamboo stick
PDA/bamboo
Pd-PDA/bamboo
Bamboo sticks
Pd NPs
(b)
UV exposure
Continuous Phase
Magnet
Dispersed Phase
Continuous Phase
(c)
50 μm
(d)
+Cd2+
+TBP
TBP
Ag2S NPs
Cation exchange
Ag doped CdS Ed-QDs
Ag+
S2-
Cd2+
Continuous flow generation of CdS: Ag Ed-QDs
PL intensity (a.u.)
Wavelength (nm)
Liquid flow rate (μL/min)

◂**Fig. 3** **a** Synthesis process of the Pd-PDA nanoparticles (Adapted with permission from [62]. Copyright (2024) Elsevier). **b** Schematic illustration of the microfluidic device used for the synthesis of magnetic iron oxide/poly(ethylene glycol) diacrylate particles production, and **c** SEM image of resultant magnetic particles (Adapted with permission from [46]. Copyright (2023) American Chemical Society). **d** Schematic demonstration of the fabrication process of the Ag-CdS quantum dots and PL intensity comparison of Ag-CdS quantum dots produced at different liquid flow rates (Adapted with permission from [31]). Copyright (2022) Springer Nature)

iron oxide nanoparticles embedded poly(ethylene glycol) diacrylate particles using a droplet-based microreactor, as shown in Fig. 3b. Importantly, the synthesized particles demonstrated a uniform particle size with a standard deviation of < 3.5% (Fig. 3c) [46].

3.2 *Organic Nanoparticles*

Organic nanoparticles can be synthesized in numerous forms, each with distinct structures, properties, and applications. The types of organic nanoparticles are generally categorized based on the materials used for their fabrication. Some examples of materials used for organic nanoparticles production include polymers, lipids, liposomes, proteins, etc. Organic nanoparticles can be synthesized using different microfluidic techniques [5]. For example, Karnik et al. have produced polymeric nanoparticles for drug delivery using a nanoprecipitation technique in hydrodynamic flow-focusing micro-channels [25]. This technique enables precise control over the drug content, shape, and size distribution of nanoparticles by adjusting the concentration, composition, and flow rates. Liposomes are also another category of organic nanoparticles that are widely used for drug delivery applications. Liposomes are composed of one or more concentric phospholipid bilayers that surround an aqueous core. This structure enables the encapsulation of hydrophilic drugs within the aqueous core and hydrophobic drugs within the lipid bilayer, enhancing the effectiveness of releasing drugs [51].

3.3 *Hybrid Nanoparticles*

Hybrid nanoparticles are composed of two or more materials that are chemically or physically combined. Typically, these nanomaterials include both organic and/or inorganic components, offering properties of both classes of materials. For instance, hybrid nanoparticles can be produced by trapping drug molecules inside the polymer core with a lipid shell. This method enhances the drug loading capacity and allows better control over the drug release [37, 43]. Furthermore, lipid nanoparticles loaded with nucleic acid are widely used in mRNA vaccines. These nanoparticles are developed in droplet-based microreactors by mixing hydrophobic lipids with a water-based solution containing mRNA [64]. In another study, Xu et al. created Cu-CuO nanocomposites by partially oxidizing copper nanoparticles in a controlled microchannel environment. This process enhanced the semiconductor properties of composite nanoparticles [17, 60]. In another study, Knauer et al. fabricated double-layered Au/Ag/Au nanoparticles with unique structural and electrical properties by concurrently reducing gold and silver precursors using the micro-segmented flow synthesis method [28].

3.4 *Other Types of Nanoparticles*

In addition to the nanoparticles mentioned above, there are other types of special nanoparticles like quantum dots, nano-clays, carbon-based nanoparticles, etc. For example, Kikker et al. showed that using continuous flow microreactors, CdSe and CdTe quantum dots can be produced safely and effectively. They successfully fabricated different sizes of quantum dots by varying the residence time [27]. ZnO quantum dots and carbon dots also can be fabricated using microfluidic devices [17]. Besides, Tang et al. demonstrated that nitrogen-doped carbon quantum dots with improved quantum yield can be fabricated in microreactors [54].

Meanwhile, special materials like fluorescent silica-coated g-Fe_2O_3 magnetic nanoparticles can be produced using microfluidic devices. The silica shell of these nanoparticles protects the inner particles and offers a biocompatible surface. The magnetic properties of the g-Fe_2O_3 nanoparticles help them to be used in target applications [3]. Köhler et al. reported that by using a two-step micro-mixer, core–shell particles like gold-core-silver-shell nanoparticles can be synthesized [29]. In addition, the production of Janus particles using microreactors has improved greatly. These Janus particles have dual functional surfaces. High monodispersity in Janus microspheres with both organic and inorganic components has been achieved by using flow-focus microfluidic devices and UV curing [64]. Janus nanoparticles that exhibit dual functionalities are important as raw materials for optical, electronic, and sensor devices fabrication.

4 Strategies for Controlling Nanoparticle Characteristics

4.1 *Nanoparticle Size, Size Distribution, and Shape*

When compared with traditional reactors, microreactors provide better control over nanoparticle synthesis owing to the precise control over reaction conditions (temperature, pressure, flow rates, heat and mass transfer etc.). These factors enable the production of nanoparticles with unique shapes, different sizes, and uniform size distribution [24]. Pan et al. reported that ultrafine Zeolite A crystals with a narrow size distribution can be produced using a water-liquid paraffin two-phase segmented microfluidic device [38]. In another study, Pillai et al. analysed the particle size distribution of Pt–Mo/C nanoparticles by utilizing a continuous flow milli-channel reactor, and reported that the increasing metal load causes a broader size distribution due to the flocculation and coagulation during the nanoparticle growth [41].

Meanwhile, in drug nanoprecipitation, artificial neural networks (ANNs) have been used to study the relationship between the prednisolone saturation levels and solvent flow rates with particle size. The anti-solvent flow rate has a strong impact on particle sizes, but prednisolone flow rates have the opposite effect [4]. In addition, the geometry of the microreactor, such as the sharper inlet angles, has a slight effect on particle sizes. Furthermore, gas bubbles can limit the distribution of particle sizes in microreactors by preventing segment mixing and reducing the distribution of residence time [36, 65]. Wang and co-workers designed a micro-porous tube-in-tube microchannel reactor with a high throughput rate (9 L/min), and they showed that the average particle size can be reduced by increasing the total flow rate and phase concentration. However, factors like the flow rate of the dispersed phase and the mixing distance have minimal impact on nanoparticle size [58]. In addition, Zhao et al. reported that segmented droplets move in a circular pattern due to the shear forces. These movements improve mixing and decrease the particles' spread along the flow. Also, these droplets allow longer nucleation stages and provide less material for later growth, which results in a smaller final particle size [64]. Adding surfactants during nanoparticle production significantly impacts the size and size distribution. For example, the size distribution of silver nanoparticles can be changed by changing the surfactant concentration, indicating the importance of surfactants in stabilizing particles and particles' growth [39].

The shape of the nanoparticles is important for various applications. Common shapes of nanoparticles include spherical, triangular, cubic, helical, cylindrical etc. For instance, Köhler et al. found that when producing hybrid Au/Ag nanoparticles, the Au/Ag ratio affects the nanoparticle shape. Lower or higher ratios tend to form spherical particles or larger aggregates, and star-like particles can be formed when the ratio is between 0.6 and 0.8 [29]. Li et al. showed that the flow patterns and droplet sizes in a two-phase flow significantly influence nanoparticle shape, where larger droplet sizes promote nucleation and lead to more uniform particle formation [30].

4.2 Crystalline Structure

The physical and chemical properties of nanoparticles are mainly determined by the crystalline structure of the nanoparticles. For the control of the nanoparticle crystal structure, traditional methods like surfactant-induced phase transitions, pressure-induced phase transitions, and thermal annealing can be used [5, 64]. However, the reaction conditions can be much more precisely controlled in microreactors, providing far better control over the crystalline structure of the nanoparticles. Temperature plays a significant role in determining the crystalline structure of nanoparticles. For example, CdS nanoparticles synthesized in aqueous solutions at room temperature often have poor crystallinity [47]. On the other hand, at room temperature, cobalt nanoparticles with fine crystal structures can be synthesized by precisely adjusting microfluidic kinetics. The distance between atoms gets shorter when high energy inputs like sonication or elevated flow rates are used. This scenario leads to the formation of face-centered cubic structures. Moreover, low kinetic energies lead to the formation of hexagonal closed-packed structures, and extremely low energy levels lead to the formation of metastable h-phase [49, 50]. Zhao et al. demonstrated that stable amorphous and cobalt-samarium nanoparticles can be produced using microreactors with rapid cooling rates [64]. Also, they showed that by using techniques like time-resolved small-angle X-ray scattering (SAXS), the crystal growth process can be monitored in real-time. In addition, the crystalline structure of the nanoparticles can be fine-tuned by altering parameters like reactant flow rates, temperatures, and mixing conditions [1, 64].

4.3 Surface Chemistry

Generally, the surface chemistry of nanoparticles refers to the chemical characteristics and interactions of the atoms, molecules, and functional groups present on the surface of the nanoparticles. Microreactors can be utilized to synthesize nanoparticles with different surface properties. The synthesis and surface functionalization of nanoparticles can be done by using the same microreactor, improving the process efficiency and reducing the risk of contamination [28]. For instance, Quantum dots with functionalized surfaces can be successfully synthesized using microreactors. Kikkeri et al. showed that quantum dots with carbohydrate-functionalized surfaces can especially interact with lectins through carbohydrate-lectin recognition [28]. Also, Valencia et al. used lipid molecules to coat nanoparticles, improving stability and preventing aggregation even at high salt concentrations [55]. Adding different additives during nanoparticle production also significantly impacts their surface properties. For instance, Huang et al. reported that adding different ligands during the production of nanoparticles allowed for the customization of nanoparticles for special applications like drug delivery and sensors [19]. In another study, Li et al. used a continuous flow microreactor to produce electronic doped quantum dots (Ed-QDs) by doping Ag to CdS quantum dots, as illustrated in Fig. 3d [31]. Furthermore, these Ed-QDs exhibited good homogeneity and stability, and can be used as a raw material for optical devices and bio-sensors.

5 Conclusions and Future Outlook

Microreactors exhibit many advantages in the field of nanoparticle synthesis, including enhanced control over the reaction conditions, the ability to produce uniform nanoparticles, reduced waste, and efficient heat and mass transfer. Although microreactors have many advantages, several challenges and limitations make them challenging to use widely in large-scale industrial production.

One of the major limitations of microreactors is their limited production rate. Microreactors are small-scale reactors, hence the flow rates of microreactors are comparatively lower than the traditional reactors. Increasing the solute concentration can mitigate this issue and improve the production rate, but this often leads to precipitation on the channel walls, resulting in particle growth inside the chip. This issue is more severe when polymerization reactions happen, especially with high molecular weights. Moreover, the narrow channel diameters in microreactors make them highly susceptible to clogging. Clogging in microreactors disrupts mixing efficiency and leads to experimental failures. Another major issue in microreactors is the formation of unwanted by-products. Even though microreactors have precise control over reaction conditions, secondary chemical reactions still can happen. These by-products may degrade the final product quality and reduce the efficiency of the microreactor. To eliminate this challenge, improved purification and extraction processes are needed.

Material compatibility is also another major issue in microreactor fabrication. Most of the materials used to fabricate microreactors have poor solvent compatibility and limited resistance to high temperatures. Therefore, it is necessary to develop new materials that exhibit high chemical resistance and thermal stability. Inconsistent throughput is another challenge that microreactors are facing. Variations in factors like operating pressure, reagent flow rates, channel geometry, and initial concentrations can significantly impact the characteristics of the produced nanoparticles. Therefore, to ensure consistent and high-quality outputs, it is critical to establish optimal operating conditions. Microreactors work well for small-scale synthesis, but scaling up microreactors for industrial applications is a difficult process. In addition, series or parallel integration of microreactors is more complicated due to the geometric variations in microreactor designs. Hence, overcoming these challenges and limitations is a must for advancing nanoparticle synthesis and fully realizing the potential of microreactors in industry-scale applications.

References

1. Abécassis B, Testard F, Spalla O, Barboux P (2007) Probing in situ the nucleation and growth of gold nanoparticles by small-angle X-ray scattering. Nano Lett 7(6):1723–1727. https://doi.org/10.1021/nl0707149
2. Abou Hassan A, Sandre O, Cabuil V, Tabeling P (2008) Synthesis of iron oxide nanoparticles in a microfluidic device: preliminary results in a coaxial flow millichannel. Chem Commun 15:1783–1785. https://doi.org/10.1039/b719550h

3. Abou-Hassan A, Bazzi R, Cabuil V (2009) Multistep continuous-flow microsynthesis of magnetic and fluorescent γ-$Fe_2O_3OSiO_2$ core/shell nanoparticles. Angew Chem Int Ed 48(39):7180–7183. https://doi.org/10.1002/anie.200902181
4. Ali HSM, Blagden N, York P, Amani A, Brook T (2009) Artificial neural networks modelling the prednisolone nanoprecipitation in microfluidic reactors. Eur J Pharm Sci 37(3–4):514–522. https://doi.org/10.1016/j.ejps.2009.04.007
5. Altammar KA (2023) A review on nanoparticles: characteristics, synthesis, applications, and challenges. Front Microbiol 14. https://doi.org/10.3389/fmicb.2023.1155622
6. Bandulasena MV, Vladisavljević GT, Benyahia B (2019) Droplet-based microfluidic method for robust preparation of gold nanoparticles in axisymmetric flow focusing device. Chem Eng Sci 195:657–664. https://doi.org/10.1016/j.ces.2018.10.010
7. Bugay CA, Caballas MC, Mercado SB, Rubio JF, Serote PK, Villarte PN, Rubi RVC (2024) A review of microreactors for process intensification. 21. https://doi.org/10.3390/engproc2024067021
8. Burlec AF, Corciova A, Boev M, Batir-Marin D, Mircea C, Cioanca O, Danila G, Danila M, Bucur AF, Hancianu M (2023) Current overview of metal nanoparticles' synthesis, characterization, and biomedical applications, with a focus on silver and gold nanoparticles. Pharmaceuticals 16(10). https://doi.org/10.3390/ph16101410
9. Bushra R, Ahmad M, Alam K, Seidi F, Qurtulen, Shakeel S, Song J, Jin Y, Xiao H (2024) Recent advances in magnetic nanoparticles: key applications, environmental insights, and future strategies. Sustain Mater Technol 40. https://doi.org/10.1016/j.susmat.2024.e00985
10. Chathuranga H, Wijesinghe I, Marriam I, Yan C (2024) Recent advances in graphene allotropes-based fire detection sensors. Critical Rev Solid State Mater Sci. https://doi.org/10.1080/10408436.2024.2385828
11. Cottam BF, Krishnadasan S, DeMello AJ, DeMello JC, Shaffer MSP (2007) Accelerated synthesis of titanium oxide nanostructures using microfluidic chips. Lab Chip 7(2):167–169. https://doi.org/10.1039/b616068a
12. Du R, Zhang K, Jia R, Zhao S, Cheng Y (2024) Continuous-flow electrochemical microreactor with three-dimensional micro-structured electrodes promoting bioalcohol upgrading coupled with hydrogen production. ACS Sustain Chem Eng 12(32):12171–12177. https://doi.org/10.1021/acssuschemeng.4c04114
13. Effenhauser CS, Bruin GJM, Paulus A, Ehrat M (1997) Integrated capillary electrophoresis on flexible silicone microdevices: analysis of DNA restriction fragments and detection of single DNA molecules on microchips
14. Feng J, Neuzil J, Manz A, Iliescu C, Neuzil P (2023) Microfluidic trends in drug screening and drug delivery. TrAC Trends Anal Chem 158. https://doi.org/10.1016/j.trac.2022.116821
15. Gañán-Calvo AM, Gordillo JM (2001) Perfectly monodisperse microbubbling by capillary flow focusing. Phys Rev Lett 87(27 I):2745011–2745014. https://doi.org/10.1103/physrevlett.87.274501
16. Günther A, Khan SA, Thalmann M, Trachsel F, Jensen KF (2004) Transport and reaction in microscale segmented gas-liquid flow. Lab Chip 4(4):278–286. https://doi.org/10.1039/b403982c
17. Hakke V, Sonawane S, Anandan S, Sonawane S, Ashokkumar M (2021) Process intensification approach using microreactors for synthesizing nanomaterials—a critical review. Nanomaterials 11(1):1–21). https://doi.org/10.3390/nano11010098
18. Hossain N, Mobarak MH, Mimona MA, Islam MA, Hossain A, Zohura FT, Chowdhury MA (2023) Advances and significances of nanoparticles in semiconductor applications—a review. Results Eng 19. https://doi.org/10.1016/j.rineng.2023.101347
19. Huang H, du Toit H, Besenhard MO, Ben-Jaber S, Dobson P, Parkin I, Gavriilidis A (2018) Continuous flow synthesis of ultrasmall gold nanoparticles in a microreactor using trisodium citrate and their SERS performance. Chem Eng Sci 189:422–430. https://doi.org/10.1016/j.ces.2018.06.050
20. Huang SY, Zhou Y, Huang YX, Wang N, Gao ZT, Sun MS, Hui W, Tao DJ (2024) Improved efficiency of continuous conversion of alcohols to chlorides in flow microreactors with DESs/HCl via predictive fluid model. J Mol Liq 413.https://doi.org/10.1016/j.molliq.2024.126041

21. Ishfaq K, Naveed R, Maqsood MA, Rehman M (2022) Analyzing laminated electrode(s) performance for the EDM of microchannel(s) in Al(6061). Int J Adv Manuf Technol 123(7–8):2941–2958. https://doi.org/10.1007/s00170-022-10360-7
22. Jahn A, Reiner JE, Vreeland WN, DeVoe DL, Locascio LE, Gaitan M (2008) Preparation of nanoparticles by continuous-flow microfluidics. J Nanoparticle Res 10(6): 925–934. https://doi.org/10.1007/s11051-007-9340-5
23. Jiang X, Li S, Sotowa KI, Tonomura O, Hoon OT (2023) High throughput continuous synthesis of size-controlled nanoFe3O4 in segmented flow. Chem Eng J 471.https://doi.org/10.1016/j.cej.2023.144546
24. Ju J, Zeng C, Zhang L, Xu N (2006) Continuous synthesis of zeolite NaA in a microchannel reactor. Chem Eng J 116(2):115–121. https://doi.org/10.1016/j.cej.2005.11.006
25. Karnik R, Gu F, Basto P, Cannizzaro C, Dean L, Kyei-Manu W, Langer R, Farokhzad OC (2008) Microfluidic platform for controlled synthesis of polymeric nanoparticles. Nano Lett 8(9):2906–2912. https://doi.org/10.1021/nl801736q
26. Khan SA, Günther A, Schmidt MA, Jensen KF (2004) Microfluidic synthesis of colloidal silica. Langmuir 20(20):8604–8611. https://doi.org/10.1021/la0499012
27. Kikkeri R, Laurino P, Odedra A, Seeberger PH (2010) Synthesis of carbohydrate-functionalized quantum dots in microreactors. Angew Chem Int Ed 49(11):2054–2057. https://doi.org/10.1002/anie.200905053
28. Knauer A, Thete A, Li S, Romanus H, Csáki A, Fritzsche W, Köhler JM (2011) Au/Ag/Au double shell nanoparticles with narrow size distribution obtained by continuous micro segmented flow synthesis. Chem Eng J 166(3):1164–1169. https://doi.org/10.1016/j.cej.2010.12.028
29. Köhler JM, Romanus H, Hübner U, Wagner J (2007) Formation of star-like and core-shell AuAg nanoparticles during two- and three-step preparation in batch and in microfluidic systems. J Nanomater 2007.https://doi.org/10.1155/2007/98134
30. Li S, Xu J, Wang Y, Luo G (2008) Controllable preparation of nanoparticles by drops and plugs flow in a microchannel device. Langmuir 24(8):4194–4199. https://doi.org/10.1021/la800107d
31. Li Y, Li Y, Li X, Hou T, Qiao C, Tai Y, Gu X, Zhao D, Sang L, Zhang J (2022) Microreactor platform for continuous synthesis of electronic doped quantum dots. Nano Res 15(10):9647–9653. https://doi.org/10.1007/s12274-022-4571-4
32. Liu K, Ding H-J, Liu J, Chen Y, Zhao X-Z (2006) Shape-controlled production of biodegradable calcium alginate gel microparticles using a novel microfluidic devicehttps://doi.org/10.1021/la061729
33. Luan W, Yang H, Fan N, Tu ST (2008) Synthesis of efficiently green luminescent CdSe/ZnS nanocrystals via microfluidic reaction. Nanoscale Res Lett 3(4):134–139. https://doi.org/10.1007/s11671-008-9125-5
34. Mekuye B, Abera B (2023) Nanomaterials: an overview of synthesis, classification, characterization, and applications. Nano Select 4(8):486–501. https://doi.org/10.1002/nano.202300038
35. Moorthy RK, Baksi S, Biswas S (2020) Process intensification—an insight. https://doi.org/10.35543/osf.io/5nghs
36. Nakamura H, Yamaguchi Y, Miyazaki M, Maeda H, Uehara M, Mulvaney P (2002) Preparation of CdSe nanocrystals in a micro-flow-reactor. Chem Commun 23:2844–2845. https://doi.org/10.1039/b208992k
37. Niculescu AG, Chircov C, Bîrcă AC, Grumezescu AM (2021) Nanomaterials synthesis through microfluidic methods: an updated overview. Nanomaterials 11(4). https://doi.org/10.3390/nano11040864
38. Pan Y, Yao J, Zhang L, Xu N (2009) Preparation of ultrafine zeolite a crystals with narrow particle size distribution using a two-phase liquid segmented microfluidic reactor. Ind Eng Chem Res 48(18):8471–8477. https://doi.org/10.1021/ie900621y
39. Patil GA, Bari ML, Bhanvase BA, Ganvir V, Mishra S, Sonawane SH (2012) Continuous synthesis of functional silver nanoparticles using microreactor: effect of surfactant and process parameters. Chem Eng Process 62:69–77. https://doi.org/10.1016/j.cep.2012.09.007

40. Phillips TW, Lignos IG, Maceiczyk RM, Demello AJ, Demello JC (2014) Nanocrystal synthesis in microfluidic reactors: where next? Lab Chip 14(17):3172–3180. https://doi.org/10.1039/c4lc00429a
41. Pillai SR, Sonawane SH, Gumfekar SP, Suryawanshi PL, Ashokkumar M, Potoroko I (2019) Continuous flow synthesis of nanostructured bimetallic Pt-Mo/C catalysts in milli-channel reactor for PEM fuel cell application. Mater Chem Phys 237.https://doi.org/10.1016/j.matchemphys.2019.121854
42. Ran J, Wang X, Liu Y, Yin S, Li S, Zhang L (2023) Microreactor-based micro/nanomaterials: fabrication, advances, and outlook. Mater Horiz 10(7):2343–2372. https://doi.org/10.1039/d3mh00329a
43. Rogach AL, Nagesha D, Ostrander JW, Giersig M, Kotov NA (2000) 'Raisin bun'-type composite spheres of silica and semiconductor nanocrystals. Chem Mater 12(9):2676–2685. https://doi.org/10.1021/cm000244i
44. Saikia A, Newar R, Das S, Singh A, Deuri DJ, Baruah A (2023) Scopes and challenges of microfluidic technology for nanoparticle synthesis, photocatalysis and sensor applications: a comprehensive review. Chem Eng Res Des 193:516–539). https://doi.org/10.1016/j.cherd.2023.03.049
45. Šalić A, Tušek A, Zelić B (2012) Application of microreactors in medicine and biomedicine. J Appl Biomed 10(3):137–153). https://doi.org/10.2478/v10136-012-0011-1
46. Saqib M, Ercan B, Erdem EY (2023) Synthesis of anisotropic magnetic polymeric Janus particles by In Situ separation of magnetic nanoparticles in a microfluidic device. Langmuir 39(48):17080–17087. https://doi.org/10.1021/acs.langmuir.3c01862
47. Shestopalov I, Tice JD, Ismagilov RF (2004) Multi-step synthesis of nanoparticles performed on millisecond time scale in a microfluidic droplet-based system. Lab Chip 4(4):316–321. https://doi.org/10.1039/b403378g
48. Song H, Tice JD, Ismagilov RF (2003) High-throughput measurements a microfluidic system for controlling reaction networks in time**. Angew Chem Int Ed 42(7)
49. Song Y, Modrow H, Henry LL, Saw CK, Doomes EE, Palshin V, Hormes J, Kumar CSSR (2006) Microfluidic synthesis of cobalt nanoparticles. Chem Mater 18(12):2817–2827. https://doi.org/10.1021/cm052811d
50. Song Y, Zhang T, Yang W, Albin S, Henry LL (2008) Fine crystal structure transition of cobalt nanoparticles formed in a microfluidic reactor. Cryst Growth Des 8(10):3766–3772. https://doi.org/10.1021/cg8003992
51. Suryawanshi PL, Gumfekar SP, Bhanvase BA, Sonawane SH, Pimplapure MS (2018) A review on microreactors: reactor fabrication, design, and cutting-edge applications. Chem Eng Sci 189:431–448. https://doi.org/10.1016/j.ces.2018.03.026
52. Takagi M, Maki T, Miyahara M, Mae K (2004) Production of titania nanoparticles by using a new microreactor assembled with same axle dual pipe. Chem Eng J 101(1–3):269–276. https://doi.org/10.1016/j.cej.2003.11.011
53. Tan YC, Fisher JS, Lee AI, Cristini V, Lee AP (2004) Design of microfluidic channel geometries for the control of droplet volume, chemical concentration, and sorting. Lab Chip 4(4):292–298. https://doi.org/10.1039/b403280m
54. Tang Y, Rao L, Li Z, Lu H, Yan C, Yu S, Ding X, Yu B (2018) Rapid synthesis of highly photoluminescent nitrogen-doped carbon quantum dots via a microreactor with foamy copper for the detection of Hg2+ ions. Sens Actuators, B Chem 258:637–647. https://doi.org/10.1016/j.snb.2017.11.140
55. Valencia PM, Basto PA, Zhang L, Rhee M, Langer R, Farokhzad OC, Karnik R (2010) Single-step assembly of homogenous lipid-polymeric and lipid-quantum dot nanoparticles enabled by microfluidic rapid mixing. ACS Nano 4(3):1671–1679. https://doi.org/10.1021/nn901433u
56. Wagner J, Tshikhudo TR, Köhler JM (2008) Microfluidic generation of metal nanoparticles by borohydride reduction. Chem Eng J 135(SUPPL. 1). https://doi.org/10.1016/j.cej.2007.07.046
57. Wang CW, Bains A, Sinton D, Moffitt MG (2013) Flow-directed loading of block copolymer micelles with hydrophobic probes in a gas-liquid microreactor. Langmuir 29(26):8385–8394. https://doi.org/10.1021/la400011n

58. Wang QA, Wang JX, Li M, Shao L, Chen JF, Gu L, An YT (2009) Large-scale preparation of barium sulphate nanoparticles in a high-throughput tube-in-tube microchannel reactor. Chem Eng J 149(1–3):473–478. https://doi.org/10.1016/j.cej.2009.02.018
59. Winterton JD, Myers DR, Lippmann JM, Pisano AP, Doyle FM (2008) A novel continuous microfluidic reactor design for the controlled production of high-quality semiconductor nanocrystals. J Nanopart Res 10(6):893–905. https://doi.org/10.1007/s11051-007-9345-0
60. Xu JH, Tan J, Li SW, Luo GS (2008) Enhancement of mass transfer performance of liquid-liquid system by droplet flow in microchannels. Chem Eng J 141(1–3):242–249. https://doi.org/10.1016/j.cej.2007.12.030
61. Yang H, Akinoglu EM, Guo L, Jin M, Zhou G, Giersig M, Shui L, Mulvaney P (2020) A PTFE helical capillary microreactor for the high throughput synthesis of monodisperse silica particles. Chem Eng J 401. https://doi.org/10.1016/j.cej.2020.126063
62. Yao S, Guo D, Han S, Fu Z, Lyu S, Li J, Lu Y (2024) Polydopamine-assisted immobilization of metallic nanoparticles confined regionally in bamboo microchannels as continuous-flow microreactors for enhanced catalysis. Chem Eng J 492. https://doi.org/10.1016/j.cej.2024.152327
63. Yao X, Zhang Y, Du L, Liu J, Yao J (2015) Review of the applications of microreactors. In: Renewable and sustainable energy reviews, vol 47. Elsevier Ltd., pp 519–539. https://doi.org/10.1016/j.rser.2015.03.078
64. Zhao CX, He L, Qiao SZ, Middelberg APJ (2011) Nanoparticle synthesis in microreactors. Chem Eng Sci 66(7):1463–1479. https://doi.org/10.1016/j.ces.2010.08.039
65. Zhao Y-F, Xie W-L, He K, Li H-P, Pan J, Xu J-H (2024) Continuous flow microreactors for the high-efficiency enzymatic synthesis of 10-hydroxystearic acid from oleic acid. ChemBioChem. https://doi.org/10.1002/cbic.202400345

High Pressure Homogenization Techniques for the Production of Nanomaterials

Gabriela Schneider Rauber, Debora Fretes Argenta, Thiago Caon, and Kateřina Poláková

Abstract High pressure homogenization is a valuable top-down technique for both the small scale and the industrial production of nanoparticles and nanovesicles. The reason for that is its ease of operation, scalability, reproducibility and high throughput. Interestingly, the technique shows a uniquely broad nanomaterial-forming capacity that can process from soft emulsions to brittle crystals. In this chapter, the use of HPH to prepare nanoemulsions was put in parallel to its application in making nanocrystals of different types. The technique appears to be the gold standard for nanoemulsion preparation, but challenges exist in the case of nanocrystals. For example, although HPH fundamentally works by breaking the vesicles/particles, the mechanical properties of the starting crystals are rarely considered during the design of the homogenizing protocols. In the case of emulsions, however, the breakage resistance originated from the system's viscosity is already well-known for affecting the outcome of the process. A series of examples and discussions are brought together to suggested that structure- and surface-specific characteristics (e.g. molecular arrangement on the crystal facets, surface adhesion, existence of cleavage planes, morphology/shape, particle aspect ratio) are to be incorporated in future studies related to HPH and nanocrystals. This may help in the rational selection of the instrument parameters, but also in understanding the use of HPH in novel technological applications, like self-stabilizing Pickering emulsions and the scalability of complex multicomponent crystal forms like cocrystals.

Keywords High pressure homogenization · Nanoemulsions · Nanocrystals · Mechanical properties · Crystal structure analysis

G. Schneider Rauber (✉) · K. Poláková
Czech Advanced Technology and Research Institute (CATRIN), Palacký University Olomouc, Olomouc, Czech Republic
e-mail: gabriela.schneider@upol.cz

D. Fretes Argenta · T. Caon
Department of Pharmaceutical Sciences, Federal University of Santa Catarina, Florianópolis, Brazil

N. M. Adassooriya et al. (eds.), *Industrial Scale Production of Nanoparticles*, Smart Nanomaterials Technology, https://doi.org/10.1007/978-981-96-9176-0_8

1 Introduction

Nanostructures can be produced by top-down (high-energy) and bottom-up (low-energy) methods. Low-energy methods use the internal chemical energy of the system/substances to prepare the particles whereas high-energy methods require mechanical devices to break them. In industry, especially in the pharmaceutical and food areas, the application of highly disruptive forces with mechanical devices has been the most used approach to break the dispersed phase into small parts within the continuous phase. High-pressure homogenization (HPH), ultrasonication and microfluidization are some examples of high-energy methods [11, 41]. In addition to the low particle sizes achieved, these preparation methods have as advantages high throughput, high reproducibility, decrease in organic solvents usage, less time consumption, and ease of industrial scalability [37].

In HPH, a high-pressure pump forces a fluid through a narrow gap valve, which greatly increases its velocity, leading to high shear stress and depressurization (Fig. 1). After the gap outlet, phenomena as turbulence and cavitation are found because of particle collision with internal surfaces of the equipment [73]. The maximum pressure achieved will depend on the equipment configuration and the valve design, which may be supplier specific. HPH allows to obtain pressures 10–15 times higher than traditional homogenizers and they usually operate between 1000 and 4000 bar [16]. When higher pressure levels are considered, they are termed as ultrahigh-pressure homogenizers [37].

Parameters such as initial fluid temperature, pressure, the number of cycles and valve design can be adjusted to control particle size and polydispersity [38]. The mechanism by which the equipment valves affect the reduction of particle size relies on the way they create obstacles to the passage of liquids [82]. Although the equipment configuration is critical, most of the applied studies in the literature have focused in the contribution of an increase in pressure as well as in the number of cycles to reduce particle size with HPH.

High levels of pressure increase shear intensity and induce the turbulent flow conditions that are needed to break the particles [17]. A concomitant increase in temperature is also observed. The consequent reduction in medium viscosity may result in a decrease of particle size and, in lipid-based systems, more uniform particles [12, 17]. There is, however, an optimal range of pressure that can be used to avoid stability problems with heat-sensitive compounds. Enzymes, for example, can be denatured when drastic operating conditions are selected [8]. This may require the operations to be carried out using heat exchangers, placed next to the homogenizing valves. Additionally, extreme conditions of pressure may be detrimental to size reduction. This can happen because particles may coalesce with the increased frequency of collision among them. Such phenomenon is especially critical in the preparation of nanocrystals and other solid nanoparticles as they can cause the blockage of the homogenizing gaps.

The use of cosolvents and surfactants can be considered when HPH equipment parameters and apparatus are adjusted but particle properties are still not suitable.

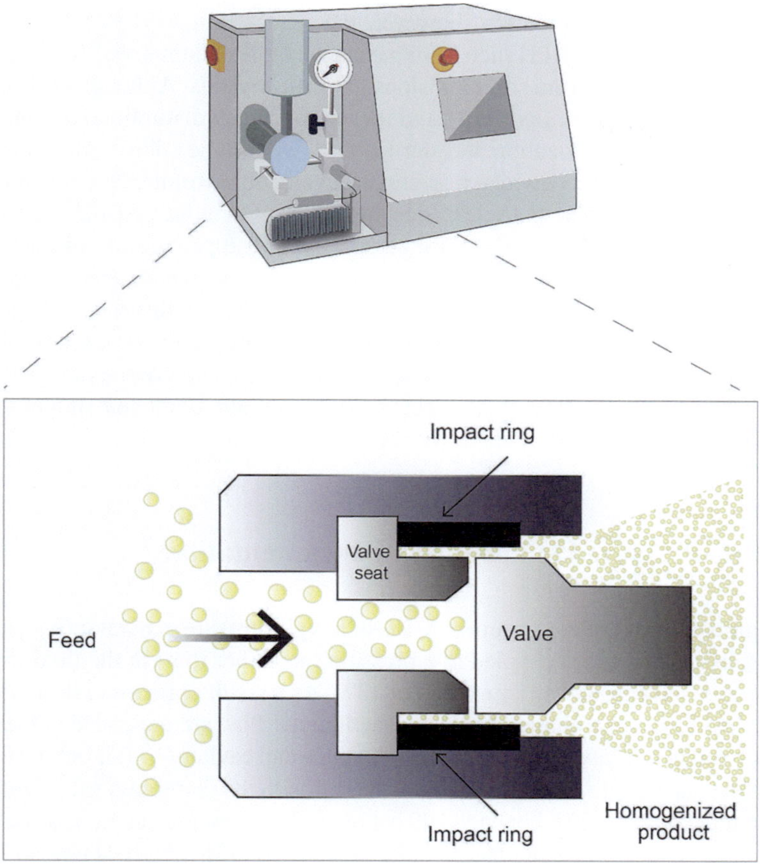

Fig. 1 Schematic representation of HPH technique (made by the authors)

For example, surfactants may reduce the interfacial tension between the oily and the aqueous phase (in the case of emulsions) and between the solid and the liquid medium (in the case of suspensions). The direct consequence is the decrease of the amount of energy needed to obtain small particles. Indirectly, surfactants may also improve particle properties, like physical stability and redispersion. In the case of cosolvents, the use of low-viscosity adjuvants in HPH can also produce smaller particles and lower the levels of pressure in the homogenizer.

Several review papers, chapters and books have already been dedicated to the HPH technique since the early 2000s. They highlight different facets of the pressure homogenization, including its instrumentation, widespread application and scalability. For the chronology of marketing of drug nanocrystals produced by HPH, as well as specific aspects of using HPH to prepare nanoemulsions and solid-lipid nanoparticles, the review written by Keck and Muller and the chapters written by Hakansson, and by Vinchhi et al., respectively, are highly recommended [18, 28,

73]. In the current contribution, however, the reader is invited to reflect on the fundamentals that make HPH such a unique but versatile nanomaterial-forming technique that can process from soft emulsions to brittle crystals. Although HPH can be used in the preparation of liposomes and for microbial cell disruption/lyse, inactivation of enzymes, pasteurization and sterilization, will not be addressed in this book chapter. The document is divided in sections devoted to instrumentation, nanoemulsions and nanocrystalline particles, focusing on providing an updated view of the current trends and closing with future perspectives. The preparation of nanoemulsions and nanocrystals were selected as models to facilitate the understanding of the technique and its gaps. While HPH is the gold standard for industrial-scale production of nanoemulsions, homogenization competes with other techniques when it comes to nanocrystallization. Both these nanomaterials show remarkably different intrinsic characteristics (Fig. 2) and, consequently, require homogenizing processes with different throughputs—which will be explored next.

2 High-Pressure Homogenizer Instrumentation

A traditional homogenizer system consists of a high-pressure generator (e.g., a positive displacement pump) coupled to a pressure intensifier to push the fluid through a disruption valve, followed by expansion [37]. As a result of pressure drop, intense mechanical forces and elongational stress in laminar flow are generated at the valve entrance and in the valve gap as well as turbulence and cavitation[1] (air bubble formation) at the gap outlet. The rapid pressure transition in the disruption valve leads to a temperature increase [72]. For each 100 bar of applied pressure, an increase between 2.0 and 3.2 °C in the product temperature is reported [79], which can be mitigated with cooling devices included in the terminal phase of the process. These devices are particularly useful during the processing of thermolabile materials such as formulations containing peptides, vitamins or essential oils. Valve gap opening adjustments ensure an efficient control of system pressure [72].

The diversity of designs/configurations (pressure intensifiers, valve types and homogenization chamber geometries) as well as high-pressure-resistant materials (e.g., diamond, ceramic, sapphire, seals) allow the equipment operation in different pressure levels and temperatures of up to 140–150 °C [9, 31]. Stainless steel of high-quality grade is often used in the production of homogenizers targeted for pharmaceutical and food as it is a corrosion resistant and wear-resistant material [73]. Commercially available units differ mainly in the high-pressure valve design, which affects the pressure range and the applicable flow rates at the laboratory, pilot, and industrial scale [9, 31].

Depending on the number of valves, they can be classified as one-stage HPH and two-stage HPH. In this last equipment configuration, the fluid passes through

[1] Formation of cavities in a fluid, their growth, and subsequent collapse at the moment in that the pressure of the liquid is suddenly reduced below its vapor pressure.

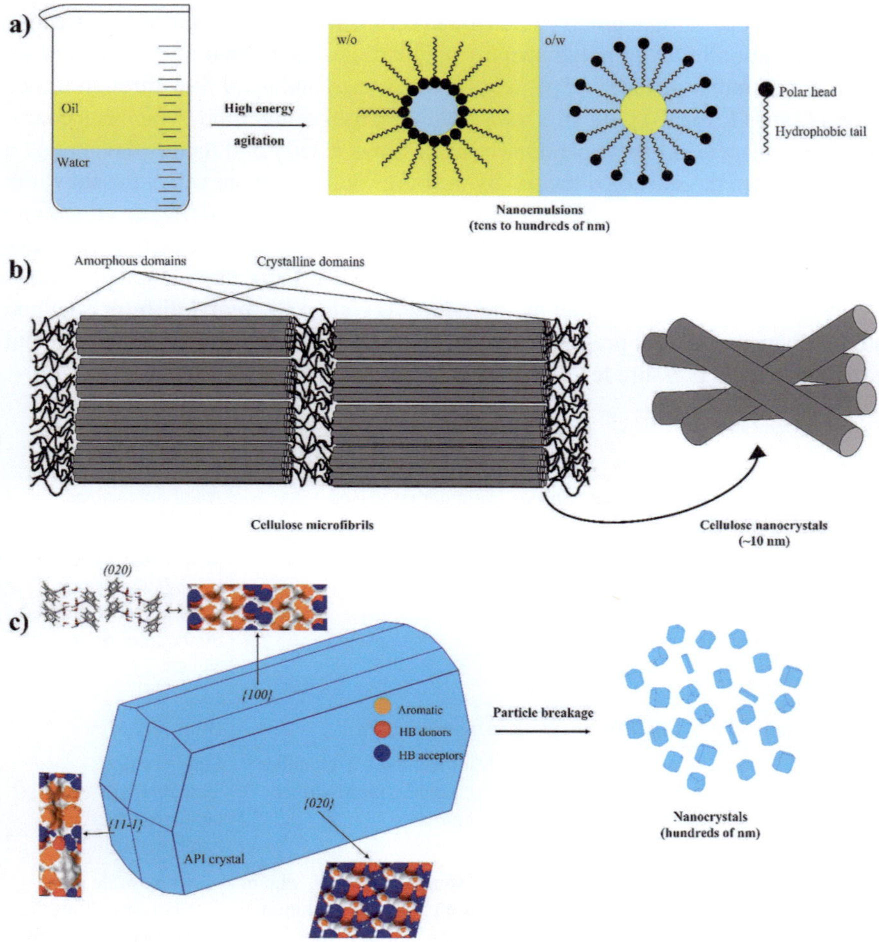

Fig. 2 Scheme comparing the microstructure and composition of **a** nanoemulsions, **b** nanocrystals of cellulose, and **c** nanocrystals of small organic molecules. In **a**, it is highlighted the need of high energy and surface-active ingredients to mix the phases and stabilize nanoemulsion droplets. The microstructure of the cellulose fibrils shown in **b**, explain the multiple steps needed to prepare nanosized crystals (e.g. hydrolysis to digest the amorphous regions and high shear to delaminate and break the crystalline domains). In **c**, the structure of carbamazepine dihydrate is used to illustrate the anisotropy of particles of small organic molecules and the different molecular arrangements over three different crystal facets (calculated with the Surface Analysis tool in CSD-Particle platform). The {*100*} surface of the crystal shows a slip/weak plane (*020*) that may anticipate breakage—a critical point in the preparation of nanocrystals

two orifices (operational units) connected in series, which results in a more effective homogenization or lower polydispersity and particle size reduction [62]. Valves are classified as radial diffuser valves, counter jet valves and axial flow through orifice valves (Table 1, Fig. 3) [14]. In a *radial diffuser*, the premix fluid, which is a coarse emulsion, suspension or dispersion, initially flows axially and then is deviated at a 90° angle from the starting direction by the plug. With a moving valve, the slit width and then the flow rate of the fluid and applied pressure are changed [29]. In *counter jet valves*, particle collision of jet streams from opposite directions is found. In *axial flow valves*, which are also called as *nozzle aggregates*, a premix flowing axially access an orifice with sharp edge inlet and outlet [73]. Unlike the radial diffusers, jet and nozzle dispersers do not present movable parts. As a consequence of this equipment variation, higher pressure levels can be achieved with these last two devices [38].

Table 1 Comparison of main valve types used in high-pressure homogenizers

	Radial diffuser valves	Counter jet valves	Axial flow valves
Flow of liquid	Premix fluid initially flows axially and then is deviated at a 90° angle from the starting direction	Two or more flow streams in opposite directions are found, which lead to particle collision	Axial entry of pressurized fluid through the orifice Flow rate is controlled by the intensifier capacity
Operating pressure limit	+ 0.1–150 MPa	++ Up to 200 MPa	+++ Up to 400 MPa
Types	Knife-edge, conical valve	Microchannel, Y-chamber, Z-chamber, multi-slotted Y-chamber	Single orifice (square-edged or quadrant edged), combined orifice
Advantages	– The gap width can be adjusted to achieve specific homogenizing pressure by moving the top plug piston	– No movable parts, which simplifies the operation and allow to obtain higher pressure levels than radial diffuser valves – More reproducible results than radial diffuser valves – For scaling up, multiple channels can be incorporated into the homogenizers to increase the throughput	– No movable parts, which simplifies the operation and allow to obtain higher pressure levels than radial diffuser valves – More reproducible results than radial diffuser valves – Pressure is controlled by gap slit
Disadvantages	– Result in larger particles when compared to other two valve types	– As premix stream using is divided into multiple channels, high flow rate is needed to operate properly and then greater energy input – Flow rate of the fluid depends on the operating pressure	– Currently available for lab-scale and pilot-scale volume streams

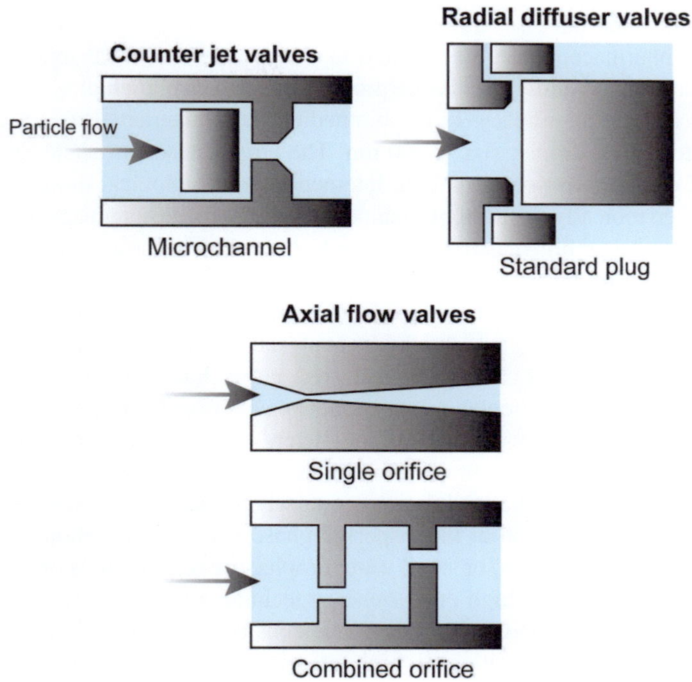

Fig. 3 Examples of valve geometries for high-pressure homogenizers. Adapted from the literature [24]. Radial diffuser valves operate mostly via impact forces, while axial flow and counter jet valves generate cavitation and shear

3 High-Pressure Homogenizer Applications

3.1 Preparation of Nanoemulsions

Nanoemulsions are dispersions of two immiscible liquids consisting of an aqueous medium and an oil phase composed of natural, synthetic, or semi-synthetic oils [68]. They are classified according to the proportion of water and oil in the dispersion. For example, nanoemulsions oil-in-water (o/w) present a small amount of oil in a large volume of water. The dispersed droplets are usually stabilized within the dispersing medium by the aid of surfactants (usually more than one type of surfactant is required). The surfactants adapt their molecular conformation to allow the interaction with water and with oil, via their polar head and lipophilic chain respectively. Surfactants reduce the rate of emulsion coalescence—a physical phenomenon of instability that occurs as the system tries to reach a state of minimum free energy. Therefore, nanoemulsions are considered thermodynamically unstable but kinetically stable, which allows them to maintain their original properties for months or years [77].

Nanoemulsions have several applications, such as in the production of food, cosmetics, pharmaceutical products and in the broader chemical industry. These systems have the potential to enhance product performance, including texture, taste and smell as well as efficacy. In the biomedical field, nanoemulsions are usually designed to have a droplet size < 500 nm. The large surface area and low surface tension of oil droplets provide a high interaction with biological tissues and cells. When applied topically, for example, nanoemulsions have the potential to improve the permeation and retention of active compounds in the skin, hence a high concentration can reach deep skin layers [70].

Droplet characteristics are based on droplet size, polydispersity index, zeta potential and interfacial tension, which may be analyzed during and after the production process. Other properties can also be analyzed and include optical and rheological parameters (gravitation, droplet aggregation, Ostwald ripening) as well as chemical and microbiological assays. All these parameters are affected by the composition and the technique used in the preparation of nanoemulsions [39].

The nanoemulsion formation process occurs when mechanical forces are applied. This leads to the separation of the dispersed phase with the formation of small oil droplets (o/w nanoemulsion) or small water droplets (w/o nanoemulsion). The shear force can be produced by high or low-energy techniques, or even by the combination of low-energy methods with the spontaneous production of droplets induced by temperature and composition. In high-energy methods, mechanical devices are used to create powerful disruptive forces to combine and break apart the oil and aqueous phases. Ease of operation, scalability, reproducibility and high throughput are the reasons why HPH is the gold standard technique for the industrial production of nanoemulsions. Also, the increased stability, the smaller droplet sizes and their homogeneity have contributed to HPH being one of the most widely used techniques in the field of nanoemulsions [54].

The preparation of nanoemulsified systems using HPH requires the prior preparation of a conventional emulsion (coarse primary emulsion), which is forced to flow through narrow passages (nozzles or valves) to reduce droplet size (Fig. 1). The emulsification process occurs in two main stages. First, the droplets are deformed and broken, which leads to an increase in the surface area of the droplets. Next, the droplets are stabilized by adsorption of the emulsifying agents at the newly formed interfaces [26]. From that, the following factors have been pointed as the most influential to the properties of droplets prepared by HPH:

- Geometry of the homogenizer;
- Homogenizer operating conditions: pressure, temperature and time;
- Nanoemulsion composition: oil type, surfactants type and their relative proportions;
- Physicochemical properties of aqueous and oil phases: interfacial tension and viscosity.

The geometry of the equipment affects the emulsification efficiency. High pressure homogenizers are equipped with plunger-like pumps and valves, nozzles, or chambers to promote cavitation, impact, shear, or their combination (Fig. 3). In general, homogenization or rupture chambers have a micrometer-sized discharge opening through which the fluid is forced, followed by an interaction chamber, where there are intense shear phenomena. The geometry of the homogenization chamber mainly affects the uniformity of the flow regimen and, consequently, the width of the droplet size distribution [82].

The vast majority of the studies, however, focus on the effect of pressure and number of cycles in the homogenizer, different types of surfactants/cosurfactants, as well as their proportions for the optimization of nanoemulsions. Table 2 shows studies that have assessed the abovementioned parameters.

The high pressure in the homogenizer can provide the necessary energy for both emulsification steps (i.e. droplet formation and stabilization) in a short time. Pressures between 500 and 3000 bar are usually applied to the coarse emulsion. In general, the mean droplet size tends to decrease with increasing homogenization pressure

Table 2 Selection of studies that have used HPH in the preparation of nanoemulsions

Equipment	Pressure (bar)	Cycles	Nanoemulsion composition	Particle size		References
				DZ[a] (nm)	PDI[b]	
Piston-orifice type homogenizer (Nano DeBEE)	500	3	8.8% olive oil 3.0% surfactant 2.0% cosurfactant	258	0.32	[86]
Panda Plus 2000 high-pressure homogenizer (GEA Mechanical)	1000	3	20% oleic acid oil 10% surfactant 5–10% cosurfactants	80.63–129.68	0.079–0.240	[15]
Motor-driven homogenizer (NLM100)	1000	5	10% limonene-soybean oil (80:20) 5–15% surfactant	40.85–99.50	0.17–0.20	[22]
EmulsiFlex-C3®	750	6	20% Copaiba oil—MCT[c] (2:1) 4% surfactants	250	0.08	[6]
Nano DeBEE	~ 2760	4	9.3% sacha inchi oil 2% surfactants at different proportions	110–199	0.04–0.27	[56]
EmulsiFlex-C3®	750	6	19% oil (mixture of *Curcuma longa* oil and MCT) 2% surfactants	~ 200	< 0.2	[45]

Particle size and polydispersity index are correlated according to the composition of the medium, pressure, number of cycles and type of equipment

[a] *DZ* droplet mean size

[b] *PDI* polydispersity index

[c] *MCT* medium chair triglycerides

and number of cycles. However, increasing the homogenization pressure should be carefully evaluated, since very high pressures can cause overprocessing of the emulsion droplets, which results in larger droplet sizes [84]. This can cause kinetic instability in the emulsion. Also, only one cycle of homogenization is usually not enough and the HPH process is usually repeated several times until size is constant [54].

The number of cycles needed to obtain stable nanoemulsions with the desired droplet sizes is mostly a function of the viscosity of the dispersed phase. The viscosity ratio (ratio between the viscosity of the dispersed phase and the viscosity of the continuous phase—ηd/ηc), however, also affects this parameter [7]. To reduce droplet size under specific pressure conditions, the product is recirculated through to re-experience all the shear, cavitation and turbulence again. The time required to complete one cycle is also dependent on the viscosity of the nanoemulsion. In general, after two or three cycles the viscosity of the nanoemulsion decreases due to reduction in the size of the droplets, which decreases the time required to complete the cycle [54]. Time is a factor that should be carefully controlled because longer emulsification times may lead to "over-processing", resulting in an increase in droplet size [26].

The optimal pressure to produce the minimum droplet size also depends on the nanoemulsion composition. However, the relationship between surfactants, oils (types and concentrations used) and the homogenizer parameters are difficult to be discriminated because they are often showed in association. Yukuyama and collaborators [86], however, applied a Design-of-Experiment (DoE) method to evaluate pressure variables, number of homogenization cycles and different concentrations of components on the olive oil nanoemulsion properties [86]. The authors found that the number of cycles did not show a significant influence on the output response of droplet size of olive oil nanoemulsions. At the pressure range evaluated (250–600 bar), minimum droplet size was achieved with pressure at 500 bar. In this specific case, pressure was considered as a critical process attribute of that nanoemulsion system. In addition, the data revealed a lowest droplet size with the highest surfactant and the lowest olive oil concentrations.

Indeed, the efficiency of HPH in forming nanoemulsions is system specific mainly because of the effect of formulation composition. Hidajat and coworkers [22], for example, have showed the effect of different concentrations of Tween 80 (an hydrophilic surfactant) at 5, 10 and 15% on nanoemulsions properties after 5 passes in the homogeneizer [22]. The authors observed a significant reduction in droplet size after 5 cycles, with a trend towards a greater reduction in droplet size as surfactant concentration increased (5% reduced 4.4-fold, 10% reduced 5.9-fold, and 15% reduced 6.5-fold in droplet size). In turn, Gawin-Mikołajewicz and coworkers [15] prepared nanoemulsions with different surfactants (Kolliphor EL, tween 80 and tween 20) by HPH [15]. After three passes through the homogenizer, the droplet size of nanoemulsions containing either Tween 80 or Tween 20 as surfactant were in the range of 80–130 nm and PDI values below 0.25. A further increase in the number of passes did not lead to a decrease in droplet size. Curiously, the combination of Tween 20 as surfactant and 10% PEG 200 as cosurfactant changed the PDI value of the formulations to above 0.3 (PDI values > 0.25 are associated with low nanoemulsion stability).

3.2 Preparation of Nanocrystals

Crystalline particles are formed when their components (i.e. atoms, ions, molecules) are close-packed and orderly arranged in the space following specific motifs of symmetry. In principle, the knowledge of the packing patterns and the intermolecular interactions within the crystal lattice allows the prediction of particle properties, namely crystal structure, morphology/shape, physical and chemical stabilities, solubility, dissolution rate, adhesion, hygroscopicity, incompatibilities with other chemicals, etc. [5]. Yet, these properties may change when the size of the crystals is reduced, especially from μm- to the nm-scale. The quantum size effect on luminescence is an example. While bulk material has a small sized band gap within its electron orbit, particles with dimension restrictions (e.g. nanowires and quantum dots) show an increased band gap and, consequently, tuneable fluorescence [81]. Another important property that changes with size is the relationship between lattice and surface energy. The implications of the increased contribution of surface energy in nanocrystals are varied, namely changes in the thermodynamic relationship between polymorphs [3], enhanced dissolution rate [13] and agglomeration [75].

Nanocrystals of organic and inorganic materials can be used for a wide range of applications, from catalysis [80] to food technology [46] and biomedicine [52]. A common challenge within all these areas, however, is the method of preparation of such small particles. The reason relies on the difficulty of stopping the crystal growth at the nanoscale after the energetic barrier of nucleation is crossed. This is especially important in the case of organic crystals, which are highly anisotropic, i.e. their molecules exhibit variations in physical properties and intermolecular interactions along different molecular axes of the substance [5]. The consequence is that, in a supersaturated solution, crystals tend to grow fast in preferred directions of the lattice and result in μm-sized particles. The main strategies that can be used to mitigate this behaviour in solution are the control of supersaturation, temperature and pressure, as well as the docking of crystal surfaces. Traditionally, surface docking can be achieved with either supramolecularly-bound additives (e.g. surfactants) or covalently-bound ligands (e.g. sulfur-substituted chemicals and polyethylene glycol) [13, 80].

Some of the nanocrystallization strategies, however, find limited application in organic crystals because these materials tend to be labile and decompose in extreme conditions of heat, pH, and oxidative stress, for instance. Therefore, top-down techniques are preferred to prepare nanocrystals of organic materials—either alone or in association. HPH is one of such techniques. In fact, most of the published works using HPH for the production of nanocrystals involve cellulosic fibres and small organic molecules of pharmaceuticals and food products.

The literature has pointed to the following factors as the most influential to the properties of nanocrystals prepared by HPH:

- Homogenizer operating conditions: pressure and cycles;
- Solution composition: presence of additives and their concentration;
- Physicochemical properties of the crystals: crystal structure, molecular arrangement on the crystal facets/surfaces, and particle morphology.

Differently from the phenomena involved in the formation of nanoemulsions, the breakage that results in nanocrystals is more dependent of the mechanical properties of the primary microparticles than of the general experimental conditions. In other words, the ability of HPH in producing nanocrystals is system-specific mainly because of the role that crystal structure plays in mechanical fracture [57, 58]. For instance, the microstructure and the highly oriented crystalline domains of cellulose fibrils and close-packed small organic molecules originate high mechanical strength and modulus, making them difficult to delaminate/break through homogenization. To overcome this characteristic, new studies have been associating HPH with other nanoparticle-forming techniques, such as ball milling [1, 51]. Process parameters can also be improved to result in efficient HPH-only methods. The particularities of the use of HPH in the preparation of nanocrystals of both cellulosic fibres and small organic molecules are illustrated below.

Nanocrystals of Small Organic Molecules

Table 3 shows some of the most cited research studies that have used HPH in the preparation of nanocrystals of small organic molecules throughout the years. Among these reports, nanocrystallization was generally used as a tool to improve dissolution rate and other biopharmaceutical properties related with drug delivery. In this context, the particle attributes that are important for the performance and the advantages of nanocrystals are the average size, particle size distribution (i.e. polydispersity), dispersibility in liquid, crystal form and physical stability.

In general, the use of HPH results in particles that range from around 100 to 500 nm and show a broader size distribution than nanoemulsions prepared with the same technique (Table 2). In regards to the effect of equipment parameters on the average particle size, the initial 5 cycles of homogenization are responsible for the biggest changes. A higher number of cycles (e.g. 20–30), in turn, contributes to lower values of polydispersity. This trend has been illustrated in the case of hesperetin [42], baicalein [87], lutein [43] and quercetin [27]. In the later, however, increased pressure levels have also significantly contributed to quicker nanocrystallization.

The usual pressure used to prepare dispersions of nanocrystals with HPH ranges from around 1000 to 2000 bar, with higher pressure levels known for accelerating the reduction of particle size. However, care should be taken upon using extreme pressures because they tend to increase the batch temperature, induce agglomeration and block the internal components of the equipment. For example, 20 cycles of HPH at 1500 bar has caused a temperature increase of 30 °C during the preparation of nifedipine nanocrystals [20]. The solution found by the authors was to use heat exchanger and keep the temperature at around 10 °C. A more prudent and efficient approach, therefore, could be to use more cycles of homogenization and reduced pressure. Additionally, pre-treatment steps can be employed to avoid agglomeration

Table 3 Selection of studies that have used HPH in the preparation of nanocrystals of small organic molecules

Material (References)	Concentration	Cycles	Pressure (bar)	Additives	Particle size (nm)	Equipment	Pre-treatment
Nifedipine [20]	5% (w/v)	20	~ 1650	HPMC (0.5%, w/w)	291[a]	EmulsiFlex C5	(i) 24,000 rpm, 10 min (ii) Low pressure homogenization cycles: 15 at ~ 480 bar, 10 at ~ 830 bar
ucb-35440-3 [21]	5% (w/v)	20	~ 1650	HPMC (2%, w/w)	182[a]	EmulsiFlex C5	(i) 500 rpm (ii) Low pressure homogenization cycles: 15 at ~ 480 bar, 10 at ~ 830 bar
				Poloxamer 407 (2%, w/w)	183[a]		
				Lauryl sulfate (2%, w/w)	10,700[a]		
				PVA (2%, w/w)	262[a]		
				Acaciae gum (2%, w/w)	407[a]		
Ascorbyl palmitate [71]	6%	20	1500	Tween 80	365[b]	MicronLAB 40	(i) 8000 rpm, 1 min (ii) Low pressure homogenization cycles: 2 at 150 bar, 2 at 500 bar, 2 at 1000 bar
				Lauryl sulfate	348[b]		
Hesperetin [42]	5% (w/w)	30	1500	Tween 80 (1%, w/v)	346[b]	MicronLAB 40	(i) 10,000 rpm, 1 min (ii) Low pressure homogenization cycles: 5
				Poloxamer 188 (1%, w/v)	301[b]		
				Plantacare 2000 (1%, w/v)	304[b]		
				Inutec SP1 (1%, w/v)	304[b]		
Rutin [40]	10% (w/w)	20	1500	Tween 80 (2%)	912[b]	MicronLAB 40	–
				Poloxamer 188 (2%)	648[b]		

(continued)

Table 3 (continued)

Material (References)	Concentration	Cycles	Pressure (bar)	Additives	Particle size (nm)	Equipment	Pre-treatment
				Lauryl sulfate (0.2%)	727[b]		
				PVA (2%)	547[b]		
Baicalein [87]	0.07% (w/v, ethanol/water)	35	1000	Tween 80 (0.5%)	293[b]	NS 1001L	(i) 1600 rpm, 5 min (ii) Low pressure homogenization cycles: 5 at 200 bar
				Poloxamer 188 (0.5%)	301[b]		
				Lauryl sulfate (0.5%)	313[b]		
Lutein [43]	5% (w/w)	25	1500	Plantacare 2000 UP (1%, w/w)	~ 429[b]	MicronLAB 40	(i) 8000 rpm, 1 min (ii) Low pressure homogenization cycles: 2 at 200 bar, 2 at 500 bar, 2 at 1000 bar
Piroxicam [30]	2.5% (w/w)	30	1500	Poloxamer 188 (0.5, 1.0 and 1.5%, w/w)	432, 553 and 414[b]	EmulsiFlex C5	(i) 8000 rpm, 1 min (ii) Sonication for 1 h (iii) Low pressure homogenization cycle: 3 at 500 bar
Quercetin [27]	2% (w/w)	20	1500	Tween 80 (0.5–2.5%, w/w)	483[b]	MicronLAB 40	–
Itraconazol [69]	1% (w/w)	15	1350	Poloxamer 188 (0.5%, w/w)	547[b]	AH100D	(i) 8000 rpm, 1 min (ii) Low pressure homogenization cycles: 2 at 150 bar, 2 at 500 bar, 2 at 1000 bar
				Poloxamer 188 (0.5%, w/w) and chitosan (0.2%, w/w)	342[b]		

(continued)

Table 3 (continued)

Material (References)	Concentration	Cycles	Pressure (bar)	Additives	Particle size (nm)	Equipment	Pre-treatment
				Poloxamer 188 (0.5%, w/w) and trimethyl chitosan (0.2%, w/w)	289[b]		
				Poloxamer 188 (0.5%, w/w) and polyethyleneimine (0.2%, w/w)	268[b]		
Quercetin [27]	5% (w/w)	20	1500	Tween 80 (1%, w/w)	338[b]	MicronLAB 40	(i) Low pressure homogenization cycles: 2 at 300 bar, 2 at 500 bar, 1 at 1000 bar
	10% (w/w)			Tween 80 (2%, w/w)	442[b]		
Clarithromicin [44]	1% (w/v)	30	1500	Tween 80 (10%, w/v)	3018[b]	APV-2000	(i) 10,000 rpm, 3 cycles of 10 min (ii) Low pressure homogenization cycles: 5 at 50 bar, 5 at 100 bar, 10 at 500 bar, 10 at 1000 bar
				Poloxamer 188 (5%, w/v)	1534[b]		
				Poloxamer 407 (2%, w/v)	761[b]		
				Lauryl sulfate (0.1%, w/v)	851[b]		
				PVA (1%, w/v)	893[b]		
Meloxicam [34]	1% (w/v)	20	1300	Tween 80 (0.5%, w/w)	432[b]	AH-Basic	(i) Low pressure homogenization cycle: 1 at 1000 bar
Atorvastatin [65]	–	20	1000	Poloxamer 188 (5:1 of drug)	258[b]	M-110P	(i) Mixing, 1 min
Baicalein [53]	2% (w/w)	20	900	Poloxamer 188 (0.4%, w/w)	207[b]	AH100D	(i) 15,000 rpm, 10 min

(continued)

Table 3 (continued)

Material (References)	Concentration	Cycles	Pressure (bar)	Additives	Particle size (nm)	Equipment	Pre-treatment
Baicalein:Nicotinamide [53]	2% (w/w)	20	900	Poloxamer 188 (0.4%, w/w)	252[b]	AH100D	(i) 15,000 rpm, 10 min
Hesperidin [35]	0.8% (w/v)	12	1000	Tea saponins (0.1%, w/v)	254[b]	AH100D	(i) 16,000 rpm, 2 min (ii) Low pressure homogenization cycles: 6 at 350 bar
				HPMC (0.1%, w/v)	> 500[b]		
				Poloxamer 188 (0.1%, w/v)	> 300[b]		
				Lauryl sulfate (0.1%, w/v)	> 300[b]		
				PVP K30 (0.1%, w/v)	> 400[b]		
				PEG 400 (0.1%, w/v)	> 300[b]		
				Gallic acid (0.1%, w/v)	> 300[b]		
4-aminosalicylic acid: sulfamethazine [61]	2% (w/w)	5	900	PVA (1%)	190[b]	Invensys APV-2000	(i) 17,500 rpm, 3 cycles of 10 min

Particle size is correlated with the number of additives and active compound, pressure, number of cycles, pre-treatment and type of equipment
[a] Measured by LD: laser diffractometry; d(v; 0.5): mean particle size of the population representing the size where 50% of the particles are below the given size
[b] Measured by PCS: photon correlation spectroscopy (also called by dynamic light scattering)

and sedimentation between cycles, consequently avoiding instrument obstruction. These steps may include the use of bench dispersers and homogenization cycles at low pressures with the HPH instrument. Interestingly, Table 3 shows that the protocols of pre-treatment steps vary more than the selected conditions of HPH per se.

Additives have also shown to improve the ability of HPH in making nano-sized crystals, especially amphiphilic substances with surfactant properties. The examples on Table 3 show that Poloxamer 188 and Tween 80 are frequently selected as HPH adjuvants and, indeed, result in good particle attributes. For example, clarithromycin has shown to reduce particle size with less cycles when surfactant (i.e. Tween, poloxamer, lauryl sulphate) was added to homogenization [44]. Also, the combination of additives seems to be an interesting strategy for reducing particle size. Cationic polymers like chitosan, N-trimethyl-chitosan, and polyethyleneimine, for instance, have shown a promising ability as steric and electrostatic stabilizers of nanocrystals of itraconazole [88].

The LightGBM algorithm has been used to identify molecular descriptors that affect the nanocrystallization through HPH [19]. The three most influential descriptors were the number of cycles, the molecular weight of the additive, and the main compound log P. This study may help raising some questions related to why, in comparison to other studies listed on Table 3, HPH has not been efficient in decreasing the particle size of rutin [40] and clarithromycin [44]. The two actives are poorly soluble glycosides with high HB donor/acceptor capacity and prone to increased molecular flexibility. Such compounds can undergo conformational changes that alter their polarity and response in a solution environment. It could be, therefore, speculated that these characteristics may affect the active's interaction with additives, impairing the performance of HPH in particle size reduction. Curiously, Esfandi and coworkers have also failed in preparing nanocrystals of clarithromycin with the aid of the same additives (i.e. Tween 80, Poloxamer 188, Lauryl sulphate and PVA), but via sonoprecipitation [10]. The authors have only succeeded when HPMC was used as an additive, instead. Indeed, a change in additive has also resulted in improved size reduction for hesperidin [35], another glycoside listed on Table 3. In hesperidin's case, the use of tea saponins consisting of hydrophilic glycosyl and hydrophobic aglycons outperformed traditional tensoactives in producing nanocrystals with HPH.

It is important to highlight that the efficiency of an additive in stabilizing nanocrystals in aqueous solutions is system specific and not directly related to the hydrophilic-lipophilicity balance between the active and the adjuvant molecules. As mentioned before, crystals tend to be anisotropic. It means that their facets are surfaces with different molecular arrangements, hydrogen bonding patterns, lipophilicity degrees and adhesion properties—especially within different crystal forms. This characteristic may change the way different additives can attach/dock on the surface of the crystals and impact growth [63, 64].

Interestingly, the facet-specific stabilization properties of additives used in the preparation of nanocrystals is still to be explored. Most of the studies on Table 3, for instance, have identified the phase of the nanocrystals prepared with HPH by means of techniques like X-ray diffraction, differential scanning calorimetry and infrared spectroscopy. None of them, however, have explored the relationship of crystal structure and particle properties (i.e. molecular arrangements on the crystal facets, existence of cleavage planes, morphology/shape, particle aspect ratio) on the products prepared by HPH. Some illustrations, however, suggest that solid form landscape is rather critical to the success of size reduction and the properties of the resulting particles. For instance, Sharma and coworkers have shown that depending on the starting polymorphic form of SN30191, same homogenization conditions resulted in different distribution of size [66]. This has also been reported in the case of meloxicam nanocrystals prepared from starting particles differing in phase, shape and size [34].

Additionally, stress-induced transformations may also occur during homogenization. In one illustration, it has been observed a partial conversion from polymorph II (stable) to I (metastable) during the HPH experiments of Clarithromycin [44]. The findings agree with the literature as this drug is known to undergo stress-induced transformations with pressure, temperature and upon contact to different polymeric matrices [25]. Also, highly energetic manufacturing processes, like HPH, tend to produce metastable crystal forms. Other examples of solid-state transformations during HPH were described for piroxicam [30], which changes from Form I to a mixture of Form III and monohydrate, and baicalein cocrystals, which amorphized in the tested conditions of homogenization [53]. Physical transformations can affect the characteristics of the nanocrystals, especially stability, re-dispersibility, particle size after reconstitution, and content uniformity. In any case, a careful understanding of the material's solid form landscape and the techniques that can identify and discriminate different crystal phases is crucial for the preparation of nanocrystals of small organic molecules, especially in highly regulated areas like in the pharmaceutical field.

Nanocrystals of Cellulose

Cellulose is a linear polysaccharide made of glucose monomers that can be isolated from the walls of plant cells (e.g. wood pulp, cotton fiber). Its microstructure is highly dominated by anisotropy—more than the crystals of small organic molecules. Cellulose is a fibrous material with high aspect ratio, mostly associated as microfibrils in its natural sources. The microfibrils, in turn, can be delaminated into nanofibers and highly oriented nanocrystals by using top-down methods like HPH [48, 55]. The applications of these materials are varied, from the preparation of iridescent materials to liquid crystals, gels, membranes, and composite films [4, 36, 50, 67]. The colloidal properties of nanocrystals of cellulose make them great candidates for the preparation of highly stable suspensions based on the formation of network of particles.

The most important particle attributes that are related to the performance of nanocrystals of cellulose are the longitudinal and cross-sectional size of the crystals, and the rheological properties of the resulting colloidal dispersions. The literature points to the following factors as the most influential to the properties of nanocrystals prepared by HPH:

- Pre-homogenization step: hydrolysis medium and other processing steps;
- Homogenizer operating conditions: pressure, number of cycles, temperature.

Table 4 compares research studies that have used HPH in the preparation of nanocrystals of cellulose from different plant sources and following different methodologies. In general, the processing of cellulose with HPH results in particles that range from around 50 to 350 nm along the longitudinal axis, and with approximately 10 nm of cross section. These particles are isolated from a multi-step protocol, mostly composed of a chemical pre-treatment with strong acids or oxidative agents, followed by homogenization. Instead of varying the number of cycles and pressure of homogenization, the studies on cellulose nanocrystals tend to focus on the pre-homogenization step as a mean to decrease particle size and change crystal form/ phase.

The pre-treatment step may vary significantly (Table 4), but it commonly consists of a hydrolysis reaction that breaks down fibril aggregates by attacking amorphous regions of the cellulose microstructure [78]. The result is a significant decrease in the particle aspect ratio, which is crucial for both the nanocrystal preparation but also for the safe subsequent use of HPH without blocking the equipment.

Although, historically, HPH has mostly been seen as secondary in the formation of nano-sized cellulose, this trend is changing. Systematic investigations have shown the effect of mechanical treatment and temperature increase provided by HPH on the separation of nanofiber bundles to form individual crystals [47]. For example, the substitution of higher pressure levels and lower cycle numbers of homogenization for repeated homogenization cycles, has been suggested. This strategy has proven to be efficient in reducing particle size and energy consumption during manufacturing, which is highly desirable in large scale production. Additionally, temperature-controlled HPH has shown to increase the crystallinity of cellulose nanocrystals in comparison to chemically extracted materials [49]. This combination of annealing and high shear is just another illustration of the versatility of HPH in the preparation of nanoparticles for a wide range of batch scales.

4 Conclusions and Future Perspectives

HPH is an extremely versatile technique with proven abilities in nanoparticle production. In this chapter, its application was illustrated by the preparation of two different types of nanostructured materials, namely nanoemulsions and nanocrystals. Both show markedly distinct characteristics and mechanisms of formation, yet, HPH has proven to be equally efficient. The flowcharts below provide a simplified visual tool to support the decision on whether HPH is the appropriate technique to be selected in the industrial scale production of the abovementioned nanomaterials.

Table 4 Selection of studies that have used HPH in the preparation of nanocrystals of cellulose

Material (References)	Concentration	Polymorph	Cycles	Pressure (bar)	Particle size (nm)		Equipment	Pre-treatment
					Long axis	Diameter		
Microcrystalline cellulose [33]	–	I	10	2070	90 ± 50	10 ± 4[a]	M-110P	(i) Sulphuric acid hydrolysis (ii) Dialysis
Cotton fiber [85]	–	I	10	2070	159 ± 5	15 ± 4[a]	M-110P	(i) Sulphuric acid hydrolysis (ii) Dialysis
	–	II		2070	76 ± 20	14 ± 3[a]		(i) Mercerization (ii) Sulphuric acid hydrolysis (iii) Dialysis
Bleached birch wood pulp [32]	0.5% (w/w)	–	5	400–950	–	10–60[b]	APV-2000	(i) Oxidative sulphonation
Microcrystalline cellulose [59]	–	–	10	2070	50–200	~ 10[a]	M-110P	(i) Sulphuric acid hydrolysis or hydrochloric acid hydrolysis (ii) Dialysis
Softwood pulp [47]	1.2% (w/w)	–	4	400–1700	–	–	M-110EH	(i) Carboxymethylation
Sugar palm fiber [23]	1.8% (w/w)		15	500	–	6 ± 1[a]	NS1001L	(i) Delignification (ii) Mercerization (iii) Bleaching (iv) Milling
Cotton fiber [49]	–	–	10–30	1400	178 ± 124	8 ± 3[a]	–	(i) Delignification

(continued)

Table 4 (continued)

Material (References)	Concentration	Polymorph	Cycles	Pressure (bar)	Particle size (nm)		Equipment	Pre-treatment
					Long axis	Diameter		
Okara fiber [78]	4.0% (w/w)	–	3	1000	343 ± 20	–[c]	D-6 L	(i) Degreasing (ii) Decolorization (iii) Delignification (iv) Bleaching (v) 10,000 rpm, 5 min
				1200	253 ± 10	–[c]		
				1400	233 ± 10	–[c]		

Particle size is correlated with the amount of cellulose, pressure, number of cycles, type of equipment and pre-treatment

[a] *TEM* transmission electron microscopy

[b] *FESEM* field emission scanning electron microscopy

[c] *PCS* photon correlation spectroscopy (also called by dynamic light scattering)

Decision tree #1: when to use HPH in the production of nanocrystals

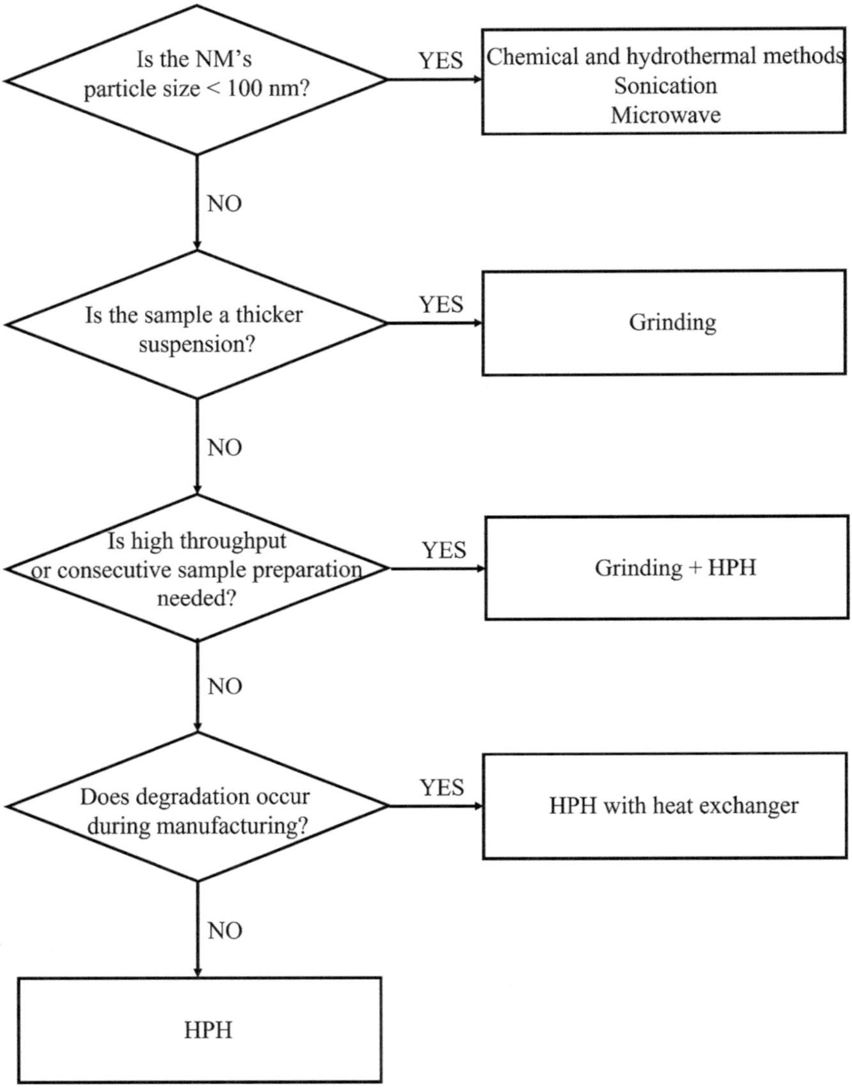

Decision tree #2: when to use HPH in the production of nanoemulsions

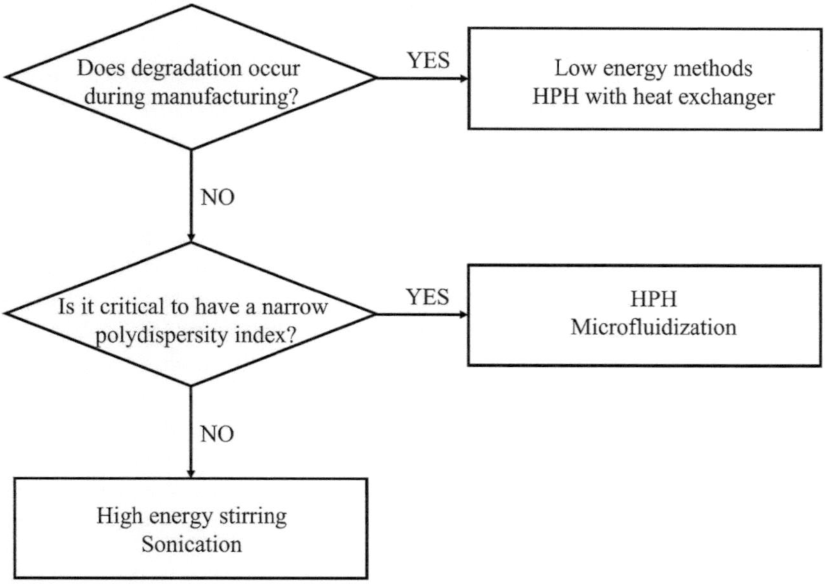

The choice of a suitable homogenizer and the appropriate working parameters, however, must take into consideration the type of nanoparticle being produced and the scale of production. This selection is based on the equipment's total throughput, which accounts for the balance among equipment internal volume, shear stress, sample size, and required number of passes. For instance, nanoemulsions require less energy and a lower number of homogenization cycles in comparison to nanocrystals. The production of nanocrystals, in turn, tend to use additional pre-homogenization steps due to the hardness of the materials and their other mechanical properties related to anisotropy. As a generalization, Scheme 1 shows the relationship between end-product attributes and operating conditions.

Although the technique is already well-stablished, especially in nanoemulsions, new trends and technological applications are noticeable. For example, emulsions and crystals have been lately combined via HPH into the so-called Pickering emulsions. These consist of droplets of one liquid phase stabilized in another liquid phase by the aid of nanocrystals. Examples of nanocrystals that could successfully stabilize Pickering emulsions through HPH are cellulose [2], quercetin [76], curcumin [74], and

Scheme 1 Process parameters that, in general, result in the decrease of particle size and in narrowing the polydispersity of nanomaterials produced by HPH. The item identified with an * refers to nanocrystals

↓ Particle concentration
↓ Viscosity
↑ Number of cycles/homogenization time
↑ Pressure
Appropriate selection of additives
Appropriate selection of crystal form and shape*
Combining HPH with other techniques

silybin [83]. The formation of such particles is based on the hydrophilic/hydrophobic balance of the nanocrystals' facets, and not of the molecule composing the crystals. Structure- and surface-specific characteristics may not only affect the formation of Pickering emulsions, but the breakage and stabilization of any nanocrystal prepared by top-down methods. For example, the shear generated by HPH may suit the preparation of materials that are susceptible to delamination, like cellulose. In the case of close-packed crystals of small organic molecules, depending on the molecular packing, impact forces may contribute more to the reduction of particle size.

This means that crystal structure should be critically assessed prior to the design of HPH methods. This can be done through different computational tools that may also grow in importance in HPH, like the group of Software available from the Cambridge Crystallographic Data Centre. Machine learning, additionally, may be used to calculate particle size and polydispersity, but also the molecular descriptors that affect the production method [19]. Design of experiments, in turn, are used to minimize the number of tests required to fine-tune particle properties [60].

Recent studies have also been using HPH to prepare nanosized cocrystals [13, 53, 61]. These are multicomponent crystal forms with a complex phase diagram that may phase-segregate if the appropriate solution composition is not selected. The use of HPH in the processing of cocrystals emphasize the need for a careful selection of methodologies to characterise the phase and other particle attributes of nanocrystals obtained through HPH. Most of the studies report the use of qualitative X-ray diffraction, scanning differential calorimetry and infrared spectroscopy. Other techniques, like Raman spectroscopy and crystallite size determination from Rietveld analysis, could also benefit future studies—especially considering materials with low crystallinity and mixed with complex matrices. In any case, the current chapter shows that, although robust, HPH is still an evolving methodology with several applications to be further explored.

Acknowledgements The work was supported by European and European Regional Development Funds—Horizon Europe Project CRYSTALS4SKIN (No. 101090272), Project "Excellence in Regenerative Medicine" (No. CZ.02.01.01/00/22_008/0004562), and project "TECHSCALE" (CZ.02.01.01/00/22_008/0004587).

References

1. Al Shaal L, Shegokar R, Müller RH (2011) Production and characterization of antioxidant apigenin nanocrystals as a novel UV skin protective formulation. Int J Pharm 420(1):133–140. https://doi.org/10.1016/j.ijpharm.2011.08.018
2. Bai L, Lv S, Xiang W, Huan S, McClements DJ, Rojas OJ (2019) Oil-in-water Pickering emulsions via microfluidization with cellulose nanocrystals: 1. Formation and stability. Food Hydrocolloids 96:699–708. https://doi.org/10.1016/j.foodhyd.2019.04.038

3. Belenguer AM, Lampronti GI, Cruz-Cabeza AJ, Hunter CA, Sanders JKM (2016) Solvation and surface effects on polymorph stabilities at the nanoscale. Chem Sci 7(11):6617–6627. https://doi.org/10.1039/c6sc03457h
4. Dalastra C, Klanovicz N, Kubeneck S, Stefanski FS, Argenta DF, Rauber GS, Caon T, Cadamuro RD, Fongaro G, Treichel H (2023) Carbohydrate-based economy: perspectives and challenges, chap 16. In: Goldbeck R, Poletto P (eds) Polysaccharide-degrading biocatalysts. Academic Press, pp 409–434. https://doi.org/10.1016/B978-0-323-99986-1.00014-4
5. Desiraju GR, Vittal JJ, Ramanan A (2011) Crystal engineering. A textbook. World Scientific
6. Dias DDO, Colombo M, Kelmann RG, Kaiser S, Lucca LG, Teixeira HF, Limberger RP, Veiga VF, Koester LS (2014) Optimization of Copaiba oil-based nanoemulsions obtained by different preparation methods. Ind Crops Prod 59:154–162. https://doi.org/10.1016/j.indcrop.2014.05.007
7. Donsì F, Sessa M, Mediouni H, Mgaidi A, Ferrari G (2011) Encapsulation of bioactive compounds in nanoemulsion-based delivery systems. Procedia Food Sci 1:1666–1671. https://doi.org/10.1016/j.profoo.2011.09.246
8. dos Santos Aguilar JG, Cristianini M, Sato HH (2018) Modification of enzymes by use of high-pressure homogenization. Food Res Int 109:120–125. https://doi.org/10.1016/j.foodres.2018.04.011
9. Dumay E, Chevalier-Lucia D, Picart-Palmade L, Benzaria A, Gràcia-Julià A, Blayo C (2013) Technological aspects and potential applications of (ultra) high-pressure homogenisation. Trends Food Sci Technol 31(1):13–26. https://doi.org/10.1016/j.tifs.2012.03.005
10. Esfandi E, Ramezani V, Vatanara A, Najafabadi AR, Pouya S, Moghaddam H (2014) Clarithromycin dissolution enhancement by preparation of aqueous nanosuspensions using sonoprecipitation technique. Shaheed Beheshti Univ Med Sci Health Serv Iran J Pharm Res 13(3)
11. Espitia PJP, Fuenmayor CA, Otoni CG (2019) Nanoemulsions: synthesis, characterization, and application in bio-based active food packaging. Comp Rev Food Sci Food Saf 18(1):264–285. https://doi.org/10.1111/1541-4337.12405
12. Fernandez-Avila C, Hebishy E, Donsì F, Arranz E, Trujillo AJ (2019) Production of food bioactive-loaded nanostructures by high-pressure homogenization. In: Jafari SM (ed) Nanoencapsulation of food ingredients by specialized equipment, vol 3. Academic Press, pp 251–340. https://doi.org/10.1016/B978-0-12-815671-1.00006-8
13. Fontana F, Figueiredo P, Zhang P, Hirvonen JT, Liu D, Santos HA (2018) Production of pure drug nanocrystals and nano co-crystals by confinement methods. Adv Drug Deliv Rev 131:3–21. Elsevier B.V. https://doi.org/10.1016/j.addr.2018.05.002
14. Gall V, Runde M, Schuchmann HP (2016) Extending applications of high-pressure homogenization by using simultaneous emulsification and mixing (SEM)—an overview. Processes 4(4). MDPI AG. https://doi.org/10.3390/pr4040046
15. Gawin-Mikołajewicz A, Nawrot U, Malec KH, Krajewska K, Nartowski KP, Karolewicz BL (2024) The effect of high-pressure homogenization conditions on the physicochemical properties and stability of designed fluconazole-loaded ocular nanoemulsions. Pharmaceutics 16(1). https://doi.org/10.3390/pharmaceutics16010011
16. Georget E, Miller B, Callanan M, Heinz V, Mathys A (2014) (Ultra) high pressure homogenization for continuous high pressure sterilization of pumpable foods—a review. Front Nutr 1. Frontiers Media S.A. https://doi.org/10.3389/fnut.2014.00015
17. Grumbach C, Krüger V, Czermak P (2022) A new control strategy for high-pressure homogenization to improve the safety of injectable lipid emulsions. Pharmaceutics 14(8). https://doi.org/10.3390/pharmaceutics14081603
18. Håkansson A (2018) Fabrication of nanoemulsions by high-pressure valve homogenization, chap 7. In: Jafari SM, McClements DJ (eds) Nanoemulsions. Academic Press, pp 175–206. https://doi.org/10.1016/B978-0-12-811838-2.00007-2
19. He Y, Ye Z, Liu X, Wei Z, Qiu F, Li HF, Zheng Y, Ouyang D (2020) Can machine learning predict drug nanocrystals? J Control Release 322:274–285. https://doi.org/10.1016/j.jconrel.2020.03.043

20. Hecq J, Deleers M, Fanara D, Vranckx H, Amighi K (2005) Preparation and characterization of nanocrystals for solubility and dissolution rate enhancement of nifedipine. Int J Pharm 299(1–2):167–177. https://doi.org/10.1016/j.ijpharm.2005.05.014
21. Hecq J, Deleers M, Fanara D, Vranckx H, Boulanger P, Le Lamer S, Amighi K (2006) Preparation and in vitro/in vivo evaluation of nano-sized crystals for dissolution rate enhancement of ucb-35440-3, a highly dosed poorly water-soluble weak base. Eur J Pharm Biopharm 64(3):360–368. https://doi.org/10.1016/j.ejpb.2006.05.008
22. Hidajat MJ, Jo W, Kim H, Noh J (2020) Article effective droplet size reduction and excellent stability of limonene nanoemulsion formed by high-pressure homogenizer. Colloids Interfaces 4(1). https://doi.org/10.3390/colloids4010005
23. Ilyas RA, Sapuan SM, Ibrahim R, Abral H, Ishak MR, Zainudin ES, Atikah MSN, Mohd Nurazzi N, Atiqah A, Ansari MNM, Syafri E, Asrofi M, Sari NH, Jumaidin R (2019) Effect of sugar palm nanofibrillated cellulose concentrations on morphological, mechanical and physical properties of biodegradable films based on agro-waste sugar palm (*Arenga pinnata* (Wurmb.) Merr) starch. J Mater Res Technol 8(5):4819–4830. https://doi.org/10.1016/j.jmrt.2019.08.028
24. Inguva P, Grasselli S, Heng PWS (2024) High pressure homogenization—an update on its usage and understanding. Chem Eng Res Des 202:284–302. https://doi.org/10.1016/j.cherd.2023.12.026
25. Ito M, Shiba R, Watanabe M, Iwao Y, Itai S, Noguchi S (2018) Phase transitions of antibiotic clarithromycin forms I, IV and new form VII crystals. Int J Pharm 547(1):258–264. https://doi.org/10.1016/j.ijpharm.2018.05.073
26. Jafari SM, Assadpoor E, He Y, Bhandari B (2008) Re-coalescence of emulsion droplets during high-energy emulsification. Food Hydrocolloids 22(7):1191–1202. Elsevier. https://doi.org/10.1016/j.foodhyd.2007.09.006
27. Kakran M, Shegokar R, Sahoo NG, Al Shaal L, Li L, Müller RH (2012) Fabrication of quercetin nanocrystals: comparison of different methods. Eur J Pharm Biopharm 80(1):113–121. https://doi.org/10.1016/j.ejpb.2011.08.006
28. Keck CM, Müller RH (2006) Drug nanocrystals of poorly soluble drugs produced by high pressure homogenisation. Eur J Pharm Biopharm 62(1):3–16. https://doi.org/10.1016/j.ejpb.2005.05.009
29. Koroleva MY, Yurtov EV (2012) Nanoemulsions: the properties, methods of preparation and promising applications. Russ Chem Rev 81(1):21–43. https://doi.org/10.1070/rc2012v081n01abeh004219
30. Lai F, Pini E, Angioni G, Manca ML, Perricci J, Sinico C, Fadda AM (2011) Nanocrystals as tool to improve piroxicam dissolution rate in novel orally disintegrating tablets. Eur J Pharm Biopharm 79(3):552–558. https://doi.org/10.1016/j.ejpb.2011.07.005
31. Levy R, Okun Z, Shpigelman A (2021) High-pressure homogenization: principles and applications beyond microbial inactivation. Food Eng Rev 13:490–508. https://doi.org/10.1007/s12393-020-09239-8/Published
32. Liimatainen H, Visanko M, Sirviö J, Hormi O, Niinimäki J (2013) Sulfonated cellulose nanofibrils obtained from wood pulp through regioselective oxidative bisulfite pre-treatment. Cellulose 20(2):741–749. https://doi.org/10.1007/s10570-013-9865-y
33. Liu H, Liu D, Yao F, Wu Q (2010) Fabrication and properties of transparent polymethylmethacrylate/cellulose nanocrystals composites. Bioresour Technol 101(14):5685–5692. https://doi.org/10.1016/j.biortech.2010.02.045
34. Liu T, Yao G, Zhang X, Zuo X, Wang L, Yin H, Möschwitzer JP (2018) Systematical investigation of different drug nanocrystal technologies to produce fast dissolving meloxicam tablets. AAPS PharmSciTech 19(2):783–791. https://doi.org/10.1208/s12249-017-0889-8
35. Long J, Song J, Zhang X, Deng M, Xie L, Zhang L, Li X (2020) Tea saponins as natural stabilizers for the production of hesperidin nanosuspensions. Int J Pharm 583. https://doi.org/10.1016/j.ijpharm.2020.119406
36. Mali P, Sherje AP (2022) Cellulose nanocrystals: fundamentals and biomedical applications. Carbohydr Polym 275:118668. https://doi.org/10.1016/j.carbpol.2021.118668

37. Malik T, Sharma R, Ameer K, Bashir O, Amin T, Manzoor S, Mohamed Ahmed IA (2023) Potential of high-pressure homogenization (HPH) in the development of functional foods. Int J Food Prop 26(1):2509–2531. Taylor and Francis Ltd. https://doi.org/10.1080/10942912.2023.2249262
38. Martínez-Monteagudo SI, Yan B, Balasubramaniam VM (2017) Engineering process characterization of high-pressure homogenization—from laboratory to industrial scale. Food Eng Rev 9(3):143–169. Springer New York LLC. https://doi.org/10.1007/s12393-016-9151-5
39. Mason TG, Wilking JN, Meleson K, Chang CB, Graves SM (2006) Nanoemulsions: formation, structure, and physical properties. J Phys Condens Matter 18(41):R635. https://doi.org/10.1088/0953-8984/18/41/R01
40. Mauludin R, Müller RH, Keck CM (2009) Kinetic solubility and dissolution velocity of rutin nanocrystals. Eur J Pharm Sci 36(4–5):502–510. https://doi.org/10.1016/j.ejps.2008.12.002
41. Mcclements DJ (2015) Nanoscale nutrient delivery systems for food applications: improving bioactive dispersibility, stability, and bioavailability. J Food Sci 80(7):N1602–N1611. https://doi.org/10.1111/1750-3841.12919
42. Mishra PR, Shaal LA, Müller RH, Keck CM (2009) Production and characterization of hesperetin nanosuspensions for dermal delivery. Int J Pharm 371(1–2):182–189. https://doi.org/10.1016/j.ijpharm.2008.12.030
43. Mitri K, Shegokar R, Gohla S, Anselmi C, Müller RH (2011) Lutein nanocrystals as antioxidant formulation for oral and dermal delivery. Int J Pharm 420(1):141–146. https://doi.org/10.1016/j.ijpharm.2011.08.026
44. Morakul B, Suksiriworapong J, Leanpolchareanchai J, Junyaprasert VB (2013) Precipitation-lyophilization-homogenization (PLH) for preparation of clarithromycin nanocrystals: influencing factors on physicochemical properties and stability. Int J Pharm 457(1):187–196. https://doi.org/10.1016/j.ijpharm.2013.09.022
45. Mossmann V, Weimer P, Nunes KAA, Rossi RC, Koester LS (2024) Essential oil extracted from the leaves of *Curcuma longa* L.: application of an agro-industrial residue in the development of anti-inflammatory nanoemulsions intended for skin delivery. Chem Afr 7(4):1849–1864. https://doi.org/10.1007/s42250-023-00872-4
46. Mu R, Hong X, Ni Y, Li Y, Pang J, Wang Q, Xiao J, Zheng Y (2019) Recent trends and applications of cellulose nanocrystals in food industry. Trends Food Sci Technol 93:136–144. https://doi.org/10.1016/j.tifs.2019.09.013
47. Naderi A, Lindström T, Sundström J (2015) Repeated homogenization, a route for decreasing the energy consumption in the manufacturing process of carboxymethylated nanofibrillated cellulose? Cellulose 22(2):1147–1157. https://doi.org/10.1007/s10570-015-0576-4
48. Nagarajan KJ, Ramanujam NR, Sanjay MR, Siengchin S, Surya Rajan B, Sathick Basha K, Madhu P, Raghav GR (2021) A comprehensive review on cellulose nanocrystals and cellulose nanofibers: pretreatment, preparation, and characterization. Polym Compos 42(4):1588–1630. Wiley. https://doi.org/10.1002/pc.25929
49. Park NM, Choi S, Oh JE, Hwang DY (2019) Facile extraction of cellulose nanocrystals. Carbohydr Polym 223. https://doi.org/10.1016/j.carbpol.2019.115114
50. Parker RM, Parton TG, Chan CLC, Bay MM, Frka-Petesic B, Vignolini S (2023) Bioinspired photonic materials from cellulose: fabrication, optical analysis, and applications. Acc Mater Res 4(6):522–535. https://doi.org/10.1021/accountsmr.3c00019
51. Patel K, Patil A, Mehta M, Gota V, Vavia P (2014) Oral delivery of paclitaxel nanocrystal (PNC) with a dual Pgp-CYP3A4 inhibitor: preparation, characterization and antitumor activity. Int J Pharm 472(1–2):214–223. https://doi.org/10.1016/j.ijpharm.2014.06.031
52. Pawar VK, Singh Y, Meher JG, Gupta S, Chourasia MK (2014) Engineered nanocrystal technology: in-vivo fate, targeting and applications in drug delivery. J Control Release 183(1):51–66. Elsevier. https://doi.org/10.1016/j.jconrel.2014.03.030
53. Pi J, Wang S, Li W, Kebebe D, Zhang Y, Zhang B, Qi D, Guo P, Li N, Liu Z (2019) A nanococrystal strategy to improve the dissolution rate and oral bioavailability of baicalein. Asian J Pharm Sci 14(2):154–164. https://doi.org/10.1016/j.ajps.2018.04.009

54. Qian C, McClements DJ (2011) Formation of nanoemulsions stabilized by model food-grade emulsifiers using high-pressure homogenization: factors affecting particle size. Food Hydrocolloids 25(5):1000–1008. https://doi.org/10.1016/j.foodhyd.2010.09.017
55. Rana AK, Frollini E, Thakur VK (2021) Cellulose nanocrystals: pretreatments, preparation strategies, and surface functionalization. Int J Biol Macromol 182:1554–1581. https://doi.org/10.1016/j.ijbiomac.2021.05.119
56. Rave MC, Echeverri JD, Salamanca CH (2020) Improvement of the physical stability of oil-in-water nanoemulsions elaborated with Sacha inchi oil employing ultra-high-pressure homogenization. J Food Eng 273. https://doi.org/10.1016/j.jfoodeng.2019.109801
57. Reddy CM, Gundakaram RC, Basavoju S, Kirchner MT, Padmanabhan KA, Desiraju GR (2005) Structural basis for bending of organic crystals. Chem Commun 31:3945–3947. https://doi.org/10.1039/B505103G
58. Reddy CM, Padmanabhan KA, Desiraju GR (2006) Structure−property correlations in bending and brittle organic crystals. Cryst Growth Des 6(12):2720–2731. https://doi.org/10.1021/cg060398w
59. Ren S, Sun X, Lei T, Wu Q (2014) The effect of chemical and high-pressure homogenization treatment conditions on the morphology of cellulose nanoparticles. J Nanomater 2014. https://doi.org/10.1155/2014/582913
60. Salazar J, Heinzerling O, Müller RH, Möschwitzer JP (2011) Process optimization of a novel production method for nanosuspensions using design of experiments (DoE). Int J Pharm 420(2):395–403. https://doi.org/10.1016/j.ijpharm.2011.09.003
61. Salem A, Takácsi-Nagy A, Nagy S, Hagymási A, Gősi F, Vörös-Horváth B, Balić T, Pál S, Széchenyi A (2021) Synthesis and characterization of nano-sized 4-aminosalicylic acid–sulfamethazine cocrystals. Pharmaceutics 13(2):1–13. https://doi.org/10.3390/pharmaceutics13020277
62. Schlender M, Minke K, Spiegel B, Schuchmann HP (2015) High-pressure double stage homogenization processes: influences of plant setup on oil droplet size. Chem Eng Sci 131:162–171. https://doi.org/10.1016/j.ces.2015.03.055
63. Schneider-Rauber G, Arhangelskis M, Bond AD, Ho R, Nere N, Bordawekar S, Sheikh AY, Jones W (2021) Polymorphism and surface diversity arising from stress-induced transformations-the case of multicomponent forms of carbamazepine. Acta Crystallogr Sect B Struct Sci Cryst Eng Mater 77:54–67. https://doi.org/10.1107/S2052520620015437
64. Schneider-Rauber G, Bond AD, Ho R, Nere N, Bordawekar S, Sheikh AY, Jones W (2021) Effect of solution composition on the crystallization of multicomponent forms of carbamazepine beyond crystal form and shape: surface as a source of diversity in the solid-form landscape. Cryst Growth Des 21(1):52–64. https://doi.org/10.1021/acs.cgd.0c00553
65. Sharma M, Mehta I (2019) Surface stabilized atorvastatin nanocrystals with improved bioavailability, safety and antihyperlipidemic potential. Sci Rep 9(1). https://doi.org/10.1038/s41598-019-52645-0
66. Sharma P, Zujovic ZD, Bowmaker GA, Denny WA, Garg S (2011) Evaluation of a crystalline nanosuspension: polymorphism, process induced transformation and in vivo studies. Int J Pharm 408(1):138–151. https://doi.org/10.1016/j.ijpharm.2011.01.032
67. Shojaeiarani J, Bajwa DS, Chanda S (2021) Cellulose nanocrystal based composites: a review. Compos Part C Open Access 5:100164. https://doi.org/10.1016/j.jcomc.2021.100164
68. Singh Y, Meher JG, Raval K, Khan FA, Chaurasia M, Jain NK, Chourasia MK (2017) Nanoemulsion: concepts, development and applications in drug delivery. J Control Release 252:28–49. Elsevier B.V. https://doi.org/10.1016/j.jconrel.2017.03.008
69. Sun W, Tian W, Zhang Y, He J, Mao S, Fang L (2012) Effect of novel stabilizers—cationic polymers on the particle size and physical stability of poorly soluble drug nanocrystals. Nanomed Nanotechnol Biol Med 8(4):460–467. https://doi.org/10.1016/j.nano.2011.07.006
70. Sutradhar KB, Amin L (2013) Nanoemulsions: increasing possibilities in drug delivery. Eur J Nanomed 5(2):97–110. Walter de Gruyter GmbH. https://doi.org/10.1515/ejnm-2013-0001
71. Teeranachaideekul V, Junyaprasert VB, Souto EB, Müller RH (2008) Development of ascorbyl palmitate nanocrystals applying the nanosuspension technology. Int J Pharm 354(1–2):227–234. https://doi.org/10.1016/j.ijpharm.2007.11.062

72. Trujillo A-J, Roig-Sagués A-X, Zamora A, Ferragut V (2016) High-pressure homogenization for structure modification, chap 12. In: Knoerzer K, Juliano P, Smithers G (eds) Innovative food processing technologies. Woodhead Publishing, pp 315–344. https://doi.org/10.1016/B978-0-08-100294-0.00012-2
73. Vinchhi P, Patel JK, Patel MM (2021) High-pressure homogenization techniques for nanoparticles. In: Patel JK, Pathak YV (eds) Emerging technologies for nanoparticle manufacturing. Springer International Publishing, pp 263–285. https://doi.org/10.1007/978-3-030-50703-9_11
74. Wang X, Liao Z, Zhao G, Dong W, Huang X, Zhou X, Liang X (2023) Curcumin nanocrystals self-stabilized Pickering emulsion freeze-dried powder: development, characterization, and suppression of airway inflammation. Int J Biol Macromol 245. https://doi.org/10.1016/j.ijbiomac.2023.125493
75. Wang Y, Zheng Y, Zhang L, Wang Q, Zhang D (2013) Stability of nanosuspensions in drug delivery. J Control Release 172(3):1126–1141. Elsevier B.V. https://doi.org/10.1016/j.jconrel.2013.08.006
76. Wang Z, Dai B, Tang X, Che Z, Hu F, Shen C, Wu W, Shen B, Yuan H (2022) Fabrication and in vitro/vivo evaluation of drug nanocrystals self-stabilized Pickering emulsion for oral delivery of quercetin. Pharmaceutics 14(5). https://doi.org/10.3390/pharmaceutics14050897
77. Wilson RJ, Li Y, Yang G, Zhao CX (2022) Nanoemulsions for drug delivery. Particuology 64:85–97. https://doi.org/10.1016/j.partic.2021.05.009
78. Wu C, McClements DJ, He M, Zheng L, Tian T, Teng F, Li Y (2021) Preparation and characterization of okara nanocellulose fabricated using sonication or high-pressure homogenization treatments. Carbohydr Polym 255. https://doi.org/10.1016/j.carbpol.2020.117364
79. Wu M, He X, Feng D, Li H, Han D, Li Q, Zhao B, Li N, Liu T, Wang J (2023) The effect of high pressure homogenization on the structure of dual-protein and its emulsion functional properties. Foods 12(18). https://doi.org/10.3390/foods12183358
80. Wu Y, Wang D, Li Y (2014) Nanocrystals from solutions: catalysts. Chem Soc Rev 43(7):2112–2124. https://doi.org/10.1039/C3CS60221D
81. Xu X, Ray R, Gu Y, Ploehn HJ, Gearheart L, Raker K, Scrivens WA (2004) Electrophoretic analysis and purification of fluorescent single-walled carbon nanotube fragments. J Am Chem Soc 126(40):12736–12737. https://doi.org/10.1021/ja040082h
82. Yadav KS, Kale K (2020) High pressure homogenizer in pharmaceuticals: understanding its critical processing parameters and applications. J Pharm Innov 15(4):690–701. Springer. https://doi.org/10.1007/s12247-019-09413-4
83. Yi T, Liu C, Zhang J, Wang F, Wang J, Zhang J (2017) A new drug nanocrystal self-stabilized Pickering emulsion for oral delivery of silybin. Eur J Pharm Sci 96:420–427. https://doi.org/10.1016/j.ejps.2016.08.047
84. Yong AP, Islam MA, Hasan N (2017) The effect of pH and high-pressure homogenization on droplet size. Int J Eng Mater Manuf 2(4):110–122. https://doi.org/10.26776/ijemm.02.04.2017.05
85. Yue Y, Zhou C, French AD, Xia G, Han G, Wang Q, Wu Q (2012) Comparative properties of cellulose nano-crystals from native and mercerized cotton fibers. Cellulose 19(4):1173–1187. https://doi.org/10.1007/s10570-012-9714-4
86. Yukuyama MN, Kato ETM, de Araujo GLB, Löbenberg R, Monteiro LM, Lourenço FR, Bou-Chacra NA (2019) Olive oil nanoemulsion preparation using high-pressure homogenization and D-phase emulsification—a design space approach. J Drug Deliv Sci Technol 49:622–631. https://doi.org/10.1016/j.jddst.2018.12.029
87. Zhang J, Lv H, Jiang K, Gao Y (2011) Enhanced bioavailability after oral and pulmonary administration of baicalein nanocrystal. Int J Pharm 420(1):180–188. https://doi.org/10.1016/j.ijpharm.2011.08.023
88. Zong L, Li X, Wang H, Cao Y, Yin L, Li M, Wei Z, Chen D, Pu X, Han J (2017) Formulation and characterization of biocompatible and stable I.V. itraconazole nanosuspensions stabilized by a new stabilizer polyethylene glycol-poly(β-benzyl-L-aspartate) (PEG-PBLA). Int J Pharm 531(1):108–117. https://doi.org/10.1016/j.ijpharm.2017.08.082

Extrusion for Nanomaterial Production

Harshani Iresha

Abstract Nanomaterial production is evolving with advanced techniques which are sometimes combined with sub-processes to develop stable and useful nanomaterials that can be compatible with diverse applications. Extrusion is one of the main nanomaterial production methods modified, scaled-down, and combined to fabricate nanomaterials in solid lipid nanoparticles, nanoemulsions, nanoparticles, liposomes, nanofibers, etc. Few classifications of nanomaterial extrusion were identified referring to literature. Among them, the hot-melt extrusion method can be identified as the classic version of extrusion. It was compacted from the plastic/rubber/food extruders for nanomaterial fabrication but innovative with advanced technologies to accommodate the favorable and critical conditions for nanomaterial production. Closely analyzing the technical aspects, for a proper nanoproduct with desired properties multi-factors are directly affected not only for the hot-melt extrusion but also for other extrusion methods referred. Therefore, for a particular type of nanomaterial fabrication for a particular application, the desired extrusion conditions are challenging to generalize. However, the trends in extrusion for nanomaterial production are being hiked toward more cutting-edge technologies.

Keywords Hot-melt extrusion · Freeze-thawing extrusion · Vesicle extrusion · Precipitation extrusion · Membrane emulsification

1 Introduction

Extrusion is a processing technique that has been evolving in macro, micro even nano-scale material processing in a wide spectrum of applications such as plastic [1], rubber [21], food [2], pharmaceutical [11], etc. Continuous operation, compatibility for different materials with vivid properties (thermoset, thermoplastic, elastomers, low–high viscous, volatile-nonvolatile), range of processing parameters (temperature,

H. Iresha (✉)
Division of Sustainable Resources Engineering, Graduate School of Engineering, Hokkaido University, Sapporo, Hokkaido, Japan
e-mail: rdhiresha@gmail.com

N. M. Adassooriya et al. (eds.), *Industrial Scale Production of Nanoparticles*, Smart Nanomaterials Technology, https://doi.org/10.1007/978-981-96-9176-0_9

pressure, flowrates, etc.), producibility of intricate shapes and sizes, lower power consumption, easiness of modification of the process complying requirements made the extrusion a handy processing technique.

The evolution of extrusion in nanotechnology is impressive when referring to the advanced research available in current literature. Production of nanoparticles such as liposomes, nanoparticles, nanofibers, nano-emulsions, nanotubes, etc. has been investigated with different advanced features added with extrusion. The processes can be classified mainly as hot-melt extrusion, freeze–thaw extrusion, vesicle extrusion, precipitation extrusion, membrane emulsification, and biological membrane extrusion, based on the different mechanisms applied in each method and types of nanomaterials produced as the final output. Specialties of these classifications will be elaborated hereafter in the coming section.

2 Types of Extrusions in Nanomaterial Production

2.1 Hot-Melt Extrusion

Hot-melt extrusion (HME) in nanomaterial production is the mini version of the industrial large-scale extruder machines used for plastic processing. It consists of a stationary extruder barrel equipped with a feeder hopper/s, mortar-driven rotating screw, heating elements, and extruder die at the end of the extruder barrel which all are in scale-down version but analogs to plastic extrusion [5, 23]. HME has emerged with numerous plus points as a modern processing technology that is much more applicable in the pharmaceutical industry in developing molecular dispersions of active pharmaceutical ingredients (API). Some of the characteristics are the obtained nanoparticles can be time-controlled, modified, extended, and targeted for drug delivery. Further, the bioavailability of drugs and taste-masking of bitter pharmaceutical ingredients was incorporated into those solid dispersions making this method a robust technology for nanomaterial production [23].

Figure 1 illustrates the twin-screw hot melt extrusion process, while Fig. 2 depicts the individual steps of the extrusion process. Initially, raw materials are introduced into the extruder through a feeding hopper located at the back of the machine. Within the extruder barrel, a continuously rotating screw drives the materials forward. The extruder barrel is equipped with controllable heating elements. These elements maintain the barrel at room temperature at the feed point and gradually increase the temperature along the barrel, reaching levels suitable for melting the feed material near the die. The screw design incorporates an increasing screw diameter along its length, which reduces the barrel volume and compresses the material. This compression, combined with friction forces and the heat supplied by the barrel, causes the material to melt. Near the end of the barrel, the screw features a kneading (metering) zone to ensure thorough mixing of the molten material. Finally, the material is pressurized through an orifice or die and directed to the final treatment stage [1, 19, 23, 29].

Product properties can be obtained by changing or setting the parameters of extrusion such as temperature control all over the extruder barrel, screw configurations, rotation speed, feeding points with side feeds, pressure, die arrangement, additives, etc. [23, 28].

Extruders can have different arrangements depending on the production rate required, complexity of formulation, etc. Mainly, single-screw [28] and twin-screw [3, 17] arrangements are commonly available in industries and research. Length (L)/diameter (D) ratio of the screw is another important factor to be considered for an extruder which determines the melting and mixing performances of the feed materials. For both the single or twin-screw arrangements L/D ratio ranges from 20/1 to 40/1 (mm). In pharmaceutical applications, the screw diameter is greater than the 50–60 mm range. Another specialty of the extruder is, the extruder barrel with screw can be generally classified into three zones called feed, melting, and metering [28]. Figure 3 is an illustration of a twin-screw setup showing multiple zones mentioned here [3]. In each zone, screw diameter, pitch dimensions, and channel dimensions (Fig. 4) change over the total length of the screws to accommodate respective feeding

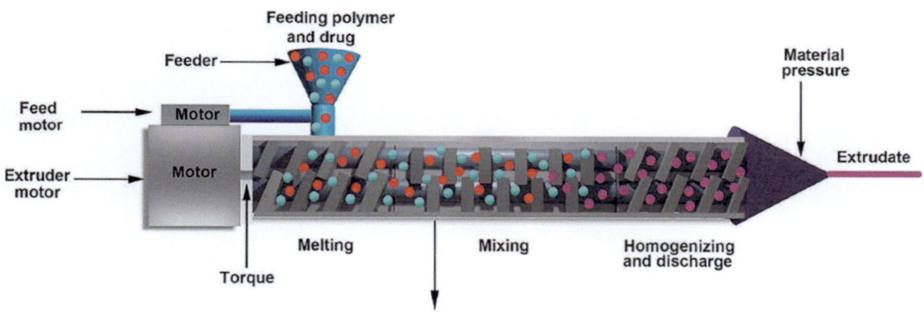

Fig. 1 Schematic representation of twin-screw HME [32]

Fig. 2 Main processing steps of HME

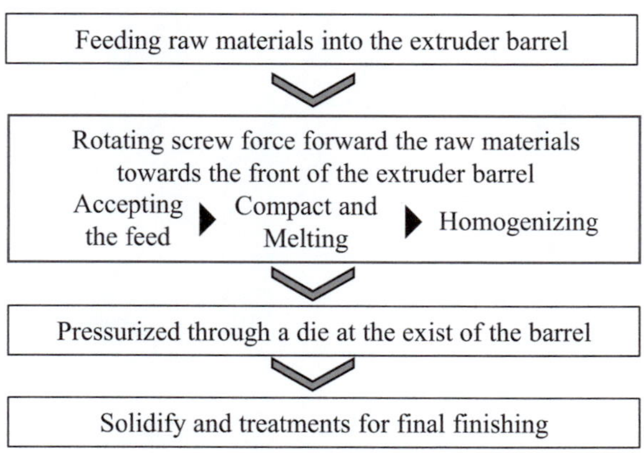

melting mixing performances corresponding to each zone and eventually accommodate the perfect mixture at the exit of the die. Considering the features and performances, twin-screw extruders can be considered a versatile selection that is applicable for a range of materials covering various applications [23].

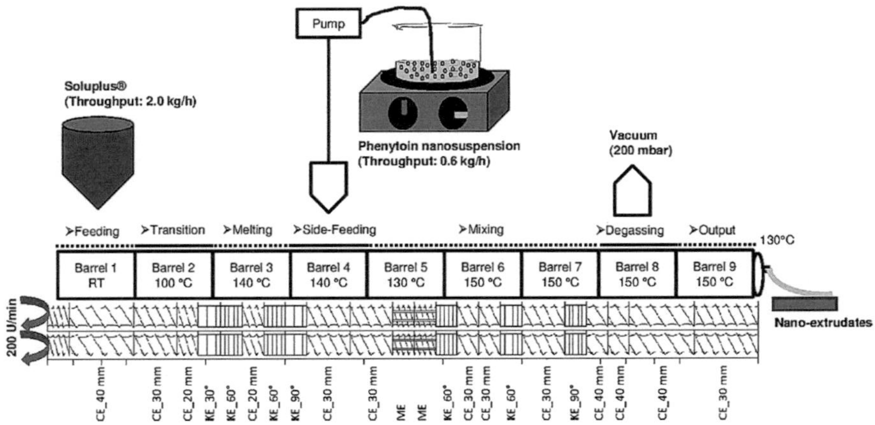

Fig. 3 Twin screw arrangement which shows different zones of the barrel, feeding points, and temperature variation over the barrel, twin screw arrangement with clarity view of screw geometry at each zone [3]

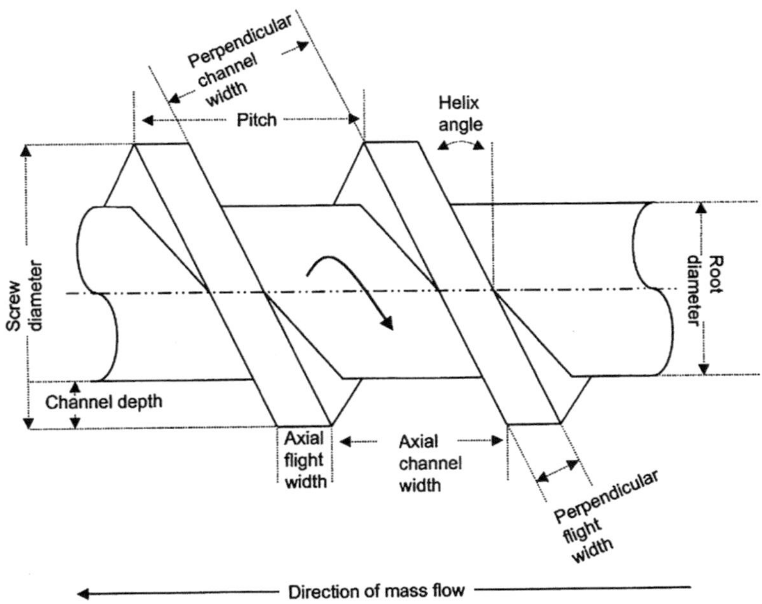

Fig. 4 Extrusion screw geometry [5]

Nanoparticle production through HME has been studied for decades. In recent history, Khinast et al. [20] reported a one-step process for solid nanoparticle of formulation from liquid stabilized TiO_2 nano-suspension using a combined twin-screw HME (screw speed 220 rpm) with devolatilization process (ZSK 18 twin screw extruder model, Coperion GmbH, Stuttgart, Germany) (nano extrusion, NANEX) [20]. There, in the combined HEM process, Soluplus® was used as the matrix polymer for formulation and fed at zone 1 at the beginning of the extruder barrel while the liquid stabilized nano-suspension was fed in the mid-way of the barrel (zone 5) via side feeding device after melting the polymer matrix. The solvent incorporated initially in liquid nano-suspension was removed by the devolatilization under vacuum at 80 mbar at zone 8. The temperature over the barrel gradually ranged from room temperature at zone 1, 100 °C at zone 2–150 °C at zone 10 at the extruder die. The extrudate was confirmed with proper embedding of TiO_2 nanoparticles in the matrix in a de-aggregated form with a size range from 200 to 400 nm range according to scanning electron microscopy (SEM) and energy dispersive X-ray spectroscopy (EDX) analysis which emphasizes the applicability of nanomaterial formulation using the twin-screw HME-devolatilization process [20].

The previously mentioned NANEX method was extended and scaled up by the research team to formulate a solid oral drug ensuring the improved solubility of the model drug phenytoin [3]. Here, a twin-screw extruder with a screw speed of 200 rpm (K-Tron twin-screw feeder K-CL-SFSKT20 model, K-Tron LLC, Switzerland) with devolatilization using a 200 mbar vacuum was used. Soluplus® was the matrix material for solid nanoparticle formulation which was fed at zone 1 while the phenytoin nanosuspension in Tween® 80 stabilizer was fed at zone 4 and solvent removal occurred at zone 8. This scale-up HME method again confirmed the phenytoin embedded in the de-aggregated form in the Soluplus® matrix with a mean nanocrystal size of 400 nm. The phenytoin release in vitro suggested controlled and enhanced release of the drug in solid nanoparticle formulation.

Another advancement of HME was studied by Patil et al. [28] for solid lipid nanoparticle (SLN) preparation by combining HME for melt-emulsification with high-pressure homogenization (HPH) for size reduction purposes [28]. Here, a co-rotating twin-screw extruder (11 mm Process 11, ThermoFisher Scientific, Karlsruhe, Germany) was used for the emulsification of Glyceryl behenate in aqueous Tween 80. The extruder barrel was maintained at temperatures above the melting point of lipids (varying from 75 to 150 °C throughout the barrel length). The study was carried out with altered experiment conditions by changing screw speed (240 or 160 rpm), the zone of aqueous stabilizer phase added to the barrel (3 or 4), changing the barrel temperature range, screw design (standard screw or modified screw arrangement), and lipid concentration (6 or 12% w/w). The outlet from the extruder (hot pre-emulsion) was fed to the HPH (Avastin Emulsiflex C5, Canada) via properly insulated tubing connected to the HME die and sample inlet at HPH where the HPH was operated in constant conditions at 75 °C and 1000 bar for all the experiments. After size reduction at HPH, emulsions were cooled at room temperature to get SLN. After analyzing SLNs, all the experiment conditions could produce SLN in nanoscale in between 80 and 850 nm whilst lipid concentration, residence time, and screw design played a major role in influencing the particle size of the SLNs.

The incorporation of biodegradable cassava starch for nanocomposite film formation with titanium dioxide nanoparticles (TiO_2NP) was carried out using extrusion at different screw speeds of 80 and 120 rpm [17]. A corotating twin screw extruder screw diameter of 16 mm and L/D = 40 (Nanjing Kerke Extrusion Equipment Co., Ltd. Jiangsu, China) was used for the research. The extrudate from the extruder was stabilized in a desiccator (room temperature, 57% relative humidity). Then the extrudate was cut into nearly 4 g of smaller pieces, sandwiched them in between Teflon sheets, fed in to the thermostatic hydraulic press to form nanocomposite films. Initially, the system was preheated heated at 130 °C for 5 min. Next, the pressure was elevated to 1 kPa and sustained for 15 min, then, further increased to 5 kPa for 15 min and finally the temperature was lowered to 40 °C at 5 kPa. At 120 rpm, the completely processed and homogenously distributed nanoparticles were observed in the film morphology which the reason could be the high mechanical forces encountered at higher speed made the perfect mixing at extrusion. Collectively, the HME is a promising advanced tool for nanomaterial preparation which can accommodate a wide spectrum of the requirements of the products [31]. In brief, HME as a process to fabricate nanoparticles the following factors are advantageous.

- Suitable for thermally stable materials plus considerably miscible components with a specific polymer
- Good as a solvent free process
- Obtain homogenous dispersion of nanoparticles
- Excellent uniformity, consistency and repeatability of the product geometry, particle size, morphology, composition and production rates
- Efficient, cost-effective and commercially viable

Despite its positive influences of HME, it is still challenging mostly in pharmaceutical industry due to heat sensitivity of drugs. Also, possibility of die clogging due to small dimensions of the equipment, difficulty in selecting a suitable polymer material compatible with the dispersion material specially in food and pharmaceutical industry, requirement of advanced technologies for proper control of processing parameters are some of the major challenges behind this handy method for nanomaterial production.

2.2 *Vesicle Extrusion and Freeze-Thawing Extrusion*

Other major versions of extrusion are vesicle extrusion and Freeze–thaw extrusion which are used to produce vesicles via nanoporous membrane structures and are considered one of the most used methods to prepare liposomes. Olson et al. [27] introduced the extrusion for multilamellar vesicle (MLVs) formation with a nanoporous membrane which reveals sound information as a prospective method in history. There, sequential extrusion with a series of nanoporous membranes with sizes of 3, 1, 0.8,

0.6, 0.4, and 0.2 μm in descending order was used for MLVs preparation. The results of negative stain and freeze-fracture electron micrographs reveal the sequential extrusion through a series of nanoporous membranes with reduced size caused to decrease in the mean vesicle diameter while creating a more homogenous size distribution of vesicles [27]. The method was then combined and applied for liposome fabrication in later research in line [15, 25].

The fact behind the title 'freeze-thawing' is the method of material preparation before the extrusion. The lipids (natural/ synthetic) are initially made into a dried lipid film using the evaporation method or inert gas purging. Then the lipid film is hydrated and simultaneously trapped with the respective solute immersed and vortexed in an aqueous solution medium. This is the step where multilamellar vesicles (MLVs) are formed, i.e., self-assembled, with a size range from 0.5 to 10 μm. These resultant MLVs undergo repeated freeze–thaw cycles depending on the requirement to optimize the lipid molecule rearrangement in lipid bilayers [12]. More specifically, ice crystal formation occurs during the freezing step will lead to the expansion and then rupture of liposomal phospholipid bilayers which eventually enhances the aqueous solute to lipid ratio, i.e., enhanced trapping efficiency of vesicles [6, 15]. The MLVs finally extrude via a single or stacked membrane/s in a nanoporous channel structure to produce unilamellar liposomes [30]. However, the routine, MLV preparation, freeze–thaw of MLV for several cycles, extrusion can also be performed by shuffling and combining the steps. Chapman et al. [7] reported the process steps as, preparation of MLV, 3 cycles of freeze–thaw-vortex, 2 cycle of extrusion followed by additional 6 freeze–thaw cycles and 8 extrusion steps to get the required large unilamellar vesicles by extrusion technique (LUVET) [7]. The reason behind the formation of downsized unilamellar vesicles is the larger sized MLVs which their size is greater than the size of nanopores of the membrane are forced to pass through the nanoporous membrane triggering rupture of the MLVs in the nanoporous channels creating unilamellar liposome formation continuously. The formed unilamellar liposomes are exit at the end of the nanoporous membrane due to the continuous flow of the extrusion. Here, to obtain liposomes with a uniform liposome size and in the required size distribution, several extrusion cycles are suggested to perform. Another key factor to be maintained is extrusion must be carried out at temperatures greater than the gel-liquid crystallization transition temperature of the lipid. Table 1 summarizes the research conducted for vesicle extrusion and combined freeze–thaw extrusion for vesicle formation. These studies have progressed to identify the most significant features that are affecting sophisticated nanomaterial, i.e., liposome fabrication and freeze–thaw extrusion became a widely used and convenient method of preparing phospholipids vesicles [7].

At the initial time, a problem associated with using organic solvents or detergent as lipid solubilizers in unilamellar vesicle formation was addressed by Hope et al. [15] incorporating repeated extrusion with freeze–thaw method to prepare MLVs in the absence of lipid solubilizers. The suitability was proved with high lipid concentrations employed with high trapping efficiency without incorporating lipid solubilizers but with the freeze–thaw method. This study was first mentioned as a protocol for freeze–thaw extrusion for liposome preparation [15].

Table 1 Summary of vesicle extrusion and combined freeze–thaw extrusion for MLV formation

MLVs composition	Freeze-thawing conditions and no. of cycles and/or extrusion conditions	Size of porous nanoporous membrane used for extrusion	Median diameter of nanomaterials obtained by extrusion	Additional remarks	References
Phosphatidylserine (PD):phosphatidylcholine (PC):cholesterol at molar ratios of 1:4:5	• Freeze–thaw: none • Extrusion pressure: 50 lb/in^2	Non-extruded	1.32 μm[a]	[a]Measured by negative stain electron microscopy and MLVs prepared by vortex mix method	[27]
		1.0 μm	0.85 μm[a]		
		0.8 μm	0.62 μm[a]		
		0.6 μm	0.44 μm[a]		
		0.4 μm	0.38 μm[a]		
		0.2 μm	0.26 μm[a]		
Egg PC	• Freezing in liquid nitrogen followed by thawing for 2 cycles • Extrusion pressure: 100–500 lb/in^2) • Extrusion flow rate: 20–60 ml/min • 10 extrusion cycles	Two stacked Polycarbonate membranes (100 nm pore size)	77 ± 16 nm	Mean trapped volume (MTV): 2.2 ± 0.5 μl/μmol of phospholipid	[15]
Soya Phosphatidylcholine (Soya PC)			94 ± 26 nm	MTV: 2.2 ± 0.1 μl/μmol of phospholipid	
Egg PC	• Freezing in liquid nitrogen followed by thawing in a 40 °C water bath for 5 cycles • Extrusion pressure: nitrogen pressure up to 800 lb/in^2 • Two stacked polycarbonate membrane • 20 times extrusion passes	400 nm	243 ± 91 nm	MTV: ≈ 3.5 μl/μmol of phospholipid	[25]
		200 nm	151 ± 36 nm	MTV: ≈ 2 μl/μmol of phospholipid	
		100 nm	103 ± 20 nm	MTV: ≈ 1.5 μl/μmol of phospholipid	

(continued)

Table 1 (continued)

MLVs composition	Freeze-thawing conditions and no. of cycles and/or extrusion conditions	Size of porous nanoporous membrane used for extrusion	Median diameter of nanomaterials obtained by extrusion	Additional remarks	References
		50 nm	68 ± 19 nm	MTV: ≈ 1 μl/μmol of phospholipid	
		30 nm	56 ± 17 nm		
ICAM-1-doxorubicin-immunoliposomes (ICAM-Dox-LPs)	• 10 freeze–thawing cycles • Extruded via a Northern Lipids Extruder	Polycarbonate membrane with 100 nm porous size	105 ± 31 nm	The lipid formulation used was DOPC: DSPE-PEGCOOH in 95:5 mol: mol	[14]
Immunoglobulin G-Dox-LPs (IgG-Dox-LPs)			101 ± 24 nm		
Soy phosphatidylcholine (SPC):cholesterol (CH):soy phosphatidylglycerol (SPG) = 6.25:3:0.75 molar ratio	• Freeze in methanol/ice at − 70 °C followed by thawing at 70 °C for 3 cycles • Polycarbonate membranes were used Continuous high-pressure extruder (CE) (Dispex Maximtor® model)	Polycarbonate with pore sizes (5.0, 1.0, 0.4, 0.2, 0.1 and 0.05 μm)	≈ 100 nm	Volume-based mean diameter was measured by photon correlation spectroscopy Extrusion was carried out sequentially through several membranes in a reduced direction of porous sizes	[4]
	Discontinuous extruder (DE) (Avestin Liposo-Fast™-50 model)	Polycarbonate with pore sizes (5.0, 1.0, 0.4 μm)	≈ 200 nm		

(continued)

Table 1 (continued)

MLVs composition	Freeze-thawing conditions and no. of cycles and/or extrusion conditions		Size of porous nanoporous membrane used for extrusion	Median diameter of nanomaterials obtained by extrusion	Additional remarks	References
			Polycarbonate with pore sizes (5.0, 1.0, 0.4, 0.2, 0.1 and 0.05 μm) Plus Drain disk (polypropylene 45 μm pore size)	≈ 100 nm		
			Polycarbonate with pore sizes (5.0, 1.0, 0.4 μm) Plus Drain disk (polypropylene 45 μm pore size)	≈ 300 nm		

In another study available inline, the same research group [25] suggested higher lipid concentrations in MLVs caused to increase in the number of extrusion cycles when using nanoporous membranes with 100, 50, and 30 nm scales to observe unilamellar vesicle characteristics, without undergoing freeze–thaw cycles in MLV preparation before the extrusion step. When MLVs endure freeze-thawing, at the number of extrusions cycles of 2 the unilamellar formation was facilitated due to the larger intermolecular spacing and larger trapped volumes of MLVs after freeze-thawing which easily downsized through extrusion [25].

Another interesting factor reported was the solute entrapment ratio of phospholipid vesicle prepared by the freeze–thaw extrusion method was > 1 [7]. In other words, the internal solute concentration of vesicles is higher than the solute concentration of the external surroundings this fact was converging for the future applications of vesicles in enhanced drug encapsulation etc. Chapman et al. [7] have proposed the mechanism of this concentrating phenomenon as the freezing-promoted expansion of the ice phase excluding the solute made the enhanced concentration in the interior thus reducing solute concentration at the external medium during the thawing-driven membrane binding which eventually created osmosis imbalance [7].

Berger et al. [4] has investigated the effect of continuous and discontinuous extrusion on liposome characteristics. There, the high-pressure continuous extruder produced comparatively smaller liposomes than the discontinuous extruder reasoning high-pressure drop and higher flow rates associated with high-pressure continuous extruder. Nevertheless, the study was supplemented with sequential extrusion through membranes with pore sizes in descending order from 5 to 0.05 μm exhibiting a decrease of mean liposome diameter via a series of filtration through decreased pore size filters for both continuous and discontinuous extrusion methods [4].

Freeze–thaw extrusion technique in targeted drug delivery was studied by Guo et al. [14] to reduce the off-target drug toxicity due to the poor recognition ability of the exact tumor locality [14]. There, the potential targets conjugated-drug encapsulated-immunoliposomes (LPs) were prepared using freeze–thaw extrusion via a nanoporous membrane with 100 nm porous size. ICAM-1 and immunoglobulin G (IgG) were incorporated as antibodies which can identify the target metastatic melanoma, the most fatal skin cancer cells. The ICAM-Dox-LPs and IgG-Dox-LPs were prepared in < 200 nm diameter using the freeze–thaw extrusion enabling them to accumulate on tumor sites leaking through pores in the tumor vasculature with greater permeability and retention ability plus extended in vivo circulation time [24]. This shows the involvement of extrusion in nanomaterial preparation and applicability in the most advanced fields of engineering plus medicine. Extended research in the literature concreting the fact of its potential in nanomaterials production is further available [9, 16, 18, 33].

2.3 Precipitation Extrusion

In the reason history, the precipitation extrusion was developed as a one-step synthesis of nanoparticles or nanofibers via nanoporous membrane [12]. In a clear view, feed solution with desired solutes is forced through a nanoporous membrane welcoming the extrudate at the end of the membrane by a solution in which the solutes are insoluble in it leading to rapid precipitate formation. The continuous flow results in carrying away the precipitates from the interface of the membrane and receiver solution allowing further solidification in the bulk of the receiver solution. The precipitation mechanism can be achieved by using antisolvent, pH-induced protonation/deprotonation, temperature changes, reaction mechanism, and self-assembly of solutes [12, 13]. The nanomaterial generation can be controlled via changing the speed through the membrane and the nature of the precipitation mechanism in this dynamic process.

Guo et al. [13] first reported this method to fabricate nanoparticles of chitosan which is a biodegradable polysaccharide polymer as the model material for the research [13]. There, the chitosan was placed in a low pH solution (chitosan in a protonated state and thus water soluble) at the feed side while sent to a high pH (deprotonation of chitosan will occur becoming water insoluble) aqueous solution as the receiver solution. The feed solution of chitosan was forced through nanoporous membranes of track-edge polycarbonate (PCTE) and anodized aluminum oxide (AAO) where their porous sizes were 10 nm (for the entire 6 μm thickness) and 20 nm (2 μm thickness at the receiver side) + 200 nm (through 58 μm thickness at the feed side), respectively. The driving force of the feed to be delivered through the membranes was the gravity flow due to two levels of height of the feed side and receiver side. The results reveal the precipitation of chitosan nanodroplets at the vicinity of the receiver-membrane interface was carried out and dispersed in the bulk due to the gravity flow. The particle size of the chitosan nanodroplets has increased exponentially with increased flow rate of gravity-driven extrusion in the range of 7.2–32 μl/min cm^2. Further, at higher flowrates hollow nanotube formation and solid nanowire formation were identified while the narrow particle size distribution was obtained at the lowest flow rate. Moreover, the viscosity effect of the feed solution affected the particle size of the chitosan nanodroplets formed. This research ensured extrusion for precipitation as a promising tool for nanomaterial production projecting more developments to the method.

The former research team later reported the formation of mineralized nanofibers from collagen fibrils containing calcium phosphate using the precipitation extrusion mechanism [22]. Here, the extrusion setup was created with track-etch polycarbonate (PCTE) membrane with pore diameters range of 50 nm to 1 μm, and the flow of feed solution was induced via 250 mbar gauge pressure created by connecting a compressed air outlet carrying a pressure reduction valve at the feed section. Here, the pH-induced hydrophilicity and hydrophobicity of collagen fibrils were utilized as the mechanism of precipitation thus the feed solution was maintained at lower pH (pH = 3) while the receiver solution was kept at higher pH values (pH =

11). The study demonstrates the formation of discrete collagen nanofibrils while controllability and the uniformity of the nanofibril diameter through the selection of nanoporous membranes with different diameters. Nonetheless, less than 200 nm diameter in the membrane caused blocking of the nanofiber formation. Mineralized collagen incorporating calcium phosphate into the collagen due to the self-assemble ability was advantageous thus the fabrication of mineralized nanofibers as one-step process in the absence of any organic solvent was achieved to be scaleup for bone drafting applications [22].

Further, the precipitation extrusion method for nanomaterial fabrication brings the advantages of shorter production time, less or no additional solvents, simple processing steps, ability to prepare both nanoparticles and nanofibers, ease to scale-up, uniformity of diameter of nanomaterials, and excellent reproducibility. However, the clogging at the receiver solution and exit of the membrane is possible if the precipitation is too fast while insufficient time for proper solidification may lead to aggregate formation if the precipitation is too slow.

2.4 Membrane Emulsification and Biological Membrane Extrusion

Membrane emulsification (ME) and biological membrane extrusions (BMEx) can be considered as two other extrusion methods for nanomaterial preparation. Both ME and BMEx are connected to use nanoporous membrane [12].

There are two types of ME methods, direct ME and premix ME. Direct ME uses a dispersed liquid phase which is forced to permeate through a nanoporous membrane into another immiscible liquid phase under a continuous phase resulting in insoluble droplets of dispersed phase in the solution. Here downsizing of droplets is not happening [12]. Direct ME is widely used in drug delivery, food, and cosmetic industries which it gives advantages of better control of droplet size and distribution, mildness of the procedure avoiding extreme energy supply for size reduction (e.g. high-pressure homogenizer), Lower energy consumption and easy to scale up. Further, lower emulsion production, difficulty in preparing high-viscosity emulsion droplets, and difficulty in achieving uniformity, unless there is a highly uniform nanoporous membrane, were the drawbacks of the direct ME [8, 12, 26, 34].

Differently, for the premix ME method, coarse emulsion droplets are first prepared by mixing two immiscible phases and forces through a nanoporous membrane to be downsized into uniform micro/nanoscale droplets [12]. This method is comparable to the vesicle extrusion method which the pre-emulsion with larger droplets is extruded via a nanoporous membrane several times to obtain emulsion droplets of nanoscale [10].

Cell membrane-cloaked nanoparticles are prepared via nanoporous membrane extrusion which is similar to vesicle extrusion yet a modified process called the BMEx. Initially, cell membranes from target cells are harvested then the extracted membranes are extruded through nanoporous membranes to form uniform cell membrane-derived vesicles (shells), and these shells are extruded again with solid nanoparticles (core) several times. At this step, these core and the shell particles fuse in the nanopore channel and fuse to form a core–shell structure [12, 35].

3 Conclusion

Extrusion for nanomaterial fabrication has been evolving for decades and has been developed to a state where different forms of nanomaterials can be formulated for a wide spectrum of applications. Hot-melt extrusion was identified as the classic version of extrusion which is compatible with more applications compared to other methods of extrusion. For example, vesicle extrusion and freeze–thaw extrusion methods are somewhat limited to liposome-based nanomaterials. Nevertheless, besides the incorporated advanced theories, mechanisms, and technologies for a particular formulation of nanomaterials extrusion conditions cannot be generalized due to the complexity of nanomaterial behaviors.

References

1. Abeykoon C, McMillan A, Nguyen BK (2021) Energy efficiency in extrusion-related polymer processing: a review of state of the art and potential efficiency improvements. Renew Sustain Energy Rev 147. https://doi.org/10.1016/j.rser.2021.111219
2. Abilmazhinov Y, Bekeshova G, Nesterenko A, Dibrova Z, Ermolaev V, Ponomarev E, Vlasova V (2023) A review on the improvement of extruded food processing equipment: extrusion cooking in food processing. Food Sci Technol 43. https://doi.org/10.5327/fst.80621
3. Baumgartner R, Eitzlmayr A, Matsko N, Tetyczka C, Khinast J, Roblegg E (2014) Nano-extrusion: a promising tool for continuous manufacturing of solid nano-formulations. Int J Pharm 477(1):1–11. https://doi.org/10.1016/j.ijpharm.2014.10.008
4. Berger N, Sachse A, Bender J, Schubert R, Brandl M (2001) Filter extrusion of liposomes using different devices: comparison of liposome size, encapsulation efficiency, and process characteristics. Int J Pharm 223(1–2):55–68. https://doi.org/10.1016/S0378-5173(01)00721-9
5. Breitenbach J (2002) Melt extrusion: from process to drug delivery technology. Eur J Pharm Biopharm 54(2):107–117. https://doi.org/10.1016/S0939-6411(02)00061-9
6. Castile JD, Taylor KMG (1999) Factors affecting the size distribution of liposomes produced by freeze-thaw extrusion. Int J Pharm 188(1):87–95. https://doi.org/10.1016/S0378-5173(99)00207-0
7. Chapman CJ, Erdahl WE, Taylor RW, Pfeiffer DR (1991) Effects of solute concentration on the entrapment of solutes in phospholipid vesicles prepared by freeze-thaw extrusion. In: Chemistry and physics of lipids, vol 60
8. Charcosset C, Limayem I, Fessi H (2004) The membrane emulsification process—a review. J Chem Technol Biotechnol 79(3):209–218. https://doi.org/10.1002/jctb.969
9. De Jesus MB, Radaic A, Zuhorn IS, De Paula E (2013) Microemulsion extrusion technique: a new method to produce lipid nanoparticles. J Nanopart Res 15(10). https://doi.org/10.1007/s11051-013-1960-3

10. El-Hawari L, Bunjes H (2021) Premix membrane emulsification: preparation and stability of medium-chain triglyceride emulsions with droplet sizes below 100 nm. Molecules 26(19). https://doi.org/10.3390/molecules26196029
11. Gallas M, Boulet P, de Margerie V (2023) Extrusion for pharma applications: an update. SPE Polym 4(1):16–23. https://doi.org/10.1002/pls2.10081
12. Guo P, Huang J, Zhao Y, Martin CR, Zare RN, Moses MA (2018) Nanomaterial preparation by extrusion through nanoporous membranes. Small 14(18). https://doi.org/10.1002/smll.201703493
13. Guo P, Martin CR, Zhao Y, Ge J, Zare RN (2010) General method for producing organic nanoparticles using nanoporous membranes. Nano Lett 10(6):2202–2206. https://doi.org/10.1021/nl101057d
14. Guo P, Yang J, Bielenberg DR, Dillon D, Zurakowski D, Moses MA, Auguste DT (2017) A quantitative method for screening and identifying molecular targets for nanomedicine. J Control Release 263:57–67. https://doi.org/10.1016/j.jconrel.2017.03.030
15. Hope MJ, Bally MB, Webb G, Cullis PR (1985) Production of large unilamellar vesicles by a rapid extrusion procedure. Characterization of size distribution, trapped volume and ability to maintain a membrane potential. Biochimica et Biophysica Acta (BBA) Biomembranes 812:55–65. https://doi.org/10.1016/0005-2736(85)90521-8
16. Hope MJ, Nayar R, Mayer LD, Cullis PR (2002) Reduction of liposome size and preparation of unilamellar vesicles by extrusion techniques. In: Liposome technology, vol 1. CRC Press, Inc., pp 123–139
17. Iacovone C, Yulita F, Cerini D, Peña D, Candal R, Goyanes S, Pietrasanta LI, Guz L, Famá L (2023) Effect of TiO_2 nanoparticles and extrusion process on the physicochemical properties of biodegradable and active cassava starch nanocomposites. Polymers 15(3):535. https://doi.org/10.3390/polym15030535
18. Jousma H, Talsma H, Spies F, Joosten JGH, Junginger HE, Crommelin DJA (1987) Characterization of liposomes. The influence of extrusion of multilamellar vesicles through polycarbonate membranes on particle size, particle size distribution and number of bilayers. Int J Pharm 35(3):263–274. https://doi.org/10.1016/0378-5173(87)90139-6
19. Khairnar SV, Pagare P, Thakre A, Nambiar AR, Junnuthula V, Abraham MC, Kolimi P, Nyavanandi D, Dyawanapelly S (2022) Review on the scale-up methods for the preparation of solid lipid nanoparticles. Pharmaceutics 14(9):1886. https://doi.org/10.3390/pharmaceutics14091886
20. Khinast J, Baumgartner R, Roblegg E (2013) Nano-extrusion: a one-step process for manufacturing of solid nanoparticle formulations directly from the liquid phase. AAPS PharmSciTech 14(2):601–604. https://doi.org/10.1208/s12249-013-9946-0
21. Lukas M, Leineweber S, Reitz B, Overmeyer L (2024) Real-time temperature control in rubber extrusion lines: a neural network approach. Int J Adv Manuf Technol 133(11–12):5233–5241. https://doi.org/10.1007/s00170-024-14061-1
22. Maas M, Guo P, Keeney M, Yang F, Hsu TM, Fuller GG, Martin CR, Zare RN (2011) Preparation of mineralized nanofibers: collagen fibrils containing calcium phosphate. Nano Lett 11(3):1383–1388. https://doi.org/10.1021/nl200116d
23. Maniruzzaman M, Boateng JS, Snowden MJ, Douroumis D (2012) A review of hot-melt extrusion: process technology to pharmaceutical products. ISRN Pharm 2012:1–9. https://doi.org/10.5402/2012/436763
24. Maruyama K (2011) Intracellular targeting delivery of liposomal drugs to solid tumors based on EPR effects. Adv Drug Deliv Rev 63(3):161–169. https://doi.org/10.1016/j.addr.2010.09.003
25. Mayer LD, Hope MJ, Cullis PR (1986) Vesicles of variable sizes produced by a rapid extrusion procedure. Biochimica et Biophysica Acta 858:161–168. https://doi.org/10.1016/0005-2736(86)90302-0
26. Nakashima T, Shimizu M, Kukizaki M (2000) Particle control of emulsion by membrane emulsification and its applications. Adv Drug Deliv Rev 45(1):47–56. https://doi.org/10.1016/S0169-409X(00)00099-5

27. Olson F, Hunt C, Szoka F, Vail W, Papahadjopoulos D (1979) Preparation of liposomes of defined size distribution by extrusion through polycarbonate membranes. Biochimica et Biophysica Acta (BBA) Biomembranes 557(1):9–23. https://www.sciencedirect.com/science/article/pii/0005273679900853
28. Patil H, Kulkarni V, Majumdar S, Repka MA (2014) Continuous manufacturing of solid lipid nanoparticles by hot melt extrusion. Int J Pharm 471(1–2):153–156. https://doi.org/10.1016/j.ijpharm.2014.05.024
29. Prabha K, Ghosh P, Abdullah S, Joseph RM, Krishnan R, Rana SS, Pradhan RC (2021) Recent development, challenges, and prospects of extrusion technology. Future Foods 3. https://doi.org/10.1016/j.fufo.2021.100019
30. Samad A, Sultana Y, Aqil M (2007) Liposomal drug delivery systems: an update review. Curr Drug Deliv 4(4):297–305. https://doi.org/10.2174/156720107782151269
31. Song D, Thio YS, Deng Y (2011) Starch nanoparticle formation via reactive extrusion and related mechanism study. Carbohyd Polym 85(1):208–214. https://doi.org/10.1016/j.carbpol.2011.02.016
32. Tambe S, Jain D, Agarwal Y, Amin P (2021) Hot-melt extrusion: highlighting recent advances in pharmaceutical applications. J Drug Deliv Sci Technol 63. https://doi.org/10.1016/j.jddst.2021.102452
33. Turanek J (1994) Fast-protein liquid chromatography system as a tool for liposome preparation by the extrusion procedure. Anal Biochem 218(2):352–357. https://doi.org/10.1006/abio.1994.1191
34. Vladisavljević GT, Williams RA (2005) Recent developments in manufacturing emulsions and particulate products using membranes. Adv Colloids Interface Sci 113(1):1–20. https://doi.org/10.1016/j.cis.2004.10.002
35. Wen Y, Fu Q, Soliwoda A, Zhang S, Zheng M, Mao W, Wan Y (2022) Cell-derived nanovesicles prepared by membrane extrusion are good substitutes for natural extracellular vesicles. Extracell Vesicle 1. https://doi.org/10.1016/j.vesic.2022.100004

Sol–Gel and Solvothermal Methods for Nanoparticles Productions

Piumika Yapa and Imalka Munaweera

Abstract The synthesis of nanoparticles has received great attention owing to their superior physical and chemical characteristics, as well as versatility. In various fabrication methods, sol–gel, and solvothermal methods are important for synthesizing high quality nanoparticles. This process involves the formation of a colloidal solution that is known as sol and conversion of this sol type into gel phase through chemical processes such as hydrolysis and condensation of metal alkoxides or metal salts. It permits the control of both the particle size as well as the composition, and it is applied to produce metal oxides and mixed-metal oxides with homogeneous characteristics. On the other hand, solvothermal is the process in which the treatment is carried out at high pressure and/or high temperature of solvent to form nucleus and grow the nanoparticles. This method offers insolubility of solvents that can be chosen at will, as well as the synthesis of various materials, including metals, semiconductors, and hybrid nanoparticles. Solvothermal process is capable to provide nanoparticles with precisely defined morphologies and dimensions together with high crystallinity. This chapter discusses on the relative benefits depending on the desired physical and chemical characteristics of the material and the intended application, thus qualifying the two methods as valuable tools in nanomaterials fabrication.

Keywords Characterization · Nanoparticles · Sol–gel · Solvothermal · Synthesis

1 Introduction

Incorporated in the new economy age, nanoparticles are the tiny building blocks that introduced the elements of precision, engineering, and material science [102]. These are particles that are smaller, normally below a hundred nanometers, and possess characteristics that would be extremely difficult, if not impossible, to achieve when using usual materials. Sol–gel and solvothermal synthesis techniques are the most

P. Yapa · I. Munaweera (✉)
Department of Chemistry, Faculty of Applied Sciences, University of Sri Jayewardenepura, Nugegoda, Sri Lanka
e-mail: imalka@sjp.ac.lk

N. M. Adassooriya et al. (eds.), *Industrial Scale Production of Nanoparticles*, Smart Nanomaterials Technology, https://doi.org/10.1007/978-981-96-9176-0_10

innovative interdisciplinary fields linking chemistry with many scientific and technical fields [123]. In material science, these techniques are essential for the creation of high-performance ceramics, nanocomposites, and functional coatings for catalysis, optical, and electronic applications [142]. In biomedical engineering, sol–gel-derived nanoparticles work on drug delivery systems and biomedical imaging contrast agents because of excellent control over the release kinetics of drugs and targeting selectivity [126]. In addition, solvothermal synthesis is useful in fabricating efficient catalysts for pollution control and renewable energy devices in environmental science which demonstrates sustainable technologies [68]. Some of the areas that have been influenced even include sensor technology since the controlled synthesis of nanocomposites improves the working of sensors for use in environmental monitoring and diagnosis. In these disciplines, sol–gel, and solvothermal processes remain essential techniques for creating nanostructures that make materials functional in ways previously unimagined to help in solving some of the world's pressing problems [3].

However, at the heart of these capabilities are some of the skills of making nanoparticles which can be said to be partly alchemy, as it is science. Out of the multitude of techniques available now, sol–gel and solvothermal are cornerstones of sol practice for the modern-day nanotechnologist [132]. Other than that, these methods aid in achieving absolute controlling of the size and form as well as the chemical nature of the nanoparticles and impose on them some functions needed for the intended uses. The synthesis methods like sol–gel and solvothermal have their roots in dissimilar chemical pathways that make it possible for the synthesis of nanoparticles that can be correctly predicted. The general method of synthesis of sol–gel involves the process of hydrolysis and condensation reaction of the precursor molecules such as metal alkoxides or metal salts in a solvent such as alcohol or water among others [21].

Sol–gel process, which is an excellent amalgamation of chemistry and colloid science helps in coagulation or condensation of solid materials from the liquid solutions through polymerization or gelation [97]. This method offers inestimable flexibility because it offers way in which we can synthesize diversified complex oxides such as silica, titania, and alumina nanoparticles [150]. Since there is a relatively low process temperature, sol–gel is useful in the dispersing of the chemicals used, the creation of intricate structures as well as the high degree of purity in the material made. This ranges from high-performance ceramics to advanced biomedical equipment, hence it indicates how much impact can be made by optimized nanoparticles [151].

Co-solvent synthesis with sol–gel is the solvothermal process, in which the chemical reactions occur in the solvent medium at high temperature and pressure to obtain desired nanoparticles with very small size distribution [45]. This immense methodological vehicle helps build different materials including metals, metal oxides, and organic/inorganic blends [7]. Concentration, temperature, pressure, and other reagent control are accurate techniques that are employed to get the required shape and crystalline form of the nanoparticle [119]. This refinement not only optimizes material characteristics but also extends the applications of these materials to fields such as energy storage, environmental remediation, and next-generation sensing systems [4].

The interest in sol–gel and solvothermal synthesis strategies comes from the possibility of creating a nanoparticle and managing its shape and dimensions at the atomic level, as well as from adaptability to the new requests addressed by technologies [12]. These techniques help scientists to steer the nanoparticles to take specific surface chemistries, get the specific size of the particle for drug delivery or to have the particle with the right activities for its intended industrial application [72]. Such a degree of control over nanoparticles can be regarded as a revolution concerning the synthesis and design of materials, so its potential for an enhanced future can be expected in the spheres including agriculture, computation, and technology, etc. [172]. Consequently, the electrochemical sol–gel and solvothermal methods are concrete examples of the discussed chemical methods with the use of the engineering approach in the synthesis of various types of nanomaterials for corresponding applications in electronics, catalysis, biomedical, and environmental technologies [33, 80, 81].

This chapter hence starts with an assessment of sol–gel and solvothermal techniques for the synthesis of nanoparticles, the recent past developments in the methods and literature concerning the techniques. Moreover, it describes the major processes of nanoparticle formation and their chemistry, especially the selection of the precursors, solvents, and the trends in nucleation. Additionally, the chapter also describes the current advancement in computational modeling to improve the synthesis process of nanoparticles by sol–gel and solvothermal techniques [176]. Real-life examples from the fields of science will show that the application of developed and targeted nanoparticles enables one to change the world using some of the biggest global challenges while at the same time building new possibilities for technology. In summary, the method of depicting nanoparticles through sol–gel and solvothermal processes can be understood as a blend of pure science and applied science. These methodologies remain essential for addressing the challenges of the twenty-first century and advancing the frontiers of nanotechnology. They play a crucial role in improving welfare, enhancing overall health, and establishing better healthcare facilities [151]. Figure 1 shows an overview of the types of processes that can be done with the sol–gel method and the products of each process.

2 Properties of Sol–Gel and Solvothermal Technology

2.1 Chemical Properties

Sol–gel processing is a well-established chemical procedure of preparing materials, especially metal oxides from a liquid-phase system referred to as sol and forms a solid gel network. Starting with the hydrolysis reactions, metal alkoxide, which is tetraethyl orthosilicate, reacts with water to form a metal hydroxide and an alcohol. The balanced chemical equation for this reaction is as follows [166]:

$$M(OR)_n + nH_2O \rightarrow M(OH)_n + nROH$$

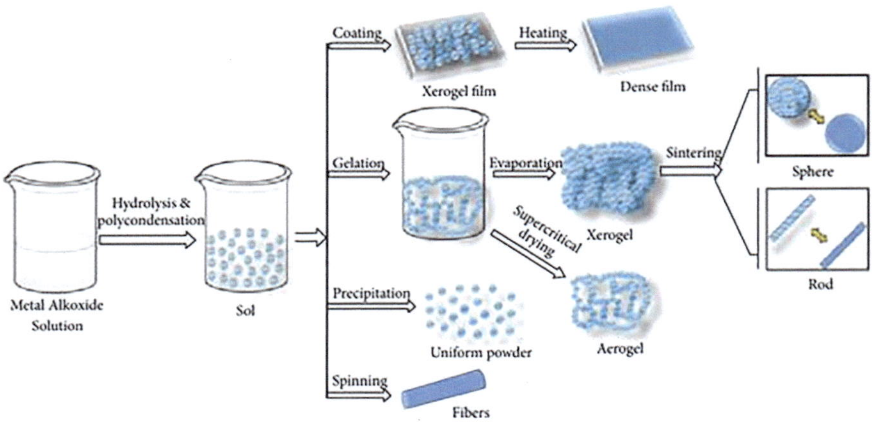

Fig. 1 An overview of the types of processes that can be done with the sol–gel method and the products of each process [21]

Further, these metal hydroxides get involved in condensation reactions to form metal–oxygen–metal cross-linked structures. This formation of the network is described by the following reaction [166]:

$$M(OH)_n + M(OH)_n \rightarrow (M - O - M) + nH_2O$$

The sol–gel method is valued for its ability to control the composition, structure, and size of materials by adjusting factors like pH and temperature, which influence the porosity and surface area of the final product [132]. This process occurs at relatively low temperatures, making it suitable for materials that are heat-sensitive or have low melting points, and allows for the creation of various material forms such as thin films, powders, monoliths, and small components. It also enables the introduction of different elements to modify material properties [132]. In contrast, the solvothermal process involves conducting reactions in a solvent at high temperatures and/or pressures within a closed system or autoclave. This method is critical for producing nanoparticles, nanostructures, and high-quality materials, with solvents like water, alcohol, and organic compounds chosen based on their solubility properties and the desired final product characteristics. Typically conducted at temperatures ranging from 100 to 300 °C and above atmospheric pressure, solvothermal conditions enhance material activity and allow for phase and structure formation that wouldn't occur under normal conditions [160]. This technique is particularly effective for controlling the size and shape of nanoscale materials like particles, rods, and plates [41]. Additionally, the high operating conditions accelerate reaction rates, shorten synthesis time, and allow for better conversion, making the solvothermal process more efficient than conventional methods [151]. It also enables the synthesis of various crystalline phases and polymorphs, broadening its application in material science research [151].

2.2 Mechanical Properties

The sol–gel method is well-regarded for its ability to create materials with specific mechanical properties, particularly for ceramic and composite materials with tailored strength and toughness [166]. It allows control over porosity and density, which is crucial for products like catalysts and thermal insulators, where a high degree of porosity is desired for lightweight and strength properties [41]. Sol–gel materials can also exhibit high mechanical strength and hardness, especially metal oxides, which benefit from strong M–O–M bonds. Post-synthesis heat treatments can enhance rigidity and stability by reducing porosity [115]. Though the process typically results in hard and brittle materials, flexibility and toughness can be achieved by incorporating polymers or using hybrid methods, making it suitable for coatings and composites [115]. Sol–gel materials are thermally stable and maintain their mechanical properties at high temperatures, making them ideal for applications in aerospace and thermal barrier coatings [29].

The solvothermal method, on the other hand, excels in producing nanostructured materials with superior mechanical properties, including nanoparticles and nanorods, which outperform bulk materials in strength due to their large surface area relative to volume [88]. The high pressure and temperature in solvothermal synthesis help to achieve high mechanical strength, hardness, and reduced flaws, creating composites with optimal crystalline structures [69]. This method also allows precise control over size and morphology, ensuring peak strength, flexibility, and impact resistance. Nanorods, for example, exhibit different mechanical performances compared to spherical nanoparticles [118]. Solvothermal materials are denser and have tightly packed particles with strong interfacial forces, enhancing toughness, hardness, and wear resistance, making them ideal for high-performance applications like coatings [118].

In summary, the sol–gel method allows for controlled mechanical properties through porosity, density, and heat treatment, making it suitable for high-strength, thermally stable, and flexible materials [11]. The solvothermal method is highly effective for synthesizing nanostructured materials with enhanced strength, hardness, and durability due to its high-pressure and high-temperature conditions, offering unique solutions for material science applications [76].

2.3 Photonic Properties

The sol–gel process is widely used in materials science for creating materials with specific photonic characteristics, particularly in optical coatings, sensors, and compact photonic devices. This method allows precise control over composition and structure, which are key to achieving desired optical properties [140]. One of its main advantages is the ability to fabricate thin, dense films with defined optical properties such as transparency, refractive index, and anti-reflection [186]. The optical properties of sol–gel materials can be tailored by adjusting parameters like precursor concentration, pH, and annealing temperature. For example, silica and titania thin films can be made highly transparent and low-scattering, useful in optical lenses and windows [2]. The sol–gel process is also used to synthesize photonic band gap structures and coatings for specific optical characteristics, such as reflective coatings and optical spacing [6]. Additionally, it allows the embedding of dopants and functionalized nanoparticles to enhance specific light properties, like fluorescence or upconversion [93].

On the other hand, the solvothermal synthesis technique excels in creating nanostructured photonic materials with highly sensitive optical characteristics that can be precisely controlled at the nanoscale. The high temperature and pressure conditions in solvothermal synthesis enable the production of quantum dots, nanowires, and nanoplates, which exhibit size-dependent optical properties, including quantum confinement effects that influence absorption, emission, or scattering [66]. For instance, solvothermal synthesis can produce semiconductor nanocrystals such as quantum dots with tunable emission colors, ideal for optoelectronic applications like light-emitting diodes, photodetectors, and fluorescence microscopy [167]. It can also enhance Surface Plasmon Resonance (SPR) characteristics, useful for Surface-Enhanced Raman Spectroscopy (SERS) and plasmonic sensors [94]. The high pressure in solvothermal processes enables the formation of novel crystalline phases and nanoscale morphologies with advanced optical properties, such as high nonlinear optical abilities or photonic band gaps, beneficial for communication and diagnostics [109, 171].

2.4 Optical Properties

Sol–gel and solvothermal methods result in materials with distinct optical characteristics, affecting their practical applications. Sol–gel processed materials typically exhibit low absorption and high transmission in the visible and near-IR spectrum, due to their homogeneous gel matrix and precise control over chemical composition and nanostructure [185]. A key advantage of sol–gel processing is its ability to control the refractive index [144], making these materials ideal for optical coatings, anti-reflective layers, and high-performance lenses. The sol–gel method also allows control over porosity and thickness, which is crucial for high-quality optical films and substrates [92].

In contrast, solvothermal synthesis involves heating solutions at high temperatures and pressures, producing materials with well-ordered crystalline structures and optical characteristics [95]. This method achieves high crystallinity and can induce quantum confinement effects, where optical properties depend on nanoscale structure size [23]. Solvothermal synthesis enables the creation of nanostructured materials with specific optical properties, fostering new optoelectronic applications and photonic devices [23].

2.5 Biological Properties

The properties of nanoparticles in medicine and biotechnology are closely linked to their preparation methods, such as sol–gel and solvothermal synthesis. Sol–gel-derived nanoparticles are biocompatible and easily functionalized, allowing for the incorporation of biological entities like antibodies or drugs, improving cell adhesion and targeting [76]. These nanoparticles are typically less toxic and suitable for drug delivery and diagnostics, with the ability to form thin, porous biocompatible films for medical implants and devices [61].

In contrast, solvothermal-synthesized nanoparticles often exhibit high crystallinity and sharp-edged nanomorphology, which enhances their biophilic reactivity [139]. These nanoparticles can have size, shape, and surface effects that improve cellular uptake and interactions with specific biological molecules. Solvothermal nanoparticles typically have a large surface area to volume ratio, benefiting drug and imaging agent delivery systems [139]. However, it is important to assess their biological safety, considering potential toxic components [188]. In summary, sol–gel nanoparticles are easier to functionalize and are generally biocompatible, while solvothermal nanoparticles offer better design flexibility for specific biological interactions and enhanced therapeutic efficacy [56]. Both methods contribute significantly to the development of nanoparticles for medical applications, though biological safety concerns must be addressed.

2.6 Electronic Properties

Sol–gel and solvothermal methods differ in their effects on the electronic properties of nanoparticles, important for electronic and optoelectronic devices. The sol–gel process allows precise control over nanoparticle size and composition, enabling tunable electronic properties [141]. It also facilitates doping and band gap adjustments, making materials more conductive or semiconductive [158], which is essential for applications like sensors [154]. The sol–gel method also influences the electronic properties of nanoparticles by controlling size and morphology [22].

Solvothermal synthesis, on the other hand, produces nanoparticles with superior electronic properties due to high crystallinity and well-defined nanostructures [133]. It allows for control of conductivity and semi - conductivity by adjusting size, shape, and crystalline phase [77]. These nanoparticles exhibit quantum confinement effects, valuable for high-performance electronic circuits and photonic devices [77]. Solvothermal synthesis also enhances chemical and electrical activity, making the nanoparticles useful for sensing and energy conversion [32]. In summary, sol–gel methods offer tunable electronic properties for various applications, while solvothermal methods provide advanced electronic qualities for high-performance devices and technologies.

3 Intrigues of Sol–Gel and Solvothermal Formation

3.1 Hydrolysis

In sol–gel synthesis, the hydrolysis process is the first and very tender process where the metal alkoxides or metal chlorides are converted into a colloidal sol of the nanoparticles. Sometimes, it begins with the combination of metal alkoxides, such as metal isopropoxides of titanium or silicon, or metal halides such as tetrachlorosilane. The general equation can describe the hydrolysis reaction:

$$M(OR)_x + xH_2O \rightarrow M(OH)_x + xROH$$

In sol–gel synthesis, metal alkoxides are dissolved in a solvent, and water is added for hydrolysis, producing metal hydroxides/oxides and alcohol. These are then polycondensed to form a gel, which undergoes further processes like gelation, aging, and drying, resulting in a solid nanoparticle network [75]. In contrast, the solvothermal method involves dissolving metal precursors like chlorides or acetates in organic solvents, then carrying out the reaction in a sealed vessel under high temperature and pressure. This enhances hydrolysis, nucleation, and nanoparticle growth, allowing control over size, shape, and crystallinity [75]. As a conclusion, the sol–gel method relies on metal alkoxides reacting with water to form metal hydroxides for nanoparticle gelation, while solvothermal synthesis uses high temperature and pressure in organic solvents to control hydrolysis and produce nanoparticles with desired properties [145].

3.2 Condensation

The hydrolysis step takes place in both methods and is crucial for the development of the nanoparticles; however, the condensation process is central to the development as well as the stabilization process of the nanoparticles. This step involves the decomposition of nitrate, sulfate, or halide salts of metal or metal–organic complexes into a network of nanoparticles [114]. In the sol–gel method, condensation is defined as reactions whereby metal hydroxides or oxyhydroxides formed during hydrolysis undergo further chemical reactions to form a three-dimensional network. The general condensation reactions can be represented as:

$$M(OH)_x + M(OH)_x \rightarrow M - O - M + xH_2O$$

In sol–gel synthesis, metal hydroxides are converted, and metal–oxygen–metal bonds form to create a gel, which is a three-dimensional network of metal-oxide or metal-hydroxy structures surrounding the particles [101]. Condensation plays a critical role in defining the gel's geometry, affecting the nanoparticles' properties. The gel is then aged, densifying further, and eventually dried and thermally treated to remove solvents, forming xerogel or aerogel with uniformly distributed nanoparticles [127]. In the solvothermal method, condensation forms metal-oxide or metal–organic frameworks under higher temperature and pressure conditions, accelerating the process [98]. Metal salts or metal–organic components are chemically transformed into metal oxide bonds, resulting in well-crystalline nanoparticles. The presence of organic solvents and additives can influence condensation, allowing control over nanoparticle size and morphology [18].

In both processes, condensation is crucial for forming a solid nanoparticle network through metal–oxygen–metal bonds. The sol–gel method requires drying and heat treatment to obtain nanoparticles, while solvothermal synthesis uses temperature and pressure to regulate condensation, influencing nanoparticle characteristics [73, 179].

3.3 *Olation*

Olation is a key step in the sol–gel process that significantly affects nanoparticle formation and characteristics, although it is less discussed than hydrolysis and condensation [48]. In this step, metal-alkoxide or metal-hydroxide species form links with other species, creating polymeric metal–oxygen–metal (M–O–M) structures. Olation plays a critical role in forming the gel network, especially in nanoparticle synthesis, as metal hydroxides or oxyhydroxides formed during hydrolysis link multiple metal centers with oxygen atoms. The general reaction for olation can be represented as:

$$M - OH + M - OH \rightarrow M - O - M + H_2O$$

In the sol–gel process, the formation of M–O–M bonds leads to the creation of a polymeric gel, pulling the network closer and forming a strong gel matrix that envelops the nanoparticles [89]. Olation defines the pore size and gel structure, which in turn determines the properties of the resulting nanoparticles [149]. In the solvothermal method, although not termed "olation," similar reactions occur under high temperature and pressure, where metal–organic precursors or salts form extended metal–oxygen structures. These reactions are faster and result in crystalline nanoparticles [86]. However, the choice of organic solvents and additives can negatively impact olation, affecting nanoparticle size, shape, and characteristics [117]. In conclusion, olation is crucial for forming composite gels and influencing nanoparticle properties in the sol–gel system. While the term is not used in solvothermal synthesis, similar reactions occur under high temperature and pressure, affecting nanoparticle structure and applications [45].

3.4 Oxolation

Oxolation is another subsequent process of sol–gel, and it is the process in which the formation of a metallic oxide framework takes place in the gel derived from the sol–gel process. This step is to add more reinforcement to the ongoing structural development and to tune the characteristics of the resultant nanoparticles [123].

In the sol–gel method of preparation, the process of oxolation occurs after the process of olation has occurred. It lead to the formation of other metal–oxygen–metal (M–O–M) bonds and thus building up a large and complex structure of metal oxides. It has been established that the oxolation process involves the following chemical equation:

$$\mathrm{M-O-M+M-OH \rightarrow M-O-M-O-M+H_2O}$$

In the sol–gel process, oxolation extends M–O–M linkages, forming a more entangled structure and enhancing the mechanical strength and heat resistance of the gel, which is crucial for nanoparticle synthesis [51]. This extended network strengthens the nanoparticles during subsequent processing stages, like drying and calcination. In the solvothermal method, though oxolation isn't directly reported, similar processes occur under high temperature and pressure. These reactions are faster, forming large metal-oxide networks and well-defined, stable nanoparticles [116]. The solvothermal method's control over this process impacts the dimensionality and crystallography of the nanoparticles.

In conclusion, oxolation is a key step in the sol–gel process, forming an extended metal-oxide structure that provides rigidity and cohesion to the gel, improving nanoparticle strength. While not named in solvothermal synthesis, similar network-forming processes occur, aiding in nanoparticle formation and enhancing their properties [70].

3.5 Formation of the Sol

The formation of the sol, which is an important step of the sol–gel process and is needed for the initial stage of the development of the nanoparticles, is suitable. In this phase, the metal precursors are in colloidal suspension and reduced to elemental form as nanoparticles, and it will be on such a matrix that all the subsequent processing will occur [44].

In this stage of the metal alkoxides or the metal chlorides are hydrolyzed and partially, condensed and forming a colloidal suspension. The sol is a standard operation of delivering nanoparticles in several solvents, for instance, water coupled with alcohol. The general reactions involved in this process can be summarized as follows:

$$\mathrm{M(OR)_x} + x\mathrm{H_2O} \rightarrow \mathrm{M(OH)_x} + x\mathrm{ROH}$$

$$M(OH)_x + M(OH)_x \rightarrow M - O - M + xH_2O$$

In the sol–gel process, the M–O–M bond forms during condensation, resulting in a system where nanoparticles are suspended in liquid media. This stage is crucial for defining the size, dispersion, and stability of the nanoparticles. The sol can be controlled by adjusting pH, temperature, and solvent choice to achieve desired properties [5]. In the solvothermal method, the concept of the sol isn't emphasized, but the metal precursors are dissolved in an organic solvent, and the reaction occurs under high temperature and pressure. This leads to a colloidal dispersion of nanoparticles, like the sol in the sol–gel process [30]. The solvothermal method deposits nanoparticles at a high, controlled rate, ensuring well-defined size and shape. Thus, sol formation is essential in the sol–gel method to create stable colloidal suspensions with desired characteristics. In solvothermal synthesis, while sol formation isn't directly observed, nanoparticles form under controlled temperature and pressure conditions to achieve specific performance parameters [190].

3.6 *Formation of the Gel*

Gelation or formation of the gel is the most crucial step in the sol–gel process and helps convert sol to a solid network of nanoparticles. It is a phase in which a colloidal suspension of the nanoparticles is transformed into a three-dimensional gel network of the nanoparticles, and further treatments on the nanoparticles are administered [151]. In the sol–gel method, after the sol formation, the process of gel formation involves the linking up of the metal-oxide bonds to give a continuous three-dimensional network. This process arises from the condensation reactions which occurs between metal hydroxides or metal alkoxy species, and as such, the gel matrix carries on thickening. The general reaction for gel formation is as follows:

$$M - (OH)_x + M - (OH)_x \rightarrow M - O - M + xH_2O$$

In the sol–gel process, condensation reactions form a network of metal-oxide bonds, gradually transforming the sol into a gel. This gel entraps nanoparticles and controls porosity and mechanical properties [151]. After aging, drying, and thermal treatment, the gel becomes a solid, typically in the form of xerogel or aerogel [189]. In solvothermal synthesis, while gelation is not central, similar structural processes occur under high temperature and pressure. Metal precursors dissolve in an organic solvent, where nanoparticles are produced. Although no traditional gel forms, metal-oxide networks indicate a solid structure. Purification through filtering or centrifugation isolates the nanoparticles [107]. Figure 2 shows an illustrational explanation of this colloidal network formation in sol–gel materials [151].

The Formation of the gel is crucial in the sol–gel process, where a colloidal sol transitions into a solid polymerizing network of metal-oxide bonds. This gel network

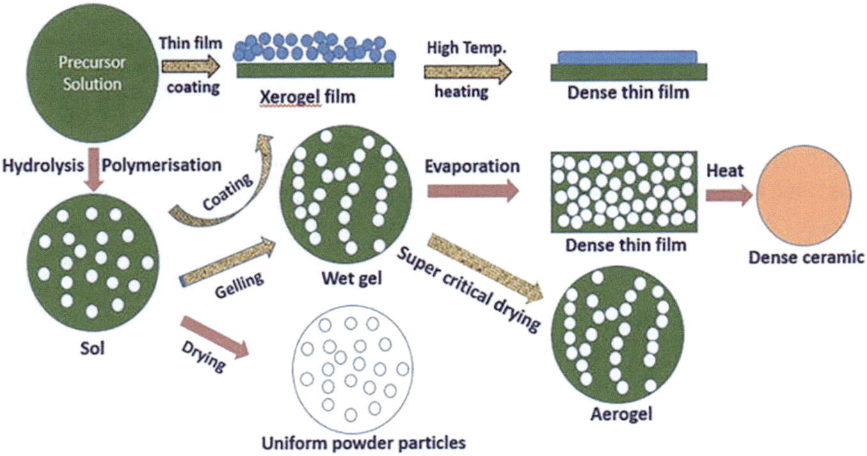

Fig. 2 Colloidal network formation in sol–gel materials [151]

plays a key role in determining the physical properties of nanoparticles, such as size, distribution, and stability. In contrast, the solvothermal method, which does not involve gel formation, relies on similar network-forming processes under specific conditions to yield well-defined nanoparticles. Both methods use this final step to ensure proper formation and purification of nanospheres tailored to their intended applications [52, 107].

3.7 Electrostatic Stabilization of the Sol

Coulter stabilization in the sol–gel process helps prevent nanoparticle agglomeration or precipitation, ensuring uniform distribution in the liquid media [10]. Electrostatic stabilization plays a key role in prolonging the stability of the sol by dispersing nanoparticles and counteracting aggregation through surface charges. This is achieved by adjusting the pH or adding charged surfactants, creating an electrostatic repulsion that prevents coalescence [103].

In the solvothermal process, electrostatic stabilization is also used, but to a lesser extent. Here, temperature, pressure, and solvent changes affect surface charges and prevent coagulation by forming a stabilizing layer on nanoparticles [125]. While the sol–gel method emphasizes controlling surface charges to avoid agglomeration, the solvothermal method uses similar stabilization techniques under varying conditions of temperature and pressure [25]. Both methods rely on electrostatic stabilization to ensure nanoparticle stability for scientific and technological applications.

3.8 Electrosteric Stabilization of the Sol

Electrosteric stabilization enhances nanoparticle dispersion by combining electrostatic and steric repulsion, preventing coagulation and ensuring uniform colloidal sol distribution [57]. In the sol–gel method, this stabilization uses surfactants or polymeric stabilizers. Electrostatic repulsion is achieved by adding ionic surfactants or charged polymers that induce surface charges, creating a double layer that prevents aggregation. Steric hindrance, through non-ionic surfactants or polymeric stabilizers, further prevents particle approach by forming a protective layer around each nanoparticle [64, 183]. This dual mechanism ensures stability and homogeneity, crucial for high-quality nanoparticle suspensions [55].

In the solvothermal process, electrosteric stabilization is also employed, though less emphasized. Stabilizers or surfactants provide electrostatic charges and steric hindrance, helping nanoparticles remain dispersed in the organic solvent despite high temperature and pressure [122, 181]. Both methods use electrostatic and steric stabilization to maintain nanoparticle dispersion, ensuring their desired size and properties for advanced materials [9, 168].

4 Influential Factors in the Sol–Gel and Solvothermal Reaction

4.1 Type of the Solvent

The solvent used in sol–gel and solvothermal reactions significantly impacts the efficiency and success of synthesis. In the sol–gel method, solvents dissolve metal alkoxides or salts, facilitating sol formation. Alcohols are commonly used due to their solubility in metal alkoxides and their role in driving hydrolysis and condensation reactions. The solvent affects the reaction rate, with solvents having high dielectric constants or forming hydrogen bonds promoting gel formation, thus altering the gel structure and properties [31]. Additionally, the solvent type influences the sol's flow behavior and rheology, affecting the homogeneity and consistency of the gel network [79].

In solvothermal reactions, the solvent determines the solubility of precursors and supports high temperature/pressure conditions. It influences nucleation and crystal growth, dictating the size, shape, and quality of crystals [92]. Cryoscopy solvents can "freeze in" specific crystal structures, affecting the final material properties. Thus, the solvent plays a crucial role in adjusting synthesis parameters and enhancing the properties of materials produced in both sol–gel and solvothermal methods [59].

4.2 Reaction Time

Reaction time plays a crucial role in both sol–gel and solvothermal processes, affecting synthesis and the final material's properties. In the sol–gel method, reaction time influences the hydrolysis and condensation of precursors. Longer durations can enhance polymerization, leading to better gel network connectivity [99]. However, extended times may cause side reactions or over-polymerization, affecting material porosity and complicating the achievement of desired mechanical properties [113]. In solvothermal reactions, reaction time impacts the formation and growth of crystalline phases. Longer reaction times can lead to well-defined, high-quality crystals with better crystallinity and phase purity, which is important for certain applications [182]. Shorter times may result in poor crystallinity or smaller particles. Therefore, the optimal reaction time is essential in both methods to control the size, morphology, and quality of the synthesized materials [67].

4.3 Agitation

Agitation/stirring is crucial in both sol–gel and solvothermal reactions, as it influences coalescence, kinetics, and yield. In the sol–gel method, stirring ensures homogeneity by facilitating equal hydrolysis and condensation reactions. Proper mixing prevents local concentration gradients or sedimentation, which can lead to inhomogeneous gels and affect product properties [8]. Stirring also aids in debubbling and impurity elimination, improving gel structure formation. In solvothermal reactions, stirring is vital for maintaining homogeneity and controlling crystal growth. For instance, adding PAA stabilizes the reaction and aids in nucleation and crystalline phase development [46]. Adequate mixing enhances heat and mass transfer, speeding up the reaction and improving crystal formation. Low agitation leads to non-uniform reactions and defects in the final product. Therefore, managing agitation is essential for reproducibility and quality in both synthesis methods [157].

4.4 pH

In the sol–gel method, pH is crucial for controlling hydrolysis and condensation of metal alkoxides or salts. Acidic conditions promote hydrolysis, while basic conditions favor condensation [39]. The gel's nature depends on the pH, and precise control is required to achieve the desired porosity and surface area. Extreme pH values may lead to incomplete reactions or unwanted crystallization, affecting material quality [161].

In solvothermal reactions, pH affects precursor solubility, nucleation, and crystalline phase development. Acidic or alkaline environments influence crystal formation rates and the size and shape of the final product. Thus, optimal pH control is essential for producing pure and well-defined materials. In both methods, pH regulation is vital for controlling reaction kinetics and ensuring high-quality, reproducible materials [15, 19].

4.5 Temperature Effects

Temperature is a critical factor in both sol–gel and solvothermal processes, influencing reaction rates, material characteristics, and outcomes. In the sol–gel method, temperature controls the hydrolysis and condensation of metal alkoxides or salts [44]. Higher temperatures promote gel formation and alter gel properties like porosity. However, excessive temperatures can lead to uncontrolled condensation or undesirable phases, affecting material characteristics [91]. In solvothermal reactions, temperature affects precursor solubility, nucleation, and crystal growth [53]. Higher temperatures enhance dissolution, promote solid-state reactions, and improve crystallization. However, excessive temperatures may cause side reactions or material transformation, negatively impacting purity and quality [147]. Thus, controlling temperature is essential to determine crystallinity, size, and morphology in both sol–gel and solvothermal synthesis [163].

4.6 Effects of the Catalysts

Catalysts are essential in controlling the rate and outcome of sol–gel and solvothermal reactions. In the sol–gel process, catalysts control hydrolysis and condensation of metal alkoxides or salts. Acidic catalysts speed up hydrolysis, while basic catalysts accelerate condensation [14]. The type and distribution of catalysts influence gelation rates, gel homogeneity, and material properties like pore size and surface area. Improper catalyst conditions can lead to incomplete reactions or undesired properties, highlighting the need for optimization [124]. In solvothermal reactions, catalysts affect solubility, nucleation, and crystal growth. They enable specific reactions, influencing phase formation, crystal size, and purity [121]. Incorrect catalysts or concentrations can lead to side reactions or phase contamination. Therefore, careful selection of catalysts is crucial for designing reaction rates and ensuring reproducibility in both sol–gel and solvothermal processes [129].

4.7 Water-to-Alkoxide Ratio

The water to alkoxide ratio is crucial in sol–gel synthesis as it controls hydrolysis and condensation of metal alkoxides, affecting gel formation and material properties. A high ratio accelerates hydrolysis, leading to quicker gelation and larger pore spaces, with better interconnection [52]. Conversely, a low ratio results in slower hydrolysis, potentially creating a less homogeneous gel with smaller pores and altered structures [162]. The ratio also impacts the chemical reaction and gel uniformity. Excess water can speed up condensation, leading to phase splitting or unwanted byproducts, while insufficient water causes poor hydrolysis and inhomogeneous gels [13]. Optimizing this ratio is essential for achieving the desired gel structure, porosity, and material quality [52].

4.8 Impact of Network Modifier

Network modifiers significantly impact sol–gel processes by altering the gel network structure. These modifiers, such as metal salts or organic compounds, influence gel connectivity, pore size, and mechanical properties by affecting M–O–M bond formation [151]. They can enhance or decrease the condensation process, thus modifying the gel's characteristics. Depending on the type and concentration of nanoparticles, network modifiers can improve properties like thermal stability, chemical resistance, and optical properties. For example, functionalizing materials with additional compounds can increase reactivity or catalytic activity [151]. However, using too many or improperly proportioned modifiers can distort the network and reduce mechanical properties. Therefore, careful consideration of the modifier type and amount is crucial for improving gel properties and controlling material performance [120].

5 Industrial Applications of Sol–Gel and Solvothermal Technology

5.1 Thermal Insulation

Sol–gel and solvothermal methods are highly effective for producing advanced thermal insulating materials. The sol–gel process is particularly known for creating macroporous materials like silica aerogels, which are valued for their low thermal conductivity and density. Silica aerogels with tunable pore sizes have achieved thermal conductivities as low as 0.015 W/m K, making them ideal for heat insulation in applications like space vehicles and energy-efficient construction [35, 36]. The sol–gel method can also improve thermal insulation by doping or functionalizing materials, such as alumina-based aerogels, which have reduced thermal conductivity and improved mechanical strength [90].

In solvothermal synthesis, nanostructured materials like ZnO nanoparticles have been used to create polymer composites with enhanced thermal insulation properties, lowering thermal conductivity [83]. Similarly, TiO_2 nanoparticles produced by solvothermal methods have been used to develop insulating thin films with stable thermal insulation characteristics [165]. Additionally, solvothermal-derived metal–organic frameworks (MOFs) incorporated into polymer matrices have shown excellent thermal insulation and high mechanical strength, further demonstrating the potential of solvothermal methods for advanced thermal barriers [74].

5.2 Acoustic Insulation

Sol–gel and solvothermal methods are crucial for developing materials that improve acoustic performance in building structures. The sol–gel process is particularly effective for synthesizing materials with high sound absorption characteristics, such as silica and carbon aerogels. Studies show that sol–gel-derived aerogels with tailored pore sizes enhance sound absorption across a broad frequency range, making them ideal for noise control in residential and commercial spaces [24]. Sol–gel also enables the creation of acoustic insulating coatings, such as titanium dioxide nanoparticle coatings, which improve sound insulation, especially reverberation time in confined areas [42, 184].

The solvothermal method contributes to acoustic insulation by forming nanocomposite compounds, such as metal–organic frameworks (MOFs) with high surface areas. These MOFs provide excellent sound absorption properties, making them suitable for building interiors and noise barriers [105]. Additionally, solvothermally synthesized graphene oxide (GO) composites offer improved sound absorption and noise attenuation compared to traditional polyurethane foam [180]. Solvothermal methods are also used to produce mesoporous ceramics with tailored pore structures for acoustic insulation [43]. In conclusion, both sol–gel and solvothermal techniques play significant roles in developing advanced acoustic insulation materials, with sol–gel focusing on high-surface-area materials and solvothermal providing nanostructured composites with tailored acoustic properties.

5.3 Protective Optical Coatings

Sol–gel and solvothermal processes are key in developing high-performance optical coatings that enhance the functionality and protection of optical devices. The sol–gel method is valued for its precision in creating coatings, such as anti-reflective layers. For example, sol–gel derived silica coatings can achieve low reflectance and high transmittance, reducing glare and improving lens performance [169]. Additionally, dopants like TiO_2 and ZrO_2 can be added to enhance optical properties. TiO_2 sol–gel coatings, for instance, offer scratch resistance while maintaining transparency [47]. Sol–gel is also used for photocatalytic coatings, such as titanium dioxide, for self-cleaning optical surfaces under UV light [17].

The solvothermal method is used to grow nanophased optical coatings, such as ZnO nanorods, which provide UV shielding and increased hardness [62]. It can also synthesize hydrophobic coatings, as seen in TiO_2 nanoparticles applied through chemical vapor deposition for water-resistant optical parts [135]. Solvothermal processes can also create multi-functional coatings, incorporating metal–organic frameworks (MOFs) that improve UV and chemical resistance [50, 156]. Both methods are crucial for creating protective optical coatings. The sol–gel process excels in anti-reflective, scratch-resistant, and photocatalytic applications, while solvothermal methods enhance UV-blocking, hydrophobicity, and multifunctionality, making them essential for improving the durability and performance of optical devices.

5.4 Lightweight Materials

Sol–gel and solvothermal methods are crucial for synthesizing lightweight materials with high strength and functionality, benefiting industries like aerospace, automotive, and construction. The sol–gel process, particularly for aerogels, generates low-density, highly porous materials with excellent thermal properties. Silica aerogels, for example, are lightweight (down to 0.003 g/cm^3) and provide heat shock protection and mechanical toughness, making them ideal for aerospace applications [104]. The sol–gel method is also used to prepare polymer matrix composites, such as alumina nanoparticle composites, which offer low density and improved mechanical properties for structural applications [108].

In the solvothermal method, lightweight nanostructured materials with high surface areas are synthesized. For instance, porous titanium dioxide (TiO_2) is produced for environmental and energy applications, offering both low density and high surface area [1]. Solvothermal techniques are also employed to create metal–organic frameworks (MOFs) with high porosity and low density, ideal for gas storage and separation [87]. Additionally, solvothermal synthesis is used for lightweight ceramics with high strength and thermal stability, suitable for structural and thermal protection uses [159]. In summary, sol–gel and solvothermal methods play a vital role in creating lightweight materials, with applications ranging from aerogels and composites to nanostructured materials and MOFs. These methods are ideal for producing high-performance materials for weight-sensitive industries.

5.5 Tough Ceramics

The synthesis of tough ceramics is critical for applications in aerospace, defense, and other industries, as these materials offer high strength, hardness, and resistance to wear and heat. Both sol–gel and solvothermal methods are advanced technologies for producing high-performance ceramics.

The sol–gel method is particularly effective for creating mechanically robust ceramics by offering fine control over microstructure and additive incorporation. For example, sol–gel-derived zirconia (ZrO_2) ceramics, often doped with yttria (Y_2O_3), exhibit increased hardness and toughness, making them suitable for high-stress applications such as cutting tools and dental implants [28]. Similarly, sol–gel-derived alumina (Al_2O_3) ceramics show enhanced wear resistance and mechanical strength due to controlled porosity and microstructure [16]. The solvothermal method is also essential for producing robust ceramics, especially nanostructured composites. For instance, solvothermal preparation of titanium carbide (TiC) nanorods results in ceramics with improved hardness and fracture toughness, ideal for cutting tools and armor [138]. Similarly, boron carbide (B_4C) and silicon carbide (SiC) nanocomposites synthesized through solvothermal processes demonstrate enhanced impact strength and high-temperature performance [40]. Solvothermal synthesis of mullite ($3Al_2O_3{\cdot}2SiO_2$) also improves toughness, making it suitable for high-temperature applications like furnace and aircraft components [34]. In summary, the sol–gel and solvothermal techniques are essential for creating ceramics with improved toughness, strength, and high-temperature resistance. These techniques enable the creation of advanced ceramics with controlled microstructures and improved mechanical properties, making them suitable for high-performance applications.

5.6 Membranes and Microfilters

Membranes and microfilters are widely used in filtration processes like water and air purification, and industrial separations. Both sol–gel and solvothermal processes are effective in synthesizing advanced membranes and microfilters.

The sol–gel method is particularly useful for creating ceramic membranes with large surface areas and precise control over pore size, enhancing filtration efficiency. For example, silica-based membranes synthesized through sol–gel processing, with nanometer-sized pores, showed excellent performance in removing bacteria and organic content from water [58]. Additionally, sol–gel-derived titania (TiO_2) membranes have been used in seawater reverse osmosis (SWRO), offering high permeability and resistance to fouling, making them suitable for desalination [78]. The solvothermal method also plays a significant role in membrane development. For instance, copper-based Metal–Organic Frameworks (MOFs) synthesized solvothermally have been used to create gas separation membranes with excellent selectivity and permeability for gases like CO_2 and CH_4 [148]. Furthermore, solvothermally synthesized graphene oxide (GO) nanomaterials incorporated into polymeric membranes have improved mechanical properties and filtration performance for water purification and industrial applications [164].

5.7 Textiles

Sol–gel and solvothermal methods are used to produce advanced functional textiles with properties such as waterproofing, self-cleaning, and bacterial resistance, which are essential for protective, medical, and all-weather wear.

The sol–gel process is particularly effective for incorporating functional properties into textiles. For instance, silica coatings applied through sol–gel techniques provide high hydrophobicity without compromising fabric moisture permeability or comfort [27]. Sol–gel is also used to create antimicrobial textiles that reduce bacterial growth, making them suitable for medical and hygiene products [63]. The solvothermal method enhances textiles by incorporating nanomaterials. For example, ZnO nanorods synthesized solvothermally have been integrated into fabrics, improving UV protection and antimicrobial performance [71]. In summary, both methods are crucial in advancing the field of functional textiles, with sol–gel excelling in coating textiles with hydrophobic and antimicrobial properties, and solvothermal methods enhancing fabric functionality with nanomaterials. Recent studies highlight their effectiveness in developing high-performance textiles and composites for various industries.

5.8 Nuclear Waste Storage

The disposal of hazardous and radioactive waste is a significant challenge, and sol–gel and solvothermal methods have proven effective for creating materials to stabilize, contain, and immobilize nuclear waste, enhancing safety and efficiency. The sol–gel process is particularly useful for synthesizing glass/ceramic matrices that immobilize radioactive materials. For example, sol–gel-derived silica glasses have been used to encapsulate radioactive elements like uranium and plutonium, offering long-term stability and resistance to leaching under environmental conditions [84]. Additionally, sol–gel titanate ceramics have been used for the immobilization of high-level radioactive wastes (HLRW), showing good chemical stability and radiation resistance, ideal for long-term storage [173]. The solvothermal method also contributes to nuclear waste disposal by producing materials with large capacity and stability. Metal–organic frameworks (MOFs) synthesized through solvothermal methods have demonstrated high uranium uptake and stability, making them suitable for trapping radioactive elements [137]. Solvothermal synthesis is also used to create phosphate ceramics and zirconium-based compounds for immobilizing radionuclides like americium, neptunium, and others, enhancing durability and radiation resistance [20, 37].

5.9 Ultra-Fine Powder Abrasives

Fine powder abrasives are crucial for precision grinding, polishing, and surface finishing across various industries. Both sol–gel and solvothermal methods are effective for producing ultra-fine abrasives with high purity and narrow particle size distribution.

The sol–gel process is ideal for synthesizing abrasives with specific particle size and shape. For example, sol–gel derived alumina (Al_2O_3) powders, with submicron sizes, offer high hardness and wear resistance, making them suitable for precision grinding and polishing [155]. Sol–gel method can also be used to synthesize silica (SiO_2) powders, enhancing the surface finish and polishing capacity of metals and ceramics [143]. The process allows precise control of particle size and distribution, critical for high - abrasive performance.

The solvothermal method also produces ultra-fine abrasives, enabling synthesis under high temperature and pressure. A notable example is cubic boron nitride (c-BN), which, when synthesized solvothermally, exhibits enhanced hardness and thermal stability, making it suitable for grinding and cutting hard materials [65]. Solvothermal synthesis of zirconia (ZrO_2) powders has also been shown to improve abrasive performance, especially in the semiconductor and aerospace industries [153]. Additionally, solvothermally synthesized titanium carbide (TiC) powders offer high hardness and wear resistance, ideal for high-precision abrasive applications [152].

5.10 Encapsulation of Biomolecules for Controlled Drug Release

The sol–gel and solvothermal methods are essential for the encapsulation of biomolecules in controlled drug delivery systems. These techniques enable the construction of carrier systems that release drugs at a controlled rate, improving therapeutic efficacy and minimizing drug toxicity.

The sol–gel process is commonly used to create porous and nanoparticle-based materials for drug delivery. A study on sol–gel synthesized mesoporous silica nanoparticles (MSNs) demonstrated their ability to effectively load and control the release of anticancer drugs, enhancing treatment effectiveness while reducing side effects [170]. Sol–gel-derived bioactive glass nanoparticles have also been used for controlled drug release in bone infection therapies, showing sustained antibiotic release in targeted areas [38]. The solvothermal method is also valuable for creating multifunctional drug delivery systems with high cargo capacity and controlled release profiles. A study demonstrated that solvothermally synthesized Metal–Organic Frameworks (MOFs) could adsorb and release anticancer drugs under physiological conditions, with controlled release rates due to adjustable pore sizes [49]. Additionally, solvothermal synthesis has been used to create polymer-based nanoparticles for drug delivery, achieving sustained release and enhanced bioavailability [131].

5.11 Antibiofouling

Antibiofouling and anti-bacterial adhesion are critical for applications in medicine and efficient foul-release systems. Both sol–gel and solvothermal processes have been used to create materials with antibacterial properties [85].

The sol–gel process can produce coatings with antimicrobial characteristics. A study by Chang et al. explored sol–gel-derived titanium dioxide (TiO_2) coatings for antibiofouling, showing that TiO_2 films exhibited photocatalytic activity under UV light, inhibiting bacterial adhesion on medical devices [31]. Additionally, silver nanoparticles incorporated into sol–gel coatings were found to exhibit long-term bactericidal properties, reducing hospital-acquired infections (HAIs) [187]. The solvothermal method is also used to synthesize antibiofouling materials. For example, solvothermally synthesized copper-based nanoparticles showed high antibacterial efficiency, preventing bacterial film formation on both marine and medical surfaces [26]. Additionally, solvothermal techniques have been used to create polymer-based antibacterial coatings, offering protection against microbial adhesion and biofilm formation in industrial and medical settings [106].

6 Scaling Up Sol–Gel Methods for Industrial Nanoparticle Production

The sol–gel method has long been recognized as a versatile and effective approach for synthesizing nanoparticles (NPs), with significant advancements being made to scale up this process for industrial applications [146, 178]. Several studies have highlighted how sol–gel-based techniques can be adapted for large-scale production, enabling the manufacturing of nanoparticles with consistent properties that meet industrial requirements [80–82, 100, 110, 111]. These scaling efforts have been demonstrated across a variety of applications, including catalyst production, photocatalysis, energy storage, and environmental remediation [134, 136].

One of the notable advancements in scaling sol–gel methods involves the production of nano-zirconia particles using supercritical hydrothermal synthesis. Liu et al. [96] successfully scaled up this process from laboratory to industrial levels, demonstrating its efficiency in producing nano-zirconia with controlled particle sizes and high conversion rates. By optimizing key synthesis parameters such as temperature, pressure, and precursor concentrations, the process enables the mass production of nano-zirconia, which has applications in fields such as catalysis and abrasives. The scalability of supercritical hydrothermal synthesis ensures that nano-zirconia can be produced in large quantities without compromising quality, addressing the increasing demand for high-performance materials in industrial sectors [96].

In a different approach, Henrique et al. explored the scaling up of sol–gel derived titanium dioxide (TiO_2) films, particularly for photocatalytic applications. The researchers developed an automated and cost-effective apparatus for depositing TiO_2 films through a spray coating technique, which could be scaled for mass production. The TiO_2 films produced via this method exhibit excellent photocatalytic activity, particularly in their anatase phase, making them highly suitable for environmental applications such as water purification and air decontamination. This automated spray coating process presents a significant step towards efficient, large-scale production of TiO_2 films, with promising implications for sustainable solutions in environmental cleanup [130].

Another key area of development involves the sol–gel synthesis of ceria nanoparticles, as demonstrated by Gilani et al. [60]. In their study, they explored pH-dependent synthesis to optimize the photocatalytic performance of ceria for the degradation of methyl orange dye in wastewater. By adjusting the pH between 6 and 12, they found that a pH of 8 resulted in the highest photocatalytic activity, achieving 75% degradation of the dye in just 80 min. This study underscores the potential for large-scale production of ceria nanoparticles, which are both cost-effective and highly efficient for industrial wastewater treatment, providing a sustainable method for dealing with dye-contaminated effluents from various industries [60].

Similarly, Liu et al. focused on scaling up the production of highly porous copper oxide, copper(I) oxide, copper, and carbon materials derived from aerogels for lithium-ion battery anodes. The sol–gel synthesis, combined with aerogel technology, has been scaled to produce these materials in large quantities, offering high capacity and excellent cycling stability. The successful scaling of this synthesis process is critical for meeting the growing demand for advanced anode materials in energy storage devices, such as batteries for electric vehicles and portable electronics, thereby supporting the transition towards more sustainable energy solutions (Liu et al. 2024).

The scaling of sol–gel methods has also been applied to the production of catalyst layers for polymer electrolyte fuel cells (PEFCs). Fikry et al. demonstrated the successful upscaling of unsupported platinum-nickel (Pt–Ni) aerogels for fuel cell cathodes. By increasing the material yield five-fold, they achieved a process that enables reproducible production of catalyst layers using an automated spray coating technique. This scalable synthesis method ensures the production of multi-gram quantities of catalyst material with consistent physicochemical properties, which is essential for commercializing fuel cell technologies that are central to clean energy production [54].

These various studies illustrate the significant progress made in scaling up sol–gel methods for the industrial production of nanoparticles. By optimizing synthesis parameters and integrating innovative techniques like supercritical hydrothermal synthesis and automated spray coating, these approaches offer viable solutions for producing high-quality nanoparticles on a large scale. The scalable production of such nanoparticles is vital for a wide range of industrial applications, from energy storage and catalysis to environmental remediation, positioning the sol–gel method as a key player in the advancement of sustainable technologies.

7 Future Directions Sol–Gel and Solvothermal Method

The sol–gel method is still evolving and is expected to undergo further development to address its current limitations, especially in industrial applications. Key future directions include improving process efficiency by reducing the lengthy gelation and drying times, potentially through microwave or supercritical drying, which could reduce processing time and increase scalability [45]. Another focus is the preparation of green precursors that are environmentally safe while maintaining performance. Improving scalability is crucial, requiring the elimination of structural complexities and more consistent tradeoffs between teams. Innovations in reactor designs and real-time monitoring could further optimize nanoparticle production [151]. Additionally, combining sol–gel methods with additive manufacturing techniques like 3D printing offers potential for designing intricate nanoparticle structures [45]. The solvothermal method, while effective, faces challenges such as high energy costs and solvent issues. Future developments are expected to enhance energy efficiency by optimizing reaction conditions and using energy-friendly reactors. Efforts to find milder solvents and improve solvent recovery are also ongoing, aiming to reduce environmental impact and operating costs [174]. Improving control over reaction conditions through advanced process analytical technologies (PATs) and feedback systems can help ensure uniform nanoparticle properties. Additionally, advancing the scalability of continuous flow solvothermal reactors is a priority to make the process suitable for large-scale industrial production [128, 151].

By addressing these challenges, both sol–gel and solvothermal methods can become more efficient, environmentally sustainable, and cost-effective, expanding the potential of nanotechnology for industrial applications [112, 175, 177].

8 Conclusion

The sol–gel and solvothermal methods are highly effective for nanoparticle synthesis, offering advantages such as precise control over size, shape, and composition. These methods have found widespread applications in fields like catalysis, energy storage, environmental control, and drug delivery.

The sol–gel process, a wet-chemical synthesis technique, is popular for producing solid nanoparticles due to its simplicity, cost-effectiveness, and ability to yield high-purity materials. Particle size and distribution can be easily controlled by adjusting precursor solutions, pH, and temperature, making it suitable for both small-scale laboratory synthesis and large-scale industrial production. Additionally, the sol–gel method can be combined with deposition techniques like spin or spray coating, expanding its use in electronics and photovoltaics. Similarly, the solvothermal method, which involves heating solvents to high temperatures and pressures, is efficient for synthesizing nanoparticles with tailored properties, including specific morphology and crystallization. This method has made significant advancements in scalability, enabling its use in industrial applications such as energy storage, catalysis, and environmental control.

While the sol–gel process excels in low-cost, high-accuracy nanoparticle production, the solvothermal process provides additional tunability for nanoparticles requiring higher temperatures and pressures. Both methods are advancing, and their integration with automation and process optimization trends holds promise for industrial applications. As demand for nanoparticles in high-tech and sustainable sectors grows, these methods will play a crucial role in shaping the next generation of materials and technologies.

References

1. Abdelnasser S, Al Sakkaf R, Palmisano G (2021) Environmental and energy applications of TiO_2 photoanodes modified with alkali metals and polymers. J Environ Chem Eng 9(1):104873. https://doi.org/10.1016/j.jece.2020.104873
2. Addonizio ML, Aronne A, Imparato C (2020) Amorphous hybrid TiO_2 thin films: the role of organic ligands and UV irradiation. Appl Surf Sci 502(144095):144095. https://doi.org/10.1016/j.apsusc.2019.144095
3. Ahire SA et al (2022) The augmentation of nanotechnology era: a concise review on fundamental concepts of nanotechnology and applications in material science and technology. Results Chem 4:100633. https://doi.org/10.1016/j.rechem.2022.100633
4. Al Mahmud MZ (2024) Investigating eco-friendly materials as energy storage solutions. Unpublished. https://doi.org/10.13140/RG.2.2.28107.57123
5. Alhalili Z (2023) Metal oxides nanoparticles: general structural description, chemical, physical, and biological synthesis methods, role in pesticides and heavy metal removal through wastewater treatment. Molecules (Basel, Switzerland) 28(7):3086. https://doi.org/10.3390/molecules28073086
6. Almeida RM, Clara Gonçalves M, Portal S (2004) Sol–gel photonic bandgap materials and structures. J Non-Cryst Solids 345–346:562–569. https://doi.org/10.1016/j.jnoncrysol.2004.08.085

7. Alshammari BH et al (2023) Organic and inorganic nanomaterials: fabrication, properties and applications. RSC Adv 13(20):13735–13785. https://doi.org/10.1039/d3ra01421e
8. Amiri S, Rahimi A (2016) Hybrid nanocomposite coating by sol–gel method: a review. Iran Polym J 25(6):559–577. https://doi.org/10.1007/s13726-016-0440-x
9. Andrade A et al (2011) Coating nanomagnetic particles for biomedical applications. In: Biomedical engineering—frontiers and challenges. InTech
10. Arcos Rosero WA et al (2024) Review of advances in coating and functionalization of gold nanoparticles: from theory to biomedical application. Pharmaceutics 16(2):255. https://doi.org/10.3390/pharmaceutics16020255
11. Babiarczuk B et al (2023) Mechanical properties of silica aerogels controlled by synthesis parameters. J Non-Cryst Solids 606:122171. https://doi.org/10.1016/j.jnoncrysol.2023.122171
12. Baig N, Kammakakam I, Falath W (2021) Nanomaterials: a review of synthesis methods, properties, recent progress, and challenges. Mater Adv 2(6):1821–1871. https://doi.org/10.1039/d0ma00807a
13. Bashir S et al (2020) Fundamental concepts of hydrogels: synthesis, properties, and their applications. Polymers 12(11):2702. https://doi.org/10.3390/polym12112702
14. Basumatary SF et al (2024) Recent advances in magnetic solid catalysts: synthesis, stabilization and application in cleaner production of biodiesel. Energy Nexus 15:100318. https://doi.org/10.1016/j.nexus.2024.100318
15. Bathe AS et al (2022) Solvothermal synthesis of soluble, surface modified anatase and transition metal doped anatase hybrid nanocrystals. Nanoscale Adv 4(24):5343–5354. https://doi.org/10.1039/d2na00640e
16. Behera PS, Bhattacharyya S (2021) Effect of different alumina sources on phase formation and densification of single-phase mullite ceramic—reference clay alumina system. Mater Today Commun 26:101818. https://doi.org/10.1016/j.mtcomm.2020.101818
17. Bergamonti L et al (2017) Enhanced self-cleaning properties of N-doped TiO_2 coating for cultural heritage. Microchem J 133:1–12. https://doi.org/10.1016/j.microc.2017.03.003
18. Bharti K, Sadhu KK (2022) Syntheses of metal oxide-gold nanocomposites for biological applications. Results Chem 4:100288. https://doi.org/10.1016/j.rechem.2022.100288
19. Biswas S, Kar S, Chaudhuri S (2007) Effect of the precursors and solvents on the size, shape and crystal structure of manganese sulfide crystals in solvothermal synthesis. Mater Sci Eng B 142(2–3):69–77. https://doi.org/10.1016/j.mseb.2007.06.019
20. Bois L et al (2001) Preliminary results on the leaching process of phosphate ceramics, potential hosts for actinide immobilization. J Nucl Mater 297(2):129–137. https://doi.org/10.1016/s0022-3115(01)00607-9
21. Bokov D et al (2021) Nanomaterial by sol-gel method: synthesis and application. Adv Mater Sci Eng 2021(1). https://doi.org/10.1155/2021/5102014
22. Borlaf M, Moreno R (2021) Colloidal sol-gel: a powerful low-temperature aqueous synthesis route of nanosized powders and suspensions. Open Ceram 8:100200. https://doi.org/10.1016/j.oceram.2021.100200
23. Borovaya M et al (2021) Synthesis, properties and bioimaging applications of silver-based quantum dots. Int J Mol Sci 22(22):12202. https://doi.org/10.3390/ijms222212202
24. Budtova T et al (2023) Acoustic properties of aerogels: current status and prospects. Adv Eng Mater 25(6). https://doi.org/10.1002/adem.202201137
25. Büyüktiryaki S et al (2022) Synthesis of core-shell magnetic nanoparticles. In: Handbook of magnetic hybrid nanoalloys and their nanocomposites. Springer International Publishing, Cham, pp 1–42
26. Calabrese C et al (2022) Antifouling and antimicrobial activity of Ag, Cu and Fe nanoparticles supported on silica and titania. Inorg Chim Acta 529:120636. https://doi.org/10.1016/j.ica.2021.120636
27. Camlibel NO, Arik B (2017) Sol-gel applications in textile finishing processes. In: Recent applications in sol-gel synthesis. InTech

28. Camposilvan E et al (2015) Enhanced reliability of yttria-stabilized zirconia for dental applications. Acta Biomater 17:36–46. https://doi.org/10.1016/j.actbio.2015.01.023
29. Catauro M, Mozzati MC, Bollino F (2015) Sol–gel hybrid materials for aerospace applications: chemical characterization and comparative investigation of the magnetic properties. Acta Astronaut 117:153–162. https://doi.org/10.1016/j.actaastro.2015.08.003
30. Chakraborty U et al (2023) Advanced metal oxides nanostructures to recognize and eradicate water pollutants. Prog Mater Sci 139:101169. https://doi.org/10.1016/j.pmatsci.2023.101169
31. Chang C et al (2023) Review of the sol–gel method in preparing nano TiO_2 for advanced oxidation process. Nanotechnol Rev 12(1). https://doi.org/10.1515/ntrev-2023-0150
32. Chaturvedi S, Dave PN, Shah NK (2012) Applications of nano-catalyst in new era. J Saudi Chem Soc 16(3):307–325. https://doi.org/10.1016/j.jscs.2011.01.015
33. Chavali MS, Nikolova MP (2019) Metal oxide nanoparticles and their applications in nanotechnology. SN Appl Sci 1(6). https://doi.org/10.1007/s42452-019-0592-3
34. Chen CY, Lan GS, Tuan WH (2000) Preparation of mullite by the reaction sintering of kaolinite and alumina. J Eur Ceram Soc 20(14–15):2519–2525. https://doi.org/10.1016/s0955-2219(00)00125-4
35. Chen D et al (2020) Shelf-stable quantum-dot light-emitting diodes with high operational performance. Adv Mater (Deerfield Beach, Fla.) 32(52). https://doi.org/10.1002/adma.202006178
36. Chen YX et al (2022) Effect of silica aerogel on thermal insulation and acoustic absorption of geopolymer foam composites: the role of aerogel particle size. Compos Part B Eng 242:110048. https://doi.org/10.1016/j.compositesb.2022.110048
37. Cheng Y et al (2019) One-pot synthesis of layered disodium zirconium phosphate: crystal structure and application in the remediation of heavy-metal-contaminated wastewater. Inorg Chem 58(19):13020–13029. https://doi.org/10.1021/acs.inorgchem.9b01938
38. Chiang Y-C et al (2021) Antibacterial silver-containing mesoporous bioglass as a dentin remineralization agent in a microorganism-challenged environment. J Dent 106:103563. https://doi.org/10.1016/j.jdent.2020.103563
39. Chiriac AP et al (2010) Sol gel method performed for biomedical products implementation. Mini Rev Med Chem 10(11):990–1013. https://doi.org/10.2174/1389557511009010990
40. Chkhartishvili L et al (2023) Advanced boron carbide matrix nanocomposites obtained from liquid-charge: focused review. Condens Matter 8(2):37. https://doi.org/10.3390/condmat8020037
41. Coradin T (2022) Sol-gel process, structure, and properties. In: Handbook of cell biosensors. Springer International Publishing, Cham, pp 497–516
42. Cucharero J et al (2021) Acoustic coatings—a discreet way to control acoustic environment. Front Built Environ 7. https://doi.org/10.3389/fbuil.2021.665332
43. Cuiyun D et al (2012) Sound absorption characteristics of a high-temperature sintering porous ceramic material. Appl Acoust 73(9):865–871. https://doi.org/10.1016/j.apacoust.2012.01.004
44. Danks AE, Hall SR, Schnepp Z (2016) The evolution of 'sol–gel' chemistry as a technique for materials synthesis. Mater Horiz 3(2):91–112. https://doi.org/10.1039/c5mh00260e
45. Demazeau G (2008) Solvothermal processes: new trends in materials chemistry. J Phys Conf Ser 121(8):082003. https://doi.org/10.1088/1742-6596/121/8/082003
46. Demazeau G (2010) Review. Solvothermal processes: definition, key factors governing the involved chemical reactions and new trends. Z Naturforsch B J Chem Sci 65(8):999–1006. https://doi.org/10.1515/znb-2010-0805
47. Dhiflaoui H, Cheikh Larbi AB (2018) Scratch resistance and wear energy analysis of TiO_2 coatings on 316L stainless steel. In: Lecture notes in mechanical engineering. Springer International Publishing, Cham, pp 405–413
48. Díez AG et al (2022) Multicomponent magnetic nanoparticle engineering: the role of structure-property relationship in advanced applications. Mater Today Chem 26:101220. https://doi.org/10.1016/j.mtchem.2022.101220

49. Ding M, Liu W, Gref R (2022) Nanoscale MOFs: from synthesis to drug delivery and theranostics applications. Adv Drug Deliv Rev 190:114496. https://doi.org/10.1016/j.addr.2022.114496
50. Ellis JE, Crawford SE, Kim K-J (2021) Metal–organic framework thin films as versatile chemical sensing materials. Mater Adv 2(19):6169–6196. https://doi.org/10.1039/d1ma00535a
51. Epifani M et al (2007) Precursors for the combustion synthesis of metal oxides from the sol–gel processing of metal complexes. J Eur Ceram Soc 27(1):115–123. https://doi.org/10.1016/j.jeurceramsoc.2006.04.084
52. Esposito S (2019) 'Traditional' sol-gel chemistry as a powerful tool for the preparation of supported metal and metal oxide catalysts. Materials 12(4):668. https://doi.org/10.3390/ma12040668
53. Feng S-H, Li G-H (2017) Hydrothermal and solvothermal syntheses. In: Modern inorganic synthetic chemistry. Elsevier, pp 73–104
54. Fikry M et al (2024) Up-scaled preparation of Pt–Ni aerogel catalyst layers for polymer electrolyte fuel cell cathodes. ACS Appl Energy Mater 7(3):896–905. https://doi.org/10.1021/acsaem.3c01930
55. Filippov AV, Starov V (2023) Interaction of nanoparticles in electrolyte solutions. J Phys Chem B 127(29):6562–6572. https://doi.org/10.1021/acs.jpcb.3c01220
56. Florensa M et al (2022) Key parameters for the rational design, synthesis, and functionalization of biocompatible mesoporous silica nanoparticles. Pharmaceutics 14(12):2703. https://doi.org/10.3390/pharmaceutics14122703
57. Fritz G et al (2002) Electrosteric stabilization of colloidal dispersions. Langmuir 18(16):6381–6390. https://doi.org/10.1021/la015734j
58. Fujii T et al (2001) The sol–gel preparation and characterization of nanoporous silica membrane with controlled pore size. J Membr Sci 187(1–2):171–180. https://doi.org/10.1016/s0376-7388(01)00338-6
59. Gao Y et al (2024) Crystal morphology prediction models and regulating methods. Crystals 14(6):484. https://doi.org/10.3390/cryst14060484
60. Gilani SAB et al (2024) pH dependent synthesis of ceria nanoparticles for efficient sunlight-driven photocatalysis of methyl orange containing wastewater. Opt Mater 148:114871. https://doi.org/10.1016/j.optmat.2024.114871
61. Gonçalves MC (2018) Sol-gel silica nanoparticles in medicine: a natural choice. Design, synthesis and products. Molecules (Basel, Switzerland) 23(8):2021. https://doi.org/10.3390/molecules23082021
62. Grüneberger F et al (2015) Nanofibrillated cellulose in wood coatings: dispersion and stabilization of ZnO as UV absorber. Prog Org Coat 87:112–121. https://doi.org/10.1016/j.porgcoat.2015.05.025
63. Gulati R, Sharma S, Sharma RK (2022) Antimicrobial textile: recent developments and functional perspective. Polym Bull (Berlin, Germany) 79(8):5747–5771. https://doi.org/10.1007/s00289-021-03826-3
64. Hang J et al (2009) Electrostatic and electrosteric stabilization of aqueous suspensions of barite nanoparticles. Powder Technol 192(2):166–170. https://doi.org/10.1016/j.powtec.2008.12.010
65. Hao X et al (2002) The effect of temperature on the synthesis of BN nanocrystals. J Cryst Growth 241(1–2):124–128. https://doi.org/10.1016/s0022-0248(02)01291-5
66. Hassan H et al (2022) Gold nanomaterials—the golden approach from synthesis to applications. Mater Sci Energy Technol 5:375–390. https://doi.org/10.1016/j.mset.2022.09.004
67. Hull R et al (2018) Stochasticity in materials structure, properties, and processing—a review. Appl Phys Rev 5(1). https://doi.org/10.1063/1.4998144
68. Huo Y et al (2023) Solvothermal synthesis and applications of micro/nano carbons: a review. Chem Eng J (Lausanne, Switzerland: 1996) 451:138572. https://doi.org/10.1016/j.cej.2022.138572

69. Hussain S, Maktedar SS (2023) Structural, functional and mechanical performance of advanced graphene-based composite hydrogels. Results Chem 6:101029. https://doi.org/10.1016/j.rechem.2023.101029
70. Inoue M (2013) Solvothermal synthesis of metal oxides. In: Handbook of advanced ceramics. Elsevier, pp 927–948
71. Irede EL et al (2024) Cutting-edge developments in zinc oxide nanoparticles: synthesis and applications for enhanced antimicrobial and UV protection in healthcare solutions. RSC Adv 14(29):20992–21034. https://doi.org/10.1039/d4ra02452d
72. Jagadeesh P, Rangappa SM, Siengchin S (2023) Advanced characterization techniques for nanostructured materials in biomedical applications. Adv Ind Eng Polym Res. https://doi.org/10.1016/j.aiepr.2023.03.002
73. Jeyaraj M et al (2019) A comprehensive review on the synthesis, characterization, and biomedical application of platinum nanoparticles. Nanomaterials (Basel, Switzerland) 9(12):1719. https://doi.org/10.3390/nano9121719
74. Jiang L et al (2021) Effect of stacking configuration on dynamic tensile properties of braided carbon fibre composites under medium-low strain rates. Compos Sci Technol 213:108924. https://doi.org/10.1016/j.compscitech.2021.108924
75. Kessler VG (2016) The synthesis and solution stability of alkoxide precursors. In: Handbook of sol-gel science and technology. Springer International Publishing, Cham, pp 1–50
76. Khan I, Saeed K, Khan I (2019) Nanoparticles: properties, applications and toxicities. Arab J Chem 12(7):908–931. https://doi.org/10.1016/j.arabjc.2017.05.011
77. Kharkwal A et al (2014) A solvothermal approach for the size-, shape- and phase-controlled synthesis and properties of $CuInS_2$. Mater Chem Phys 144(3):252–262. https://doi.org/10.1016/j.matchemphys.2013.12.023
78. Kim J et al (2016) A high-performance and fouling resistant thin-film composite membrane prepared via coating TiO_2 nanoparticles by sol-gel-derived spray method for PRO applications. Desalination 397:157–164. https://doi.org/10.1016/j.desal.2016.07.002
79. Kokol V et al (2021) Rheological properties of gelatine hydrogels affected by flow- and horizontally-induced cooling rates during 3D cryo-printing. Colloids Surf A Physicochem Eng Asp 616:126356. https://doi.org/10.1016/j.colsurfa.2021.126356
80. Kumarage S et al (2022a) A comprehensive review on electrospun nanohybrid membranes for wastewater treatment. Beilstein J Nanotechnol 13(1):137–159. https://doi.org/10.3762/bjnano.13.10
81. Kumarage S et al (2022b) Contemporary, multidisciplinary roles of mesoporous silica nanohybrids/nanocomposites. ChemistrySelect 7(21):e202200574. https://doi.org/10.1002/slct.202200574
82. Kumarage S et al (2023) Electrospun amine-functionalized silica nanoparticles–cellulose acetate nanofiber membranes for effective removal of hardness and heavy metals (As(V), Cd(II), Pb(II)) in drinking water sources. Environ Sci Water Res Technol 9(10):2664–2679. https://doi.org/10.1039/d3ew00312d
83. Lee S-H et al (2001) Synthesis of TiO_2 photocatalyst thin film by solvothermal method with a small amount of water and its photocatalytic performance. J Photochem Photobiol A Chem 146(1–2):121–128. https://doi.org/10.1016/s1010-6030(01)00553-6
84. Lee KR et al (2019) Investigation of physical and chemical properties for upgraded SAP ($SiO_2Al_2O_3P_2O_5$) waste form to immobilize radioactive waste salt. J Nucl Mater 515:382–391. https://doi.org/10.1016/j.jnucmat.2019.01.004
85. Li B, Logan BE (2005) The impact of ultraviolet light on bacterial adhesion to glass and metal oxide-coated surface. Colloids Surf B Biointerfaces 41(2–3):153–161. https://doi.org/10.1016/j.colsurfb.2004.12.001
86. Li J et al (1999) Low temperature route towards new materials: solvothermal synthesis of metal chalcogenides in ethylenediamine. Coord Chem Rev 190–192:707–735. https://doi.org/10.1016/s0010-8545(99)00107-1
87. Li B et al (2014) Porous metal–organic frameworks for gas storage and separation: what, how, and why? J Phys Chem Lett 5(20):3468–3479. https://doi.org/10.1021/jz501586e

88. Li J, Wu Q, Wu J (2015) Synthesis of nanoparticles via solvothermal and hydrothermal methods. In: Handbook of nanoparticles. Springer International Publishing, Cham, pp 1–28
89. Li H et al (2017) Recent advances in metal-containing polymer hydrogels. Macromol Rapid Commun 38(14). https://doi.org/10.1002/marc.201700109
90. Li C et al (2021) A review of silicon-based aerogel thermal insulation materials: performance optimization through composition and microstructure. J Non-Cryst Solids 553:120517. https://doi.org/10.1016/j.jnoncrysol.2020.120517
91. Li M et al (2022) The effect of heating method on the gel structures and properties of surimi prepared from Bombay duck (*Harpadon nehereus*). Front Nutr 9. https://doi.org/10.3389/fnut.2022.1060188
92. Li T et al (2024) Fabrication of a porous low refractive index anti-reflective coating with high transmittance by using the porogen of Laureth-5 carboxylic acid. Coatings 14(4):440. https://doi.org/10.3390/coatings14040440
93. Liang G et al (2020) Recent progress in the development of upconversion nanomaterials in bioimaging and disease treatment. J Nanobiotechnol 18(1). https://doi.org/10.1186/s12951-020-00713-3
94. Lin Y et al (2019) Size and dimension dependent surface-enhanced Raman scattering properties of well-defined Ag nanocubes. Appl Mater Today 14:224–232. https://doi.org/10.1016/j.apmt.2018.12.012
95. Liu L et al (2022a) Solvothermal synthesis of zirconia nanomaterials: latest developments and future. Ceram Int 48(22):32649–32676. https://doi.org/10.1016/j.ceramint.2022.07.290
96. Liu L et al (2022b) From modification to mechanism: supercritical hydrothermal synthesis of nano-zirconia. Ceram Int 48(4):4401–4423. https://doi.org/10.1016/j.ceramint.2021.11.028
97. Livage J et al (1989) The chemistry of the sol-gel process. Solid State Ionics 32–33:633–638. https://doi.org/10.1016/0167-2738(89)90338-x
98. Łuczak J et al (2023) Morphology control through the synthesis of metal-organic frameworks. Adv Colloid Interface Sci 314:102864. https://doi.org/10.1016/j.cis.2023.102864
99. Luković Golić D et al (2011) Structural characterization of self-assembled ZnO nanoparticles obtained by the sol–gel method from $Zn(CH_3COO)_2{\cdot}2H_2O$. Nanotechnology 22(39):395603. https://doi.org/10.1088/0957-4484/22/39/395603
100. Madhusha C et al (2023) Synthesis and structural characterization of copper nanoparticles doped activated carbon derived from coconut coir for drinking water purification. Mater Today Chem 27:101312. https://doi.org/10.1016/j.mtchem.2022.101312
101. Madhushika IP et al (2024) The antimicrobial synergy of polymer based nanofiber mats reinforced with antioxidants intercalated layered double hydroxides as a potential active packaging material. Nano Express 5(2):025018. https://doi.org/10.1088/2632-959x/ad4a95
102. Malik S, Muhammad K, Waheed Y (2023) Nanotechnology: a revolution in modern industry. Molecules (Basel, Switzerland) 28(2):661. https://doi.org/10.3390/molecules28020661
103. Matter F, Luna AL, Niederberger M (2020) From colloidal dispersions to aerogels: how to master nanoparticle gelation. Nano Today 30:100827. https://doi.org/10.1016/j.nantod.2019.100827
104. Mekonnen BT et al (2021) Preparation of aerogel and its application progress in coatings: a mini overview. J Leather Sci Eng 3(1). https://doi.org/10.1186/s42825-021-00067-y
105. Miller QRS et al (2018) Microporous and flexible framework acoustic metamaterials for sound attenuation and contrast agent applications. ACS Appl Mater Interfaces 10(51):44226–44230. https://doi.org/10.1021/acsami.8b19249
106. Mirghani R et al (2022) Biofilms: formation, drug resistance and alternatives to conventional approaches. AIMS Microbiol 8(3):239–277. https://doi.org/10.3934/microbiol.2022019
107. Mo L-E et al (2019) Facile synthesis of TiO_2 microspheres via solvothermal alcoholysis method for high performance dye-sensitized solar cells. Sol Energy (Phoenix, Ariz.) 177:448–454. https://doi.org/10.1016/j.solener.2018.10.013
108. Mohamad SNS et al (2019) Synthesis of alumina nanoparticles by sol-gel method and their applications in the removal of copper ions (Cu^{2+}) from the solution. IOP Conf Ser Mater Sci Eng 701(1):012034. https://doi.org/10.1088/1757-899x/701/1/012034

109. Munaweera I, Yapa P (2024) Principles and applications of nanotherapeutics. CRC Press, Boca Raton
110. Munaweera I et al (2014) Chemoradiotherapeutic wrinkled mesoporous silica nanoparticles for use in cancer therapy. APL Mater 2(11). https://doi.org/10.1063/1.4899118
111. Munaweera I et al (2015) Novel wrinkled periodic mesoporous organosilica nanoparticles for hydrophobic anticancer drug delivery. J Porous Mater 22(1):1–10. https://doi.org/10.1007/s10934-014-9897-1
112. Munaweera I, Pathiraja AS, Kottegoda N (2023) Surface functionalized mesoporous silica nanoparticles for enhanced removal of heavy metals: a review. Vidyodaya J Sci 1(s1). https://doi.org/10.31357/vjs.v1is1.6705
113. Mustapha AN et al (2022) A systematic study on the reaction mechanisms for the microencapsulation of a volatile phase change material (PCM) via one-step in situ polymerisation. Chem Eng Sci 252:117497. https://doi.org/10.1016/j.ces.2022.117497
114. Narayan R et al (2018) Mesoporous silica nanoparticles: a comprehensive review on synthesis and recent advances. Pharmaceutics 10(3):118. https://doi.org/10.3390/pharmaceutics10030118
115. Navas D et al (2021) Review on sol-gel synthesis of perovskite and oxide nanomaterials. Gels (Basel, Switzerland) 7(4):275. https://doi.org/10.3390/gels7040275
116. Ndlwana L et al (2021) Sustainable hydrothermal and solvothermal synthesis of advanced carbon materials in multidimensional applications: a review. Materials 14(17):5094. https://doi.org/10.3390/ma14175094
117. Nedylakova M et al (2024) Iron oxide magnetic aggregates: aspects of synthesis, computational approaches and applications. Adv Colloid Interface Sci 323:103056. https://doi.org/10.1016/j.cis.2023.103056
118. Nguyen MD et al (2021) Fe_3O_4 nanoparticles: structures, synthesis, magnetic properties, surface functionalization, and emerging applications. Appl Sci (Basel, Switzerland) 11(23):11301. https://doi.org/10.3390/app112311301
119. Nyabadza A et al (2023) A review of physical, chemical and biological synthesis methods of bimetallic nanoparticles and applications in sensing, water treatment, biomedicine, catalysis and hydrogen storage. Adv Colloid Interface Sci 321:103010. https://doi.org/10.1016/j.cis.2023.103010
120. Oey T et al (2019) The role of the network-modifier's field-strength in the chemical durability of aluminoborate glasses. J Non-Cryst Solids 505:279–285. https://doi.org/10.1016/j.jnoncrysol.2018.11.019
121. Ooi MDJ, Aziz AA, Adnan R (2021) Solvent effects on the structural and catalytic properties of Pt nano- and microstructures synthesised in solvothermal system. Int J Hydrogen Energy 46(8):5926–5937. https://doi.org/10.1016/j.ijhydene.2020.11.218
122. Osman AI et al (2024) Synthesis of green nanoparticles for energy, biomedical, environmental, agricultural, and food applications: a review. Environ Chem Lett 22(2):841–887. https://doi.org/10.1007/s10311-023-01682-3
123. Owens GJ et al (2016) Sol–gel based materials for biomedical applications. Prog Mater Sci 77:1–79. https://doi.org/10.1016/j.pmatsci.2015.12.001
124. Palmisano G, Jitan SA, Garlisi C (2022) Surface area and porosity. In: Heterogeneous catalysis. Elsevier, pp 101–140
125. Paszkiewicz M et al (2016) The ILs-assisted solvothermal synthesis of TiO_2 spheres: the effect of ionic liquids on morphology and photoactivity of TiO_2. Appl Catal B Environ 184:223–237. https://doi.org/10.1016/j.apcatb.2015.11.019
126. Patra JK et al (2018) Nano based drug delivery systems: recent developments and future prospects. J Nanobiotechnol 16(1). https://doi.org/10.1186/s12951-018-0392-8
127. Payanda Konuk O et al (2023) The effect of synthesis conditions and process parameters on aerogel properties. Front Chem 11. https://doi.org/10.3389/fchem.2023.1294520
128. Peiris SE et al (2022) In vitro evaluation of antibacterial activity of copper and sulfur nanoparticles for controlling bacterial blight caused by *Xanthomonas* sp. in *Anthurium andraeanum* Lind. SLIIT J Humanit Sci 3(1):46–55. https://doi.org/10.4038/sjhs.v3i1.50

129. Perveen F et al (2022) Catalytic conversion of agricultural waste biomass into valued chemical using bifunctional heterogeneous catalyst: a sustainable approach. Catal Commun 171:106516. https://doi.org/10.1016/j.catcom.2022.106516
130. Pinton JHB et al (2024) Development of an automated and cost-effective apparatus for sol-gel solution deposition using spray coating technique and its application for TiO_2-based photocatalytic films. Mater Chem Phys 318:129213. https://doi.org/10.1016/j.matchemphys.2024.129213
131. Pontes-Quero GM et al (2021) Modulation of inflammatory mediators by polymeric nanoparticles loaded with anti-inflammatory drugs. Pharmaceutics 13(2):290. https://doi.org/10.3390/pharmaceutics13020290
132. Pretto T et al (2023) A sol-gel/solvothermal synthetic approach to titania nanoparticles for Raman thermometry. Sensors (Basel, Switzerland) 23(5):2596. https://doi.org/10.3390/s23052596
133. Rai P, Kwak W-K, Yu Y-T (2013) Solvothermal synthesis of ZnO nanostructures and their morphology-dependent gas-sensing properties. ACS Appl Mater Interfaces 5(8):3026–3032. https://doi.org/10.1021/am302811h
134. Rajapaksha S, Yapa P, Munaweera I (2025) Innovation management and nanotechnology: a PRISMA-based analysis and research implications. Int J Innov Sci. https://doi.org/10.1108/ijis-08-2024-0215
135. Ramakrishnan VM et al (2018) Size controlled synthesis of TiO_2 nanoparticles by modified solvothermal method towards effective photo catalytic and photovoltaic applications. Mater Res Bull 97:351–360. https://doi.org/10.1016/j.materresbull.2017.09.017
136. Ranathunga K et al (2024) Preparation and characterization of Fe–ZnO cellulose-based nanofiber mats with self-sterilizing photocatalytic activity to enhance antibacterial applications under visible light. RSC Adv 14(26):18536–18552. https://doi.org/10.1039/d4ra03136a
137. Rani L et al (2023) Significance of MOF adsorbents in uranium remediation from water. Environ Res 236:116795. https://doi.org/10.1016/j.envres.2023.116795
138. Rasaki SA et al (2018) Synthesis and application of nano-structured metal nitrides and carbides: a review. Prog Solid State Chem 50:1–15. https://doi.org/10.1016/j.progsolidstchem.2018.05.001
139. Rathod S et al (2024) Exploring synthesis and applications of green nanoparticles and the role of nanotechnology in wastewater treatment. Biotechnol Rep (Amsterdam, Netherlands) 41:e00830. https://doi.org/10.1016/j.btre.2024.e00830
140. Righini GC et al (2023) Sol–gel photonic glasses: from material to application. Materials 16(7):2724. https://doi.org/10.3390/ma16072724
141. Saleem S et al (2023) Analysis and characterization of opto-electronic properties of iron oxide (Fe_2O_3) with transition metals (Co, Ni) for the use in the photodetector application. J Mater Res Technol 25:6150–6166. https://doi.org/10.1016/j.jmrt.2023.07.065
142. Saleh HM, Hassan AI (2023) Synthesis and characterization of nanomaterials for application in cost-effective electrochemical devices. Sustainability 15(14):10891. https://doi.org/10.3390/su151410891
143. Salinas AJ, Vallet-Regí M (2008) The sol-gel production of bioceramics. Key Eng Mater 391:141–158. https://doi.org/10.4028/www.scientific.net/kem.391.141
144. Sasani Ghamsari M, Bahramian AR (2008) High transparent sol–gel derived nanostructured TiO_2 thin film. Mater Lett 62(3):361–364. https://doi.org/10.1016/j.matlet.2007.05.053
145. Schubert U (2015) Chemistry and fundamentals of the sol–gel process. In: The sol-gel handbook. Wiley, pp 1–28. https://doi.org/10.1002/9783527670819.ch01
146. Selvaraj S et al (2024) In vitro analysis of XLAsp-P2 peptide loaded cellulose acetate nanofiber for wound healing. J Pharm Sci. https://doi.org/10.1016/j.xphs.2024.10.050
147. Sequeira CAC (2013) Role of solid state chemistry in high temperature corrosion. Chem Eng Res Des 91(2):318–324. https://doi.org/10.1016/j.cherd.2012.09.021
148. Shekhah O et al (2018) Metal–organic framework membranes: from fabrication to gas separation. Crystals 8(11):412. https://doi.org/10.3390/cryst8110412

149. Shimoga G, Shin E-J, Kim S-Y (2019) Thermal, dielectric and catalytic behavior of palladium doped PVC films. Polímeros 29(2). https://doi.org/10.1590/0104-1428.08718
150. Sivakumar S et al (2004) Nanoporous titania–alumina mixed oxides—an alkoxide free sol–gel synthesis. Mater Lett 58(21):2664–2669. https://doi.org/10.1016/j.matlet.2004.03.050
151. Sopan Mahato S et al (2023) Perspective chapter: sol-gel science and technology in context of nanomaterials—recent advances. In: Sol-gel method—recent advances. IntechOpen
152. Srivastava AK, Das K (2010) The abrasive wear resistance of TIC and (Ti, W)C-reinforced Fe–17Mn austenitic steel matrix composites. Tribol Int 43(5–6):944–950. https://doi.org/10.1016/j.triboint.2009.12.057
153. Subuki I et al (2020) Study of the synthesis of zirconia powder from zircon sand obtained from zircon minerals Malaysia by caustic fusion method. Indones J Chem 20(4):782. https://doi.org/10.22146/ijc.43936
154. Sudheep CV et al (2024) Revolutionizing gas sensors: the role of composite materials with conducting polymers and transition metal oxides. Results Chem 7:101255. https://doi.org/10.1016/j.rechem.2023.101255
155. Svetlizky D, Eliaz N (2021) Sol–gel encapsulation of ZnAl alloy powder with alumina shell. Coatings 11(11):1389. https://doi.org/10.3390/coatings11111389
156. Tan TL et al (2022) One-pot solvothermal synthesis of Zr-based MOFs with enhanced adsorption capacity for Cu^{2+} ions removal. J Solid State Chem 315:123429. https://doi.org/10.1016/j.jssc.2022.123429
157. Thompson LH, Doraiswamy LK (2000) The rate enhancing effect of ultrasound by inducing supersaturation in a solid–liquid system. Chem Eng Sci 55(16):3085–3090. https://doi.org/10.1016/s0009-2509(99)00481-9
158. Tsay C-Y, Lee W-C (2013) Effect of dopants on the structural, optical and electrical properties of sol–gel derived ZnO semiconductor thin films. Curr Appl Phys 13(1):60–65. https://doi.org/10.1016/j.cap.2012.06.010
159. Vakifahmetoglu C, Zeydanli D, Colombo P (2016) Porous polymer derived ceramics. Mater Sci Eng R Rep 106:1–30. https://doi.org/10.1016/j.mser.2016.05.001
160. Valencia S et al (2013) Sol–gel and low-temperature solvothermal synthesis of photoactive nano-titanium dioxide. J Photochem Photobiol A Chem 251:175–181. https://doi.org/10.1016/j.jphotochem.2012.10.025
161. Valliant EM et al (2012) Role of pH and temperature on silica network formation and calcium incorporation into sol–gel derived bioactive glasses. J Mater Chem 22(4):1613–1619. https://doi.org/10.1039/c1jm13225c
162. Venkateswara Rao A, Bhagat SD (2004) Synthesis and physical properties of TEOS-based silica aerogels prepared by two step (acid–base) sol–gel process. Solid State Sci 6(9):945–952. https://doi.org/10.1016/j.solidstatesciences.2004.04.010
163. Walton RI (2011) Solvothermal synthesis of cerium oxides. Prog Cryst Growth Charact Mater 57(4):93–108. https://doi.org/10.1016/j.pcrysgrow.2011.10.002
164. Wang X et al (2018) Graphene oxide-based polymeric membranes for water treatment. Adv Mater Interfaces 5(15). https://doi.org/10.1002/admi.201701427
165. Wang Y et al (2021) Solvothermal synthesis of ZnO nanoparticles for photocatalytic degradation of methyl orange and p-nitrophenol. Water 13(22):3224. https://doi.org/10.3390/w13223224
166. Wannaborworn M, Praserthdam P, Jongsomjit B (2015) A comparative study of solvothermal and sol-gel-derived nanocrystalline alumina catalysts for ethanol dehydration. J Nanomater 2015(1). https://doi.org/10.1155/2015/519425
167. Wei J et al (2021) Direct photoinduced synthesis of lead halide perovskite nanocrystals and nanocomposites. Nano Today 39:101179. https://doi.org/10.1016/j.nantod.2021.101179
168. Worthen AJ et al (2016) Steric stabilization of nanoparticles with grafted low molecular weight ligands in highly concentrated brines including divalent ions. Soft Matter 12(7):2025–2039. https://doi.org/10.1039/c5sm02787j
169. Xu SH et al (2021) Sol-gel preparation of high transmittance of infrared antireflective coating for TeO_2 crystals. Infrared Phys Technol 118:103881. https://doi.org/10.1016/j.infrared.2021.103881

170. Xu B et al (2023) Multifunctional mesoporous silica nanoparticles for biomedical applications. Sig Transduct Target Ther 8(1). https://doi.org/10.1038/s41392-023-01654-7
171. Yadav A et al (2020) Fabrication of 3D polymeric photonic arrays and related applications. Mater Today Chem 15:100208. https://doi.org/10.1016/j.mtchem.2019.100208
172. Yadav N et al (2023) Recent advances in nanotechnology for the improvement of conventional agricultural systems: a review. Plant Nano Biol 4:100032. https://doi.org/10.1016/j.plana.2023.100032
173. Yang D et al (2013) Titanate-based adsorbents for radioactive ions entrapment from water. Nanoscale 5(6):2232. https://doi.org/10.1039/c3nr33622k
174. Yang H, Zhu M, Li Y (2023) Sol–gel research in China: a brief history and recent research trends in synthesis of sol–gel derived materials and their applications. J Sol-Gel Sci Technol 106(2):406–421. https://doi.org/10.1007/s10971-022-05750-y
175. Yapa PN et al (2024a) Metal doped silica nanohybrids with extensive bacterial coverage for antibacterial applications exhibit synergistic activity. Biomater Adv 157:213753. https://doi.org/10.1016/j.bioadv.2023.213753
176. Yapa PN et al (2024b) Nanoarchitectonics for synergistic activity of multimetallic nanohybrids as a possible approach for antimicrobial resistance (AMR). J Biol Inorg Chem (JBIC) 29(5):477–498. https://doi.org/10.1007/s00775-024-02066-w
177. Yapa P et al (2024a) Potential antifungal applications of heterometallic silica nanohybrids: a synergistic activity. Biomater Adv 162:213930. https://doi.org/10.1016/j.bioadv.2024.213930
178. Yapa P et al (2024b) Synergistic antimicrobial nanofiber membranes based on metal incorporated silica nanoparticles as advanced antimicrobial layers. RSC Adv 14(46):33919–33940. https://doi.org/10.1039/d4ra05052e
179. Yarbrough R et al (2020) A sol–gel synthesis to prepare size and shape-controlled mesoporous nanostructures of binary (II–VI) metal oxides. RSC Adv 10(24):14134–14146. https://doi.org/10.1039/d0ra01778g
180. Ye H, Liu Y, Zhao X (2024) Recent research progress on graphene-based sound-absorbing materials. Mater Today Chem 38:102139. https://doi.org/10.1016/j.mtchem.2024.102139
181. Yeap SP et al (2012) Electrosteric stabilization and its role in cooperative magnetophoresis of colloidal magnetic nanoparticles. Langmuir 28(42):14878–14891. https://doi.org/10.1021/la303169g
182. Yoshimura M, Byrappa K (2008) Hydrothermal processing of materials: past, present and future. J Mater Sci 43(7):2085–2103. https://doi.org/10.1007/s10853-007-1853-x
183. Yu W, Xie H (2012) A review on nanofluids: preparation, stability mechanisms, and applications. J Nanomater 2012(1). https://doi.org/10.1155/2012/435873
184. Zahrouni A et al (2023) Effect of sol–gel derived TiO_2–SiO_2 binary nanoparticles on thermomechanical property of polymer matrix composite. J Inorg Organomet Polym Mater 33(12):4052–4067. https://doi.org/10.1007/s10904-023-02637-5
185. Zayat M et al (2011) Optical and electro-optical materials prepared by the sol-gel method. Adv Mater (Deerfield Beach, Fla.) 23(44):5318–5323. https://doi.org/10.1002/adma.201101927
186. Zayat M et al (2018) Sol-gel materials for optics and electrooptics. In: Handbook of sol-gel science and technology. Springer International Publishing, Cham, pp 2065–2092
187. Zhang S et al (2021) A sol–gel based silver nanoparticle/polytetrafluorethylene (AgNP/PTFE) coating with enhanced antibacterial and anti-corrosive properties. Appl Surf Sci 535:147675. https://doi.org/10.1016/j.apsusc.2020.147675
188. Zhang Q et al (2023) Recent advances in metal-organic frameworks: synthesis, application and toxicity. Sci Total Environ 902:165944. https://doi.org/10.1016/j.scitotenv.2023.165944
189. Zuo L et al (2015) Polymer/carbon-based hybrid aerogels: preparation, properties and applications. Materials 8(10):6806–6848. https://doi.org/10.3390/ma8105343
190. Nguyen MD et al (2021) Fe3O4 nanoparticles: structures, synthesis, magnetic properties, surface functionalization, and emerging applications. Appl Sci (Basel, Switzerland) 11(23):11301. https://doi.org/10.3390/app112311301

Supercritical Fluid Technologies for Nanoparticle Production

C. S. Kalpage and R. A. T. Isuranga

Abstract Supercritical fluid (SCF) technologies offer a versatile and sustainable approach to synthesizing nanoparticles with precise control over size, morphology, and functionality, suitable for various industrial applications. SCFs, particularly carbon dioxide and water above their critical temperature and pressure, exhibit liquid-like solvating power and gas-like diffusivity, making them ideal for producing particles with tunable properties. This paper explores the principles and processes of SCF technologies, emphasizing methods like Rapid Expansion of Supercritical Solutions (RESS), Supercritical Anti-solvent (SAS), and Particles from Gas-Saturated Solutions (PGSS), each demonstrating unique advantages in nanoparticle production. RESS involves rapid depressurization, allowing fine control over particle size, while SAS uses SCFs to precipitate nanoparticles from solutions, offering enhanced bioavailability and solubility for pharmaceutical applications. The PGSS technique encapsulates various compounds without requiring solubility in SC-CO_2, supporting applications in food and pharmaceuticals. Advancements in SCF dynamics, such as SAS with enhanced mass transfer (SAS-EM), have enabled efficient, high-quality production of nanoparticles while minimizing organic solvent use. SCF technology has proven advantageous for industrial processes due to its ability to meet regulatory and safety requirements, particularly in the pharmaceutical and materials sciences sectors. This review underscores SCF's role in clean nanoparticle production, exploring its application in drug delivery, catalysis, and electronic materials, and highlights ongoing research for optimizing process variables and scaling up operations. The environmental and economic benefits of SCF technologies demonstrate their potential to revolutionize industries by aligning with green chemistry principles and sustainable development goals.

Keywords Supercritical fluids · Nanoparticles · Rapid Expansion of Supercritical Solutions · Supercritical anti-solvent · Green chemistry · Particle size control · Drug delivery · Nanomaterials · Sustainable synthesis

C. S. Kalpage (✉) · R. A. T. Isuranga
Department of Chemical and Process Engineering, Faculty of Engineering, University of Peradeniya, Peradeniya, Sri Lanka
e-mail: csk@eng.pdn.ac.lk

N. M. Adassooriya et al. (eds.), *Industrial Scale Production of Nanoparticles*, Smart Nanomaterials Technology, https://doi.org/10.1007/978-981-96-9176-0_11

Nomenclature and Notations

API	Active pharmaceutical ingredient
ASES	Aerosol Solvent Extraction System
DCM	Dichloromethane
DME	Dimethyl ether
DMSO	Dimethyl sulfoxide
ELS	Expanded liquid anti-solvent
PE	Polyethylene
PEG	Polyethylene glycol
PGSS	Particles from Gas-Saturated Solutions
RESOLV	Rapid Expansion of a Supercritical Solution into a Liquid Solvent
RESS	Rapid Expansion of Supercritical Solutions
SAILA	Supercritical assisted injection in a liquid anti-solvent
SAS	Supercritical Anti-solvent
SAS-EM	Supercritical Anti-solvent with Enhanced Mass Transfer
SC-CO_2	Supercritical carbon dioxide
SCF	Supercritical fluid
SCFs	Supercritical fluids
SC-H_2O	Supercritical water
SEDS	Solution-Enhanced Dispersion System
SFEE	Supercritical Fluid Extraction of Emulsions
SpEDS	Suspension-Enhanced Dispersion by Supercritical Fluids
VLE	Vapor-liquid equilibria
$AgNO_3$	Silver nitrate
CO_2	Carbon dioxide
H_2O	Water
P_c	Critical pressure
T_c	Critical temperature

1 Introduction

Supercritical fluids (SCFs) are substances above their critical temperature and pressure, exhibiting properties of both liquids and gases. They possess liquid-like solvating power and gas-like diffusion and viscosity, enabling rapid penetration into crevices and effective contaminant removal [11, 83]. The critical point of a substance is characterized by its critical temperature (T_c) and critical pressure (P_c), above which the distinction between liquid and gas phases disappears [22]. This phenomenon was first observed by Cagniard de la Tour in 1822, who noted the cessation of splashing in a sealed cannon at a certain temperature. The critical point marks the end of the gas–liquid coexistence curve on a phase diagram, where the densities of both phases become identical [22] (Fig. 1).

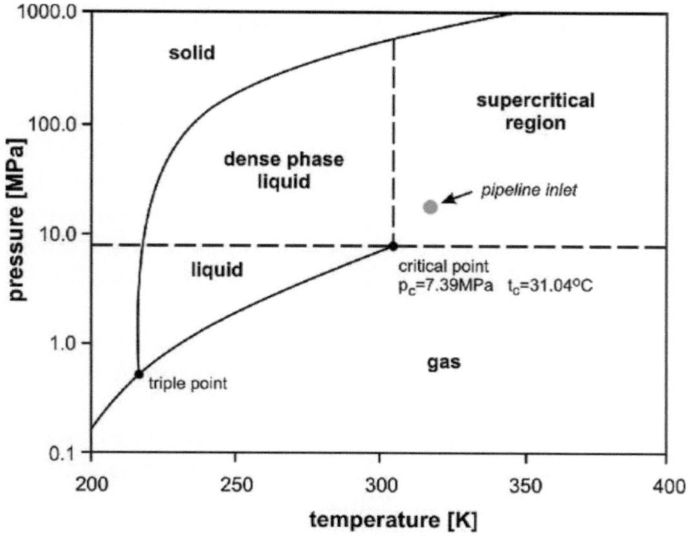

Fig. 1 Phase diagram of CO_2

SCFs, particularly CO_2, are widely used in extraction and purification processes due to their tunable properties, which can be adjusted by altering pressure and temperature [68, 83]. Their unique characteristics, such as negligible surface tension and high diffusivity, make them suitable for extracting solutes from various plant materials and as alternatives to organic solvents in industrial application [68]. While CO_2 is the most commonly used SCF, others like propane, butane and water are also employed. Co-solvents and modifiers can enhance extraction performance, especially for polar molecules [68].

SCFs offer unique advantages for nanoparticle synthesis and processing due to their tunable properties. By adjusting temperature and pressure near the critical point, researchers can precisely control solvent strength, density, and diffusivity [100]. This tunability enables size-selective precipitation and dispersibility of nanocrystals, as demonstrated with gold nanoparticles in near-critical and supercritical ethane and propane [37]. SCF technology, particularly using CO_2 and H_2O, has become crucial for fabricating inorganic and hybrid nanomaterials for biomedical applications, such as phosphors, magnetic materials, and carbon nanotubes [14]. The ability to control nanoparticle size and shape is essential for determining their properties and application potential. SCF synthesis offers a potentially greener route for producing metal oxide nanoparticles with precise control over morphology, making it a promising technique for various applications [66].

Recent advances in supercritical fluid dynamics have enhanced our understanding of their unique flow behavior and applications. Supercritical fluids exhibit higher diffusivity than liquids and lower viscosity than gases, enabling efficient penetration of porous materials [8]. They demonstrate transitional behavior between laminar and turbulent flows at lower Reynolds numbers compared to traditional fluids, facilitating better mixing and homogenization [8].

These properties are particularly beneficial in nanoparticle production and pharmaceutical applications [58, 118]. The zero surface tension of supercritical fluids makes them ideal for nanotechnology processes, such as lithography in the electronic industry [20]. Improved computational fluid dynamics modeling has allowed manufacturers to better predict supercritical fluid behavior during nanoparticle synthesis, leading to more consistent product quality in scaled-up processes [8]. These advancements have contributed to the design and understanding of catalysts, crystals, sol-gels, and composite materials [20].

SCF technology enables the creation of nanoparticles with narrow size distributions and specific shapes, which are essential for applications in drug delivery, catalysis, and electronics [101]. SCF methods allow for the production of functionalized and composite nanoparticles, enhancing performance in targeted drug delivery and related applications [14]. The process is advantageous due to its mild conditions, which benefit labile compounds, and its environmentally benign materials that reduce the use of organic solvents [70]. SCF technology has been applied to numerous pharmaceutical processes, such as crystallization, particle size reduction, and the development of drug delivery systems [70, 118]. Recognized as a sustainable approach to nanoparticle synthesis, SCF processes—especially those utilizing CO_2 and H_2O—reduce or eliminate toxic organic solvents, promoting cleaner production with reduced waste [44]. The technology enables the controlled production of nanomaterials with specific properties tailored for diverse applications in catalysis, electronics, and pharmacology [6]. Controlling the size, morphology, and composition of nanoparticles is particularly important for biomedical uses, including bio-imaging, cancer therapy, and drug delivery. Recent advancements in SCF technology have focused on in situ surface modification and the formation of single nanocrystals, further enhancing its versatility in nanomaterial design [14].

Supercritical carbon dioxide (SC-CO_2) has emerged as a versatile and environmentally friendly medium for various applications. Its unique properties, including zero surface tension, low viscosity, and high solubility, make it ideal for materials processing and synthesis [123]. In the pharmaceutical industry, SC-CO_2 technologies can improve drug solubility, bioavailability, and therapeutic effects, leading to the development of new formulations with reduced dosage requirements [71]. SC-CO_2 plays a critical role in nanotechnology processes, particularly in the synthesis of nanoengineered materials with enhanced mechanical, optical, chemical, and electrical properties [20]. Its applications extend to drug solubility studies, extraction of valuable materials, and nanoparticle production for pharmaceutical processes [102]. The growing interest in SC-CO_2 technologies across multiple industries suggests a promising future for their integration in advanced material processing and drug development. In pharmaceutical applications, SC-CO_2 facilitates particle formation and clean synthesis of drug compounds, although challenges remain in scaling up these processes [103]. Recent developments include the design of surfactants for SC-CO_2, enabling the production of metal nanoparticles and porous polymers [117]. The technology has also been applied to deposit nanocatalysts on various support materials, resulting in improved performance for energy storage systems [26]. While SC-CO_2 offers numerous advantages in materials processing, further research is needed to optimize process variables such as temperature, pressure, and contact time to enhance the synthesis of nanostructured materials [26].

Supercritical water (SC-H_2O) has emerged as a promising medium for various applications, including wastewater treatment, biomass conversion, and materials synthesis for solar energy as an alternative to SC-CO_2 [4, 13, 61, 122]. Its unique properties at high temperatures and pressures enable efficient organic waste destruction and hydrogen production from biomass [4, 122]. However, challenges such as corrosion, salt deposition, and high energy consumption have hindered widespread commercialization [122]. Recent advancements in reactor design, materials modification, and energy recovery systems have addressed some of these issues [13, 122]. The development of compact heat exchangers has improved overall energy efficiency. Despite progress, technical challenges like corrosion and solid handling persist [4]. Ongoing research focuses on optimizing reactor designs, exploring catalysis, and investigating SC-H_2O's potential in generating novel materials for solar energy applications [61, 122]. Supercritical water oxidation (SCWO) has emerged as a promising method for destroying explosive materials and other energetic compounds. Studies have demonstrated that SCWO can achieve destruction efficiencies exceeding 99.9% for various explosives, including RDX, HMX, and TNT, within 30 s at temperatures above 600 °C. The process primarily produces N_2, N_2O, and CO_2, with some nitrate and nitrite ions [9]. SCWO has also shown effectiveness in treating colored dyes and pyrotechnic compositions, achieving destruction and removal efficiencies typically above 99.5% in less than 10 s [93]. Researchers have investigated the chemistry and reaction products of SCWO for various explosives and propellants [35]. While SCWO offers advantages over traditional disposal methods like open burning or incineration, challenges such as salt deposition and corrosion of reactor materials require further investigation [93].

Recent research explores eco-friendly alternatives to traditional supercritical fluids. Dimethyl ether (DME) has emerged as a promising green solvent for extracting bioactive compounds from microalgae and other natural sources, offering high extraction rates without the need for drying [115]. DME also shows potential as an environmentally friendly fuel, with reduced emissions compared to petroleum-based fuels [67]. Other supercritical fluids being investigated include propane and butane, alongside the dominant CO_2 [68]. These alternative fluids offer specific advantages for certain extractions, but their commercial application remains limited. Most research is still conducted at laboratory scale, with little information on scale-up strategies [68]. Despite these challenges, supercritical fluid technology continues to advance, with applications expanding in various industries, including power generation, food processing, and materials science [84].

2 Principles of Supercritical Fluid Technology

Supercritical fluid (SCF) technology leverages the distinctive properties of fluids above their critical point, where the substance exhibits both liquid-like density and gas-like diffusivity [2]. Understanding the phase behavior and solubility of SCFs is essential for various applications, such as extraction, particle production, and crystallization [2, 60]. The phase behavior of SCFs is intricate, with solubility undergoing significant variations based on pressure changes [60]. Several SCF technologies have been developed to harness these unique properties, including the Rapid Expansion of Supercritical Solutions (RESS), Gas Antisolvent Process (GAS), and Supercritical Fluid Extraction [76]. Factors influencing solubility in SCFs include the selection of solvent, the chemical functionality of the solute, and the temperature and pressure conditions [38]. The ability to control phase behavior and equilibrium is central to SCF separation, extraction, and reaction processes, making SCFs a promising technology for future industrial applications [2].

Accurate predictions of critical properties are vital for SCF applications. Schmidt and Wenzel [98] proposed a cubic equation of state to estimate critical temperature (T_c) and pressure (P_c) using the liquid density at room temperature and boiling temperatures under specific pressures. This method demonstrated good agreement with experimental data for non-associating substances and was applied to nearly 1700 compounds. Furthermore, the authors suggested that this approach could be extended to associating substances [107]. Table 1 provides the critical temperatures and pressures of several substances, which are essential for understanding their behavior in SCF applications.

Recent studies have significantly enhanced the understanding of critical conditions and high-pressure phenomena in materials. Research on metal superhydrides has shown potential for near-room-temperature superconductivity under extreme pressures [82]. The supercritical state exhibits a universal transition between liquid-like and gas-like behaviors across various fluids, challenging earlier conceptions of a featureless supercritical region [23]. Hydrogen, in particular, has been extensively

Table 1 Critical temperatures and pressures of some substances [22]

Substance	Critical temperature (K)	Critical pressure (bar)
Carbon dioxide	304	74
Water	647	221
Ethane	305	49
Ethane	282	50
Propane	370	43
Xenon	290	58
Ammonia	406	114
Nitrous oxide	310	72
Fluoroform	299	49

studied under extreme conditions, revealing novel phases and properties through advanced experimental and theoretical techniques [43]. Major breakthroughs in high-pressure science, such as room-temperature superconductivity and metallic hydrogen, have been reported [69]. However, claims of metallic hydrogen discovery require further substantiation and the enhancement of experimental capabilities to confirm its exotic behaviors [69]. These advancements contribute to a refined understanding of critical conditions and their effects on material properties.

Advancements have also focused on developing equations of state for multicomponent mixtures, including polymer solutions, and measuring critical properties of binary mixtures [79]. Ternary phase diagrams and computer programs designed to calculate critical mixture and phase border curves have been developed to assist in process design [79]. Numerous SCF processes, such as RESS, PGSS, GAS, and SAS, have been explored for applications in pharmaceuticals, natural products, and chemicals [41]. While these technologies show promise in areas like solubility enhancement, polymer processing, and extraction, challenges persist in implementing them at a commercial scale [41, 79].

The solubility of solids in SCFs can be significantly enhanced by adjusting the temperature, pressure, and using co-solvents [57]. Common polar co-solvents, such as ethanol, as well as nonpolar co-solvents like alkanes, can increase hydrocarbon solubility in SC-CO_2 [49]. Interestingly, the mechanism of solubility enhancement by co-solvents appears to be driven by entropy rather than enthalpy, as previously believed [121]. This suggests that co-solvents enhance solubility by altering the local density of the SC solvent around the solute and preferentially associating with it [121]. This understanding is crucial for optimizing SCF processes across various industries, including pharmaceuticals, food processing, and pollution remediation [57, 60].

The addition of co-solvents significantly increases solubility, particularly for polar solutes [57, 121]. This enhancement is primarily attributed to an increase in the entropy of solution, rather than enthalpy changes [121]. Common co-solvents include dimethylacetamide, dimethyl sulfoxide, ethanol, glycerin, polyethylene glycol, and propylene glycol [113]. Table 2 illustrates the solubility profile of rifabutin in pure solvents, while Table 3 presents the solubility of rifabutin in solvent mixtures, highlighting the significant improvements that co-solvents can offer.

The phase behavior and solubility in SCFs play a critical role in these applications, as solubility depends on temperature, pressure, and molecular interactions [86]. In supercritical fluid extraction, mass transfer is influenced by external mass transfer

Table 2 Solubility profile of rifabutin in pure solvents [54]

Solvent	Solubility (mg/ml)
Water	304
Ethanol	647
Propylene glycol (PG)	305
Polyethylene glycol 400 (PEG 400)	282

Table 3 Solubility profile of rifabutin in mixture of solvents [54]

Solvent blend	Solubility (mg/ml)
20% PEG 400 + 80% water	1.27
40% PEG 400 + 60% water	6.78
60% PEG 400 + 40% water	12.56
80% PEG 400 + 20% water	19.87
50% PEG 400 + 20% water + 30% PG	24.89
90% PEG 400 + 10% ethanol and PG	31.56

coefficients and effective diffusivity, which can be up to 100,000 times higher than binary diffusion in CO_2 [32]. Key factors affecting SCF processes, such as diffusivity and low viscosity, contribute to efficient mass transfer [99]. The controllability of fluid properties like density, viscosity, and diffusivity through pressure and temperature adjustments makes SCFs versatile solvents for various industrial applications, including the extraction of plant essential oils and high-value compounds [2, 32].

Retrograde behavior in SCFs is characterized by decreased solubility with increasing temperature at constant pressure [48]. This phenomenon results from the formation of large solvent molecule clusters around solute molecules near the solvent's critical point [30]. Retrograde behavior is relevant for separation processes, such as retrograde crystallization for isomer separation [48], as well as in enhanced molecular resolution in supercritical fluid chromatography [56]. In confined nanopore environments, retrograde behavior differs from that of bulk mixtures, with confined dew and bubble points occurring at lower pressures than their bulk counterparts [106]. Furthermore, confined mixtures exhibit a unique multiphase equilibrium curve at moderate pressures, beyond which equilibrium occurs between bulk and confined phases, resulting in the absence of critical points and cricondenbar in confined mixtures [106].

Recent advancements in SCF technologies have significantly enhanced nanoparticle production by improving mass transfer. The Supercritical Antisolvent with Enhanced Mass Transfer (SAS-EM) technique, a modification of the conventional SAS process, employs ultrasonic vibrations to atomize solution jets and improve mixing. This results in smaller particles with narrower size distributions [18]. This method has been successfully applied to produce protein nanoparticles while maintaining biological activity [17]. Studies have shown that by optimizing parameters such as flow rate, ultrasonic amplitude, and drug concentration, nanoparticles of quercetin as small as 120–450 nm can be produced, with improved dissolution rates [51]. These methods offer distinct advantages over traditional techniques like spray drying and milling, especially for producing particles in the submicron range (<300 nm), which are crucial for drug targeting and controlled release [17]. Additionally, the Rapid Expansion of Supercritical Solutions (RESS) method is a promising technique for particle formation, as it enables rapid attainment of uniform conditions and high supersaturations [112].

Table 4 Kinetics models proposed in SCF technology for nano particle synthesis

Model type	Focus area	Key factors considered	References
Computational Fluid Dynamics (CFD) + Population Balance Equation (PBE)	Particle formation, size prediction, and nucleation	Hydrodynamics, population balance, scale-up process	Cardoso et al. [16]
Supercritical Antisolvent (SAS) Process Models	Micronization, solubility, and bioavailability	Temperature, pressure, solvent type, flow rates	Abuzar et al. [1]

Kinetic modeling in SCF processes is essential for optimizing particle formation and characteristics. The complex interplay of solubility, mass transfer, and nucleation rates influences the kinetics of nanoparticle formation [36]. Mathematical models that incorporate fluid dynamics, mass transfer, and population balances can predict particle size and nucleation rates across different scales [16]. These models take into account the unique properties of SCFs, which differ from gases or liquids, affecting theoretical modeling and equipment design [8]. The Supercritical Antisolvent (SAS) technique is particularly effective in enhancing the solubility and bioavailability of poorly water-soluble drugs by controlling micronization and crystal morphology [1]. Understanding and modeling the kinetics of SCF processes allow for precise control over particle characteristics, making them valuable tools for pharmaceutical development and nanotechnology applications [1, 36]. Table 4 summarizes various kinetic models used in SCF technology for nanoparticle synthesis, highlighting the key factors considered and their focus areas.

Recent advancements in thermodynamics and high-pressure equipment design have significantly contributed to the commercial applications of SCF technology, including extraction processes, reaction chemistry, and particle formation for pharmaceuticals [8, 50]. The ability to optimize and control SCF properties has opened up new opportunities across various industries, enhancing processes and products at both laboratory and commercial scales.

3 Supercritical Fluid Processes for Nanoparticle Production

Supercritical fluid (SCF) technology, particularly using CO_2 and H_2O, has emerged as a valuable tool for nanoparticle production in the pharmaceutical industry [14, 44]. This approach offers advantages such as reduced particle size, narrow size distribution, and minimal solvent residue [29]. Various SCF-based techniques, including Rapid Expansion of Supercritical Solutions (RESS), Super Critical Anti-Solvent (SAS), and Particles from Gas-Saturated Solutions (PGSS) any many other methods have been developed for nanoparticle synthesis, each with specific merits and applications [44, 101]. These methods are particularly useful for improving the bioavailability of poorly soluble drugs and have potential applications in bio-imaging, cancer therapy, and targeted drug delivery [14, 92]. Despite its advantages, SCF technology has limitations, including the poor solvent power of CO_2 and high costs [101]. Nevertheless, it remains a promising approach for producing nanoparticles with controlled size, morphology, and surface properties for pharmaceutical and biomedical applications [14, 39].

3.1 Rapid Expansion of Supercritical Solutions

The Rapid Expansion of Supercritical Solutions (RESS) is an innovative technique for producing micro- to nanoparticles with narrow size distributions, as highlighted by Kumar et al. [63] and Tom and Debenedetti [112]. This process leverages supercritical fluids, predominantly carbon dioxide (CO_2), as solvents to precipitate a range of materials, including pharmaceuticals, polymers, and organic compounds [25, 42]. RESS offers notable advantages, such as rapid attainment of uniform conditions and high supersaturation, which promote the formation of small, monodisperse particles [112]. The particle size and distribution are significantly influenced by factors such as solute solubility in supercritical CO_2, supersaturation levels, and key process parameters like temperature, pressure, and nozzle dimensions [25, 62].

The RESS process involves saturating a suitable supercritical fluid with a solid substrate, such as a polymer or drug, followed by depressurization of the solution through a heated nozzle into a low-pressure chamber. This rapid expansion facilitates the nucleation of the substrate into very small particles, which are subsequently collected from the gaseous stream. A schematic diagram of the test rig used by Tom and Debenedetti [112] is shown in Fig. 2.

The morphology of the resulting solid material—whether crystalline or amorphous—depends on several factors, including the chemical structure of the material and RESS parameters like temperature, pressure drop, jet impact distance, and nozzle geometry [46]. The quick release of solute into the gaseous medium ensures the production of very small nanoparticles, a key advantage of RESS due to the absence of organic solvents in the process [88].

Particle size control in RESS is primarily governed by solubility and supersaturation [63]. Parameters such as extraction temperature, pre-expansion temperature, and pressure play critical roles in determining particle size. For instance, increasing pre-expansion temperature typically results in larger particles, whereas higher extraction temperatures and pressures tend to produce smaller particles due to retrograde solubility behavior in supercritical solutions [64]. These conditions allow the RESS process to be applied across various materials, including pharmaceuticals, polymers, and inorganics [112]. Optimization efforts, such as modifying active pharmaceutical

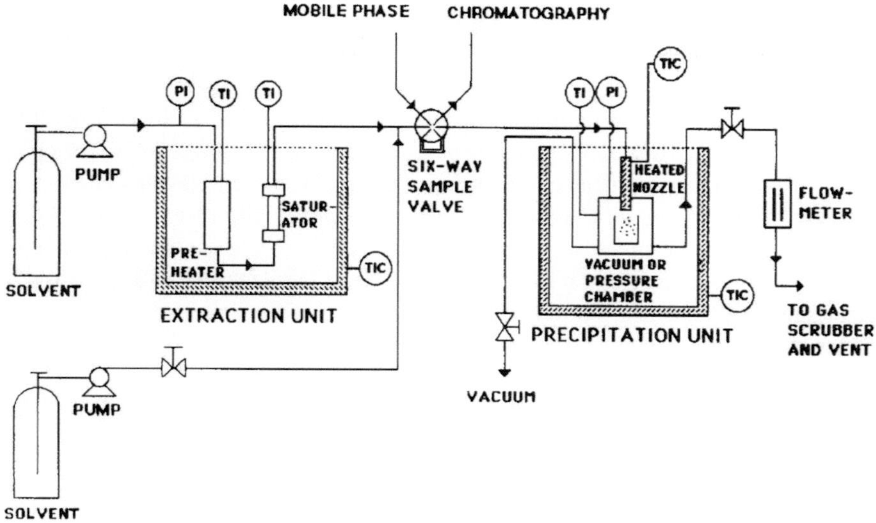

Fig. 2 Schematic diagram for Rapid Expansion of Supercritical Solutions [112]

ingredient (API) solubility, employing co-solvents, and implementing process modifications, have enhanced the technique's effectiveness in producing nanoparticles for drug delivery and other applications [63].

A notable variant of RESS is the Rapid Expansion of a Supercritical Solution into a Liquid Solvent (RESOLV). In this approach, the supercritical solution is atomized into a liquid, which quenches particle growth in the precipitator and improves the RESS technique's performance. The interaction between nucleating solid particles and the liquid phase can also facilitate chemical reactions, potentially enhancing nanoparticle properties [27].

Despite its advantages, the RESS process faces challenges in consistently producing high-quality nanoparticles. A primary issue is particle size control, as particles may coalesce in the supersonic free jet generated during expansion, often forming needle-like particles. This phenomenon can be attributed to electrostatic charges induced by the rapid relative motion between particles and the gas in the expansion chamber [89]. The RESOLV technique, however, has demonstrated greater efficacy in preventing particle agglomeration. By atomizing the jet into a liquid containing a stabilizing agent, particle growth can be suppressed, and stabilization is improved [27].

Both RESS and RESOLV techniques are limited by the solubility of materials in supercritical CO_2. Many compounds with high molecular weights and polar bonds, ideal for nanoparticle synthesis, exhibit low or negligible solubility in SC-CO_2, restricting the range of processable materials [88]. Additionally, the RESOLV method presents challenges in recovering particles from the liquid phase, which compromises the solventless advantage typically associated with supercritical processes.

Numerous studies underscore the potential of RESS in nanoparticle synthesis. Turk et al. [114] successfully utilized RESS to produce nanoparticles of β-sitosterol, an anticholesteremic agent, achieving an average diameter of approximately 200 nm by varying pre-expansion temperatures and pressures. Similarly, Sun and Rollins [104] employed the RESOLV process to synthesize cadmium sulfide (CdS) nanoparticles with an average diameter of 3.3 nm and ibuprofen nanoparticles with an average diameter of 40 nm and a standard deviation of 8.5 nm. Modifications to RESOLV, such as using a water-in-supercritical CO_2 microemulsion to dissolve $AgNO_3$, have yielded silver nanoparticles with a mean size of 7.8 nm. While the RESS and RESOLV techniques hold promise for nanoparticle synthesis, their successful application hinges on addressing challenges related to solubility and particle recovery.

3.2 *Particles from Gas-Saturated Solutions*

The Particles from Gas-Saturated Solutions (PGSS) process is a versatile technique for producing microparticles using supercritical carbon dioxide (SC-CO_2). This method has been effectively applied to encapsulate various compounds, including proteins, edible oils, and algal pigments, producing microparticles ranging in size from submicron to several micrometers [5, 120]. The ability of PGSS to modulate particle properties by incorporating additives such as polyethylene glycol (PEG) or PEGylated copolymers has been demonstrated to influence drug release behavior and particle morphology [80, 120]. Furthermore, PGSS-produced microparticles often exhibit porous structures, which enhance solubility and flow properties, making them suitable for diverse applications [7].

One of the significant advantages of PGSS is its high encapsulation efficiency for various compounds, including edible oils and algal pigments [5, 7]. This capability positions the PGSS process as a promising approach for applications in food, pharmaceutical, and cosmetic industries, particularly due to its ability to preserve bioactive compounds and control their release [120].

The PGSS technique overcomes the challenge of solubilizing polar compounds or those with higher molecular weights in SC-CO_2, a nonpolar solvent, by incorporating co-solvents when necessary. SC-CO_2's ability to penetrate organic substances, such as polymers, reduces their melting points and viscosity, facilitating the encapsulation process [101]. During PGSS operations, a physical mixture of the drug and polymer is subjected to SC-CO_2 under pressure, dissolving the solid components. Additional SC-CO_2 further reduces viscosity, forming a gas-saturated solution. This solution is then atomized through a nozzle into a receiver, where rapid depressurization causes SC-CO_2 to escape, resulting in the formation of composite microparticles [72].

The ability to form gas-saturated solutions is crucial to the PGSS process. SC-CO_2's solubility in melted or liquid-suspended materials forms a solution that, when expanded through a nozzle, produces solid particles or droplets. This capability broadens the range of materials that can be processed, as solubility in SC-CO_2 is not required. Consequently, the PGSS method is suitable for suspensions of active substrates in polymers or other carrier substances, enabling the formation of composite microspheres [14, 28, 46, 119].

The process's versatility extends beyond pharmaceutical applications, as it has been employed to develop inorganic powders and organic compounds, including albuterol sulfate, calcium antagonists, cromolyn sodium, DL-alanine, glucose, glutathione, and nifedipine, among others. It has also been used for materials such as glycerides, metal oxides, pigments, phosphors, and spinels [14, 28, 46, 119].

Despite its advantages, the PGSS process faces challenges, particularly ingredient segregation during the pressure drop, which can result in the independent formation of drug and polymer particles. This limitation complicates the creation of polymer microparticles encapsulating the drug. To address these issues, modifications to the PGSS technique have been proposed, such as incorporating two separate mixing compartments. In the first compartment, the drug and polymer are homogeneously mixed and melted in SC-CO_2. This mixture is then transferred to a second compartment, where additional SC-CO_2 further reduces viscosity before being sprayed and expanded to form uniform microparticles of the polymer-drug mixture [19, 101].

While the PGSS process shares benefits with the RESS technique, such as the absence of organic solvents and lower pressures, its challenges include monitoring particle size and achieving uniform encapsulation. These limitations underscore the need for continued refinement of the PGSS technique to fully realize its potential in producing microparticles for diverse industrial applications.

3.3 Supercritical Anti-solvent Process

The Supercritical Antisolvent (SAS) process is a versatile technique that leverages supercritical fluids, particularly carbon dioxide (CO_2), to precipitate nanoparticles from liquid solutions [90]. This method has found applications across various fields, including the production of superconductor precursors [90], pharmaceutical preparations [108], and nano catalysts for biodiesel production [97]. The key advantages of SAS include precise control over particle size, morphology, and distribution, which are critical for applications such as drug bioavailability [108]. Factors such as pressure, temperature, and solution concentration play significant roles in determining the characteristics of the final product [90]. Furthermore, studies on ampicillin solutions highlight the importance of understanding hydrodynamic conditions to achieve the desired particle properties [109].

The SAS process involves introducing a liquid solution containing the solute to be micronized into a supercritical fluid. For successful precipitation, the supercritical fluid must be fully miscible with the liquid solvent while the solute remains insoluble. The contact between the liquid solution and the supercritical fluid results in the formation of a supersaturated solution, which triggers solute precipitation. This rapid process is facilitated by the enhanced mass transfer rates characteristic of supercritical fluids, enabling efficient nanoparticle production [52].

This technique is particularly suited for compounds with poor solubility in supercritical fluids. Organic solvents such as acetone, dichloromethane (DCM), and dimethyl sulfoxide (DMSO) are commonly employed to dissolve materials, while the supercritical fluid acts as a nonsolvent for the solute or active pharmaceutical ingredient (API) [52]. As the liquid solution is exposed to the supercritical fluid, rapid formation of a supersaturated solution occurs, leading to solute precipitation. This process's efficiency is largely driven by the high mass transfer rates between the liquid and supercritical phases [52].

Depending on operational requirements, SAS can be conducted in batch or semi-continuous modes [90]. In batch operations, the precipitation chamber is first filled with a known quantity of the liquid solution before introducing the supercritical anti-solvent to achieve the desired pressure. In contrast, semi-continuous operations involve continuous delivery of both the liquid solution and the supercritical anti-solvent, either in co-current or counter-current configurations. The design of the liquid solution injection device is crucial, as it facilitates liquid jet breakup and generates small-sized droplets, maximizing the mass transfer surface area [116].

High-pressure vapor–liquid equilibria (VLE) and mass transfer between the liquid and the supercritical fluid are critical determinants of SAS efficiency. The ternary system comprising solute, solvent, and supercritical anti-solvent exhibits complex phase behaviors. Achieving a single supercritical region—a prerequisite for effective nanoparticle production—depends on precise control of VLE conditions. The positioning of the working point relative to these equilibria significantly influences process success [91]. At the final precipitation stage, washing the precipitate with pure supercritical anti-solvent is essential to prevent liquid medium condensation, which could compromise the final product's characteristics.

The SAS process poses challenges due to the difficulty of predicting VLE adjustments caused by solute presence in the binary liquid-supercritical fluid system. These adjustments can lead to intricate phase behaviors and higher mixture critical points (MCP), necessitating extremely high pressures to achieve single-phase conditions [91]. Phase separation during SAS can yield various morphologies, depending on whether precipitation occurs from a liquid-rich state (yielding crystals) or a supercritical fluid-rich state (yielding amorphous particles). The partitioning of the solute between the phases dictates the relative proportions of these morphologies [90].

The SAS process provides precise control over particle size by manipulating key parameters. For instance, ultrasonic vibration can significantly reduce particle size and distribution by enhancing mass transfer and minimizing agglomeration [18]. Higher solubility of the solute in the organic solvent-supercritical fluid mixture generally results in larger particles, as does increased operating pressure [95]. Temperature and pressure adjustments can also influence particle size: higher temperatures tend to increase mean particle size due to plasticizing effects, while higher pressures typically reduce particle size by enhancing molecular interactions and solubility [53]. For polymer materials, liquid-side mass transfer controls particle size, with low temperatures and high pressures favoring smaller particles [87].

The SAS technique's outcome is highly dependent on the sequence of introducing the solvent, supercritical fluid, and other substrates. Optimizing parameters such as solute and solvent composition, temperature, and pressure is crucial for achieving the desired results. SAS has demonstrated superior drug loading capacity compared to the Rapid Expansion of Supercritical Solution (RESS) technique, enabling the production of fine particles [72]. Innovations in SAS, such as the expanded liquid anti-solvent (ELS) technique [85] and supercritical-assisted injection in a liquid anti-solvent (SAILA) [15], have enhanced its capabilities. Moreover, tailored SAS methods, such as the supercritical-assisted extraction of emulsions (SFEE), have shown promise in specific applications [33]. Despite its advantages, the SAS process requires careful optimization of operational conditions and an in-depth understanding of the underlying phase behaviors to fully exploit its potential for nanoparticle production.

3.4 Aerosol Solvent Extraction System

The Aerosol Solvent Extraction System (ASES) is an emerging process for producing fine pharmaceutical particles using supercritical gases [10, 62]. In ASES, a drug-polymer solution is atomized into an antisolvent environment, typically supercritical CO_2, causing rapid supersaturation and particle formation [62]. This process allows for the production of microparticles with very low residual solvent levels, below 30 ppm [10]. ASES has been successfully used to encapsulate various drugs, including prednisolone, with particle sizes as small as 230 nm when using a ternary system of drug, polymer, and surfactant [73]. Process parameters such as spraying rate and CO_2 pump rate can influence particle characteristics, including residual solvent content and yield [94]. However, ASES may not be suitable for non-polar drugs due to complete extraction with the organic solvent. Because it prevents the desired particle formation and encapsulation of the drug [10].

3.5 Solution-Enhanced Dispersion by Supercritical Fluids

Solution-Enhanced Dispersion by Supercritical Fluids (SEDS) represents a sustainable and effective approach for producing micro- and nanoparticles of biomaterials and pharmaceuticals [55]. This method utilizes supercritical fluids, particularly CO_2, to extract solvents from sprayed solutions, resulting in particle formation [40].

Initially, as the solution is introduced into the chamber, supercritical CO_2 rapidly interacts with the solvent, decreasing its ability to dissolve the compound. This leads to the compound precipitating out of the solution as fine particles. A coaxial nozzle ensures controlled interaction between supercritical CO_2 and the solution, creating a high-supersaturation zone that fosters uniform particle nucleation. Subsequently, as the solvent continues to mix with supercritical CO_2, it is extracted from the precipitated particles, leaving behind dry particles with tailored size and morphology. The precision in nozzle design and flow rate control ensures a narrow particle size distribution, enabling consistent and high-quality particle production suitable for drug delivery and other applications [55].

SEDS provides significant advantages over conventional methods, such as enhanced control, reproducibility, and scalability [40]. Key parameters—temperature, pressure, and solution flow rate—greatly impact particle characteristics [124]. The technique has demonstrated efficacy in micronizing poorly water-soluble drugs like curcumin, enhancing their solubility and dissolution rates [124]. Furthermore, SEDS has been applied to biodegradable polymers such as polyhydroxybutyrate-co-hydroxyvalerate (PHBV), yielding spherical particles with sizes ranging from 210 to 720 nm [3]. Its versatility makes SEDS highly valuable for incorporating proteins and peptides into biodegradable microparticles, particularly for drug delivery applications [40].

3.6 Supercritical Anti-solvent with Enhanced Mass Transfer

The Supercritical Anti-solvent with Enhanced Mass Transfer (SAS-EM) technique improves upon the traditional supercritical anti-solvent method by integrating ultrasonic vibration to enhance mass transfer and limit particle growth [110]. This approach has been effectively used to produce nanoparticles of compounds such as hydrocortisone [110], quercetin [51], lysozyme [18], and β-carotene [47].

Key process variables, including flow rate, ultrasonic power, drug concentration, and flow duration, enable precise control over particle size and morphology. Compared to conventional SAS methods, SAS-EM can generate significantly smaller particles, typically in the range of 120–450 nm. These nanoparticles often exhibit improved dissolution rates and lower crystallinity, making them more effective for various applications [51].

3.7 Suspension-Enhanced Dispersion by Supercritical Fluids

The Suspension-Enhanced Dispersion by Supercritical Fluids (SpEDS) technique modifies the Suspension-Enhanced Dispersion System (SEDS) to address certain operational challenges [21]. While both methods share similar principles, SpEDS incorporates an auxiliary injector system to efficiently deliver loaded suspensions into the chamber. This enhanced approach is particularly suited for producing core–shell micro- and nanoparticles, which exhibit superior drug encapsulation efficiency and sustained release properties compared to other supercritical fluid-based co-precipitation methods. Such advancements are especially beneficial in pharmaceutical applications, where controlled release is critical for therapeutic success.

3.8 *Supercritical Fluid Extraction of Emulsions*

Supercritical Fluid Extraction of Emulsions (SFEE) is a novel method for creating micro- and nanoparticles tailored for drug delivery and food applications. By utilizing supercritical CO_2 to extract the organic phase from oil-in-water emulsions, SFEE enables the precipitation of dissolved substances into small particles suspended in water [19, 111]. SFEE offers distinct advantages, including precise particle size control, high encapsulation efficiency, and minimal residual solvent content [19, 96]. The technique is well-suited for encapsulating poorly water-soluble and food-grade bioactive substances [31, 96]. Operational setups such as spray columns, bubble columns, and countercurrent columns enable batch, semi-continuous, or continuous processing. The properties of the resulting nanocapsules are influenced by variables like emulsion composition, pressure, temperature, and flow rates [31]. Recent advancements have focused on scaling up the process and integrating solvent recovery systems, enhancing the technique's commercial viability [111].

4 Industrial Applications

SCFs have been successfully employed in various processes, including extraction, impregnation, particle formation, and reactive systems [12]. SCF technology has shown promise in synthesizing iron oxide nanoparticles for multidisciplinary industrial applications [65]. Additionally, SCFs have been utilized to produce metal and semiconductor nanoparticles and nanowires, addressing the limitations of traditional lithography-based methods [45]. The ability of SCFs to create nanostructured materials with precise control over their properties makes them particularly valuable for biomedical applications, such as bio-imaging, hyperthermia, and targeted drug delivery [14]. As research progresses, SCF technology is expected to play an increasingly important role in nanomaterial synthesis and processing across various industries.

4.1 *Pharmaceutical Industry*

SCF technologies have emerged as promising methods for enhancing drug delivery and formulation in the pharmaceutical industry. These techniques offer advantages in particle size control, solubility improvement, and clean processing [77]. Various SCF processes, including SAS and RESS, have been developed for producing drug-loaded nanoparticles with improved bioavailability and targeted delivery [101]. Supercritical CO_2-based technologies are particularly effective in generating small particles with smooth surfaces and narrow size distributions [101]. These methods have applications in particle design, drug purification, and innovative formulations for controlled

Table 5 Applications of SCF in pharmaceutical industry

Application	SCF process	Description	References
Particle size control and solubility improvement	Supercritical Anti-Solvent (SAS)	SAS is used to produce fine particles with improved solubility and bioavailability by precipitating drugs from a solution upon contact with supercritical CO_2	Pathak et al. [77]
Nanoparticle production for drug delivery	Rapid Expansion of Supercritical Solutions (RESS)	RESS generates nanoparticles with controlled particle size and narrow distribution, enhancing targeted drug delivery and bioavailability	Sheth et al. [101]
Drug purification and particle design	Supercritical CO_2-based purification	Supercritical CO_2 is used for purifying drugs, removing impurities, and designing particles with controlled morphologies and surfaces suitable for drug formulations	Perrut [81]
Controlled release and formulations	Cyclodextrin complex formation	SCF technology enables the formation of drug-cyclodextrin complexes for improved stability, solubility, and controlled drug release, which are beneficial for various formulations	Pasquali and Bettini [75]
Biotechnological compound formulation	Supercritical CO_2 for biotech compounds	Supercritical CO_2 is applied in producing biotechnological compounds with precise particle characteristics, enhancing their therapeutic efficacy and stability	Pasquali and Bettini [75]

release systems [81]. Despite their potential, SCF technologies face challenges such as the poor solvent power of CO_2 and high costs [101]. Nevertheless, ongoing research continues to explore SCF applications in areas like particle engineering, cyclodextrin complexes, and biotechnological compound formulation [75] (Table 5).

4.2 Materials Science

Supercritical fluids, particularly CO_2, have emerged as versatile tools in materials science and engineering. They offer unique properties for synthesizing and processing advanced materials, including nanoparticles, catalysts, and coatings [24, 105]. These fluids enable the production of functional materials with controlled morphologies, sizes, and compositions, enhancing their performance in various applications such as catalysis, electronics, and pharmaceuticals [6]. Supercritical fluid technology has also shown promise in graphene processing, allowing for the scalable production of graphene-based composites, aerogels, and foams [74]. This approach offers advantages over conventional liquid-phase processing, including improved

Table 6 Applications of SCF in material science

Traditional method	Green chemistry approach	Description	References
Organic solvent-based synthesis of nanoparticles	Supercritical CO_2 in nanoparticle synthesis	Conventional synthesis uses volatile, toxic organic solvents (e.g., toluene), causing environmental pollution and health risks. Supercritical CO_2 enables solvent-free synthesis, reducing waste	Cooper [24]
Chemical vapor deposition (CVD) of graphene	Supercritical fluid processing of graphene	CVD uses high-energy, toxic gases like methane or acetylene, producing hazardous emissions. Supercritical CO_2 provides milder, safer processing for graphene-based composites	Padmajan Sasikala et al. [74]
Heavy metal catalysts in polymer synthesis	Supercritical water or CO_2 catalysis for polymer synthesis	Heavy metals (e.g., mercury, cadmium) in traditional polymer synthesis are toxic and persistent. Supercritical fluids reduce or eliminate toxic catalysts, lowering environmental and health risks	Aymonier et al. [6]
Solvent-intensive extraction of bioactives	Supercritical fluid extraction (SFE) of bioactive compounds	Organic solvents used in traditional extraction are toxic and leave residues. SFE with supercritical CO_2 achieves higher purity, is non-toxic, and preserves bioactivity	Aymonier et al. [6]
Acid etching for surface treatments	Supercritical CO_2 for surface treatments	Acid etching (e.g., using sulfuric acid) releases toxic fumes and produces acidic waste. Supercritical CO_2 enables controlled etching with minimal emissions and waste	Cooper [24]

material properties and reduced environmental impact. As the field evolves, supercritical fluids are increasingly recognized as a sustainable alternative for designing multifunctional nanomaterials, aligning with the principles of green chemistry and sustainable development [6] (Table 6).

4.3 *Electronics and Energy*

SCF technologies have emerged as a sustainable approach for synthesizing nanomaterials, particularly for energy storage applications. SCF methods offer advantages in producing well-designed nanomaterials with controlled properties, including phase purity, morphology control, and size tuning down to 5 nm [6, 34]. These techniques have been successfully applied to synthesize various cathode materials for lithium-ion batteries, such as layered transition metal oxides, spinel structures, and lithium metal phosphates [34]. SCF synthesis can significantly impact battery performance by enabling the production of nanoparticles and nanowires with tailored characteristics [45]. The use of SCF methods aligns with green chemistry principles, reducing environmental impact compared to traditional solution chemistry approaches [6, 118]. This sustainable nanomaterial production technique shows promise for advancing energy storage technologies in electronic devices and electric vehicles.

5 Regulatory and Safety Considerations

Supercritical fluid extraction (SFE) is a versatile technology with applications in food, pharmaceutical, and chemical industries. It offers environmental, health, and safety benefits compared to conventional extraction methods [59]. The solvating power of supercritical fluids depends on their density, which is controlled by pressure rather than intermolecular attraction [78]. Carbon dioxide is commonly used in SFE due to its safety profile, reducing the need for flammable organic solvents [78]. Industrial-scale SFE plants are used for various applications, including hop extraction, decaffeination, and lecithin separation [59]. The unique properties of supercritical fluids also enable integrated extractions and in situ formulations [59]. Research in this field encompasses thermodynamics, phase equilibria, and process design [50]. As the technology advances, SFE continues to find new applications in areas such as natural product extraction and pollutant removal [50].

5.1 Regulatory Frameworks

The use of supercritical fluids (SCFs) in industrial applications is governed by a range of regulations established by various regulatory bodies, including the U.S. Food and Drug Administration (FDA), Environmental Protection Agency (EPA), and international standards organizations such as the International Organization for Standardization (ISO). Recent updates to these regulations emphasize the need for thorough safety and efficacy evaluations of processes that utilize SCFs, particularly in the pharmaceutical and food industries. For instance, the FDA has issued guidelines for the use of supercritical CO_2 in the extraction and purification of food products and active pharmaceutical ingredients (APIs), requiring manufacturers to demonstrate compliance with safety standards, including limits on residual solvents and contaminants. Similarly, the EPA has increased scrutiny on the environmental impact of SCF operations, mandating assessments that ensure SCF processes do not adversely affect air and water quality.

These regulations significantly impact research and development efforts within the supercritical fluid technology sector. Companies must adapt their operational procedures to align with updated guidelines, which can involve re-evaluating existing processes or investing in new technologies that meet stringent safety and environmental standards. This may include conducting comprehensive risk assessments and modifying equipment to comply with pressure and safety requirements set forth by regulatory agencies. The need for compliance can slow down innovation cycles as companies must ensure that any new developments or modifications are thoroughly tested and validated to meet regulatory expectations.

The requirement for environmental impact assessments (EIAs) is a critical aspect of regulatory compliance for supercritical fluid operations. These assessments evaluate the potential environmental consequences of SCF processes, including emissions, waste generation, and resource consumption. Regulatory bodies are increasingly prioritizing sustainability in their guidelines, pushing companies to adopt practices that minimize ecological footprints. Operators must conduct detailed analyses of their processes to identify potential environmental impacts and implement strategies to mitigate adverse effects. This emphasis on sustainability not only ensures compliance but also enhances the public perception of supercritical fluid technologies as environmentally responsible alternatives.

Safety assessments for high-pressure systems are essential to ensure the protection of workers and the surrounding environment. Regulations require that companies conduct thorough safety evaluations, including hazard analyses, risk assessments, and emergency preparedness plans. These assessments focus on identifying potential hazards associated with SCF operations, such as leaks, equipment failures, or operational mishaps. Companies must adhere to established guidelines for safe operation, which include regular maintenance and inspections of high-pressure equipment, training for personnel, and the implementation of safety management systems. These safety protocols are vital to prevent accidents and ensure a safe working environment.

5.2 *Safety Protocols*

Recent advancements in safety equipment for supercritical fluid processes have significantly improved operational safety. This includes the development of advanced pressure relief systems designed to prevent overpressure incidents, ensuring that equipment operates within safe limits. Safety interlocks have also become a standard feature in modern SCF systems, automatically shutting down operations when critical safety parameters are exceeded. Furthermore, real-time monitoring systems allow for continuous surveillance of pressure, temperature, and other vital process parameters, enabling operators to respond quickly to any deviations from normal operating conditions. These innovations contribute to a safer working environment by minimizing the risk of accidents and facilitating proactive maintenance.

The integration of automation and control systems in supercritical fluid operations enhances safety by reducing the potential for human error in high-pressure environments. Automated systems can control process parameters such as temperature and pressure with high precision, minimizing the risk of mishaps due to manual handling. Additionally, advanced data analytics and machine learning technologies can predict potential failures before they occur, allowing for preemptive maintenance and timely interventions. This level of automation not only improves safety but also increases operational efficiency, as consistent monitoring and adjustments can lead to better process optimization.

Effective emergency response plans are crucial for the safe operation of supercritical fluid processes. These plans should outline procedures for various emergency scenarios, including equipment failures, leaks, and other incidents that could pose risks to personnel and the environment. Key components of an emergency preparedness plan include evacuation protocols, containment measures for spills or leaks, and communication strategies to ensure that all personnel are informed and can respond effectively in a crisis. Regular drills and training sessions should be conducted to familiarize staff with these protocols, ensuring a swift and coordinated response during emergencies.

Training programs for personnel working with supercritical fluids are essential to ensure safety and compliance with regulatory requirements. These programs should cover a range of topics, including the properties and behavior of supercritical fluids, safety procedures, and emergency response strategies. Employees should be well-versed in the operational protocols specific to SCF processes, including the proper handling of equipment and materials. Regular refresher courses and updates on new regulations or technological advancements are also crucial to maintain a high level of safety awareness among workers. By investing in comprehensive training, organizations can foster a culture of safety and preparedness, significantly reducing the likelihood of accidents in high-pressure environments.

6 Conclusion

The exploration of SCF technologies has revealed their transformative potential in nanoparticle synthesis and industrial applications. These technologies leverage the unique properties of SCFs—gas-like diffusivity, liquid-like solvating power, and tunable density—to achieve remarkable precision in controlling particle size, morphology, and functionality. Such attributes have positioned SCFs as pivotal tools for meeting the growing demands of industries like pharmaceuticals, materials science, and energy storage.

Through innovative processes such as Rapid Expansion of Supercritical Solutions (RESS), Supercritical Anti-solvent (SAS), and Particles from Gas-Saturated Solutions (PGSS), SCF technologies address longstanding challenges in nanoparticle production. These methods enable the creation of nanoparticles with narrow size distributions and tailored properties, essential for applications ranging from drug delivery to catalysis and electronic materials. Additionally, advanced techniques like Supercritical Anti-solvent with Enhanced Mass Transfer (SAS-EM) have improved efficiency by enhancing mixing and minimizing agglomeration, making the processes even more reliable and environmentally friendly.

The environmental and economic advantages of SCF technologies are particularly noteworthy. By significantly reducing the use of harmful organic solvents and aligning with green chemistry principles, these methods contribute to sustainability goals while maintaining regulatory compliance. This dual focus on performance and environmental stewardship makes SCF technologies a critical asset for industries striving for sustainable innovation.

Table 7, which can be placed here, encapsulates the technological progress made in this field. It provides a comparative analysis of the efficiencies and applications of various SCF processes, highlighting their advancements in reducing solvent usage, optimizing particle size control, and enhancing bioavailability. The data in Table 7 underscores how SCF methods outperform traditional techniques in both environmental impact and product quality.

Despite these achievements, challenges remain in scaling up SCF processes to industrial levels and optimizing key variables such as temperature, pressure, and flow rates. Ongoing research into computational modeling, reactor design, and the exploration of alternative SCFs, such as dimethyl ether or supercritical water, is expected to overcome these barriers. Such advancements will not only enhance the versatility of SCF technologies but also ensure their broader adoption across diverse industries.

In conclusion, supercritical fluid technologies are poised to revolutionize nanoparticle production by combining precision, efficiency, and sustainability. With continued innovation, these methods will play a pivotal role in shaping a cleaner, greener, and more efficient future for industrial manufacturing.

Table 7 Summary on SCF process in nanoparticle production

Decision factors	RESS	RESOLV	PGSS	SAS	ASES	SEDS	SAS-EM	SFEE
Material type	Soluble in SCF	Soluble in SCF	Soluble or insoluble in SCF	Poor solubility in SCF	Soluble in organic solvents	Soluble in organic solvents	Soluble in organic solvents	Oil-in-water emulsions
Particle size range	Micro to nanoparticles	Nanoparticles	Submicron to several micrometers	Nanoparticles	Microparticles	Micro- and nanoparticles	Nanoparticles	Micro- and nanoparticles
Particle morphology	Crystalline or amorphous	Improved control with liquid quenching	Porous structures; dependent on additives	Crystalline or amorphous	Smooth, fine particles	Uniform, tailored morphology	Controlled morphology, smaller size	Controlled morphology
Key advantages	Solvent-free, rapid nucleation, precise particle control	Enhanced stabilization, suppression of particle growth	High encapsulation efficiency, versatility for polymers and inorganics	Precise control over size and morphology, scalable	Low residual solvent content, rapid supersaturation	Sustainable, eco-friendly, consistent quality	Enhanced mass transfer, reduced particle size	High encapsulation efficiency, low residual solvents
Key limitations	Limited to SCF-soluble materials, particle agglomeration	Particle recovery from liquid phase can compromise solventless advantage	Challenges in uniform encapsulation; ingredient segregation during pressure drop	Complex phase behaviors, solubility prediction challenges	Limited to polar drugs, requires suitable surfactants	Dependency on nozzle design, flow rates, and mixing efficiency	Requires ultrasonic optimization, higher energy usage	Sensitive to emulsion composition and operational variables

(continued)

Table 7 (continued)

Decision factors	RESS	RESOLV	PGSS	SAS	ASES	SEDS	SAS-EM	SFEE
Best for …	Pharmaceuticals, polymers, inorganics	Nanoparticles requiring surface modification	Pharmaceuticals, cosmetics, food	Superconductor precursors, pharmaceutical applications	Pharmaceuticals requiring minimal residual solvents	Biomaterials, poorly water-soluble drugs	Pharmaceuticals needing enhanced dissolution rates	Food-grade bioactives, poorly water-soluble compounds
Scalability	Moderate	Moderate	High	Moderate	Moderate	High	Moderate	Moderate
Cost efficiency	High setup cost; low operating cost	Higher due to additional liquid recovery steps	High due to versatility and low operating costs	Moderate to high; depends on optimization	Moderate	High due to reduced solvent use	Moderate due to energy requirements	Moderate due to solvent recovery and setup
Typical applications	Drug delivery, catalysis, nanomaterial synthesis	Cadmium sulfide (CdS) nanoparticles, ibuprofen	Microparticles for encapsulation of oils, pigments, and drugs	Biodiesel nano catalysts, pharmaceutical drugs	Fine pharmaceutical particles	Poorly water-soluble drugs, biodegradable polymers	Hydrocortisone, β-carotene, protein nanoparticles	Nanoencapsulation of food bioactives
Customization flexibility	High, based on process parameters	Moderate; requires liquid-phase compatibility	High, customizable through co-solvents and mixing	High; dependent on VLE and phase behavior	Moderate; parameters like CO_2 rate and spray rate are influential	High due to nozzle and flow control	High due to ultrasonic vibration customization	High, influenced by emulsion and flow parameters

References

1. Abuzar SM, Hyun SM, Kim JH, Park HJ, Kim MS, Park JS, Hwang SJ (2018) Enhancing the solubility and bioavailability of poorly water-soluble drugs using supercritical antisolvent (SAS) process. Int J Pharm 538(1–2):1–13. https://doi.org/10.1016/j.ijpharm.2017.12.041
2. Adschiri T (2019) Supercritical fluid technology. The expanding world of chemical engineering. https://api.semanticscholar.org/CorpusID:199184332
3. Aguiar GPS, Magro CD, Oliveira JV, Lanza M (2018) Poly(hydroxybutyrate-co-hydroxyvalerate) micronization by solution enhanced dispersion by supercritical fluids technique. Braz J Chem Eng 35(4):1275–1282. https://doi.org/10.1590/0104-6632.20180354s20170501
4. Andrea Kruse FK (2008) Supercritical water gasifi cation Andrea. Biofuels Bioprod Biorefin 8(6):23. https://doi.org/10.1002/bbb.93. Wiley InterScience
5. Aredo V, Bittencourt GM, de Jesus Agnolon Pallone EM, Guimarães FEC, de Oliveira AL (2021) Formation of edible oil-loaded beeswax microparticles using PGSS—particles from gas-saturated solutions. J Supercrit Fluids 169. https://doi.org/10.1016/j.supflu.2020.105106
6. Aymonier C, Erriguible A, Marre S, Serani A, Cansell F (2007) Processes using supercritical fluids: a sustainable approach for the design of functional nanomaterials. Int J Chem React Eng 5. https://doi.org/10.2202/1542-6580.1424
7. Banožić M, Čolnik M, Škerget M, Cikoš AM, Aladić K, Jokić S (2022) Formation and characterization of *Fucus virsoides* J. Agardh pigment–polyethylene glycol microparticles produced using PGSS process. Appl Sci (Switzerland) 12(22). https://doi.org/10.3390/app122211496
8. Bałdyga J, Henczka M, Shekunov B (2004) Fluid dynamics, mass transfer, and particle formation in supercritical fluids. https://api.semanticscholar.org/CorpusID:199081445
9. Beulow SJ, Dyer RB, Harradine DM, Robinson JM, Oldenborg RC, Funk KA, McInroy RE, Sanchez JA, Spontarelli T (1993) Destruction of energetic materials by supercritical water oxidation. https://api.semanticscholar.org/CorpusID:93890023
10. Bleich J, Müller BW (1996) Production of drug loaded microparticles by the use of supercritical gases with the Aerosol Solvent Extraction System (ASES) process. J Microencapsul 13(2):131–139. https://doi.org/10.3109/02652049609052902
11. Bok E, Automation I, Kelch D, Robot M (1992) Supercritical fluids for single wafer cleaning. June, pp 117–120
12. Brunner G (2010) Applications of supercritical fluids. Annu Rev Chem Biomol Eng 1:321–342. https://doi.org/10.1146/annurev-chembioeng-073009-101311
13. Brunner G (2014) Hydrothermal and supercritical water processes. https://api.semanticscholar.org/CorpusID:92689887
14. Byrappa K, Ohara S, Adschiri T (2008) Nanoparticles synthesis using supercritical fluid technology—towards biomedical applications. Adv Drug Deliv Rev 60(3):299–327. https://doi.org/10.1016/j.addr.2007.09.001
15. Campardelli R, Adami R, Della Porta G, Reverchon E (2012) Nanoparticle precipitation by Supercritical Assisted Injection in a Liquid Antisolvent. Chem Eng J 192:246–251. https://doi.org/10.1016/j.cej.2012.04.010
16. Cardoso FAR, Vogel EM, Souza MF, Cardozo-Filho L (2019) Mathematical modeling to predict the size and nucleation rate of micro and nanoparticles using the scale-up process with supercritical CO_2. J Supercrit Fluids 154:104608. https://doi.org/10.1016/j.supflu.2019.104608
17. Chattopadhyay P, Gupta RB (2001) Production of griseofulvin nanoparticles using supercritical CO_2 antisolvent with enhanced mass transfer. Int J Pharm 228(1–2):19–31. https://doi.org/10.1016/S0378-5173(01)00803-1
18. Chattopadhyay P, Gupta RB (2002) Protein nanoparticles formation by supercritical antisolvent with enhanced mass transfer. AIChE J 48(2):235–244. https://doi.org/10.1002/aic.690480207

19. Chattopadhyay P, Huff RW, Shekunov B (2006) Drug encapsulation using supercritical fluid extraction of emulsions. J Pharm Sci 95(3):667–679. https://api.semanticscholar.org/CorpusID:2968752
20. Chehroudi B (2006) Supercritical fluids: nanotechnology and select emerging applications. Combust Sci Technol 178(1–3):555–621. https://doi.org/10.1080/00102200500294247
21. Chen AZ, Wang GY, Wang SB, Li L, Liu YG, Zhao C (2012) Formation of methotrexate-PLLA-PEG-PLLA composite microspheres by microencapsulation through a process of suspension-enhanced dispersion by supercritical CO_2. Int J Nanomed 7:3013–3022. https://doi.org/10.2147/IJN.S32662
22. Clifford AA, Williams JR (2000) Introduction to supercritical fluids and their applications. Supercrit Fluid Methods Protoc 13(1):1–16. https://doi.org/10.1385/1-59259-030-6:1
23. Cockrell C, Trachenko K (2022) Double universality of the transition in the supercritical state. Sci Adv 8(32):1–8. https://doi.org/10.1126/sciadv.abq5183
24. Cooper AI (2001) Recent developments in materials synthesis and processing using supercritical CO_2. Adv Mater 13(14):1111–1114. https://doi.org/10.1002/1521-4095(200107)13:14%3c1111::AID-ADMA1111%3e3.0.CO;2-L
25. Corazza ML, Cardozo Filho L, Dariva C (2006) Modeling and simulation of rapid expansion of supercritical solutions. Braz J Chem Eng 23(3):417–425. https://doi.org/10.1590/S0104-66322006000300015
26. Costa JM, de Almeida Neto AF (2021) Nanocatalysts deposition assisted by supercritical carbon dioxide technology: a review. Synthet Met 271:116627. Elsevier B.V. https://doi.org/10.1016/j.synthmet.2020.116627
27. Dalvi S, Azad M, Dave R (2013) Precipitation and stabilization of ultrafine particles of Fenofibrate in aqueous suspensions by RESOLV. Powder Technol 236:75–84. https://doi.org/10.1016/j.powtec.2012.05.038
28. Date AA, Patravale VB (2004) Current strategies for engineering drug nanoparticles. Curr Opin Colloid Interface Sci 9(3–4):222–235. https://doi.org/10.1016/j.cocis.2004.06.009
29. De Marco I (2022) Supercritical fluids and nanoparticles in cancer therapy. Micromachines 13(9). https://doi.org/10.3390/mi13091449
30. Debenedetti PG, Kumar SK (1988) The molecular basis of temperature effects in supercritical extraction. AIChE J 34(4):645–657. https://doi.org/10.1002/aic.690340414
31. del Cerro DR, Rojas A, Torres A, Villegas C, Galotto MJ, Guarda A, Romero J (2023) Nanoencapsulation of food-grade bioactive compounds using a supercritical fluid extraction of emulsions process: effect of operational variables on the properties of nanocapsules and new perspectives. LWT. https://api.semanticscholar.org/CorpusID:260267053
32. del Valle JM, de la Fuente JC, Uquiche E, Zetzl C, Brunner G (2011) Mass transfer and equilibrium parameters on high-pressure CO_2 extraction of plant essential oils. In: Food engineering series. https://doi.org/10.1007/978-1-4419-7475-4_17
33. Della Porta G, Falco N, Reverchon E (2010) NSAID drugs release from injectable microspheres produced by supercritical fluid emulsion extraction. J Pharm Sci 99(3):1484–1499. https://doi.org/10.1002/jps.21920
34. Devaraju MK, Truong QD, Tomai T, Honma I (2014) Supercritical fluid methods for synthesizing cathode materials towards lithium ion battery applications. RSC Adv 4(52):27452–27470. https://doi.org/10.1039/c4ra01772b
35. Dyer RB, Buelow SJ, Harradine DM, Robinson JM, Foy B, Atencio JH, DellOrco PC, Funk KA, McInroy RE, Rofer CK, Counce DA, Trujillo PE, Wander JD (1992) Destruction of explosives and rocket fuels by supercritical water oxidation. https://api.semanticscholar.org/CorpusID:94215116
36. Erkey C, Türk M (2021) Modeling of particle formation in supercritical fluids (SCF). In: Synthesis of nanostructured materials in near and/or supercritical fluids—methods, fundamentals and modeling. https://api.semanticscholar.org/CorpusID:244864719
37. Fernandez CA, Hoppes EM, Bekhazi JG, Wang C, Wiacek RJ, Warner MG, Fryxell GE, Bays JT, Addleman RS (2008) Tuning and quantifying the dispersibility of gold nanocrystals in liquid and supercritical solvents. J Phys Chem C 112(36):13947–13957. https://doi.org/10.1021/jp8038237

38. Foster NR, Lucien FP, Mammucari R (2010) Basic physical properties, phase behavior and solubility. In: Handbook of green chemistry, vol 4, pp 77–100. https://doi.org/10.1002/9783527628698.hgc036
39. Franco P, De Marco I (2021) Nanoparticles and nanocrystals by supercritical CO_2-assisted techniques for pharmaceutical applications: a review. Appl Sci (Switzerland) 11(4):1–27. https://doi.org/10.3390/app11041476
40. Ghaderi R (2000) A supercritical fluids extraction process for the production of drug loaded biodegradable microparticles
41. Girotra P, Singh SK, Nagpal K (2013) Supercritical fluid technology: a promising approach in pharmaceutical research. Pharm Dev Technol 18(1):22–38. Informa Healthcare. https://doi.org/10.3109/10837450.2012.726998
42. Gomes MTMS, Santana ÁL, Santos DT, Meireles MAA (2018) Trends on the rapid expansion of supercritical solutions process applied to food and non-food industries. Recent Pat Food Nutr Agric 10(2):82–92. https://doi.org/10.2174/2212798410666180925160459
43. Goncharov A (2020) Phase diagram of hydrogen at extreme pressures and temperatures; updated through 2019 (review article). Low Temp Phys 46(2). https://doi.org/10.1063/10.0000526
44. Hamidreza AB, Reza BS, Mahdi CS, Mahdi DY, Fatemeh EF, Hassan FH (2016) Nanotechnology and supercritical fluids. J Fundam Appl Sci 8(2):839. https://doi.org/10.4314/jfas.8vi2s.138
45. Holmes JD, Lyons DM, Ziegler KJ (2003) Supercritical fluid synthesis of metal and semiconductor nanomaterials. Chem Eur J 9(10):2144–2150. https://doi.org/10.1002/chem.200204521
46. Jennifer Jung MP (2001) Introduction. J Supercrit Fluids 20:179–219. http://linkinghub.elsevier.com/retrieve/pii/S089684460100064X
47. Jin HY, Hemingway M, Xia F, Li SN, Zhao YP (2011) Production of β-carotene nanoparticles by the solution enhanced dispersion with enhanced mass transfer by ultrasound in supercritical CO_2 (SEDS-EM). Ind Eng Chem Res 50(23):13475–13484. https://doi.org/10.1021/ie2011565
48. Johnston KP, Barry SE, Read NK, Holcomb TR (1987) Separation of isomers using retrograde crystallization from supercritical fluids. Ind Eng Chem Res 26(11):2372–2377. https://doi.org/10.1021/ie00071a033
49. Johnston KP, Dobbs JM, Wong JM (1986) Nonpolar co-solvents for solubility enhancement in supercritical fluid carbon dioxide. J Chem Eng Data 31(3):303–308. https://doi.org/10.1021/je00045a014
50. Johnston KP, Penninger JML (1989) Supercritical fluid science and technology. https://api.semanticscholar.org/CorpusID:92957157
51. Kakran M, Sahoo NG, Antipina MN, Li L (2013) Modified supercritical antisolvent method with enhanced mass transfer to fabricate drug nanoparticles. Mater Sci Eng C 33(5):2864–2870. https://doi.org/10.1016/j.msec.2013.03.002
52. Kalani M, Yunus R (2011) Application of supercritical antisolvent method in drug encapsulation: a review. Int J Nanomed 6:1429–1442. https://doi.org/10.2147/ijn.s19021
53. Kalani M, Yunus R (2012) Effect of supercritical fluid density on nanoencapsulated drug particle size using the supercritical antisolvent method. Int J Nanomed 7:2165–2172. https://doi.org/10.2147/IJN.S29805
54. Kalra JM, Kalra K, Pokhriyal V, Husaain K (2022) Solubility enhancement of rifabutin by co-solvency approach. J Pharm Negat Results 13(9):1748–1754. https://doi.org/10.47750/pnr.2022.13.s09.210
55. Kankala RK, Chen BQ, Liu CG, Tang HX, Wang SB, Chen AZ (2018) Solution-enhanced dispersion by supercritical fluids: an ecofriendly nanonization approach for processing biomaterials and pharmaceutical compounds. Int J Nanomed 13:4227–4245. https://doi.org/10.2147/IJN.S166124
56. Kelley FD, Chimowitz EH (1990) Near-critical phenomena and resolution in supercritical fluid chromatography. AIChE J 36(8):1163–1175. https://doi.org/10.1002/aic.690360806

57. Knez Ž, Cör D, Knez Hrnčič M (2011) Solubility of solids in sub- and supercritical fluids: a review 2010–2017. J Chem Eng Data 63(4):860–884. https://doi.org/10.1021/acs.jced.7b00778
58. Knez Ž, Hrnčič MK, Škerget M (2015) Particle formation and product formulation using supercritical fluids. Annu Rev Chem Biomol Eng 6:379–407. https://doi.org/10.1146/annurev-chembioeng-061114-123317
59. Knez Ž, Škerget M, Knez Hrnčič M (2010) Principles of supercritical fluid extraction and applications in the food, beverage and nutraceutical industries. In: Separation, extraction and concentration processes in the food, beverage and nutraceutical industries, pp 3–38. https://doi.org/10.1533/9780857090751.1.3
60. Knox DE (2005) Solubilities in supercritical fluids. Pure Appl Chem 77(3):513–530. https://doi.org/10.1351/pac200577030513
61. Kozinski JA, Thomas ML, Butler IS, Dalai AK (2015) Supercritical water (SCW): a new medium to generate novel materials for solar energy applications. In: SASEC 2015 third Southern African solar energy conference, Kruger National Park, South Africa, 11–13 May 2015. https://api.semanticscholar.org/CorpusID:54024113
62. Kumar R (2021) Numerical simulation to estimate the droplet size in aerosol solvent extraction system. Mater Today Proc. https://api.semanticscholar.org/CorpusID:245462579
63. Kumar RI, Thakur AK, Banerjee N, Chaudhari P (2021) A critical review on the particle generation and other applications of rapid expansion of supercritical solution. Int J Pharm 121089. https://api.semanticscholar.org/CorpusID:237545449
64. Kwauk X, Debenedetti PG (1993) Mathematical modeling of aerosol formation by rapid expansion of supercritical solutions in a converging nozzle. J Aerosol Sci 24:445–469. https://api.semanticscholar.org/CorpusID:95577873
65. Lam UT, Mammucari R, Suzuki K, Foster NR (2008) Processing of iron oxide nanoparticles by supercritical fluids. Ind Eng Chem Res 47(3):599–614. https://doi.org/10.1021/ie070494+
66. Lane MKM, Zimmerman JB (2019) Controlling metal oxide nanoparticle size and shape with supercritical fluid synthesis. Green Chem 21(14):3769–3781. Royal Society of Chemistry. https://doi.org/10.1039/c9gc01619h
67. Makos P, Słupek E, Sobczak J, Zabrocki D, Hupka J, Rogala A (2019) Dimethyl ether (DME) as potential environmental friendly fuel. E3S Web Conf 116. https://doi.org/10.1051/e3sconf/201911600048
68. Manjare SD, Dhingra K (2019) Supercritical fluids in separation and purification: a review. Mater Sci Energy Technol 2(3):463–484. https://doi.org/10.1016/j.mset.2019.04.005
69. Mao HK, Chen B, Gou H, Li K, Liu J, Wang L, Xiao H, Yang W (2021) 2020—transformative science in the pressure dimension. Matter Radiat Extrem 6(1):2020–2022. https://doi.org/10.1063/5.0040607
70. Martin Del Valle EM, Galan MA (2005) Supercritical fluid technique for particle engineering: drug delivery applications. Rev Chem Eng 21(1):33–69. https://doi.org/10.1515/revce.2005.21.1.33
71. Milovanović S, Lukić I (2022) An overview on the application of supercritical carbon dioxide for the processing of pharmaceuticals. Arh Farm 72(6):566–590. https://doi.org/10.5937/ARHFARM72-39999
72. Mishima K (2008) Biodegradable particle formation for drug and gene delivery using supercritical fluid and dense gas. Adv Drug Deliv Rev 60(3):411–432. https://doi.org/10.1016/j.addr.2007.02.003
73. Moribe K, Fukino M, Tozuka Y, Higashi K, Yamamoto K (2009) Prednisolone multicomponent nanoparticle preparation by aerosol solvent extraction system. Int J Pharm 380(1–2):201–205. https://doi.org/10.1016/j.ijpharm.2009.06.030
74. Padmajan Sasikala S, Poulin P, Aymonier C (2016) Prospects of supercritical fluids in realizing graphene-based functional materials. Adv Mater 28(14):2663–2691. https://doi.org/10.1002/adma.201504436
75. Pasquali I, Bettini R (2008) Are pharmaceutics really going supercritical? Int J Pharm 364(2):176–187. https://doi.org/10.1016/j.ijpharm.2008.05.014

76. Patel P, Patel KV (2013) Supercritical fluid technology. https://api.semanticscholar.org/CorpusID:65639807
77. Pathak P, Meziani MJ, Sun YP (2005) Supercritical fluid technology for enhanced drug delivery. Expert Opin Drug Deliv 2(4):747–761. https://doi.org/10.1517/17425247.2.4.747
78. Payne KM, King JW (2020) Supercritical fluid extraction/chromatography. https://api.semanticscholar.org/CorpusID:213776493
79. Penninger JML, McHugh, Radosz M, Krukonis VJ (1985) Supercritical fluid technology. East Pharm 44:35–43. https://api.semanticscholar.org/CorpusID:93192576
80. Perinelli DR, Cespi M, Bonacucina G, Naylor A, Whitaker MJ, Lam JKW, Howdle SM, Casettari L, Palmieri GF (2016) PEGylated biodegradable polyesters for PGSS microparticles formulation: processability, physical and release properties. Curr Drug Deliv 13(5):673–681. https://api.semanticscholar.org/CorpusID:46709263
81. Perrut M (2004) Supercritical fluids; applications in the pharmaceutical industry. ChemInform 35. https://api.semanticscholar.org/CorpusID:96849244
82. Pinsook U (2020) In search for near-room-temperature superconducting critical temperature of metal superhydrides under high pressure: a review. J Met Mater Miner 30(2):31–41. https://doi.org/10.55713/jmmm.v30i2.858
83. Prausnitz EP (2004) What is a supercritical fluid. 976(102):976
84. Proctor JE, Maynard-Casely HE (2020) Applications of supercritical fluids. https://api.semanticscholar.org/CorpusID:225221724
85. Prosapio V, Reverchon E, De Marco I (2016) Polymers' ultrafine particles for drug delivery systems precipitated by supercritical carbon dioxide + organic solvent mixtures. Powder Technol 292:140–148. https://doi.org/10.1016/j.powtec.2016.01.033
86. Rad HB, Sabet JK, Varaminian F (2019) Study of solubility in supercritical fluids: thermodynamic concepts and measurement methods—a review. Braz J Chem Eng 36(4):1367–1392. https://doi.org/10.1590/0104-6632.20190364s20170493
87. Rantakylä M (2004) Particle production by supercritical antisolvent processing techniques, issue 76. http://lib.hut.fi/Diss/
88. Reverchon E, Adami R (2006) Nanomaterials and supercritical fluids. J Supercrit Fluids 37(1):1–22. https://doi.org/10.1016/j.supflu.2005.08.003
89. Reverchon E, Delia Porta G, Taddeo R, Pallado P, Stassi A (1995) Solubility and micronization of griseofulvin in supercritical CHF_3. Ind Eng Chem Res 34(11):4087–4091. https://doi.org/10.1021/ie00038a051
90. Reverchon E, Della Porta G, Sannino D, Ciambelli P (1999) Supercritical antisolvent precipitation of nanoparticles of a zinc oxide precursor. Powder Technol 102(2):127–134. https://doi.org/10.1016/S0032-5910(98)00207-1
91. Reverchon E, Spada A (2004) Crystalline microparticles of controlled size produced by supercritical-assisted atomization. Ind Eng Chem Res 43(6):1460–1465. https://doi.org/10.1021/ie034111t
92. Reverchon E, De Marco I (2006) Supercritical fluid extraction and fractionation of natural matter. J Supercrit Fluids 38(2):146–166. https://doi.org/10.1016/j.supflu.2006.03.020
93. Rice SF, LaJuenesse CA, Hanush RG, Aiken JD, Johnston SC (1994) Supercritical water oxidation of colored smoke, dye, and pyrothechnic compositions. Sandia report SAND94-8209
94. Ruchatz F, Kleinebudde P, Müller BW (1997) Residual solvents in biodegradable microparticles. Influence of process parameters on the residual solvent in microparticles produced by the aerosol solvent extraction system (ASES) process. J Pharm Sci 86(1):101–105. https://doi.org/10.1021/js960136o
95. Sacha G, Schmitt W, Nail S (2006) Identification of physical-chemical variables affecting particle size following precipitation using a supercritical fluid. Pharm Dev Technol 11(2):195–205. https://doi.org/10.1080/10837450600561299
96. Santos DT (2019) Supercritical antisolvent precipitation process fundamentals, applications and perspectives. Springer International Publishing

97. Santos S, Puna J, Gomes J (2022) A brief review of the supercritical antisolvent (SAS) technique for the preparation of nanocatalysts to be used in biodiesel production. Energies 15(24). https://doi.org/10.3390/en15249355
98. Schmidt G, Wenzel H (1981) Estimation of critical data by equation of state. Can J Chem Eng 59(4):527-531. https://doi.org/10.1002/cjce.5450590421
99. Schneider GM (1983) Fluid Ph Equilib 10:141–157
100. Shah PS, Hanrath T, Johnston KP, Korgel BA (2004) Nanocrystal and nanowire synthesis and dispersibility in supercritical fluids. J Phys Chem B 108(28):9574–9587. https://doi.org/10.1021/jp049827w
101. Sheth P, Sandhu H, Singhal D, Malick W, Shah N, Serpil Kislalioglu M (2012) Nanoparticles in the pharmaceutical industry and the use of supercritical fluid technologies for nanoparticle production. Curr Drug Deliv 9(3):269–284. https://doi.org/10.2174/156720112800389052
102. Sodeifian G, Usefi MMB (2022) Solubility, extraction, and nanoparticles production in supercritical carbon dioxide: a mini-review. ChemBioEng Rev. https://api.semanticscholar.org/CorpusID:254438096
103. Subramaniam B, Rajewski RA, Snavely K (1997) Pharmaceutical processing with supercritical carbon dioxide. J Pharm Sci 86(8):885–890. https://doi.org/10.1021/js9700661
104. Sun YP, Atorngitjawat P, Meziani MJ (2001) Preparation of silver nanoparticles via rapid expansion of water in carbon dioxide microemulsion into reductant solution. Langmuir 17(19):5707–5710. https://doi.org/10.1021/la0103057
105. Sun Y (2002) Supercritical fluid technology in materials science and engineering: syntheses: properties, and applications. https://api.semanticscholar.org/CorpusID:94005112
106. Tan SP, Piri M (2017) Retrograde behavior revisited: implications for confined fluid phase equilibria in nanopores. Phys Chem Chem Phys 19(29):18890–18901. https://doi.org/10.1039/c7cp02446k
107. Taylor CMV, Jacobson GB (1981) Supercritical fluids. EMagRes 2007(10). https://doi.org/10.1002/9780470034590.emrstm0544
108. Tenorio A, Gordillo MD, Pereyra CM, De La Ossa EJM (2010) Supercritical antisolvent process applied to the pharmaceutical industry. Part Sci Technol 28(3):262–266. https://doi.org/10.1080/02726351.2010.481589
109. Tenorio A, Jaeger P, Gordillo MD, Pereyra CM, De La Ossa EJM (2009) On the selection of limiting hydrodynamic conditions for the Supercritical AntiSolvent (SAS) process. Ind Eng Chem Res 48(20):9224–9232. https://doi.org/10.1021/ie801940p
110. Thakur R, Gupta RB (2006) Production of hydrocortisone micro- and nano-particles using supercritical anti-solvent with enhanced mass transfer. Chem Eng Commun 193(3):293–305. https://doi.org/10.1080/009864490949143
111. Tirado DF, Cabañas A, Calvo L (2023) Modelling and scaling-up of a supercritical fluid extraction of emulsions process. Processes 11(4). https://doi.org/10.3390/pr11041063
112. Tom JW, Debenedetti PG (1991) Particle formation with supercritical fluids—a review. J Aerosol Sci 22(5):555–584. https://doi.org/10.1016/0021-8502(91)90013-8
113. Trivedi JS, Yue Z (2008) Solubilization using cosolvent approach. https://api.semanticscholar.org/CorpusID:202053096
114. Türk M, Upper G, Hils P (2006) Formation of composite drug-polymer particles by co-precipitation during the rapid expansion of supercritical fluids. J Supercrit Fluids 39(2):253-263. https://doi.org/10.1016/j.supflu.2006.04.004
115. Wang T, Zhu L, Mei L, Kanda H (2024) Extraction and separation of natural products from microalgae and other natural sources using liquefied dimethyl ether, a green solvent: a review. Foods 13(2). https://doi.org/10.3390/foods13020352
116. Warwick B, Dehghani F, Foster NR, Biffin JR, Regtop HL (2002) Micronization of copper indomethacin using gas antisolvent processes. Ind Eng Chem Res 41(8):1993–2004. https://doi.org/10.1021/ie010760y
117. Woods HM, Silva MMCG, Nouvel C, Shakesheff KM, Howdle SM (2004) Materials processing in supercritical carbon dioxide: surfactants, polymers and biomaterials. J Mater Chem 14(11):1663–1678. https://doi.org/10.1039/b315262f

118. Ye X, Wai CM (2003) Making nanomaterials in supercritical fluids: a review. J Chem Educ 80(2):198–204. https://doi.org/10.1021/ed080p198
119. York P, Palakodaty S, Kordikowski A, Sloan R (2002) Pharmaceutical and biological materials processing with supercritical fluids. https://doi.org/10.1201/9780203909362.ch12
120. Yun JH, Haque ASMT, Chun BS (2015) Characterization of wheat germ oil particles formed by gas-saturated solutions process with polyethylene glycol. J Food Process Preserv 39(6):1720–1728. https://doi.org/10.1111/jfpp.12403
121. Zhang X, Han B, Hou Z, Zhang J, Liu Z, Jiang T, He J, Li H (2002) Why do co-solvents enhance the solubility of solutes in supercritical fluids? New evidence and opinion. Chem Eur J 8(22):5107–5111. https://doi.org/10.1002/1521-3765(20021115)8:22%3c5107::AID-CHEM5107%3e3.0.CO;2-0
122. Zhang S (2017) A review of challenges and recent progress in supercritical water oxidation of wastewater. Chem Eng Commun 6
123. Zhang X, Heinonen S, Levänen E (2014) Applications of supercritical carbon dioxide in materials processing and synthesis. RSC Adv 4(105):61137–61152. https://doi.org/10.1039/c4ra10662h
124. Zhao Z, Xie M, Li Y, Chen A, Li G, Zhang J, Hu H, Wang X, Li S (2015) Formation of curcumin nanoparticles via solution-enhanced dispersion by supercritical CO_2. Int J Nanomed 10:3171–3181. https://doi.org/10.2147/IJN.S80434

Plasma and Laser Technology for Nanoparticles Production

W. A. D. M. Jayathilaka

Abstract The synthesis of nanoparticles while preserving their inherited characteristics and size-dependent properties is still a challenge that pushes researchers to their limits. With the understanding of plasma engineering and the development of laser and laser-based material processing, researchers could absorb them into nanoparticle synthesis. As a result, different plasma synthesis techniques and laser ablation techniques were introduced and developed over the years. As the understanding of the principle improves over time, the techniques and methodologies have evolved to match the specific requirements. In this chapter, principles of plasma synthesis techniques and laser-based nanoparticle synthesis techniques will be presented. Along with that, some key research will be discussed with an overview and a highlight of their contributions to the development.

Keywords Plasma synthesis · Laser synthesis · Nanoparticles

1 Introduction

Over the years, nanotechnology and nanoparticles have become of key interest among researchers owing to their unique size-dependent properties. A vast range of nanomaterials like metal nanoparticles [65], carbon-based nanoparticles [7, 68], ceramic nanoparticles, semiconductor nanoparticles [19] and polymeric nanoparticles [28] have been used in a broader range of applications like wearable electronics [29], medicine [30], energy storage, catalysts and many more [6, 14, 21, 62]. The uniqueness of nanoparticles is often attributed to the large surface-to-volume ratio and the specific characteristics like crystalline structure, size, high-purity levels and precise composition. Since uniqueness is often associated with the synthesis of nanoparticles, the synthesis of nanoparticles, especially on a large scale while maintaining the physical properties, is highly significant [63].

W. A. D. M. Jayathilaka (✉)
Department of Mechanical Engineering, University of Moratuwa, Katubedda, Sri Lanka
e-mail: dumithj@uom.lk

N. M. Adassooriya et al. (eds.), *Industrial Scale Production of Nanoparticles*, Smart Nanomaterials Technology, https://doi.org/10.1007/978-981-96-9176-0_12

The nanoparticle synthesis falls into two categories: the top-down approach, where nano-sized particles are obtained by reducing the bulk material and the bottom-up approach, where atoms or molecules are used as the starting blocks [27, 52]. Mainly, nanoparticle synthesis is carried out through traditional methods like chemical vapour deposition (CVD), solvothermal methods and sol–gel methods [38]. By following traditional wet chemical methods, nanoparticles must go through additional processing steps to remove any liquid effluents, eventually resulting in costly and environmentally harmful results. Hence, the plasma and laser synthesis techniques have become environmentally benign since they can produce highly pure, pristine nanoparticles through cleaner routes. The absence of chemical ligands and dispersants has been highlighted [40]. In addition, they have shown great control over the final particle size [43], the composition, crystallinity, morphology [17] and surface properties [32, 38, 47].

This chapter will discuss plasma synthesis techniques and laser ablation synthesis techniques, with an overview of their fundamental working principle. In each section, the introduction will be followed by a brief discussion on key research highlighting the influencing parameters and presenting some key concepts in the processes. In the final section, the chapter will be summarised, with a short perspective on the challenges in the scaling-up process of lab-level research to the industrial demands.

2 Plasma-Based Nanoparticle Synthesis

In brief, plasma is defined as the fourth state of matter, which consists of energetic group matter like electrons, ions, atoms, charged particles and excited molecules mixed with neutral gas [38]. Plasma synthesis, in general, is used as a term that defines the nanoparticle synthesis process where charged particles are observed [67]. Unlike most liquid-based nanoparticle synthesis methods, gas-based synthesis techniques like plasma synthesis offer a cleaner and more controllable approach to producing nanoparticles. Plasma-based synthesis can offer high throughput production with flexibility in controlling material morphology by tuning the plasma parameters. Tu et al. [65] have synthesised spherical Ti nanoparticles with an average diameter of 64.6 nm and a standard deviation of 1.15. Also, an environmentally friendly method was proposed in [68], where spherical carbon nanoparticles (SCN) and graphene nanoflakes (GNF) were synthesised with the presence of different buffer gases.

2.1 The Principle

Figure 1 shows two basic arrangements for nanoparticle synthesis via plasma techniques [67]. The plasma originated between two electrodes in both arrangements, with DC or AC electric power. A powerful stream of gas is used to blow away the plasma from the electrodes, where the precursor is introduced either co-axially or radially to the plasma flame directly. The gas flow serves the cooling purposes of the electrodes and provides a means of supplying reactive gasses to the plasma. The presence of precursors (or particles) can be described as dusty plasma, and it is associated with the synthesis process. The precursor can either be in the form of powders or liquid form in an aqueous solution. While the axial feed nozzles can accommodate both forms of precursors, the radial feed system can accommodate only powders as the precursor.

The temperature of the high-temperature plasma flame can reach temperatures above 4000 K very easily, and most of the precursor feed will immediately evaporate, and particles will move in the direction of gas flow. As the particles randomly flow at very high speeds, they promote particle-to-particle collisions and agglomeration, forming particle clusters. In order to control the particle agglomeration and to gain control over the particle size distribution, a rapid quenching zone is introduced just after the plasma flame, usually with the help of Ar or inert gas [71]. After the quenching, the yield is recovered and isolated by means like filtration and cyclone separator, while the unreacted feedstock and the plasma are recycled back into the process.

The nanoparticle synthesis or the particle growth associated with plasma synthesis can be briefly described in three steps [65]. First, the precursor is separated and

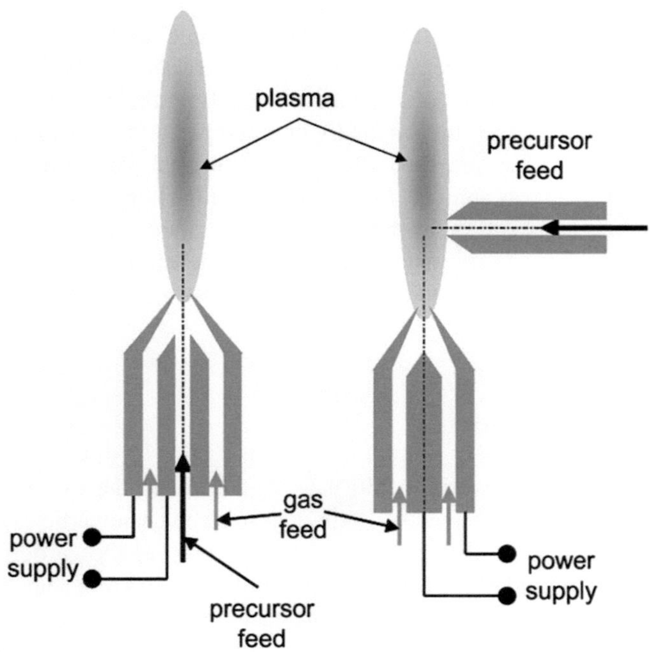

Fig. 1 Two configurations of the basic high-temperature plasma-based nanoparticle synthesis types. Reproduced via licence CC By 4.0 from [67]

nucleated into initial clusters. Then, these clusters get agglomerated into larger particles. Thirdly, these particles will continue to grow through the surface deposition of the precursor. At a certain point in the growth, the agglomerated particles will become large enough such that the charge of the particles themselves prevents further agglomeration.

The plasma synthesis can be categorised based on different approaches. The basic or fundamental categorisation is based on the thermal equilibrium of the plasma, where two categories can be identified: equilibrium plasma (thermal plasma) and non-equilibrium plasma (nonthermal plasma) [67]. Secondly, based on plasma generation, the synthesis can be categorised into classes such as electrical power systems (AC or DC), microwave, radiofrequency (RF), plasma torch, electron beam, etc. The third categorisation is based on the process pressure, as in low-pressure plasma synthesis and atmospheric-pressure plasma synthesis.

2.2 *Thermal Plasma Process*

The thermal plasma process, also known as the equilibrium plasma process, operates at very high temperatures at ambient pressure conditions. The power requirement for the plasma is generally produced by AC, DC or RF power sources. As shown in Fig. 1, the electric power (AC or DC) is supplied through electrodes and the plasma is generated at the tip of the electrode with temperatures above 4000 K. With the gas blowing at high speeds, the plasma is carried away from the nozzle tip, preventing electrode meltdown. In fact, the extremely high temperatures generated during the process are considered an inherited advantage of thermal plasma techniques over other bottom-up synthesis approaches since they result in high-quality nanoparticles [51]. All the reactants get vaporised at high temperatures and promote the high chemical activity of precursor materials. The rapid quenching offers extremely fine nanoparticles with well-controlled size distribution and a cleaner working window for high-purity yield.

2.2.1 AC Plasma Synthesis

In their work, Liu and coworkers [39] have demonstrated the use of the AC arc plasma technique to produce low-cost, high-quality carbon nanoparticles with highly controllable structures. In their work, Ar gas was used as the carrier gas, while Methane (CH_4) was used as the carbon source. With an AC power source of voltage 380 V and 50 Hz, the process was carried out under different electrode materials with different electric currents while varying the gas flow rate of both Ar and Methane. With W and Cu electrodes, spherical carbon particle synthesis was confirmed through scanning electron microscopy (SEM) images. The presence of Cu elements was visible through energy-dispersive spectrum (EDS) results. When Fe was used as the electrode material, the synthesis of nanofiber-like carbon structures was visible with the presence of some Fe elements as well. The presence of Cu and Fe elements in carbon nanoparticle collection was expected since the plasma can generally reach beyond their melting points. As the Fe electrode melted and sputtered under the plasma, authors concluded that the iron nanoparticles had catalysed the growth of carbon nanofiber-like structure, which represented the typical morphology of carbon nanotube growth under plasma. The effect of gas flowrate over the nanoparticle production has also been discussed, and it was concluded that the average diameter of spherical carbon nanoparticles was decreased with the increase of Ar gas flow rate while the same was increased with the increase of Methane gas flow rate [39]. With an increased Methane gas flow rate, the carbon nucleation at the plasma flame was increased, promoting a much higher rate of agglomeration of carbon sub-particles into large nanoparticles (see Fig. 2).

One of the key advantages of plasma synthesis is its capability to scale up production with continuous and high-throughput yield. A design proposed by Hamady et al. [22] highlighted the approach of high-volume carbon nanoparticle synthesis with an AC 3-phase plasma reactor with specialised electrode design. In their design, they used three graphite electrodes for three phases with a divergence from the feedstock inlet. The divergence promotes a self-stabilising and yet scalable plasma. Under

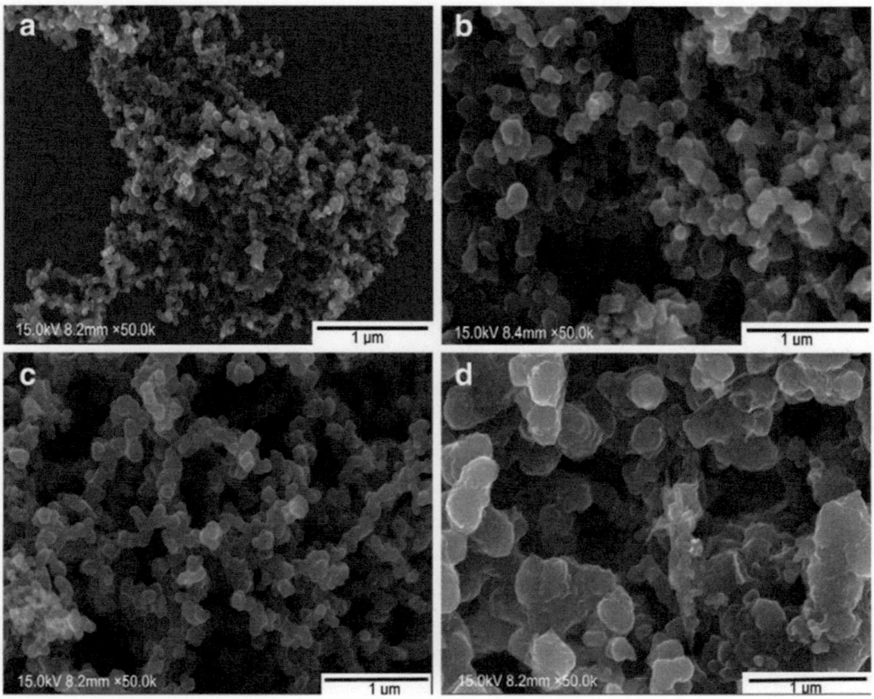

Fig. 2 The SEM images of the carbon nanoparticles produced under different methane flow rates: **a** 15 ml/min, **b** 30 ml/min, **c** 60 ml/min, **d** 80 ml/min. Reprinted with permission from Elsevier from [39]

helium gas supply and the plasma power of 6–9 kW, 'onion-like' carbon nanoparticles of diameters 10–20 nm were obtained.

2.2.2 DC Plasma Synthesis

DC power supply has also been used for nanoparticle synthesis in many applications. Under atmospheric pressure conditions, silicon oxide (SiO_2) nanoparticles were synthesised from hexamethyldisiloxane (HMDSO) using a DC plasma torch under the presence of oxygen (reactant) and nitrogen (carrier) gas supply [13]. A 1 kW DC plasma torch was supplied with Oxygen and Nitrogen gas mix while the precursor was bubbled using a pre-heated Nitrogen gas supply as the carrier gas. After the plasma and a rapid quenching step, nanoparticles of size 8–14 nm were collected at a bag house. They also observed the formation of agglomerated nanoparticles of size 130–260 nm. Further, they have studied the influence of plasma power over nanoparticle synthesis. As the power of the DC plasma torch was increased from 800 to 1600 W, a sharp drop in mean nanoparticle size was observed from 14.5 to 10 nm. The change was attributed to the higher fragmentation, higher nuclei production and the oxidation of precursors under the high plasma power.

Lin and coworkers [36] have studied the synthesis of hexagonal wurtzite ZnO nanoparticles using the DC thermal plasma technique with commercially available ZnO powders (average particle size of 10 μm). The SEM images have shown two types of nanoparticle morphology: spherical ZnO nanoparticles under pure N_2 plasma and tetrapod-like ZnO nanoparticles under the plasma of a mixture of Ar and N_2. The diameter of the spherical ZnO particles was found to be about 31 nm, while the tetrapod-like nanoparticles had four crystalline rods of lengths 100–200 nm.

2.2.3 RF Plasma Synthesis

In both AC and DC plasma methods, the electrodes are directly in contact with the plasma flame. Though the function of the carrier gas is to blow away the plasma and protect the electrodes, it is inevitable that the electrodes get melted at extreme temperatures and can contaminate nanoparticle production. Though it may be useful in some cases as a catalyst for the synthesis [39], it is not the optimal condition for most of the scenarios.

In such premises, induction thermal plasma can provide an alternative approach to producing contamination-free nanoparticles. With the design, the plasma power is generated through induction without any internal electrodes in contact with the plasma. They can reach very high temperatures, with higher purity and high production rate with this method.

Zhang and coworkers [71] have synthesised silicon nanoparticles via the inductive thermal plasma synthesis method and successfully coated them with a carbon coating by introducing a carbon source at a relatively low-temperature stage of the process. Here, the plasma is generated in the plasma torch with the electromagnetic induction of a water-cooled induction coil operated at 4 MHz frequency and 17.5 kW power supply. Micro-sized silicon power is used as the precursor, and Ar is the carrier gas. As the precursor enters the plasma region, it immediately gets vaporised and transported to the tail-end of the plasma torch, where nucleation and solidification happen. Here, the authors introduced ethylene gas when the temperature drops below 1680 K to coat the silicon nanoparticles with an extremely fine coating of carbon over the particles. Silicon nanoparticles were in the range of 67–78 nm, while the carbon coating was measured in the range of 2–8 nm.

As shown in Fig. 3, an inductively coupled thermal plasma (ICTP) process was used to synthesise silicon nanoparticles. With Ar gas as the carrier gas, micro-sized silicon particles were introduced as the precursor of the system. First, nanoparticles with a mean diameter of 150 nm and a standard deviation of 82 nm were synthesised without introducing any quenching medium [60]. When an Ar gas flow was introduced as a quenching medium, the mean diameter of the synthesised silicon nanoparticles was dropped significantly to 27 nm with a standard deviation of just 9 nm. Through a computational analysis defined alongside the experimental analysis, authors have claimed that the quenching increases the number of small nuclei in the quenching region. As the number of nuclei increases, the remaining few vapour atoms condense on nuclei surfaces, resulting in a large number of smaller nanoparticles.

2.3 Low-Temperature Plasma Processes

Unlike the processes explained in the previous section, the low-temperature plasma processes are associated with the low-pressure systems typically operating at temperatures below 1000 K. As a result, the materials like functional materials and temperature-sensitive materials that are not capable of being processed under high-temperature environments can be processed into nano-sized particles [28, 38].

In low-temperature plasma setups, the power is generally provided by microwaves or RF supplies. In either case, the energy transfer is happening to the charged particles in an oscillating field. The amount of energy transferred is proportional to the electric charge of the field, while it is inversely proportional to the mass of the charged particle and the frequency of the oscillation. Since the mass of an electron is less than that of an ion by a thousandfold, the amount of energy transferred to electrons is a thousandfold larger than that of ions. Hence, the particles within the plasma are at different temperatures during the process, making this process a non-equilibrium plasma (nonthermal plasma) process [67].

Heavy ions usually are at temperatures near room temperature, while free electrons can reach temperatures well above 10,000 K. And hence, the collisions between these two elements are very effective in producing highly reactive radicals and ions by ionising the precursors. When compared with other forms of thermal plasma processes, the nonthermal plasma process offers some key advantages: the synthesis of high-purity nanocrystals due to the absence of solvents, synthesis of nanocrystals

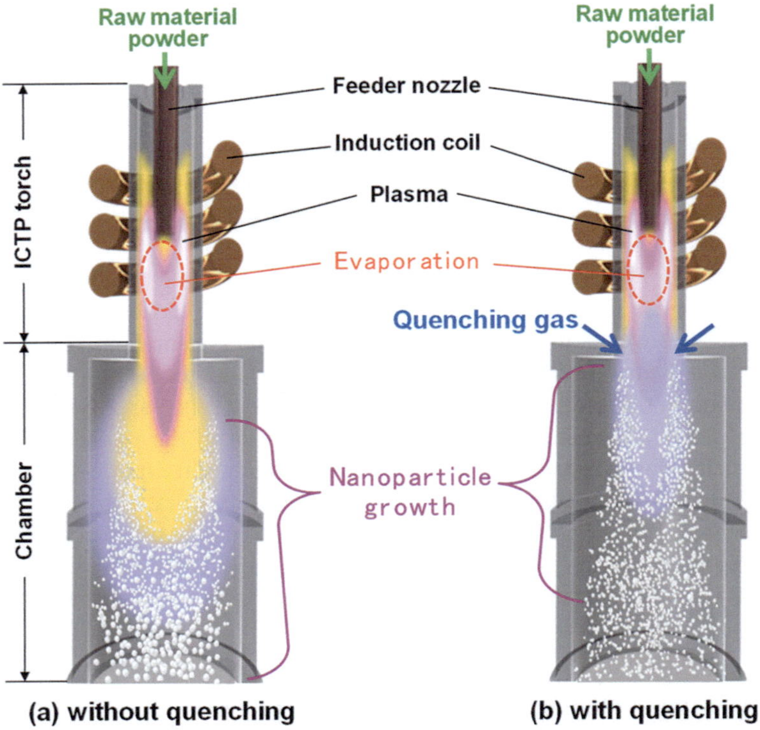

Fig. 3 The schematic of the particle variation in inductively coupled thermal plasma (ICTP) method for the synthesis of silicon nanoparticles **a** without quenching and **b** with quenching using an additional Ar gas supply. Reproduced via licence CC By 4.0 from Shigeta et al. [60]

of high-melting point materials and prevention of particle agglomeration due to the negatively charged plasma [20, 33].

Nonthermal plasma synthesis of carbon nanoparticles is discussed in [68], with a plasma source of magnetically stabilised gliding arc discharge (MSGAD) plasma. Here, two co-centric stainless-steel electrodes are used as the cathode and the anode of the system, with the cathode as the inner electrode. The outer electrode: anode is surrounded by an annular magnet that provides an axial magnetic field. When a DC voltage of 400–700 V is supplied under a fixed magnetic field, a rotating plasma from the cathode is observed. With an electron energy of 1 eV, the heavy particles like ions reached a temperature value of 3000 K. By introducing different buffer gasses to the system they were able to manipulate the crystal structure of the carbon nanoparticle synthesis. With the introduction of hydrogen as the plasma atmosphere, the dangling carbon bonds are terminated, and the growth of planar graphene sheets is encouraged. As a result, few-layer graphene nanoflakes (GNFs) were synthesised with high crystallinity and high purity yield.

Titanium nanocrystals were synthesised using nonthermal plasma methods in a capacitively coupled plasma (CCP) reactor [65]. With titanium tetrachloride ($TiCl_4$)

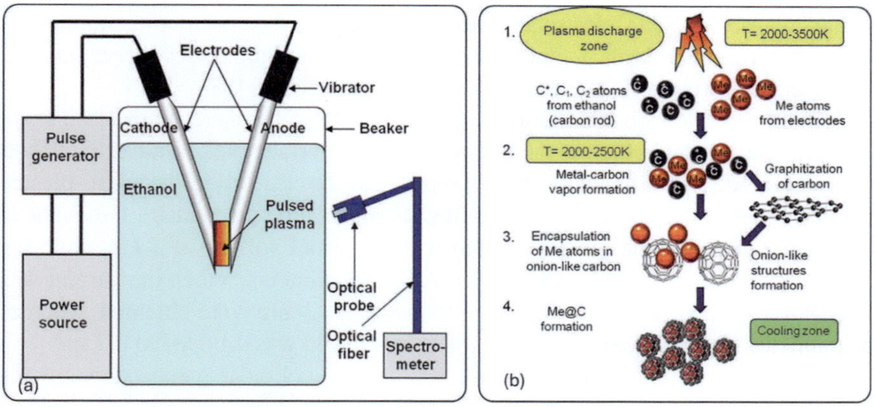

Fig. 4 **a** The schematic diagram of the pulsed plasma synthesis in a liquid medium, **b** The encapsulation process. Reproduced with permission of Elsevier from Abdullaeva et al. [1]

as the precursor, Ti nanoparticles with an average diameter of 64.6 nm were obtained. Since the as-prepared Ti nanoparticles were highly sensitive to oxidation, an H_2 plasma post-treatment method is introduced with a secondary capacitively coupled copper electrode. The post-treatment removed residual chlorides from the surface of the nanoparticles, resulting in lower oxidation of the yield. The X-ray diffraction (XRD) results summarised that the synthesised Ti nanoparticles resembled a non-equilibrium face-centred cubic (FCC) crystal structure.

Synthesis of Co, Ni and Fe encapsulated within a carbon shell structure was proposed by [1]. As shown in Fig. 4a, carbon-encapsulated metal nanoparticles were synthesised with different types of electrodes using the plasma-in-liquid method. Here, two electrodes are submerged in a liquid medium, and a pulsed electrical discharge plasma is generated between the electrodes. In their work, [1] used ethanol as the dielectric medium and copper and nickel-doped electrodes for Co–C encapsulation, pure Ni electrodes for Ni–C encapsulation and Fe electrodes for Fe–C encapsulation. With high-resolution transmission electron microscopy (HRTEM) images, the core–shell nature of carbon encapsulation was confirmed for all three systems. The average diameter for the Co–C encapsulation was 8.5 nm, while 15.1 and 7.9 nm were measured for the encapsulations of Ni–C and Fe–C systems. With the encapsulation, the toxicity of Co, Ni and Fe nanoparticles was lowered, suggesting the applications in biological treatments.

2.4 *Micro-plasma Techniques*

A special category of plasma technique is available, as in the micro-plasma technique, with a sub-millimetre length scale in at least one dimension [41]. As the dimensions get reduced below the millimetre range, the surface-to-volume ratio gets increased, and the electrode spacing gets decreased, promoting high operating pressures, microscale geometry, high-yield and continuous-flow operation, and self-assembly through Coulombic and van der Waals forces [38].

Crystalline Co_3O_4 quantum dots of a diameter of 2–5 nm were synthesised using the micro-plasma technique at the helium-ethanol interface [45]. With a thin cobalt foil as the anode and an absolute ethanol solution as the electrolyte, the micro-plasma was developed at the cathode: a helium-flowing nickel tube suspended 1.5 mm above the ethanol surface. The process started with an initial voltage of 3 kV while the current was set as either 0.7 mA or 4.6 mA for experiments. When the current was set as 4.6 mA, quantum dots of a mean diameter of 1.9 nm were obtained, whereas the diameter was increased to 4.6 nm when the current was dropped to 0.7 mA.

Richmonds and Sankaran [55] have synthesised colloidal metal nanoparticles through the micro-plasma method. The authors have submerged a metal foil inside an acidic liquid medium. The acidic medium was selected considering the conductivity of the solution. A stainless-steel tube with Ar gas flow was suspended above the liquid surface to form the cathode of the micro-plasma. As the micro-plasma developed directly above the liquid surface, a well-dispersed colloidal nanoparticle suspension was obtained within minutes. The average particle size was measured to be around 10 nm. As they used a metal foil, no metal cations were initially available within the solution. Then, with the dissolution of the metal foil, cations travel to the micro-plasma. Also, it was concluded that the particle growth is controlled by parameters like current density, acid concentration and the metal cation concentration in the liquid medium.

2.5 *Nanoparticle Production with Plasma Techniques*

As described above, plasma-based nanoparticle synthesis has been widely adopted by researchers and a number of industrial setups to produce high-purity, size-controllable nanoparticles. Including the limited number of methods explained in the previous sub-sections, plasma synthesis follows a number of processes and design variations to gain the best results possible. Table 1 summarises some of such approaches in producing nanoparticles via the plasma synthesis route.

3 Laser-Based Nanoparticle Synthesis

Recently, the laser synthesis of nanoparticles or pulsed laser-assisted nanoparticle synthesis has received the attention of both researchers and commercial manufacturers. In laser synthesis or laser ablation method, a solid target is 'ablated' with a laser (light amplification by stimulated emission of radiation) as the energy source. With the concentration of extremely high energy at a single point on the target material,

Table 1 Selected research work for the synthesis of nanoparticles via plasma methods

References	Nanomaterial	Method	Description
[68]	Spherical carbon nanoparticles, graphene nanoflakes	Nonthermal plasma (magnetically stabilised gliding arc discharge)	Different crystallinity under different duffer gasses (Ar, He, H_2, N_2), size variation from 50 to 200 nm
[56]	Ag nanoparticles	Micro-plasma synthesis	Ag NP capped with sucrose, PVA and PVP were synthesised; average particle size: 65 nm with PVA capping and 120 nm with sucrose capping
[45]	Co_3O_4 quantum dots	Micro-plasma synthesis	With absolute ethanol as the electrolyte liquid, particle size in the range of 2–5 nm
[40]	Au Nanoparticles	Droplet-plasma synthesis within a microfluidic reactor	Micro-fluidic based reactor, Au NP with an average diameter of about 4 nm, continuous operation with a high rate of synthesis
[8]	Ag nanoparticles	Micro-plasma synthesis	Ag NP of diameters around 14–26 nm, Silver nitrate as the precursor solution and He plasma for the synthesis
[2]	Ag nanoparticles	Submerged plasma synthesis	Ar gas plasma is submerged in silver nitrate solution, with particle diameter in the range of 5–7.3 nm based on the plasma gas, a very high synthesis rate than micro-plasma
[61]	Ag nanoparticles, Au nanoparticles, Au–Ag core–shell nanoparticles	Dual plasma synthesis	Dual plasma at anode and cathode in liquid mediums, different particle sizes and morphologies at anode and cathode, Ag NP at cathode plasma with Ag precursors, Au Np at both anode and cathode with Au precursors, core–shell (Au–Ag) NP when precursors are mixed
[66]	Silica nanoparticles	Plasma assisted synthesis	Deposition of Silicon NP generated via plasma methods on a polymer scaffold, average particle size 200 ± 20 nm, clustered deposition
[44]	Au nanoparticles with polydopamine functionalisation (Au@PDA NP)	Micro-plasma synthesis	Au@PDA NP with an average diameter of 44 nm, single-step synthesis with precursor solution, Ar plasma at the liquid–air interface
[9]	Iron oxide, Iron oxide @ graphene oxide composite	Nonthermal plasma	FeO_x diameter at 5.4 nm, Graphene oxide encapsulated iron oxide (FeO_x@GO) diameter around 20 nm, Ar plasma, Direct deposition on the substrate

(continued)

Table 1 (continued)

References	Nanomaterial	Method	Description
[65]	Ti nanoparticles	Nonthermal plasma with capacitively coupled plasma (CCP)	FCC crystalline Ti NP with diameter 64.6 nm, Hydrogen plasma for Ti NP synthesis from Titanium tetrachloride solution, secondary H_2 plasma to remove residual chlorides, reduce oxidation
[35]	Silicon nanoparticles	Microwave plasma reactor (pilot-plant-scale)	Pilot-plant-scale reactor, Ar and H_2 gas flows, axial and tangential gas flows to improve the plasma, Nanoparticle diameter between 1 and 43 nm depending on precursor concentration,
[15]	Silicon nanoparticles	Microwave plasma synthesis	Ar as the carrier gas, Si NP with an average diameter of 58 nm, in-line coating of NP with different precursors, coating thickness of 6 nm, carbon and non-carbon based coating (HMDSO, OMCTS and TEOS)
[71]	Silicon nanoparticles	Induction thermal plasma	Silicon powder as the precursor, ethylene gas supply for a carbon coating over silicon NP, Mean diameter around 67–78 nm
[37]	Ag nanoparticles	Micro-plasma synthesis	Silver nitrate electrolyte, Pt and SS electrodes with Ar plasma, Spherical nanoparticles, Average Ag NP diameter of 9.64 nm
[24]	Ag, Au, tungsten oxide nanoparticles	Plasma synthesis in water driven by RF	Electrode material based on the choice of nanoparticles, 27.12 MHz Rf input for the electrodes, spherical and rectangular particles for tungsten oxide NP, 9 nm spherical particles, 5 * 19 nm for rectangular NP, spherical Ag NP of 5.5 nm diameter and spherical Au NP of 4.5 nm
[22]	Carbon nanoparticles	Plasma synthesis	3-Phase AC plasma, three-electrode design, large volume plasma, 10–20 nm carbon nanoparticles
[31]	Silica-coated iron oxide nanoparticles	DC plasma synthesis	Hot-wall and cold-wall reactor designs, 200 nm particles in cold-wall reactor and 20 nm particles in hot-wall reactor, iron oxide core and amorphous silica shell

the target gets evaporated immediately, producing a high-purity yield [32]. Figure 5 represents the basic steps of nanoparticle synthesis under the laser synthesis route.

When a focused laser beam 'hits' a target surface, which can be in either a gaseous or liquid medium, the temperature of the target spot increases rapidly. The photon energy from the laser is transferred to the target material lattice in the form of thermal energy. Here, the ablation can follow two routes based on the temporal width of the laser pulse. When the temporal width of the laser pulse is longer than the electron-photon coupling constant of the target material, the incident laser energy will be dissipated over the target spot via heat energy. It will result in material melting, and the ablation will be due to the melting and vaporisation of the target material. On the other hand, when the temporal width of the incident laser is shorter than the electron-photon coupling constant, the heat dissipation will be minimal. Instead, the ablation of nanoparticles will be due to the direct solid-to-vapour transition and sublimation of the material [58]. As the material gets vaporised, the released atoms and clusters get energised, forming a plasma, which is known as a laser-induced plasma plume. In a gaseous environment, this plume can grow larger, while in a liquid medium, the plume is suppressed by the surrounding liquid pressure, limiting it to a small and confined area [32].

The laser synthesis depends on the control of several parameters like laser wavelength, fluence, reaction time, the medium of operation, pulse repetition rate, solvent and, of course, the target material. Though it complicates the process, it also provides more control over the final output and its properties, such as size, composition, crystalline structure, surface properties, etc. Since the process requires no surfactants or capping agents, it is known for its 'green nature'. Also, it provides a much cleaner yield with very high purity within an extremely short duration. Also, the process can be adjusted to produce metal complexes and coated nanoparticles as well [54, 63].

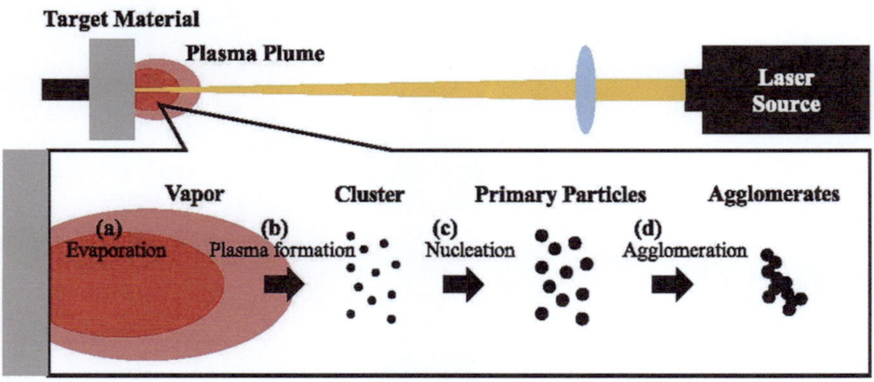

Fig. 5 Schematic representation of the nanoparticle synthesis process in the laser synthesis process. Reproduced via licence CC By 4.0 from Kim et al. [32]

3.1 Laser Ablation in Liquid Mediums

Laser ablation in liquid mediums has vastly been employed in a number of different research and industrial settings. As the synthesis is happening within the liquid medium, the resultant nanoparticles are confined within the liquid itself without any workplace contamination and health risks. Also, it provides some control over the heat dissipation during the process as well [70].

The laser ablation of nanoparticles with liquid medium can be categorised into four types based on their working principle (see Fig. 6) [5]. The first category is known as laser ablation in liquid (LAL), where a solid target is immersed in a liquid medium, and nanoparticles are synthesised by focusing a laser beam on the target surface. Since most LAL processes are performed with pulsed laser beams, it is also known as pulsed laser ablation in liquid (PLAL). The second method is known as laser fragmentation in liquid (LFL), where the precursor material is provided by means of a liquid dispersion. Here, the laser is focused on the liquid medium, inducing fragmentation in the colloidal solution. It has also been observed that the laser-induced fusion of nanoparticles into larger particles is also possible in a liquid medium. Such a synthesis approach is known as laser melting in liquid (LML) [70]. Also, the laser beam can be used to introduce defects into the nanoparticles, such that the structural properties of the nanoparticles are altered without significant changes in size. The method is known as laser defect-engineering in liquid (LDL) and has been used to alter crystalline structure, band gaps and oxidation or reduction of nanoparticles [5].

Laser ablation in liquid (LAL) was used to synthesise pure silver nanoparticles in distilled water to be used in medical treatments [42]. As explained by the authors, the use of distilled water was necessary to avoid the decomposition of any organic solvent and the contamination of silver nanoparticles. A pulsed laser generated from an Nd:YAG laser with a frequency of 2–15 Hz was used to ablate the silver granular

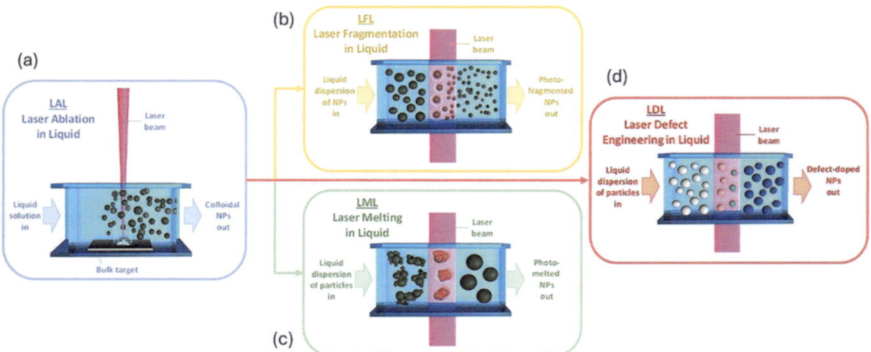

Fig. 6 Four main categories of laser ablation method in liquid mediums. **a** Laser ablation in liquid, **b** laser fragmentation in liquid, **c** laser melting in liquid and **d** laser defect-engineering in liquid. Reproduced with licence CC BY from Amendola et al. [5]

in distilled water. With exposure times over one hour, they have obtained silver nanoparticles in the range of 16–30 nm.

As discussed briefly in previous sections, the liquid medium can influence the synthesis process of the nanoparticles in laser ablation using the liquid method. A study [47] on the effect of the liquid medium has summarised that the choice of liquid medium can influence the size of the nanoparticle and its properties, such as the electrical conductivity and the chemical composition. The synthesis of carbon nanoparticles was examined using three mediums: DI water, ethanol, and a gel medium based on magnesium, sodium, and potassium. As it was claimed, the resultant NP size depended on the amount of ion concentration in the electrolyte solution and the viscosity of the solution. Under constant laser fluence and scan speed, the nanoparticle diameter in ethanol and gel-based solution has an almost tenfold increase compared with the LAL in DI water. The charged ions available in the solution have provided additional nucleation sites, increasing the NP size. Also, the charged ions have agglomerated with the carbon nanoparticles, resulting in a highly conductive yield compared with the resultant based on the DI water sample.

In addition, the final yield's production rate and properties can be controlled by the laser fluence and the ablation time [17]. A copper target submerged within DI water was ablated using a Ce:Nd:YAG laser source with a pulse duration of 10 ns. The laser fluence and ablation time were varied to determine the influence of those factors. It was found that the ablation time and fluence have an effect on the crystallinity of the nanoparticles as a phase shift from Cu to Cu_2O was visible with the increased fluence and ablation time. Out of them, the fluence was more effective in the synthesis of Cu_2O nanoparticles since the higher fluence has synthesised much smaller nanoparticles, which are easier to oxidise in a charged aqueous environment. Also, higher fluence and longer ablation time have resulted in smaller nanoparticles than in other scenarios, where the results were attributed to the fact that the particles tend to collide more with high energy input.

As reported by Intartaglia and coworkers, laser ablation in a liquid method is capable of producing bimetallic nanoparticles as well [26]. Bimetallic nanoparticles of Au–Ag were synthesised using an Ag target submerged within an Au NP colloidal suspension, which was previously synthesised following the LAL method. First, Au NPs were synthesised using a 60 ps Nd:YAG laser system with three different harmonic wavelengths. The obtained Au NPs had a nearly spherical shape with an average diameter of 8 nm. Then, an Ag target was submerged in the Au colloidal solution, and the same procedure was followed to obtain the bimetallic samples. The resultant bimetallic nanoparticles were less than 20 nm in diameter and were found to have 15–25% of Ag content. The silver content can be controlled by the concentration of Au NPs in the colloidal solution. Here, the authors reported the presence of plasma shielding phenomenon when the target was ablated with the harmonics of the original laser wavelength. When the laser of 1064 nm was used, the Ag NP yield was unaffected by the concentration of the Au NP in the solution. However, with lasers of 532 and 355 nm, the incident laser tended to interact with the Au NP colloidal solution, weakening the strength of the laser reaching the target.

This has resulted in a drop in Ag NP synthesis and nano-alloying percentage, thus providing control over the composition of the final bimetallic sample.

Alternatively, a hybrid method has been proposed for the synthesis of ZnO@NiO core–shell nanoparticles [25]. NiO nanoparticles were first synthesised via the LAL method by immersing a Ni target in distilled water. Then, the colloidal solution was subjected to an Ar plasma jet with a submerged Zn target. During the second step, Zno NP was formed with the plasma, and the NiO nanoparticles presented within the colloidal solution were ionised. With the high energy from the plasma, the ZnO nanoparticle surface accommodates the ionised NiO nanoparticles over the outer surface, forming a core–shell structure.

The synthesis of single-sized gold nanoparticles in a liquid–liquid microdroplet dispersion system was presented and discussed in [48]. A femtosecond laser pulse (40 fs) generated from a Ti:sapphire was focused on a quartz cuvette filled with aqueous $HAuCl_4$ and n-hexane mixture to synthesise gold nanoparticles with an average diameter of 3.4 nm. The mixture of n-hexane has provided an oil-in-water type emulsion with n-hexane droplets, which have successfully prevented the Au nanoparticle agglomeration and further growth. When the nanoparticles were synthesised with just aqueous $HAuCl_4$, the nanoparticles grew into larger nanoparticles with more than 10 nm in diameter. Authors have claimed that the synthesis of such finer nanoparticles is due to the laser fragmentation in the liquid (LFL) route, and the diameter is rarely affected by the laser irradiation time.

Though PLAL has been widely used for the synthesis of nanoparticles with controlled sizes and crystallinity, the exact working sequence is an interesting point to discuss. As discussed by Dell'Aglio et al. [16], the nanoparticle synthesis is due to the plasma cooling phase happening right after the laser-induced breakdown of the target (see Fig. 7). When the laser 'hits' the target, the surface breaks down, and a shock wave surrounds the spot. The plasma expands and is subjected to a cooling phase from the outside to the inside of the plasma. Hence, the temperature within the shockwave remains constant and enables nanoparticle formation within the plasma plume. As this plasma and the adjacent liquid layer exchange energy quickly, the liquid at the frontier changes its phase by forming a thin vapour layer. With the temperature and the pressure, the vapour expands until it reaches the saturation point before collapsing. The nanoparticles get diffused within the surrounding liquid as the cavitation bubble collapses.

In fact, the cavitation bubble can interfere with the incident laser beam and hinder nanoparticle synthesis as well [50]. When the cavitation bubble doesn't collapse before the subsequent laser pulse, the laser irradiation will not reach the metal target and might hinder the nanoparticle synthesis and, eventually, the size of the nanoparticles.

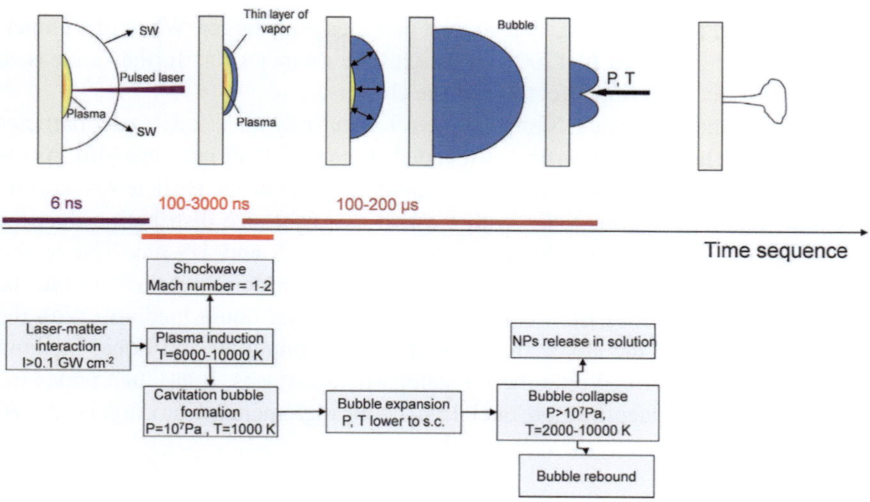

Fig. 7 The process sequence of the laser ablation process in a liquid medium. Reproduced with permission from Elsevier from Dell'Aglio et al. [16]

3.2 *Laser Ablation in Air*

Though it was not as popular as LAL methods, laser ablation in air has also been used for the synthesis of nanoparticles. Though LAL methods have provided a good option to synthesise nanoparticles efficiently compared to traditional wet-chemical methods, additional treatments are required to isolate the synthesised nanoparticles. Hence, the process lacks the productivity demand that may be required for certain applications.

Boutinguiza and the team have discussed a method to synthesise Ag nanoparticles by ablating a solid Ag plate in open-air conditions [10]. The laser ablation of solids by open-air conditions (LASOA) method has produced Ag nanoparticles with an average diameter of 15 nm. Unlike in LAL methods, since no liquid-based cooling is available in the LASOA method, the expansion of the laser-induced plasma plume is unrestricted. Hence, the nanoparticle production was higher than that of the LAL methods. However, the choice of surrounding air is important as it can prevent any surface oxidation of the nanoparticles. Also, it has been noted that the air pressure significantly influences the particle agglomeration. With low air pressure, more clusters were observed, while the presence of isolated nanoparticles was obvious with high air pressure.

A better analysis of the surrounding air conditions on the particle synthesis process is available in [64]. The synthesis of Ti nanoparticles under different air temperatures was carried out within a tube furnace. A clear increase in particle size with the increase in air temperature was observed through different imaging tools. At room temperatures, the nanoparticle's diameter was 11 nm, while at 800 °C, the particle diameter increased to 24 nm. With the temperature rise, the number of nuclei in the plume dropped, the quenching slowed, and particles grew larger. When the temperature is further increased to 1200 °C, the particle diameter has further increased, mainly due to the sintering effect within the clusters.

Following a new approach, Nishi and coworkers have synthesised Pd nanoparticles by focusing the laser on the particle suspended at the liquid–air interface [46]. As the incident laser beam is focused on the air–liquid interface of the particle suspension, they have obtained nanoparticles with a much narrower size distribution than the PLAL method. The average diameter was between 0.5 and 1.5 nm. The reason for such a smaller diameter and much narrower distribution was attributed to the existence of charged ions from both surrounding air and liquid medium. With the laser-induced plume, the atoms from both the surrounding air and liquid medium have ionised and adsorbed onto the Pd nanoparticle surfaces. It has then prevented nanoparticles from attaching more nuclei and being agglomerated into larger clusters.

3.3 Nanoparticle Synthesis in Laser Ablation Process

Due to the 'green' synthesis process with the absence of surfactants and stabilising chemicals and the ability to synthesise high-purity nanoparticles with better control over the size distribution, the laser synthesise process has become popular to produce different nanoparticles. The process has offered a number of controls over the particle size distribution and has resulted in different classes of nanoparticles like metals, metal oxides, alloys and core–shell nanoparticles [59]. Table 2 summarises some selected research work with a brief overview of some noteworthy aspects of each selected work.

4 Conclusions and Future Perspectives

This chapter discusses and reviews thermal and nonthermal plasma techniques and laser techniques in both liquid and open-air techniques for nanoparticle synthesis. When compared with traditional wet chemical methods, these two approaches have offered routes to synthesise nanoparticles with much cleaner methods, as they eliminate the need for post-cleaning steps to remove chemical effluents. Also, they have offered better control over the yield with the control of particle size, morphology, crystallinity, composition, surface properties and productivity. Both processes have been proven to have different metal nanoparticles, semiconductor nanoparticles, magnetic nanoparticles, carbon-based nanoparticles, and many more with different process variations and modifications. As the researchers start to understand the principles and process sequences clearly, more process variations will be introduced to target specific material groups and process conditions.

However, it was quite evident through literature that most of these techniques are developed on a lab scale and still haven't gone beyond that level. Very few works [34,

Table 2 Selected research work on the synthesis of nanoparticles via laser ablation routes

References	Nanoparticle	Method	Description
[48]	Au nanoparticles	Laser fragmentation in liquid (LFL)	Ti:sapphire laser, femtosecond pulses, diameter of 3.4 nm, n-hexane microdroplets to prevent agglomeration
[42]	Ag nanoparticles	Laser ablation in liquid (LAL)	Nd:YAG laser, long exposure time, nanoparticle diameter of 16–30 nm,
[26]	Au–Ag bimetallic nanoparticles	Laser ablation in liquid (LAL)	Nd:YAG laser with three different harmonic wavelengths, bimetallic samples with Ag solid submerged in Au NP solution, Alloy diameter less than 20 nm
[47]	Carbon nanoparticles	Laser ablation in liquid (LAL)	Nd:YAG laser, synthesis with three different liquid mediums, diameter variation from ~ 10–1300 nm based on the liquid and fluence,
[25]	ZnO@NiO core–shell nanoparticles	Laser ablation in liquid (LAL) and laser-induced plasma hybrid technique	Nd:YAG laser and Ar gas plasma, Average particle size about 50 nm, ZnO core and NiO shell structure
[17]	Cu/Cu_2O nanoparticles	Laser ablation in liquid (LAL)	Ce:Nd:YAG laser with 10 ns pulse, solid copper target within DI water, different ablation time and laser fluence, average diameter from 40 to 79 nm depending on the fluence and ablation time
[10]	Ag nanoparticles	Laser ablation of solids in open air conditions (LASOA)	Nd:YVO_4 laser with 50 ns pulse duration, Solid Ag plate in Ar air flow, Average diameter of 15 nm, high production rate
[64]	TiO_2 nanoparticles	Laser ablation in open air	Nd:YAG laser with 6 ns pulse, 11 nm nanoparticles at room temperature, up to 45 nm with increased air temperature
[4]	Ag nanoparticles	Laser ablation in liquid (LAL)	Nd:YAG laser with 7 ns pulses, solid Ag target is DI water, Spherical particles with less than 100 nm diameter
[57]	Si@Au and Si@Ag nanoparticles	Laser ablation in liquid (LAL)	Silicon target in metal ion solution, laser beam to form semiconductor core @ metal shell, Average diameter ~ 500 nm
[53]	Au:ZnO core–shell nanoparticles	Laser ablation in liquid (LAL)	Nd:YAG laser with 9 ns pulses, Au pellet in DI water followed by Zn pellet in Au NP colloidal solution, Average diameter ~ 53 nm

(continued)

Table 2 (continued)

References	Nanoparticle	Method	Description
[18]	Au nanoparticles	Laser ablation in liquid (LAL)	Ti:sapphire laser, use of femtosecond pulses, simultaneous spatial and temporal focusing (SSTF) for laser focusing, average diameter 17 nm, 2.4 times increased productivity
[49]	Ag nanoparticles	Laser ablation in liquid (LAL)	Nd:YAG laser, Ag plate submerged in DI water, spherical nanoparticles with an average diameter of 50 nm
[23]	Au-graphene quantum dots composite nanoparticles	Laser ablation in liquid (LAL)	Nd:YAG laser with 7 ns pulses, GQD and Au nanoparticles separately with graphite target and gold foil in DI water, mixing of precursors and laser ablation to form composite, Au NP diameter ~ 10–18 nm, GQD diameter ~ 7–9 nm, composite diameter ~ 15 nm
[50]	Ni nanoparticles	Laser ablation in liquid (LAL)	Nd:YAG laser with a Ni target submerged in DI water, three different laser powers, 49, 19 and 8 nm particles for increasing laser powers
[69]	Be nanoparticles	Laser ablation in liquid (LAL)	Nd:YAG laser, Be target in liquid mediums, No NP formation with acetone as liquid (protective carbon layer formation), BE NP formation with bi-distilled water, particle diameter 54–90 nm, Be in heavy water resulted in NP of 80 nm diameter

35] have even discussed the topic. Although nanomaterials have shown multitudes of advantages and potential to revolutionise the current medical, energy, agricultural, and engineering fields, the application of industry technology is still a challenge. As discussed in the editorial, the issue can be due to two main reasons [3]. Firstly, the characteristics of nanomaterials can be altered during any scaled-up synthesis route. Secondly, the capital investment required for such a large-scale facility could be huge, and a healthy profit must be ensured before securing such investment. Scaling up lab research to a level of industry would demand analytical characterisation, batch-to-batch consistency analysis, prototype and pilot productions, stress testing to understand the limits of the process, propriety methodologies, cost optimisation, raw material quality assurance and many more [11]. Charitidis and the team highlighted five key factors to be considered during a transition from lab to industry: product orientation, continuous industry interaction, spiral product development method, attraction and retention of industry partners, and mitigation of associated risks [12].

In a nutshell, the future of nanomaterial synthesis requires more effort to reap the true benefits of the technology. There are plenty of opportunities available for new projects, mainly in two main aspects: one in the technology and knowledge aspect and the second in industry collaboration aspects. Though much research is available on laser and plasma synthesis routines, we have yet to claim success in the knowledge and technology aspects. New research must focus more on the production aspects, focusing on the applications. The standard practices on technology readiness levels can be used to identify the gaps and potential for development. This may involve a series of different characterisation results, stress analyses, process optimisation techniques, consistency analyses and many more. Another aspect that can be included here is the green manufacturing aspects and the associated risk. As quantum mechanics governs the properties of nanomaterials at the nanoscale, the health impacts like the toxicology and associated health impacts are still unclear. Also, the push for the circular economy, sustainable manufacturing processes, supply of raw materials, cleaner production, and waste management can open up new avenues. It can be impactful with investments as well.

Secondly, industrial collaboration is crucial for successful technology transfer, from the point of understanding the industrial requirement to the point where the final product completes its full cycle. First, industry partners should be attracted to the project by sharing their knowledge and the value addition aspects. The requirement should be assessed with their input to transform the lab technology into a marketable product. The input will optimise the process and will identify drawbacks easily. Collaborative partnerships are crucial as they can provide the necessary resources, expertise, and financial support required. As most industries are oriented towards sustainability development goals, the project and product development should also be aligned with them.

References

1. Abdullaeva Z, Omurzak E, Iwamoto C et al (2012) Onion-like carbon-encapsulated Co, Ni, and Fe magnetic nanoparticles with low cytotoxicity synthesised by a pulsed plasma in a liquid. Carbon N Y 50:1776–1785. https://doi.org/10.1016/j.carbon.2011.12.025
2. Acharya TR, Jang M, Lee GJ, Choi EH (2023) A comprehensive study on the synthesis, characteristics, and catalytic applications of submerged hydrogen-mixed argon plasma-synthesised silver nanoparticles. Curr Appl Phys 56:36–46. https://doi.org/10.1016/j.cap.2023.09.003
3. Ahmed H (2022) Editorial feature: the challenges behind scaling up nanomaterials. In: AZO NANO. https://www.azonano.com/article.aspx?ArticleID=6126. Accessed 8 Oct 2024
4. Alhamid MZ, Hadi BS, Khumaeni A (2019) Synthesis of silver nanoparticles using laser ablation method utilising Nd:YAG laser. AIP Conf Proc 2202. https://doi.org/10.1063/1.5141626
5. Amendola V, Amans D, Ishikawa Y et al (2020) Room-temperature laser synthesis in liquid of oxide, metal-oxide core-shells, and doped oxide nanoparticles. Chem A Eur J 26:9206–9242. https://doi.org/10.1002/chem.202000686
6. Baig N, Kammakakam I, Falath W, Kammakakam I (2021) Nanomaterials: a review of synthesis methods, properties, recent progress, and challenges. Mater Adv 2:1821–1871. https://doi.org/10.1039/d0ma00807a
7. Barberio M, Antici P (2017) Laser-plasma driven synthesis of carbon-based nanomaterials. Sci Rep 7:12009. https://doi.org/10.1038/s41598-017-12243-4
8. Bisht A, Deen GR, Ilyas U et al (2013) Synthesis of nanoparticles using atmospheric microplasma discharge. Singapore

9. Bjelajac A, Phillipe AM, Guillot J et al (2024) Iron oxide/graphene oxide nanocomposite synthesis using atmospheric cold plasma. RSC Adv 14:1750–1756. https://doi.org/10.1039/d3ra05560d
10. Boutinguiza M, Comesaña R, Lusquiños F et al (2015) Production of silver nanoparticles by laser ablation in open air. Appl Surf Sci 336:108–111. https://doi.org/10.1016/j.apsusc.2014.09.193
11. Cerionnano (2023) Scale up nanomaterial production—cerion nanomaterials. https://cerionnano.com/scale-up/. Accessed 8 Oct 2024
12. Charitidis CA, Georgiou P, Koklioti MA et al (2014) Manufacturing nanomaterials: from research to industry. Manuf Rev 1:11. https://doi.org/10.1051/mfreview/2014009
13. Chau-Nan Hong F, Yan CJ (2018) Synthesis and characterisation of silicon oxide nanoparticles using an atmospheric DC plasma torch. Adv Powder Technol 29:220–229. https://doi.org/10.1016/j.apt.2017.11.006
14. Choi S, Lee H, Ghaffari R et al (2016) Recent advances in flexible and stretchable bio-electronic devices integrated with nanomaterials. Adv Mater 28:4203–4218
15. Dasgupta M, Fortugno P, Wiggers H (2020) Plasma-assisted gas-phase synthesis and in-line coating of silicon nanoparticles. Plasma Process Polym 17
16. Dell'Aglio M, Gaudiuso R, De Pascale O, De Giacomo A (2015) Mechanisms and processes of pulsed laser ablation in liquids during nanoparticle production. Appl Surf Sci 348:4–9. https://doi.org/10.1016/j.apsusc.2015.01.082
17. Dizajghorbani Aghdam H, Azadi H, Esmaeilzadeh M et al (2019) Ablation time and laser fluence impacts on the composition, morphology and optical properties of copper oxide nanoparticles. Opt Mater (Amst) 91:433–438. https://doi.org/10.1016/j.optmat.2019.03.027
18. Doñate-Buendía C, Fernández-Alonso M, Lancis J, Mínguez-Vega G (2019) Overcoming the barrier of nanoparticle production by femtosecond laser ablation in liquids using simultaneous spatial and temporal focusing. Photonics Res 7:1249. https://doi.org/10.1364/prj.7.001249
19. Dołan I, Van De Sanden MCM (2016) Gas-phase plasma synthesis of free-standing silicon nanoparticles for future energy applications. Plasma Process Polym 13:19–53. https://doi.org/10.1002/ppap.201500197
20. Dou S, Tao L, Wang R et al (2018) Plasma-assisted synthesis and surface modification of electrode materials for renewable energy. Adv Mater 30:1–24. https://doi.org/10.1002/adma.201705850
21. Gong S, Cheng W (2017) One-dimensional nanomaterials for soft electronics. Adv Electron Mater 3:1600314
22. Hamady M, Sheppard D, Seddighi K, et al (2015) Synthesis of nanocarbons using a large volume AC plasma reactor. MRS Proc 1747:mrsf14-1747-hh08-06. https://doi.org/10.1557/opl.2015.341
23. Hameed R (2021) Synthesis and physical properties of nanocomposite gold nanoparticles attached to graphene quantum dots prepared by laser ablation. Mater Today Proc 42:1829–1833. https://doi.org/10.1016/j.matpr.2020.12.195
24. Hattori Y, Nomura S, Mukasa S et al (2013) Synthesis of tungsten oxide, silver, and gold nanoparticles by radio frequency plasma in water. J Alloys Compd 578:148–152. https://doi.org/10.1016/j.jallcom.2013.05.032
25. Imran HJ, Hubeatir KA, Aadim KA (2023) A novel method for ZnO@NiO core–shell nanoparticle synthesis using pulse laser ablation in liquid and plasma jet techniques. Sci Rep 13:1–13. https://doi.org/10.1038/s41598-023-32330-z
26. Intartaglia R, Das G, Bagga K et al (2013) Laser synthesis of ligand-free bimetallic nanoparticles for plasmonic applications. Phys Chem Chem Phys 15:3075–3082. https://doi.org/10.1039/C2CP42656K
27. Jamkhande PG, Ghule NW, Bamer AH, Kalaskar MG (2019) Metal nanoparticles synthesis: an overview on methods of preparation, advantages and disadvantages, and applications. J Drug Deliv Sci Technol 53:101174. https://doi.org/10.1016/j.jddst.2019.101174
28. Jang HJ, Jung EY, Parsons T et al (2021) A review of plasma synthesis methods for polymer films and nanoparticles under atmospheric pressure conditions. Polymers (Basel) 13. https://doi.org/10.3390/polym13142267

29. Jayathilaka WADM, Qi K, Qin Y et al (2019) Significance of nanomaterials in wearables: a review on wearable actuators and sensors. Adv Mater 31:1805921. https://doi.org/10.1002/adma.201805921
30. Kaushik NK, Kaushik N, Linh NN et al (2019) Plasma and nanomaterials: fabrication and biomedical applications. Nanomaterials 9:1–19. https://doi.org/10.3390/nano9010098
31. Kim D-W, Kim T-H, Choi S et al (2012) Preparation of silica coated iron oxide nanoparticles using non-transferred arc plasma. Adv Powder Technol 23:701–707. https://doi.org/10.1016/j.apt.2011.09.001
32. Kim M, Osone S, Kim T et al (2017) Synthesis of nanoparticles by laser ablation: a review. KONA Powder Part J 34:80–90. https://doi.org/10.14356/kona.2017009
33. Kortshagen UR, Sankaran RM, Pereira RN et al (2016) Nonthermal plasma synthesis of nanocrystals: fundamental principles, materials, and applications. Chem Rev 116:11061–11127. https://doi.org/10.1021/acs.chemrev.6b00039
34. Kunze F, Kuns S, Hülser T et al (2023) Validation of the silicon nanoparticle production on the pilot plant scale via long-term gas-phase synthesis using a microwave plasma reactor. Appl Energy Combust Sci 15:100195. https://doi.org/10.1016/j.jaecs.2023.100195
35. Kunze F, Kuns S, Spree M et al (2019) Synthesis of silicon nanoparticles in a pilot-plant-scale microwave plasma reactor: Impact of flow rates and precursor concentration on the nanoparticle size and aggregation. Powder Technol 342:880–886. https://doi.org/10.1016/j.powtec.2018.10.042
36. Lin HF, Liao SC, Hung SW (2005) The DC thermal plasma synthesis of ZnO nanoparticles for visible-light photocatalyst. J Photochem Photobiol A Chem 174:82–87. https://doi.org/10.1016/j.jphotochem.2005.02.015
37. Lin L, Li X, Zhou J et al (2021) Plasma-aided green and controllable synthesis of silver nanoparticles and their compounding with Gemini surfactant. J Taiwan Inst Chem Eng 122:311–319. https://doi.org/10.1016/j.jtice.2021.04.061
38. Lin L, Wang Q (2015) Microplasma: a new generation of technology for functional nanomaterial synthesis. Plasma Chem Plasma Process 35:925–962. https://doi.org/10.1007/s11090-015-9640-y
39. Liu XY, Hong RY, Feng WG, Badami D (2014) Synthesis of structure controlled carbon nanomaterials by AC arc plasma process. Powder Technol 256:158–165. https://doi.org/10.1016/j.powtec.2014.01.096
40. Maguire P, Rutherford D, Macias-Montero M et al (2017) Continuous In-flight synthesis for on-demand delivery of ligand-free colloidal gold nanoparticles. Nano Lett 17:1336–1343. https://doi.org/10.1021/acs.nanolett.6b03440
41. Mariotti D, Sankaran RM (2010) Microplasmas for nanomaterials synthesis. J Phys D Appl Phys 43. https://doi.org/10.1088/0022-3727/43/32/323001
42. Mwenze NM, Juma M, Maaza M et al (2023) Laser liquid ablation for silver nanoparticles synthesis and conjugation with hydroxychloroquine for Covid-19 treatment. Mater Today Proc. https://doi.org/10.1016/j.matpr.2023.08.195
43. Nancy P, James J, Valluvadasan S et al (2018) Laser–plasma driven green synthesis of size controlled silver nanoparticles in ambient liquid. Nano-Struct Nano-Objects 16:337–346. https://doi.org/10.1016/j.nanoso.2018.09.006
44. Nguyen LN, Kaushik N, Lamichhane P et al (2020) In situ plasma-assisted synthesis of polydopamine-functionalised gold nanoparticles for biomedical applications. Green Chem 22:6588–6599. https://doi.org/10.1039/D0GC01348J
45. Ni C, Carolan D, Rocks C et al (2018) Microplasma-assisted electrochemical synthesis of Co_3O_4 nanoparticles in absolute ethanol for energy applications. Green Chem 20:2101–2109. https://doi.org/10.1039/C8GC00200B
46. Nishi T, Suzuki N, Takahashi N, Yano K (2013) Preparation of monodispersed Pd nanoparticles by laser ablation at air-suspension interface. J Nanoparticle Res 15. https://doi.org/10.1007/s11051-013-1569-6
47. Nyabadza A, Vázquez M, Fitzpatrick B, Brabazon D (2022) Effect of liquid medium and laser processing parameters on the fabrication of carbon nanoparticles via pulsed laser ablation in

liquid towards paper electronics. Colloids Surf A Physicochem Eng Asp 636. https://doi.org/10.1016/j.colsurfa.2021.128151
48. Okamoto T, Nakamura T, Sakota K, Yatsuhashi T (2019) Synthesis of single-nanometer-sized gold nanoparticles in liquid-liquid dispersion system by femtosecond laser irradiation. Langmuir 35:12123–12129. https://doi.org/10.1021/acs.langmuir.9b01854
49. Qayyum H, Ahmed W, Hussain S et al (2020) Laser synthesis of surfactant-free silver nanoparticles for toxic dyes degradation and SERS applications. Opt Laser Technol 129:106313. https://doi.org/10.1016/j.optlastec.2020.106313
50. Rahman A, Guisbiers G (2024) Synthesis of nickel-based nanoparticles by pulsed laser ablation in liquids: correlations between laser beam power, size distribution and cavitation bubble lifetime. Metals (Basel) 14:224. https://doi.org/10.3390/met14020224
51. Rai P, Park JS, Park GG et al (2014) Influence of carbon precursors on thermal plasma assisted synthesis of SiC nanoparticles. Adv Powder Technol 25:640–646. https://doi.org/10.1016/j.apt.2013.10.004
52. Rane AV, Kanny K, Abitha VK, Thomas S (2018) Methods for synthesis of nanoparticles and fabrication of nanocomposites. Elsevier Ltd.
53. Rashid TM, Nayef UM, Jabir MS, Mutlak FAH (2021) Synthesis and characterisation of Au:ZnO (core:shell) nanoparticles via laser ablation. Optik (Stuttg) 244:167569. https://doi.org/10.1016/j.ijleo.2021.167569
54. Reza Sadrolhosseini A, Adzir Mahdi M, Alizadeh F, Abdul Rashid S (2019) Laser ablation technique for synthesis of metal nanoparticle in liquid. In: Laser technology and its applications. IntechOpen
55. Richmonds C, Sankaran RM (2008) Plasma-liquid electrochemistry: rapid synthesis of colloidal metal nanoparticles by microplasma reduction of aqueous cations. Appl Phys Lett 93:129–132. https://doi.org/10.1063/1.2988283
56. Saleem MT, Bashir S, Bashir M (2021) Microplasma assisted synthesis of silver nanoparticles capped with PVA, PVP and sucrose. Nano Express 2. https://doi.org/10.1088/2632-959X/ac0965
57. Saraeva IN, Van LN, Kudryashov SI et al (2018) Laser synthesis of colloidal Si@Au and Si@Ag nanoparticles in water via plasma-assisted reduction. J Photochem Photobiol A Chem 360:125–131. https://doi.org/10.1016/j.jphotochem.2018.04.004
58. Semaltianos NG (2010) Nanoparticles by laser ablation. Crit Rev Solid State Mater Sci 35:105–124. https://doi.org/10.1080/10408431003788233
59. Semaltianos NG, Karczewski G (2021) Laser synthesis of magnetic nanoparticles in liquids and application in the fabrication of polymer-nanoparticle composites. ACS Appl Nano Mater 4:6407–6440. https://doi.org/10.1021/acsanm.1c00715
60. Shigeta M, Hirayama Y, Ghedini E (2021) Computational study of quenching effects on growth processes and size distributions of silicon nanoparticles at a thermal plasma tail. Nanomaterials 11. https://doi.org/10.3390/nano11061370
61. Shirai N, Uchida S, Tochikubo F (2014) Synthesis of metal nanoparticles by dual plasma electrolysis using atmospheric DC glow discharge in contact with liquid. Jpn J Appl Phys 53:046202. https://doi.org/10.7567/JJAP.53.046202
62. Simon P, Gogotsi Y (2008) Materials for electrochemical capacitors. Nat Mater 7:845–854. https://doi.org/10.1038/nmat2297
63. Theerthagiri J, Karuppasamy K, Lee SJ et al (2022) Fundamentals and comprehensive insights on pulsed laser synthesis of advanced materials for diverse photo- and electrocatalytic applications. Light Sci Appl 11:250. https://doi.org/10.1038/s41377-022-00904-7
64. Tsuji M, Seto T, Otani Y (2012) Effect of surrounding gas temperature on the morphological evolution of TiO_2 nanoparticles generated by laser ablation in tubular furnace. J Nanoparticle Res 14. https://doi.org/10.1007/s11051-011-0674-7
65. Tu Q, Poerschke DL, Kortshagen UR (2024) Nonthermal plasma synthesis of metallic Ti nanocrystals. Nanomaterials 14. https://doi.org/10.3390/nano14030264
66. Vijayan VM, Tucker BS, Dimble PS et al (2020) Dusty-plasma-assisted synthesis of silica nanoparticles for in situ surface modification of 3D-printed polymer scaffolds. ACS Appl Nano Mater 3:7392–7396. https://doi.org/10.1021/acsanm.0c01734

67. Vollath D (2007) Plasma synthesis of nanoparticles. KONA Powder Part J 25:39–55. https://doi.org/10.14356/kona.2007007
68. Wang C, Li D, Lu ZS et al (2020) Synthesis of carbon nanoparticles in a nonthermal plasma process. Chem Eng Sci 227:115921. https://doi.org/10.1016/j.ces.2020.115921
69. Yehia SA, Carpen LG, Stokker-Cheregi F et al (2022) Laser ablation of a solid target in liquid medium for beryllium nanoparticles synthesis. Nucl Mater Energy 31. https://doi.org/10.1016/j.nme.2022.101160
70. Zhang D, Gökce B, Barcikowski S (2017) Laser synthesis and processing of colloids: fundamentals and applications. Chem Rev 117:3990–4103. https://doi.org/10.1021/acs.chemrev.6b00468
71. Zhang X, Hayashida R, Tanaka M, Watanabe T (2020) Synthesis of carbon-coated silicon nanoparticles by induction thermal plasma for lithium ion battery. Powder Technol 371:26–36. https://doi.org/10.1016/j.powtec.2020.05.084

Sonochemical Methods for Nanoparticle Production

Dylan G. Ramanan and Nadeesh M. Adassooriya

Abstract Sonochemical methods are an environmentally clean and an energy-efficient method of synthesizing nanoparticles. These methods can be utilized to produce them through various means such as radical formation, sonofragmentation, and nucleation. Most of these mechanisms are made possible due to a phenomenon known as acoustic cavitation, which is the nucleation, growth and implosion of bubbles. During this process, very high local temperatures and pressure are generated, while not influencing the bulk temperature significantly. Sonochemical methods are propelled by ultrasonic transducers of various forms which include the bath, probe and cup horn sonicators. Although ultrasonication is a popular method for the synthesis of metal nanoparticles, ceramic nanoparticles, polymeric and crystalline nanoparticles, industrially it has only been used widely to disperse agglomerates. However, during dispersion microscale agglomerates can be made in to nanoscale particles with long term storage capability and stability. In this book chapter we discuss about the mechanisms of acoustic cavitation, nanoparticle synthesis through pathways such as nucleation, radical formation and sonofragmentation and scalability of the process. Synthesis and modification of certain nanoparticles have also been looked into with examples. Furthermore, scalability is of extreme importance for large scale production and industrialization. Here, we discuss ways in which sonochemical methods can be introduced in to already existing systems to efficiently synthesize nanoparticles.

Keywords Sonochemical · Ultrasonication · Nanoparticles · Acoustic cavitation · Scalability

D. G. Ramanan · N. M. Adassooriya (✉)
Department of Chemical and Process Engineering, Faculty of Engineering, University of Peradeniya, Peradeniya, Sri Lanka
e-mail: nadeeshm@eng.pdn.ac.lk

N. M. Adassooriya et al. (eds.), *Industrial Scale Production of Nanoparticles*, Smart Nanomaterials Technology, https://doi.org/10.1007/978-981-96-9176-0_13

1 Introduction

Sonochemistry is defined as the usage of ultrasound in chemistry and processing. The ultrasound waves lie in the region where it is not audible for the human ear, usually in the range of 20 kHz to 10 MHz. Furthermore, ultrasonic waves can be subdivided in to three main regions: Low frequency high power, intermediate frequency medium power and high frequency low power. However, usually in sonochemistry applications, a frequency of 20 kHz to 1 MHz is utilized [27]. Chemical reactions require energy to take place, where most reaction rates are limited by temperature and pressure, which is determined by the energy source. In comparison to other modes of energy supply, ultrasonication varies in the time of interaction, pressure and energy per molecule (eV) (Fig. 1) [39].

When ultrasonic waves pass through a solution, it creates regions with high and low pressure, dependent on the periodic expansion and compression of the source. This leads to a scenario known as acoustic cavitation, which is the nucleation, growth, and collapse of bubbles in the solution. The air molecules dissolved in the medium diffuse during the low pressure cycle to form bubbles, subsequently during the high pressure cycle the external pressure compresses the bubble until it collapses releasing high amounts of energy leading to extreme local heating and pressures for a fraction of time [31]. The formed bubbles usually have a diameter of 100 μm and implode

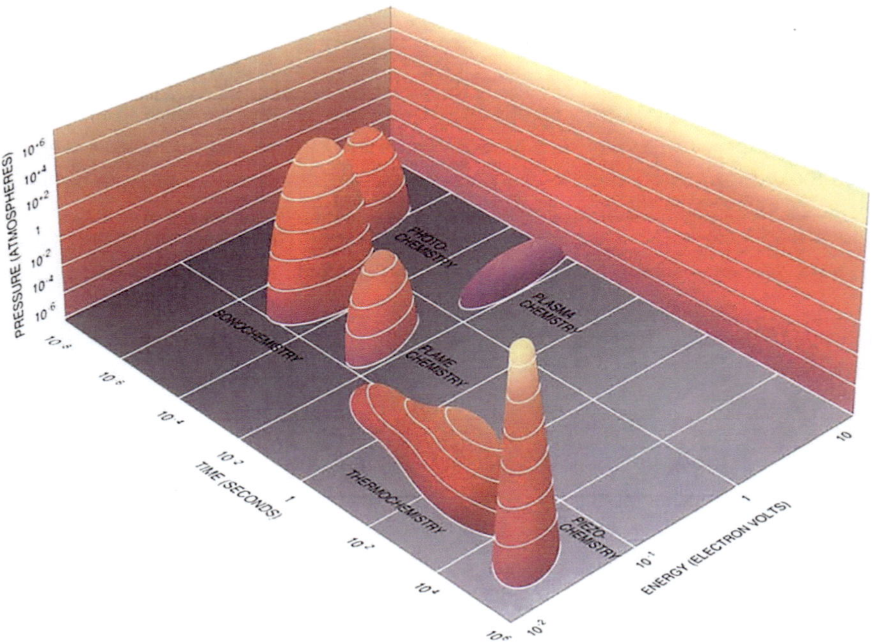

Fig. 1 The interaction of energy and matter [38]

violently in a duration of a microsecond, while temperatures can reach up to 5500 °C and pressures exceeding 1000 atm [39].

History of acoustic cavitation dates back to the nineteenth century, where the adverse effects were studied. Cavitation was observed in all kinds of hydraulic machinery, including the blades on ships. This phenomenon led to the damage and erosion of the ships' propellers which led to inefficiencies during long journeys across the sea [30]. This was predominantly due to the collapse of pre-existing air bubbles rather than those which are formed due to acoustics. In the 1920s, with the development of ultrasonics, the uses of sonochemistry came in to light in industries such as chemical and processing, industrial cleaning and tinning-soldering.

In the last two decades' nanoparticles have garnered much attention due to its revolutionary properties. Industries such as pharmaceuticals, electronics, food and catalysis have welcomed research towards nano-sizing particles. Nanoparticles are defined as objects with dimensions in the range of 1–100 nm [34]. The most exceptional characteristic of nanoparticles is its' large surface area to volume ratio, allowing distinct properties of those to the bulk.

Nanoparticles are synthesized using two methods which are top down and bottom up. Top down synthesis is the breaking down of large particles to smaller ones using mechanical means, whereas bottom up synthesis is the formation of nanoparticles from a solution, which is known as a precursor, where the chemicals nucleate from it to from nanoparticles [2]. Even though ultrasonication is known for the dispersion of agglomerated nanoparticles [36], it is not an effective method for top down synthesis of nanoparticles. Thus it is classified as a bottom up method. However, there exists scenarios where sonofragmentation has been utilized to prepare nanoparticles in certain circumstances, but literature is limited.

This chapter delves into the fundamentals of sonochemical synthesis of nanoparticles which includes, fundamental principles, mechanisms, current uses, scalability and future prospects. The fundamental principles explain how acoustic cavitation takes place and the mechanisms followed during top down and bottom up synthesis of nanoparticles. Comparatively ultrasonication is more environmentally friendly than other techniques, thus current uses, scalability and future prospects are essential in identifying real world applications of research conducted in laboratory scale.

2 Fundamental Principles

2.1 Acoustic Cavitation

Acoustic cavitation is the driving force behind sonochemistry and its applications. The kinetic energy of liquid motion converts to heat, when the bubbles collapse in the solution [40]. The high local heat and pressure generated during this process combined with rapid cooling, drives chemical reactions under extreme conditions. The process of acoustic cavitation involves bubble growth and bubble collapse, while properties such as bubble temperature, bubble size and bubble lifetime are essential parameters which determine the influence on the constituents of the solution [4].

Bubble growth happens through two mechanisms, rectified diffusion and bubble coalescence [12]. Rectified diffusion is defined as the net gas flow into the bubble system. Gas diffuses in to the bubble during the expansion half cycle, and leaves the bubble during compression half cycle, due to the oscillating pressure field, below and above the mean pressure value (Fig. 3) [14]. Techniques such as light scattering, stroboscopic or fast video recording can be utilized for understanding rectified diffusion growth of single bubble systems, it is difficult to study multi bubble systems which are found in non-ideal systems.

Bubble coalescence is the merging of two or more microbubbles to form a single bubble [32]. Initially, two bubbles approach each other while they are expanding, subsequently, the surfaces of contact of the bubbles flatten (Fig. 2b), trapping liquid between them. Flattening of the surfaces induced by the liquid inertia which overcomes the capillary pressure. The trapped liquid exits the gap until a critical separation is reached (Fig. 2c). Thereafter, due to thermal fluctuation which leads to small perturbations in the micrometer range [37], the film thins until it reaches a critical thickness, rupturing to form a merged bubble [6] (Fig. 2d). When the bubbles merge they form an ellipsoidal shape, finally reaching a spherical shape due to surface tension. Coalescence of bubbles can simultaneously take place while rectified diffusion happens, however smaller bubbles have a higher probability to coalesce than larger bubbles [32]. Bubbles which have a high velocity, have a higher chance to bounce, as the contact time is less than the time required for the trapped fluid to escape.

From the findings of Ashokkumar, it can be noted that the contributions to bubble growth from coalescence and rectified diffusion are nearly equal, but highly dependent on power, number of cycles and the solution used [4].

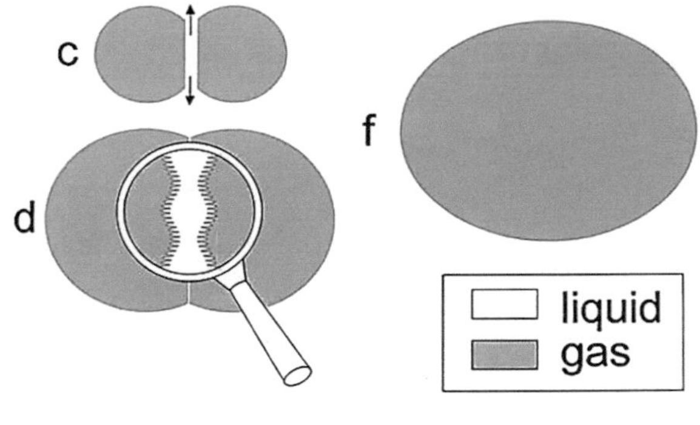

Fig. 2 Schematic of the stages of bubble growth from coalescence. Adapted from Postema et al. [32]

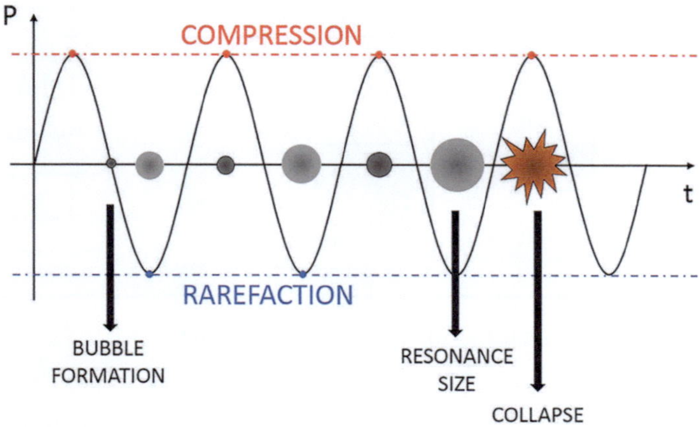

Fig. 3 Schematic of acoustic cavitation process due to high intensity ultrasound. Adapted from Nalesso et al. [29]

For the preparation of nanoparticles, an important phenomenon is the implosion of the bubbles, which provides micro-jets and shock waves, which also generate high temperatures and pressure which in return drives reactions. When the pressure wave changes to a high pressure phase the bubbles implode violently (Fig. 3) creating high local temperatures and pressures. The formation of micro-jets occurs when the bubbles burst close to surfaces larger than the bubbles radius [12]. The collapse near surfaces create non-spherical shock waves, driving the fast jets into the surface, creating damage and cavitation. The shock waves are also able to accelerate particles to high speeds, inducing change to the reactivity, surface morphology and composition [40].

2.2 *Nanoparticle Synthesis Through Radical Formation*

Ultrasonication can be used drive reactions in the range of oxidation, reduction, dissolution and decomposition, due to its high intensity. During ultrasonication three different regions have been identified namely, the inner environment, interfacial region and the bulk solution (Fig. 4). The inner environment consists of the gas phase where extremely high temperature and pressures are observed. Whereas the interfacial region has a lower temperature but still the conditions are able induce sonochemical reactions. The bulk of the solution is at ambient temperature [45]. When ultrasonic waves travel through a solution, its velocity and wavelength are both greater than the molecular scale, thus the change created in the reactants are through the local pressure and temperature spikes. The nanoparticles are formed from a solution, thus categorizing it as a bottom up technique.

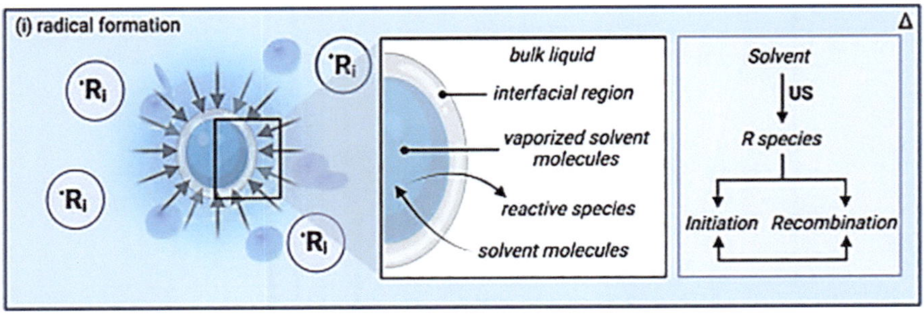

Fig. 4 Radical formation in bubbles. Adapted from Devos et al. [8]

The primary driving force of the reactions is the radicals that are formed during sonication. The region where the chemical reactions occur depends on the physio-chemical properties of the chemicals involved. Volatile molecules are able to traverse the liquid gas interface into the bubble, encountering very high temperatures and pressures that lead to the breakage of chemical bonds, forming radicals, inside the bubble cavity. This is known as primary sonochemical reactions [25]. The radicals formed may undergo further reaction within the gas phase, or may return in to the bulk of the liquid to react with less volatile substances, which is known as secondary sonochemical reactions [25]. Organic components undergo homolytic bond breakage to form radicals [22]. At frequencies in the range of 200–800 kHz, majority of the reactions occur at the interface where the efficiency is the highest [5].

An experiment to identify radical formation was conducted, by sonolysis of water, where the formation of H_2O_2 was studied. Through photography the formation of highly reactive OH radicals was identified through the addition of a luminol. Furthermore, the dissociation of O_2 and N_2 was also confirmed at high frequencies, which confirms the generation of very high energy release during ultrasonication [19].

The formed radicals react with the constituents in the bulk to create nanoparticles. An example of this is the formation of ZnO nanoparticles in a study conducted by Raghvendra et al., where the reaction steps were found to be different during the presence of sonication [45]. Because of the limited volatility of zinc nitrate and sodium hydroxide during normal conditions, the dissociation or ionization might have happened within the interfacial region, powered by sonication. The formation of ZnO could be due to the dominance of OH radicals in sonicated samples which takes place in aqueous samples [21].

In the absence of ultrasound, the following reactions take place:

$$H_2O \leftrightarrow H^+ + OH^- \tag{1}$$

$$NaOH \leftrightarrow Na^+ + OH^- \tag{2}$$

$$Zn(NO_3)_2 \leftrightarrow Zn^{2+} + NO_3^{2-} \tag{3}$$

whereas in the presence of ultrasound, the reactions proceed as follows,

$$Zn^{2+} + 4OH^{-} \leftrightarrow [Zn(OH)_4]^{2-} \tag{4}$$

$$2[Zn(OH)_4]^{2-} \leftrightarrow 2ZnO + 4H_2 + 3O_2 \tag{5}$$

Another examples of radical formation were explained by Arami et al. during the preparation of TiO_2 from a solution containing NaOH. It is explained that the formation of TiO_2 has intermediates such as Ti–O–Na and Ti–OH, which are converted to Ti–O–Ti bonds through hydroxyl radicals (OH), eventually forming the TiO_6 octahedral which is the building blocks of TiO_2 crystal structure [3].

2.3 *Nanoparticle Formation Through Sonofragmentation*

Sonofragmentation is the process, where high intensity ultrasound is used to break down larger particles to smaller ones, often in the nanoscale. This is a phenomenon where sonication is considered as a top down synthesis method. The causes for this phenomenon can be segmented into 4 categories, namely, (a) Inter particle collisions, (b) Particle–wall collision, (c) Particle-shockwave interaction and (d) particle-sonicator interaction [31]. The theory of micro-jets will not affect particle production in the nanoscale, because the initial starting material is in the micrometer range, where micro-jets are unable to form as there should be a large enough surface. However, micro-jetting can aid in the fragmentation of larger particles above the 200 μm range [40].

Zeiger et al. conducted research to identify the primary driving force of sonofragmentation (Fig. 5) in molecular crystals, among the four possibilities identified, the particle wall collision and particle sonicator interactions were controlled by decoupling them. It was discovered that their contribution for fragmentation was negligible. To identify the effects of inter particle collisions, the particle concentration was increased. If the rate of fragmentation was primarily dependent on inter particle collisions, an increase in concentration would lead to a higher rate of fragmentation. Similarly, if the inter particle collisions were the dominant mechanism then the rate of fragmentation would be directly proportional to the square of the particle concentration. From experiments conducted, an absence of a relationship between particle concentration and rate of reaction was noted, helping to identify that the inter particle collisions played no role in the fragmentation, concluding that the primary driving force of crystal fragmentation was due to particle-shockwave interactions [46].

However, In the case of metal nanoparticle synthesis using ultrasonication, inter particle collisions pose a threat. Due to acoustic cavitation, the particles are propagated at high velocities and upon collision they have sufficient energy to induce melting. Doktycz et al. did experiments with different transition metal powders to identify the maximum speeds and temperatures achieved during collisions. It was

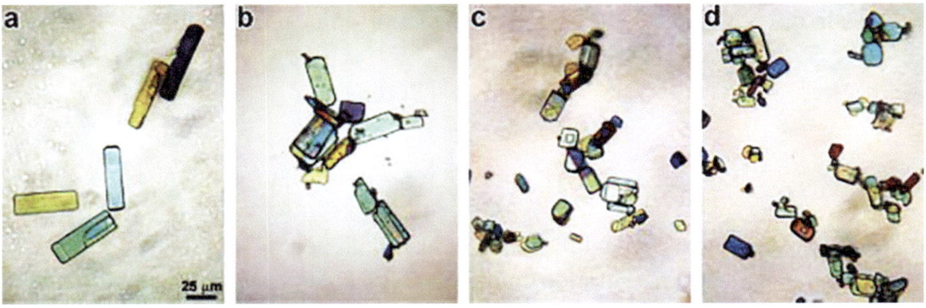

Fig. 5 Cross polarized optical micrographs of sonofragmented aspirin crystals; **a** before sonication, **b** sonicated for 1 min, **c** sonicated for 3 min, **d** sonicated for 10 min. Adapted from Zeiger and Suslick [46]

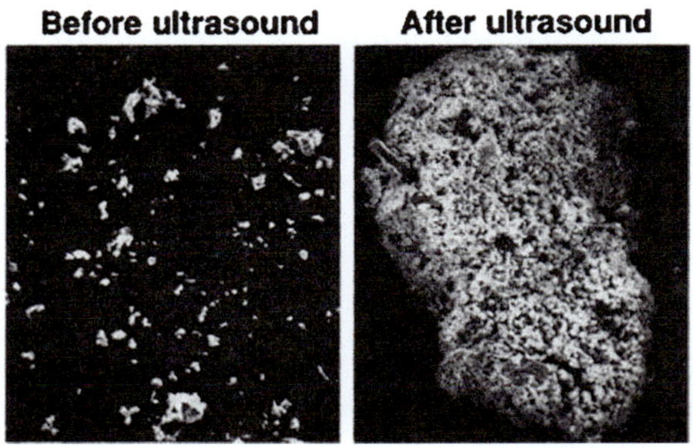

Fig. 6 Before and after the effects of ultrasonication of chromium particles with average diameter of 3 μm. Adapted from Doktycz and Suslick [9]

deduced that the metal particles (average diameter of 10 μm) sonicated in a hydrocarbon liquid reached velocities of up to 500 m s^{-1} and generated localized temperatures between 2600 and 3400 °C at the point of impact, creating aggregates induced by melting and rapid cooling (Fig. 6) [9]. This scenario was only observed with metals which had melting points less than the range indicated above, posing a threat to metal nanoparticle synthesis during prolonged ultrasonication.

2.4 *Nanocrystal Formation Through Nucleation*

Nanocrystals formed from sonication are known to have uniform shapes with high crystallinity and purity. A definite crystal formation mechanism hasn't been still identified for ultrasonication stimulated crystal growth, which is currently explained through various theories and pathways (Fig. 7). Primary homogenous nucleation during ultrasonication is said to be induced by extreme bubble temperatures and pressures. Usually high temperatures increase solubility, which in return decreases

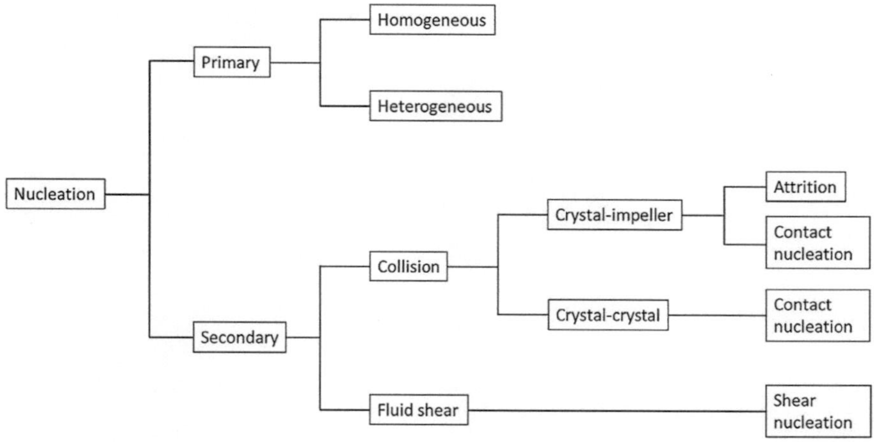

Fig. 7 Nucleation classification adapted from Nalesso et al. [29] and Agrawal and Paterson [1]

super saturation. However, it is explained that the rapid cooling rates (10^7 to 10^{10} K/s) and high pressures observed during ultrasonication can enhance nucleation.

Another factor that propagates nucleation could be the evaporation of solvent on the surface in to the bubbles interior. As the solvent evaporates it absorbs latent heat, creating a localized cooling effect. Furthermore, the evaporation of the solvent into the bubble increases the concentration of the solute in the bulk of the liquid, leading to super saturation.

Another theory promoting heterogeneous nucleation is the reduction in crystallization time during sonication could be due to cavitation bubble surfaces acting as nucleation sites decreasing the effective surface energy. Furthermore, the heterogeneous nucleation theory mentions the possibility of sonocrystallization taking place without the presence of cavitation bubbles, as non-cavitation bubbles can also act as sites [29].

Secondary nucleation theories suggest, particles separated due to erosion and abrasion enhanced by particle–particle collisions and particle shock wave interactions (sonofragmentation), can act as nuclei for the formation of new crystals [20].

3 Experimental Setup for Nanoparticle Synthesis

The experimental apparatus involves one of three types of sonicators which are, bath sonicators, probe sonicators and cup horn sonicators (Fig. 8).

Bath sonicators are the simplest form of sonication equipment found in almost all chemistry/chemical engineering laboratories. They are generally low power, usually in the range of 1–5 W cm^{-1}. Major drawback of bath sonicators are its reproducibility of experiments, as finding the area of the highest intensity is a critical issue. The discovery of the highest intensity is done by a method known as an "aluminium

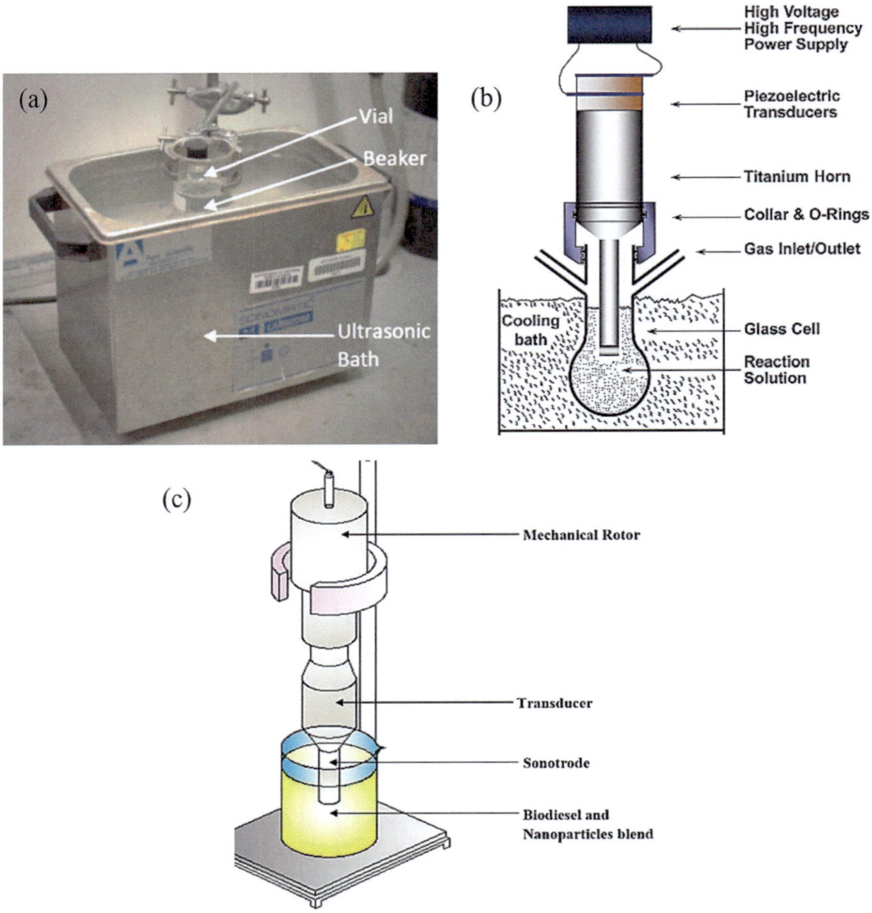

Fig. 8 Schematic of **a** bath sonicator [16], **b** cup-horn sonicator [12], **c** probe sonicator [15]

foil test" [35]. For initial experiments bath sonication may be the best option due to its availability and cost. While bath sonicators are known to disperse agglomerated nanoparticles, they are not recommended for nanoparticle synthesis from a precursor solution. Even though there are examples such as production of spherical gold particles in the nanometer range, it requires a higher duration of time [23].

The probe sonicator comes directly in contact with the medium and the power supplied by the probe may be 100 times more powerful than the bath sonicator. Due to its high energy, the probe sonicator has the tendency to increase the bulk temperature driving unnecessary oxidation and transformations, which should be carefully monitored during nanoparticle synthesis [41]. Another important point to consider is the degradation of materials due to high intensity waves. A study conducted on SiO_2 using both the bath and probe sonicator, helped identify how the probe sonicator was able to disperse particles with an average size of 141 nm in 5 min compared to a bath sonicator which dispersed particles to an average size 796 nm in 30 min [10].

Similar to bath sonicators, cup-horn sonicators also provide indirect sonication. Indirect sonication is when the waves are to travel through the wall of the sample container. The major benefit of the cup-horn sonicator is the ability to process nanoparticles without cross contamination as they are usually sealed in vials. However, they are less powerful compared to probe sonicators, but are more capable than the bath sonicators [35].

4 Sonochemical Synthesis and Modification of Selected Nanoparticles

The effects of sonication on nanoparticles are due to both chemical (radical formation) and physical (sonofragmentation) causes [33]. Components undergo both chemical and physical effects, but by optimization a better control can be acquired.

4.1 Carbon Nanoparticle Synthesis and Surface Modification.

Sonochemical surface modifications of carbon nanoparticles is a popular technique [33]. In an experiment conducted by Xing et al. on carbon nanotubes (CNT) of 30 nm in diameter, in a solution of HNO_3 and H_2SO_4, the morphology of the surface of the CNTs were studied using TEM. With increasing time of sonication, they were able to discover the roughness of the surface increasing, making it heterogeneous (Fig. 9). The attack of the acids on the surface is known to be slow during normal conditions, whereas under sonication it is expedited. It is also mentioned that significant structural change of CNTs were not observed because low power bath sonicators were used, rather than probe sonicators. Furthermore, when settling time was studied, comparatively sonicated samples had a higher settling time, which is expected [44].

By performing such morphological changes using ultrasonication, the catalytic properties of carbon can be enhanced by deposition of metal nanoparticles on them [24]. This process promotes the density of functional groups on the surface.

Another interaction between carbon nanoparticles and sonication is in the study conducted by Jeong et al., to produce single walled carbon nanotubes (SWCNT). A bottom up synthesis route using sonication removes multi-step purification and the use of specialized equipment. The prepared SWCNT portrayed an average diameter of 1.5 nm under TEM characterization [18].

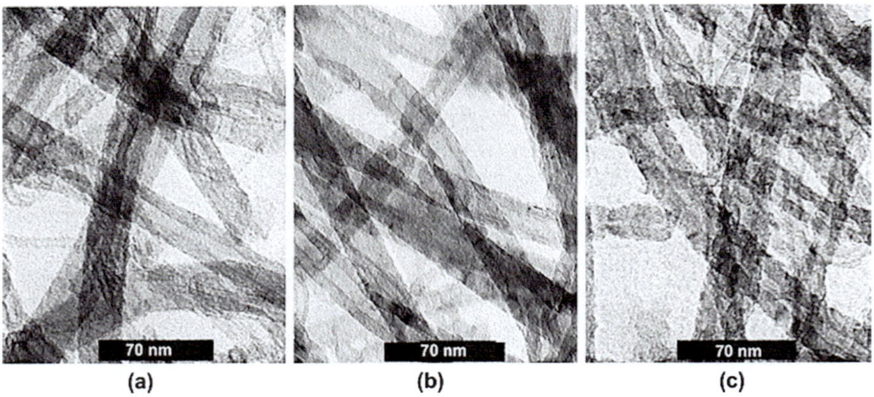

Fig. 9 TEM images of CNT, surface morphology after (a) 2 h, (b) 4 h, and (c) 8 h Adapted from Xing et al. [44]

4.2 Ceramic Nanoparticle Synthesis from Sonofragmentation

Compared to traditional synthesis path ways of nanoceramics which involve grinding, the usage of low frequency (20–35 kHz) high intensity ultrasound is more energy efficient. Acoustic cavitation induces inter particle collisions and shockwave particle interactions which ultimately reduces particle sizes. In an experiment conducted on sonofragmentation of nanoceramics, the ceramics were dispersed in 100 ml of water. Then the slurry was introduced to a probe sonicator operating at 20 kHz frequency and 500 W power. From the experiment it was concluded, for rounded nanoceramic synthesis in the size range of 1–300 nm (Fig. 10), sonofragmentation for a duration of 30–45 min is optimum [11].

Sato et al., made a comparison between bead milling and ultrasonication for the synthesis of TiO_2, where a similar particle size reduction was observed when the

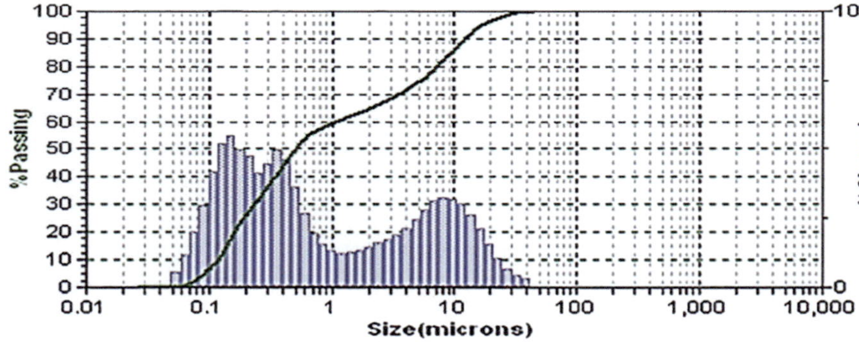

Fig. 10 Size distribution of ceramics after 30 min exposure to ultrasound (20 kHz and 500 W) adapted from Gopi and Nagarajan [11]

Fig. 11 SEM of TiO_2 nanoparticles formed by sonication. Adapted from Arami et al. [3]

bead mill was operated for a duration of 150 min, whereas ultrasonic synthesis took only 30 min. With ultrasonication a highly dispersed dense solution was obtained [36].

A bottom up synthesis of TiO_2 was conducted by dissolving TiO_2 pellets in an NaOH solution, which was irradiated using an ultrasonic bath. The precipitated sample underwent characterization and was confirmed to be TiO_2 nanoparticles (Fig. 11) [3]. Thus the usage of ultrasonication is versatile.

4.3 Pharmaceutical Nanocrystals from Ultrasonication

In the last two decades with the increase in research related to nanocrystal drug delivery, ultrasonication has gained popularity as an effective standalone and combination method due to its efficiency.

Sonia et al. fabricated nanocrystals of meloxicam by sonicating a nanosuspension, while cooling the sample to limit the temperature rise. They concluded the factors which affected the most are increase in sonication time and an increase in amplitude. The higher amplitude and time also improved the uniformity with a lower polydispersity index of the nanocrystals formed. Ultimately, it was confirmed ultrasonicated nanocrystals had an improved dissolution rate of up to 93% whereas crystals from the stock solution had a dissolution of 41% [17].

Polymorphism was observed in certain crystal structures when exposed to ultrasonication. In an experiment conducted by Mori et al. different polymorphs of the drug Acetaminphen were observed when synthesized with and without sonication. When produced by stirring, the crystals were in form I, whereas sonicated samples were in needle shaped form II (Fig. 12) [28]. Extreme care should be taken when nanocrystals are produced through sonication due to high energy intensity, rise to unexpected crystal forms with very different therapeutic effects might occur.

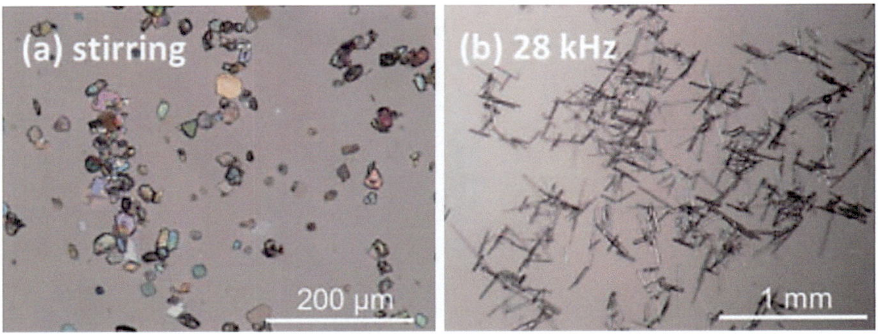

Fig. 12 Different polymorphic forms during **a** stirring, **b** sonication at 28 kHz [28]

5 Agglomeration Control of Nanoparticles Using Sonochemical Methods

Ultrasonication is a well-known technique for separation of agglomerates in nanoparticle synthesis. Due to the high surface area to volume ratio, nanoparticles possess high surface energy. To minimize this energy, they tend to cluster together to reduce the overall surface area to reach a more thermodynamically stable state [26] at microscale. Agglomeration reduction can be done using the usage of stabilizers but further dispersion without usage of chemicals can be done using ultrasonication.

Experiments were conducted using platinum nanoparticles at different ultrasonic frequencies (488, 408, 326 and 210 kHz) to understand the agglomeration size dependence on frequency. Form Fig. 13, it is understood that lower frequency ultrasound reduces agglomeration [13].

Another study showed the agglomeration dispersion of hydrophilic fumed silica with a primary particle diameter of 12 nm. Different samples sizes were loaded to study the effect of loading on dispersion. It was found that an increase of sonication time led to a higher volume of finer particles. The average particle size was fairly consistent with a variation of less than 15 nm, thus indicating a high uniformity.

Fig. 13 Average agglomerate sizes at different frequencies. Adapted from [13]

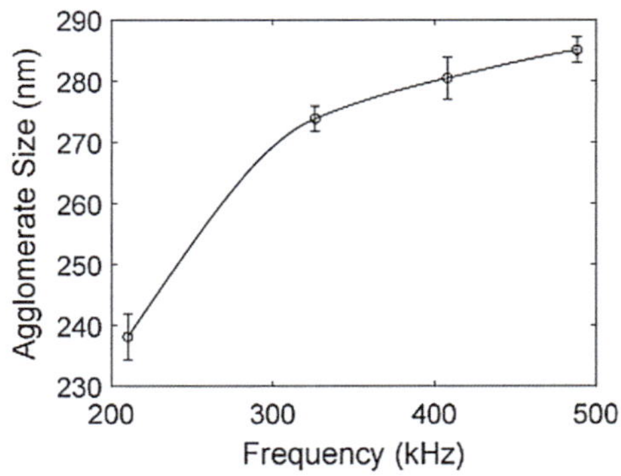

Furthermore the stability of the nanoparticles studied fir a duration of 1 month indicated no agglomeration over time, but some aggregation was observed over longer duration of storage [43].

6 Scalability and Future Prospects of Sonochemical Methods in Nanoparticle Synthesis

Ultrasonication has been used in the industry on a large scale for applications such as homogenization of emulsions, dispersion of paints, cleaning and degreasing, however the use of sonochemical methods for nanoparticle synthesis in large scale applications is still limited [7]. For solutions undergoing crystallization, sonication and the physical effects such as temperature and pressure, reduce the metastable zone width and the induction time and increase the crystal formation [20]. During the scale up of ultrasonication applications it is essential to know the path way of the reaction, whether it is a chemical reaction or is it undergoing sonofragmentation or dispersion. Thompson et al., describes many ways in which chemical reactions can be induced using ultrasonication in the large scale. Batch and continuous reactors can be coupled with transducers in different ways such as with hexagonal bath systems (Fig. 15) or external flow loop systems (Fig. 16). This would enable large scale sonication for nanoparticle synthesis [42]. Despite effective large scale scalability, there are certain considerations to be noted as visualized in Fig. 14.

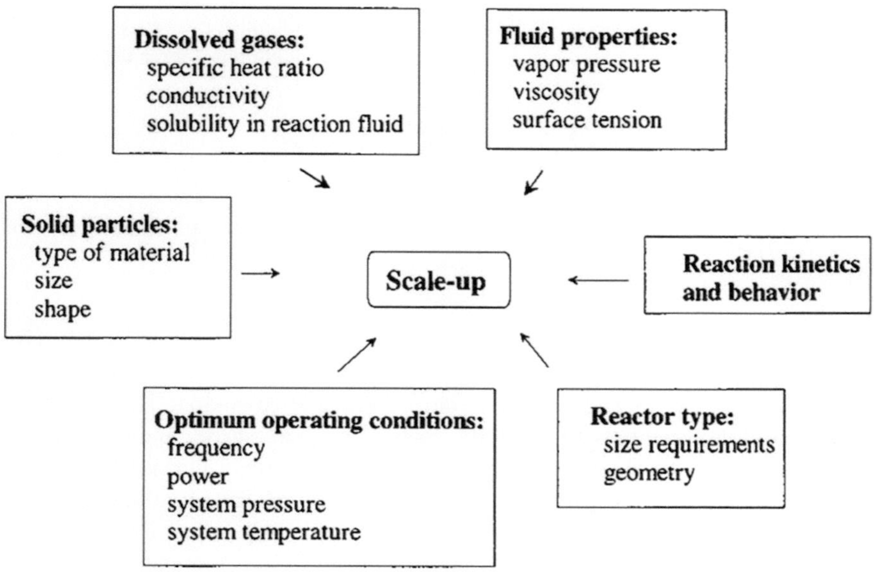

Fig. 14 Factors to be considered during scale up of sonochemical systems. Adapted from Thompson and Doraiswamy [42]

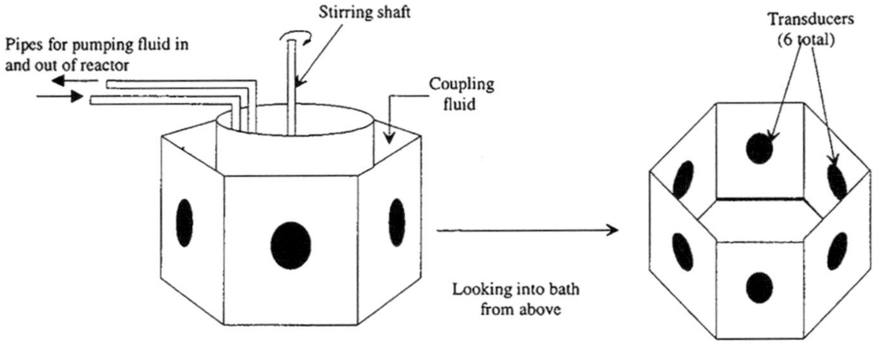

Fig. 15 Hexagonal bath system. Adapted from Thompson and Doraiswamy [42]

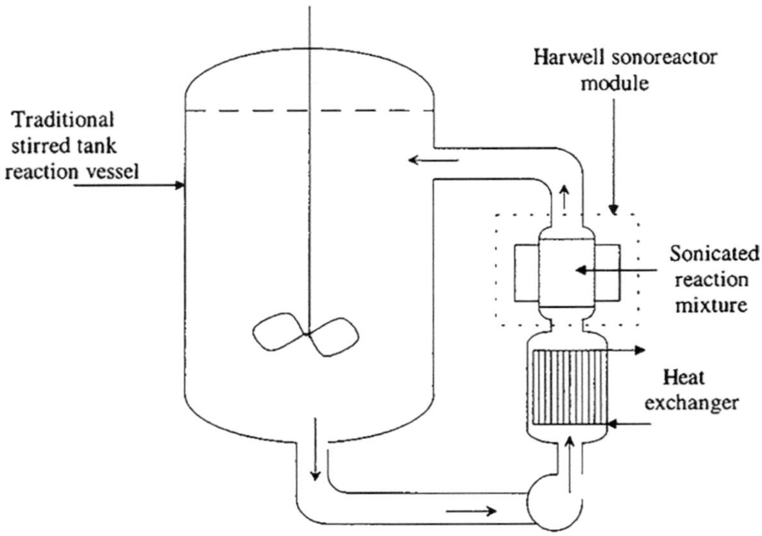

Fig. 16 External loop system installed to traditional reactor. Adapted from Thompson and Doraiswamy [42]

Sonochemical methods for nanoparticle synthesis offers exciting opportunities for advancing nanotechnology because they create unique reaction conditions, such as very high temperatures and pressures, during acoustic cavitation. These extreme conditions allow scientists to carefully control the size, shape, and structure of nanoparticles, making the method useful for many applications. One promising area is the study of nucleation pathways, which is key to understanding how particles start to form at the molecular level. This knowledge can help improve the design of nanomaterials with specific properties as the pathway isn't theoretically sound.

Sonochemical methods are also increasingly being used to synthesize nanococrystals, which combine two or more components into a single crystal structure, improving properties like solubility and stability for pharmaceutical and industrial uses. Beyond this, sonochemical techniques have potential in creating advanced materials for drug delivery, energy storage, and environmental cleanup, such as efficient catalysts or materials for removing pollutants during water treatment.

Another significant advantage is the ability to alter surface morphologies even at the nanoscale, allowing to fine tune certain characteristics of nanoparticles. Additionally, the usage of extra chemicals and high energy is reduced during sonication, making it an effective method of nanoparticle synthesis while being environmentally friendly. With further research and improvements, sonochemical synthesis can unlock even more possibilities in producing innovative materials for a wide range of industries.

7 Conclusion

In conclusion, sonochemical methods are an effective way of fabricating nanoparticles, using different path ways such as through radical formation, sonofragmentation and promotion of nucleation. These methods are initiated through the fundamental phenomena known as acoustic cavitation.

Radical formation takes place through high temperature and pressure zones initiated by ultrasonication. The formed radicals react with the molecules in the bulk to form nanoparticles. The phenomenon known as sonofragmentation breaks down larger particles to the nanoscale. Sonofragmentation takes place through (a) Inter particle collisions, (b) Particle–wall collision, (c) Particle-shockwave interaction and (d) particle-sonicator interaction. Lastly, initiation of nucleation happens primarily due to local high temperature, which increases the solubility, promoting super saturation. Another mechanism through which nucleation is promoted is by bubble surfaces formed during acoustic cavitation behaving as nucleation sites. These two mechanisms are considered as primary nucleation. Furthermore, sonochemical methods can be also used to fine tune the properties of nanoparticles which may have useful applications and to prevent agglomeration of nanoparticles.

Sonication can be achieved through different apparatuses such as bath, probe or cup-horn sonicators. Bath sonicators are simple and of lower power, whereas probe sonicators are more powerful but can cause heating and material degradation. Cup-horn sonicators offer indirect sonication, preventing cross contamination, while providing moderate power.

Sonochemical methods offer unique advantages such as the ability to precisely control nanoparticle size, shape and structure through extreme conditions created during acoustic cavitation. Moreover, sonochemical methods require lesser energy and chemicals, making it environmentally friendly for industrial scale applications. Industrial scale systems can be achieved through careful consideration to reaction pathways and innovative reactor designs such as the hexagonal bath systems and external flow loop systems.

References

1. Agrawal SG, Paterson AHJ (2015) Secondary nucleation: mechanisms and models. Chem Eng Commun 202(5):698–706. https://doi.org/10.1080/00986445.2014.969369
2. Anand C, Yamaguchi Y, Liu Z, Ibe S, Elangovan SP, Ishii T, Ishikawa T, Endo A, Okubo T, Wakihara T (2016) Pioneering in situ recrystallization during bead milling: a top-down approach to prepare zeolite a nanocrystals. Sci Rep 6:1–7. https://doi.org/10.1038/srep29210
3. Arami H, Mazloumi M, Khalifehzadeh R, Sadrnezhaad SK (2007) Sonochemical preparation of TiO_2 nanoparticles. Mater Lett 61(23–24):4559–4561. https://doi.org/10.1016/j.matlet.2007.02.051
4. Ashokkumar M (2011) The characterization of acoustic cavitation bubbles—an overview. Ultrason Sonochem 18(4):864–872. https://doi.org/10.1016/j.ultsonch.2010.11.016
5. Asiva Noor Rachmayani (2015) Synthetic organic sonochemistry
6. Chaudhari RV, Hofmann H (1994) Coalescence of gas bubbles in liquids. Rev Chem Eng 10(2):131–190. https://doi.org/10.1515/REVCE.1994.10.2.131
7. Destaillats H, Lesko TM, Knowlton M, Wallace H, Hoffmann MR (2001) Scale-up of sonochemical reactors for water treatment. Ind Eng Chem Res 40(18):3855–3860. https://doi.org/10.1021/ie010110u
8. Devos C, Bampouli A, Brozzi E, Stefanidis GD, Dusselier M, Van Gerven T, Kuhn S (2024) Ultrasound mechanisms and their effect on solid synthesis and processing: a review. Chem Soc Rev. https://doi.org/10.1039/d4cs00148f
9. Doktycz SJ, Suslick KS (1987) Interparticle collisions driven by ultrasound_Suslick_Science_1990.pdf. Sci Rep 2(67):1987–1989
10. Glaubitz C, Rothen-Rutishauser B, Lattuada M, Balog S, Petri-Fink A (2022) Designing the ultrasonic treatment of nanoparticle-dispersions via machine learning. Nanoscale 14(35):12940–12950. https://doi.org/10.1039/d2nr03240f
11. Gopi KR, Nagarajan R (2008) Application of power ultrasound in cavitation erosion testing of nano-ceramic particle/polymer composites. Solid State Phenom 136:191–204. https://doi.org/10.4028/www.scientific.net/SSP.136.191
12. Hangxun X, Zeiger BW, Suslick KS (2013) Sonochemical synthesis of nanomaterials. Chem Soc Rev 42(7):2555–2567. https://doi.org/10.1039/c2cs35282f
13. Hansen HE, Seland F, Sunde S, Burheim OS, Pollet BG (2022) Frequency controlled agglomeration of Pt-nanoparticles in sonochemical synthesis. Ultrason Sonochem 85. https://doi.org/10.1016/j.ultsonch.2022.105991
14. Hsieh D-Y, Plesset MS (1961) Theory of rectified diffusion of mass into gas bubbles. J Acoust Soc Am 33(2):206–215. https://doi.org/10.1121/1.1908621
15. Hussain F, Soudagar MEM, Afzal A, Mujtaba MA, Rizwanul Fattah IM, Naik B, Mulla MH, Badruddin IA, Yunus Khan TM, Raju VD, Gavhane RS, Ashrafur Rahman SM (2020) Enhancement in combustion, performance, and emission characteristics of a diesel engine fueled with Ce-ZnO nanoparticle additive added to soybean biodiesel blends. Energies 13(17):1–20. https://doi.org/10.3390/en13174578
16. Ibrahim I, Rümmeli MH, Warner JH, Bachmatiuk A, Gorantla S (2012) Methods for obtaining graphene. In: Graphene: fundamentals and emergent applications. https://doi.org/10.1016/B978-0-12-394593-8.00004-7
17. Iurian S, Tomuţa I, Rus L, Achim M, Leucuta SE (2015) Optimization of the sonication process for meloxicam nanocrystals preparation. Clujul Med 88(3):366–372. https://doi.org/10.15386/cjmed-445
18. Jeong SH, Ko JH, Park JB, Park W (2004) A sonochemical route to single-walled carbon nanotubes under ambient conditions. J Am Chem Soc 126(49):15982–15983. https://doi.org/10.1021/ja0451867
19. Ji R, Pflieger R, Virot M, Nikitenko SI (2018) Multibubble sonochemistry and sonoluminescence at 100 kHz: the missing link between low- and high-frequency ultrasound. J Phys Chem B 122(27):6989–6994. https://doi.org/10.1021/acs.jpcb.8b04267

20. Kim HN, Suslick KS (2018) The effects of ultrasound on crystals: sonocrystallization and sonofragmentation. Crystals 8(7). https://doi.org/10.3390/cryst8070280
21. Kimmel E (2010) Acoustic cavitation. In: Therapeutic ultrasound: mechanisms to applications. https://doi.org/10.1201/9780367853495-6
22. Kuna E, Behling R, Valange S, Chatel G, Colmenares JC (2017) Sonocatalysis: a potential sustainable pathway for the valorization of lignocellulosic biomass and derivatives. Top Curr Chem 375(2). https://doi.org/10.1007/s41061-017-0122-y
23. Lee JH, Choi SUS, Jang SP, Lee SY (2012) Production of aqueous spherical gold nanoparticles using conventional ultrasonic bath. Nanoscale Res Lett 7:1–7. https://doi.org/10.1186/1556-276X-7-420
24. Levêque JM, Duclaux L, Rouzaud JN, Reinert L, Komatsu N, Desforges A, Afreen S, Sivakumar M, Kimura T (2017) Ultrasonic treatment of glassy carbon for nanoparticle preparation. Ultrason Sonochem 35:615–622. https://doi.org/10.1016/j.ultsonch.2016.02.004
25. Li Z, Dong J, Zhang H, Zhang Y, Wang H, Cui X, Wang Z (2021) Sonochemical catalysis as a unique strategy for the fabrication of nano-/micro-structured inorganics. Nanoscale Adv 3(1):41–72. https://doi.org/10.1039/d0na00753f
26. Loza K, Epple M, Maskos M (2019) Stability of nanoparticle dispersions and particle agglomeration. In: Nanoscience and technology (Issue Imm). Springer. https://doi.org/10.1007/978-3-030-12461-8_4
27. Mason TJ, Cintas P (2007) Sonochemistry. Handbook of green chemistry and technology, pp 372–396. https://doi.org/10.1002/9780470988305.ch16
28. Mori Y, Maruyama M, Takahashi Y, Ikeda K, Fukukita S, Yoshikawa HY, Okada S, Adachi H, Sugiyama S, Takano K, Murakami S, Matsumura H, Inoue T, Yoshimura M, Mori Y (2015) Selective crystallization of metastable phase of acetaminophen by ultrasonic irradiation. Appl Phys Express 8(6). https://doi.org/10.7567/APEX.8.065501
29. Nalesso S, Bussemaker MJ, Sear RP, Hodnett M, Lee J (2019) A review on possible mechanisms of sonocrystallisation in solution. Ultrason Sonochem 57:125–138. https://doi.org/10.1016/j.ultsonch.2019.04.020
30. Neppiras EA (1984) Acoustic cavitation series: part one. Acoustic cavitation: an introduction. Ultrasonics 22(1):25–28. https://doi.org/10.1016/0041-624X(84)90057-X
31. Pokhrel N, Vabbina PK, Pala N (2016) Sonochemistry: science and engineering. Ultrason Sonochem 29:104–128. https://doi.org/10.1016/j.ultsonch.2015.07.023
32. Postema M, Marmottant P, Lancée CT, Hilgenfeldt S, Jong ND (2004) Ultrasound-induced microbubble coalescence. Ultrasound Med Biol 30(10):1337–1344. https://doi.org/10.1016/j.ultrasmedbio.2004.08.008
33. Price GJ, Nawaz M, Yasin T, Bibi S (2018) Sonochemical modification of carbon nanotubes for enhanced nanocomposite performance. Ultrason Sonochem 40:123–130. https://doi.org/10.1016/j.ultsonch.2017.02.021
34. Saba H (2014) A review on nanoparticles: their synthesis and types. Res J Recent Sci Uttar Pradesh (Lucknow Campus) 4:1–3
35. Santos HM, Capelo JL (2007) Trends in ultrasonic-based equipment for analytical sample treatment. Talanta 73(5):795–802. https://doi.org/10.1016/j.talanta.2007.05.039
36. Sato K, Li JG, Kamiya H, Ishigaki T (2008) Ultrasonic dispersion of TiO_2 nanoparticles in aqueous suspension. J Am Ceram Soc 91(8):2481–2487. https://doi.org/10.1111/j.1551-2916.2008.02493.x
37. Sharma A, Ruckenstein E (1987) Critical thickness and lifetimes of foams and emulsions: role of surface wave-induced thinning. J Colloid Interface Sci 119(1):14–29. https://doi.org/10.1016/0021-9797(87)90240-2
38. Suslick KS (1989) The chemical effects of ultrasound. Sci Am 260(2):80–86. https://doi.org/10.1038/scientificamerican0289-80
39. Suslick KS (1990) Sonochemistry
40. Suslick KS, Didenko Y, Fang MM, Hyeon T, Kolbeck KJ, McNamara WB, Mdleleni MM, Wong M (1999) Acoustic cavitation and its chemical consequences. Philos Trans Royal Soc A: Math Phys Eng Sci 357(1751):335–353. https://doi.org/10.1098/rsta.1999.0330

41. Tang SY, Qiao R, Yan S, Yuan D, Zhao Q, Yun G, Davis TP, Li W (2018) Microfluidic mass production of stabilized and stealthy liquid metal nanoparticles. Small 14(21):1–8. https://doi.org/10.1002/smll.201800118
42. Thompson LH, Doraiswamy LK (1999) Sonochemistry: science and engineering. Ind Eng Chem Res 38(4):1215–1249. https://doi.org/10.1021/ie9804172
43. Vikash, Kumar V (2020) Ultrasonic-assisted de-agglomeration and power draw characterization of silica nanoparticles. Ultrasonics Sonochem 65, 105061. https://doi.org/10.1016/j.ultsonch.2020.105061
44. Xing Y, Li L, Chusuei CC, Hull RV (2005) Sonochemical oxidation of multiwalled carbon nanotubes. Langmuir 21(9):4185–4190. https://doi.org/10.1021/la047268e
45. Yadav RS, Mishra P, Pandey AC (2008) Growth mechanism and optical property of ZnO nanoparticles synthesized by sonochemical method. Ultrason Sonochem 15(5):863–868. https://doi.org/10.1016/j.ultsonch.2007.11.003
46. Zeiger BW, Suslick K (2013) Sonofragmentation of molecular crystals: observations and modeling. Proc Meetings Acoust 19:1–5. https://doi.org/10.1121/1.4800930

Nanolithographic Techniques for Scaled-Up Manufacturing of Nanostructures

Benxuan Li

Abstract Nanolithography, a cornerstone of modern nanotechnology, has evolved significantly with techniques ranging from traditional photolithography to advanced methods like extreme Ultraviolet lithography (EUVL), electron beam lithography (EBL), ion beam lithography, and soft lithography. Each technique offers unique advantages in resolution, throughput, or versatility while facing distinct challenges in scaled-up manufacturing. Photolithography remains industry-dominant due to its high throughput and scalability, while EUVL and EBL push resolution boundaries but struggle with large-scale production speed or cost-effectiveness. soft lithography offers versatility for specialized applications and the potential for roll-to-roll manufacturing. Future advancements in nanolithography will likely focus on balancing high resolution with increased throughput for scaled-up production, expanding material compatibility, and developing more cost-effective and environmentally friendly processes. Emerging trends include the integration of artificial intelligence for process optimization in large-scale manufacturing, the development of three-dimensional (3D) nanofabrication techniques for complex structures, and the exploration of quantum and biomimetic approaches for novel patterning methods. As nanotechnology expands into fields like quantum computing and flexible electronics, nanolithography will continue to adapt, driving innovations in scaled-up manufacturing across various technological domains.

1 Introduction

Nanolithography, the art and science of fabricating structures at the nanoscale, has become a cornerstone of modern technology. As demand grows for smaller, faster, and more efficient devices, the ability to manufacture nanostructures at scale has become increasingly crucial [1, 2]. Nanolithographic techniques for scaled-up manufacturing of nanostructures represent a frontier where cutting-edge science

B. Li (✉)
Electrical Engineering Division, Engineering Department, University of Cambridge, Cambridge, UK
e-mail: benxuan1013@outlook.com

N. M. Adassooriya et al. (eds.), *Industrial Scale Production of Nanoparticles*, Smart Nanomaterials Technology, https://doi.org/10.1007/978-981-96-9176-0_14

meets industrial application, owing to their remarkable capability to create intricate micro- and nanoscale patterns with precise detail while allowing for selective material placement across different substrate areas [3]. A primary advantage of nanolithographic techniques lies in their exceptional spatial resolution capabilities, enabling the creation of features with dimensions as small as 10 nm or less—a level of precision that far exceeds the capabilities of conventional printing-based approaches and unlocks unprecedented opportunities for advancing nanostructure development in electronic and optoelectronic applications [4–6]. Nanolithographic techniques encompass a diverse array of methods, each with its strengths and limitations [7]. While still widely used, traditional photolithography faces fundamental limitations as feature sizes approach the wavelength of light. The demand for greater precision has driven the advancement of cutting-edge lithographic methods, such as extreme ultraviolet lithography (EUVL), which uses much shorter wavelengths to achieve higher resolution [8].

Beyond photon-based methods, electron beam lithography offers exceptional resolution but has historically been limited by low throughput. Recent advances in multi-beam electron lithography systems aim to address this limitation, potentially enabling large-scale production of highly detailed nanostructures [9, 10]. Nanoimprint lithography (NIL) presents another promising approach, offering high-resolution patterning with the potential for high-throughput production [11]. By physically deforming a resist layer with a pre-patterned mold, this technique can rapidly replicate nanoscale features over large areas. Emerging techniques like directed self-assembly leverage the intrinsic properties of materials to form ordered nanostructures, potentially offering a bottom-up complement to traditional top-down approaches [12–14].

Several lithographic techniques excel in industrial scalability by delivering large patterned areas (> 10 cm^2) with high throughput (> 100 $cm^2\ h^{-1}$), crucial factors that ensure cost-effectiveness and manufacturability for large-scale production of micro- and nanostructured devices. There is a clear trade-off between nanolithographic technique resolution (nm) and throughput ($cm^2\ h^{-1}$). Figure 1 presents a comparative analysis of various lithographic and printing processes, plotting their feature resolution capabilities against fabrication throughput. Notably, photolithography and soft lithography, represented by large overlapping ovals, offer a balance of moderate to high resolution (10–10^4 nm) and throughput. In contrast, inkjet and screen printing techniques, shown by a blue circle on the far right, provide the highest throughput but at the cost of lower resolution. Note that the discussion of nanoimprint lithography and other printing techniques is not included in this chapter, but related comprehensive reviews can be found [11, 15, 16]. At the opposite end of the spectrum, advanced lithographic techniques like EBL and SPL (Scanning Probe Lithography) achieve the highest resolutions, down to 1 nm, but with significantly lower throughput. NIL occupies a middle ground, offering moderate resolution and throughput. The challenge lies not only in developing these individual techniques but in integrating them into cohesive manufacturing processes capable of producing nanostructures at industrial scales. This requires overcoming issues of repeatability, defect control, and cost-effectiveness.

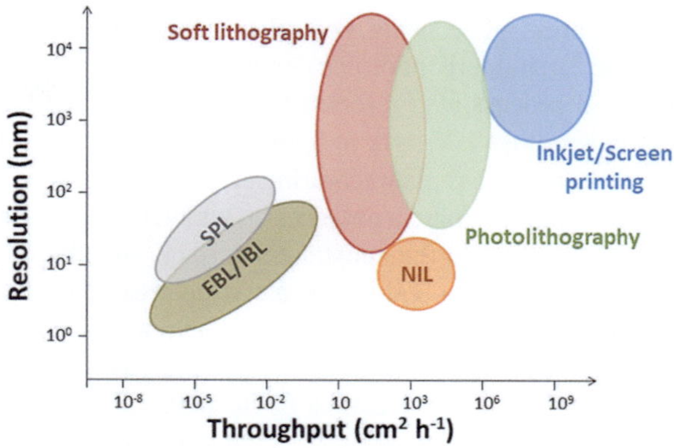

Fig. 1 Nanolithographic techniques feature resolution capabilities against fabrication throughput, illustrating trade-offs between precision and production speed across different techniques [17]. Copyright 2021 American Chemical Society

In this chapter, we begin by examining traditional photolithography, which remains the industry standard due to its high throughput and scalability. We then delve into advanced methods such as EUVL and EBL, which push resolution boundaries but face challenges in large-scale production. Ion beam lithography and soft lithography are also discussed, highlighting their unique advantages and potential for specialized applications of nanostructures. Throughout our analysis, we focus on the critical balance between achieving high resolution and maintaining efficient, cost-effective production on a scale. We evaluate each technique's strengths, limitations, and potential for integration into large-scale manufacturing processes. Additionally, we explore emerging trends and future directions in the field, including the integration of artificial intelligence for process optimization and the development of novel three-dimensional (3D) nanofabrication techniques.

2 Nanolithographic Techniques for Scaled-Up Nanostructure Manufacturing

2.1 *Photolithography*

The etymology of "photolithography" reveals its fundamental principles, as the term derives from Greek roots—"photo" signifying light, "litho" denoting stone, and "graphy" meaning writing, which aptly encapsulates the essence of this technique where light serves as the crucial element [18]. Photolithography is a traditional and well-established method with a history dating back to the 1820s [19]. It was first developed by Nicéphore Niépce, who developed the first photoresist using natural Syrian asphalt to create a photographic process, essentially inventing the first photoresist and creating the world's first permanent photographic image. Subsequently, two chemists, Oskar Süß and Louis Plambeck Jr. created the first positive photoresist and a letterpress plate using Dycryl polymers, respectively, which became the foundation for modern photolithography in the semiconductor industry. Photolithography, a process also referred to as optical lithography or ultraviolet (UV) lithography, employs UV light to transfer intricate patterns from a photomask—typically a transparent glass plate featuring strategically placed opaque regions that define the desired pattern—onto a photoresist, which is a light-sensitive organic polymer meticulously applied to the substrate's surface, thereby enabling the precise replication of complex micro- and nanoscale structures [20, 21]. Photolithography is a multi-step process applied to create micro- and nanopatterns on the substrate, typically a silicon wafer. The main steps involved are as follows, as illustrated in Fig. 2a: The wafer surface is first chemically cleaned with isopropyl alcohol (IPA) and distilled water to remove impurities and contaminants, followed by the deposition of silicon dioxide as a barrier layer. Then, the light-sensitive photoresist is spin-coated over the barrier layer. There are two types of photoresists. When exposed to UV light, positive photoresists increase solubility, whereas negative photoresists undergo the opposite reaction, becoming less soluble upon UV exposure. The process begins with soft baking, where gentle heating removes excess solvent from the photoresist coating. Precise alignment between the photomask and wafer is crucial before UV light projects the pattern through the mask onto the photoresist layer. The final step involves applying a developer solution, which dissolves and washes away the soluble portions of the exposed photoresist, revealing the intended pattern on the wafer surface. The patterned photoresist is baked at relatively high temperatures, usually above 100 °C, to obtain chemical and thermal stability for the subsequent etching process, where the exposed areas of the substrate are selectively removed by the etching solution while the photoresist-covered regions remain protected. Finally, the remaining photoresist is stripped away using a liquid-resist stripper and plasma oxidation.

Photolithography has evolved over the years, with three main UV light exposure methods emerging: contact photolithography, proximity photolithography, and photolithography (as shown in Fig. 2b). Contact and proximity photolithography were the pioneering methods, with each technique offering distinct benefits and limitations based on the spatial configuration between the mask, photoresist layer, and UV light source. Their broad application remains in the industry but suffers from the relatively poor resolution of around 1 μm [19]. In contact photolithography, the photomask is placed directly on the photoresist-coated wafer, providing high resolution but risking mask damage and contamination. Proximity photolithography maintains a small gap between the mask and wafer to minimize mask degradation, though this spacing inevitably reduces resolution due to light diffraction. Taking the technology further, projection photolithography overcame these limitations by using specialized lenses to focus and project the mask pattern onto the wafer, achieving both enhanced resolution and prolonged mask life. Using the Rayleigh equation, the resolution R and the corresponding depth of focus (DOF) of the exposure system can be calculated:

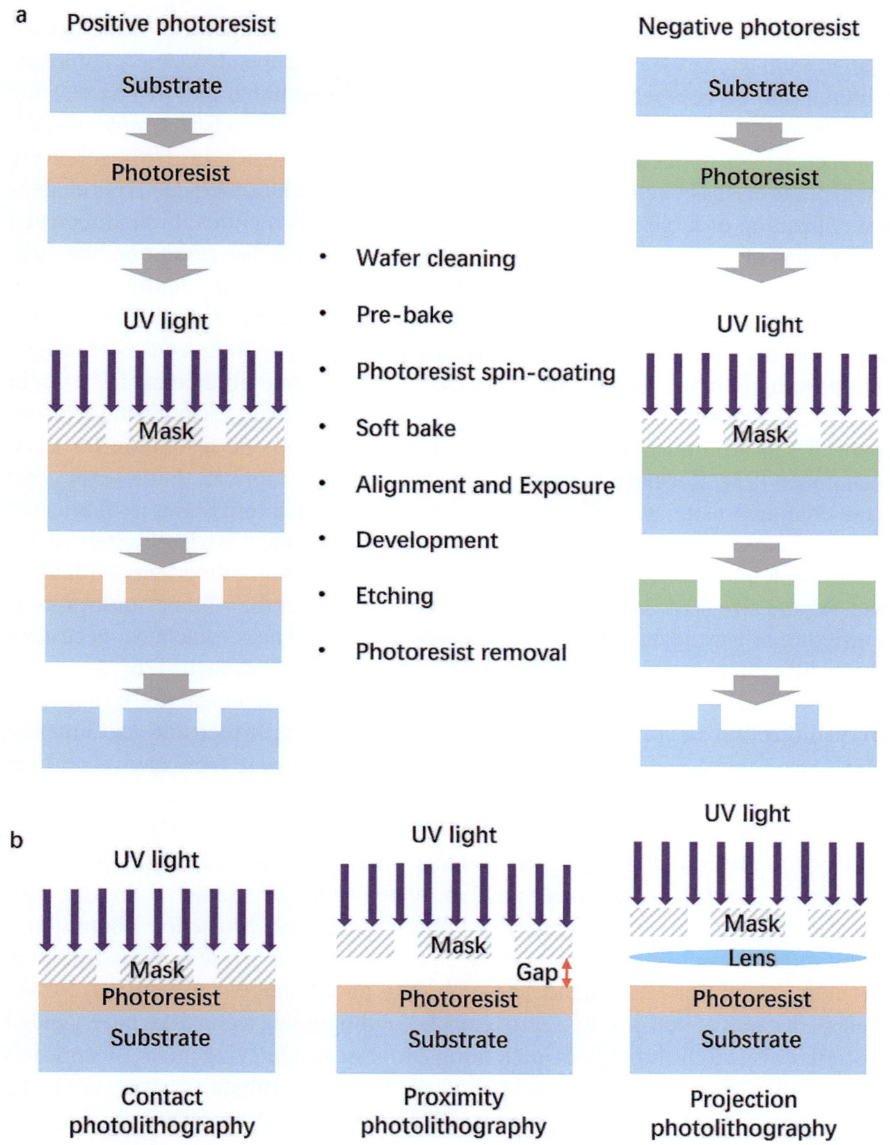

Fig. 2 Basics of photolithography. **a** Schematic of the fabrication process of photolithography using positive and negative photoresists. **b** Exposure methods used in photolithography

$$R = \frac{k_1 \cdot \lambda}{\mathrm{NA}} \tag{1}$$

$$DOF = \frac{k_2 \cdot \lambda}{\mathrm{NA}^2} \tag{2}$$

where λ is the exposure wavelength, NA is the numerical aperture of the optical system, and k_1 and k_2 are Rayleigh constants which depend on the photoresist material and process technology [22]. Projection photolithography overcomes the micrometre-scale limitations of contact and proximity methods. It employs additional projection optics to create high-resolution and large-scale arrays from relatively small photomasks by manipulating parameters in the resolution Eq. (1). This enables the realization of large-scale and high-throughput nanostructures through repeated projections of light through the same photomask [23]. Further resolution enhancements can be achieved by increasing the numerical aperture of the additional optical system and utilizing the immersion method, which replaces air with water as the background medium to increase the refractive index (n) [24]. State-of-the-art projection printing employs deep ultraviolet (DUV) and extreme ultraviolet (EUV) light sources, with wavelengths of 193 nm and 13.5 nm, respectively, enabling the achievement of sub-40 nm resolution for DUV and sub-10 nm node resolution for EUV techniques [25]. Recently, several groups have tried to fabricate large-scale optic nanostructures using photolithography. For example, photolithography-fabricated large-scale metasurfaces offer solutions to traditional optical device limitations, as demonstrated by a half-wave plate created using DUV photolithography with a 12-inch feature size (Fig. 3a and b). This approach overcomes the volume constraints of conventional waveplates, which require substantial thickness to achieve necessary phase differences for polarization control in various optical systems [26].

In addition, a centimetre-scale, all-glass metalenses was demonstrated by applying DUV projection stepper lithography for visible wavelength focusing and imaging [27]. On a single 4-inch fused-silica wafer, 45 metalenses were fabricated, each 1 cm in diameter, showcasing size-scalability and mass-production potential. The scanning electron microscope (SEM) image and photograph are provided in Fig. 3c–e.

Modern projection lithography systems have reached remarkable precision, capable of creating features as small as a few tens of nanometers in size [28, 29]. Despite offering exceptional resolution capabilities, projection photolithography requires sophisticated and costly equipment, making it less accessible than the more straightforward contact and proximity photolithography methods. Nevertheless, decades of optimization across all photolithographic techniques have yielded cost-effective, high-throughput patterning processes, achieving rates of approximately [30]. Photolithography's technological maturity and ubiquity have established it as a cornerstone technique for the mass production of nanostructures.

2.2 Extreme Ultraviolet Lithography

The increasing demand for enhanced performance, combined with the resolution limits of conventional wavelength technologies, drove semiconductor companies to pioneer new exposure methods. Extreme Ultraviolet Lithography (EUVL) emerged as the leading solution for microelectronic chip production among various nanostructure fabrication approaches. Its unique combination of high throughput and process

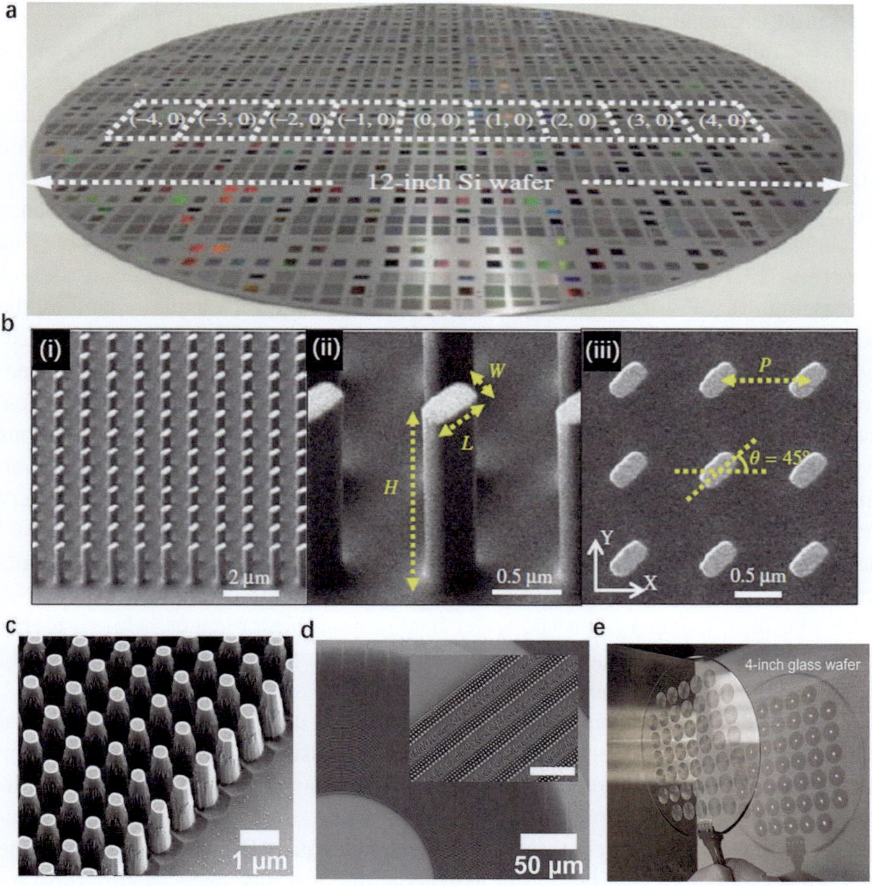

Fig. 3 **a** Photograph of the fabricated 12-inch Si metasurface wafer. **b** SEM image (tilted) of the Si pillar array at position (0, 0). (ii) Zoomed-in view of a single Si pillar. The scale bar is 0.5 μm. (iii) Top view of the Si pillars. The scale bar is 0.5 μm. Copyright 2019 De Gruyter. **c** Zoom-in SEM image of nanopillars of the metalens. **d** Top-view SEM image near the centre (inset shows the edge of the lens). **e** Photograph of 45 fabricated metalenses on a 4-inch diameter SiO_2 wafer. Copyright 2019 American Chemical Society

compatibility has established EUVL as the definitive technology for next-generation lithographic manufacturing. The self-aligned double patterning (SADP) technique achieves doubled line density through a two-step process: first creating a core pattern using conventional photolithography, then applying chemical vapor deposition to grow sidewall spacers, resulting in a pitch-halving hard mask [31] (Fig. 4a). Furthermore, EUVL enhances pattern fidelity, as illustrated in Fig. 4b, enabling greater design flexibility, improved performance, and increased scalability from a design perspective [32]. In 2006, Advanced Semiconductor Materials Lithography (ASML) built and launched the first commercial tool featuring this innovative technology, marking the beginning of its market dominance. The widespread adoption of cost-effective immersion ArF scanners and multi-patterning technology resulted in the 10 nm/7 nm FinFET technology node with acceptable cost, die yield and high efficiency in the semiconductor industry [8]. The development of the EUV light source, a multi-coated reflection mirror system, EUV photoresist, and EUV mask enabled High-Volume Manufacturing (HVM) [33]. For example, the ASML NXE EUVL module is equipped with several advanced components that significantly enhance its capabilities. At its core, the system centers on a powerful EUV light source that generates 13.5 nm wavelength radiation at 250 W, while next-generation sources surpassing 400 W are being engineered. Operating within a vacuum chamber, the scanner employs a full-reflection mirror system that accomplishes dual critical objectives: preventing high absorption along the EUV optical path and maximizing reflectivity. Additionally, the system features a redesigned reflection mirror and reticle system with a 0.33 NA, developed by Zeiss SMT. These cutting-edge innovations work in concert to enable an impressive throughput of up to 140 wafers per hour, marking a significant advancement in semiconductor manufacturing efficiency. Very recently, The NXE:3800E EUV system was released for cost-efficient 2 nm logic and advanced dynamic random access memory (DRAM) production, which complements immersion lithography capabilities. Its new bottom module, shared with the 0.55 NA EXE platform, features improved wafer handling and faster stages, while a higher power source boosts productivity. Figure 4c demonstrates a structural illustration of the NXE:3800E EUV system.

EUVL's transition to replace 193 nm lithography cost-effectively faces critical challenges, including line edge roughness (LER) and resist resolution limitations. The technique's high-vacuum requirement and resist outgassing issues restrict its applicability, while EUV absorption affects resist profiles, complicating device manufacturing. Moreover, insufficient throughput due to inadequate EUV source power at the wafer level remains a significant hurdle to economic viability, necessitating substantial power increases for widespread adoption. The widespread adoption of EUVL for nanostructure manufacturing hinges on overcoming several critical challenges: significantly reducing defect density on reflective masks, improving pattern generation and transfer, addressing difficulties with movable pellicles and thermophoretic protection as alternatives to non-pellicle masks, and enhancing resist performance for higher resolution, increased productivity, and reduced line edge roughness. Substantial advancements in these areas are essential for EUVL to become a practical and widely adopted technique.

2.3 Electron Beam Lithography

The advancement of photolithography faces the major challenge of resolution enhancement, while EUVL suffers from LER and EUV absorption in photoresist

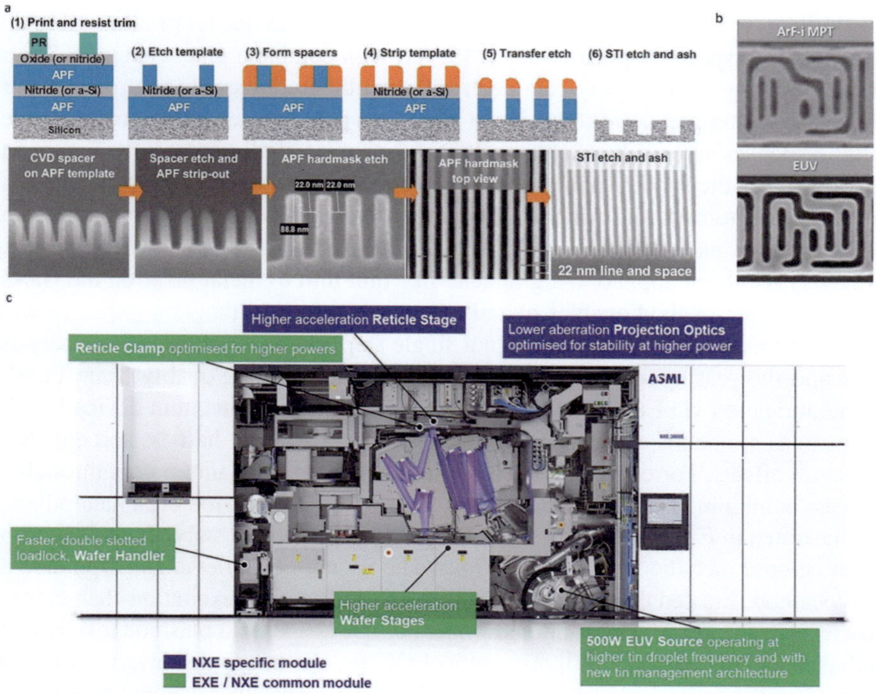

Fig. 4 **a** The scheme of SADP based on photolithography scanners using 193 nm wavelength lasers and an immersion system. Copyright 2008 Society of Photo-Optical Instrumentation Engineers (SPIE). **b** The Metal1 layer's standard cell patterns in 7 nm logic are fabricated using two distinct methods: multiple patterning with ArF immersion lithography and single-exposure EUV lithography. Copyright 2017 Society of Photo-Optical Instrumentation Engineers (SPIE). **c** NXE:3800E EUV system. It presents that the EUVL system contains a powerful EUV source and sophisticated optics and control systems. *Source* ASML

materials. As semiconductor devices continue to shrink in size and improve performance, their feature dimensions are constantly decreasing. Electron beam lithography (EBL) emerged as a breakthrough technology by harnessing the extraordinarily short wavelength of electrons—less than 0.1 nm at 10–50 keV electrons—to achieve unprecedented levels of practical resolution [34]. EBL traces its origins to the early 1960s, evolving from SEM technology. The first practical EBL systems emerged in the late 1960s and early 1970s, utilizing modified SEM to write patterns on electron-sensitive resists. After decades of development, EBL became crucial for photomask production in the semiconductor industry, enabling the creation of increasingly smaller circuit features. The evolution of electron beam lithography has witnessed a dramatic reduction in e-beam spot size, progressing from approximately 10 nm in the 1980s to an exceptional 0.1 nm through the implementation of aberration-corrected scanning transmission electron microscopy (STEM) systems [35, 36].

Williams et al. developed a wafer-scale fabrication technique for producing highly efficient multispectral optical filters (MSFAs) with complete pixel coverage. Their method employs a single grayscale electron beam lithography step to control insulator thickness within metal–insulator-metal structures precisely, enabling customizable filter responses across ultraviolet, visible, and near-infrared spectra [37]. Figure 5a demonstrates the MSFA photolithography process flow schematic comparing a grayscale photomask and a binary photomask. A 3-inch wafer was fabricated with $\sim$ 32 MSFAs, each containing nine spectral bands (3×3), as shown in Fig. 5b. EBL has also been applied to a graphene-like thin film of metal layer on the GaAs substrate. This method enables quantitative evaluation of lithography quality and critical dimensional measurements in a single step, significantly reducing analysis time and showcasing its potential for efficient nano-patterning quality control [38]. Nanofabrication via EBL at sub-10 nm scales is crucial for quantum device development, as structures at this near-atomic level can effectively harness and enhance quantum effects. For example, Valenta et al. fabricated Si quantum dots through a process combining EBL, plasma etching, and two-stage oxidation of Si nanopillars, as illustrated in Fig. 5c [39]. Recently, Researchers performed stable photoemission from tailored metallic nanoparticle arrays. The vacuum tube device incorporated a photocathode created through EBL, with the emitter's optical excitation characteristics revealing crucial insights into how nanoparticle shapes and plasmon resonances influence device response [40]. The optical microscope image displayed in Fig. 5d showcases the complete device alongside SEM views revealing detailed features of both nanorod and nanotriangle emitter structures.

Currently, the limitation threshold for EBL is more dependent on the point spread function (PSF), which characterizes how energy is spatially distributed around the small electron beam. A photoresist undergoes exposure only when the total energy density absorbed exceeds a critical threshold value. However, even when the energy density falls below this threshold, subsequent nearby exposures can adversely contribute to this partial exposure, potentially leading to an accumulation that crosses the threshold and results in unintended exposure regions. This phenomenon, known as the proximity effect, poses a significant challenge to the precision of EBL [41]. Electron beam lithography resolution is limited by various scattering phenomena that extend resist exposure beyond the focused beam spot. These include forward scattering of electrons entering the resist, backscattering from the substrate, generation of low-energy secondary electrons, and volume plasmons altering chemical bonds, as depicted in Fig. 5e [34]. Collectively, these effects contribute to the proximity effect, where partially exposed areas may accumulate enough energy from nearby exposures to cross the threshold for resist modification, resulting in unintended pattern alterations. Using chromatic aberration-corrected energy-filtered transmission electron microscopy (EFTEM), Berggren and colleagues conducted detailed measurements and analysis of the PSF at dimensions below 10 nm [42]. The researchers utilized a 10 nm silicon nitride (SiN_x) substrate, which effectively minimizes, nearly eliminating, the backscattered component. Specifically, the effects of secondary electrons and volume plasmons persist as significant factors influencing the overall beam spread. It should be noted that the SiN_x membrane is not a usual substrate, and an

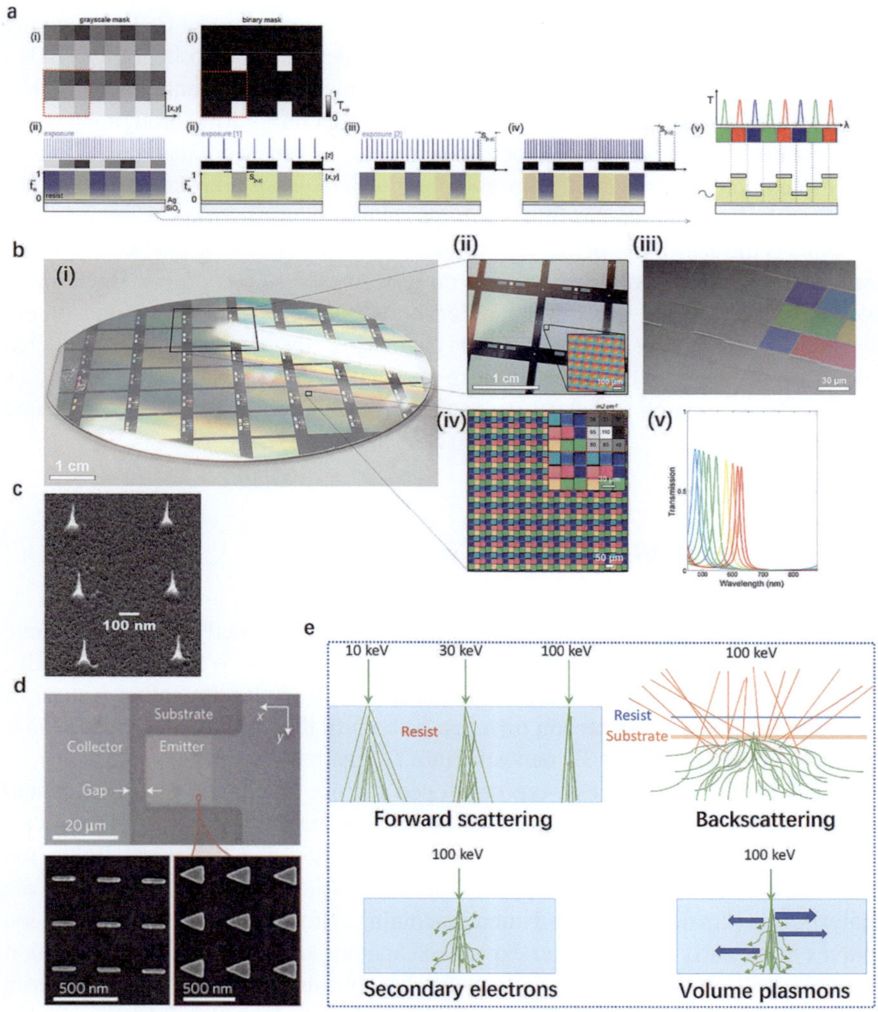

Fig. 5 **a** Schematic of photolithography-based MSFA fabrication process flow. Grey mask: (i) 3 × 3 grayscale photomask. (ii) A single exposure can be performed by imparting a spatially varying dose profile into the photoresist. Binary mask: (i) 3 × 3 binary photomask. (ii–iv) The mask is translated in-plane for each spectral pixel, with varying exposure levels. (v) The identical spectral response of final MSFA. (b) Photograph of a 3-inch wafer with ~32 9-band MSFAs. (i) A zoomed-in image of the region was captured with a macro lens. (ii–iii) SEM micrograph. (iv) Optical image (transmission) of a different region of the wafer, with labeled equivalent exposure pattern (inset). (v) corresponding transmission spectra for each spectral band. Copyright 2019 American Chemical Society. **c** SEM images (45° tilted view) of Si quantum dots. Copyright 2002 Elsevier. **d** Top optical microscope image of an emitter array. Bottom SEM of example nanotriangle and nanorod arrays. Copyright 2016 Nature Publishing Group. **e** Components of the point spread function of an electron beam. Copyright 2022 MDPI

electron beam of 200 keV is much higher than that in conventional EBL. Following this, Ehud Fuchs et al. demonstrated that hydrogen depassivation lithography (HDL) enables significantly higher resolution and precision over the conventional EBL [41].

The fundamental limitation of EBL lies in its inherently low throughput, as its serial patterning process requires hours to complete the same work that optical lithography accomplishes in minutes on complete wafers. This time-intensive process significantly limits its applicability in high-volume manufacturing. Compounding this issue is the sensitivity of electron beams to electrical and magnetic noise, which can cause disruptive perturbations. To mitigate these effects, proper shielding is essential, necessitating careful separation between the beam-column and potential interference sources. Despite these challenges, the technique's unparalleled resolution capabilities continue to make it valuable for specialized applications, particularly in research and development settings where precision outweighs speed considerations.

2.4 Ion Beam Lithography

Ion beam lithography harnesses focused ion beams to directly pattern substrate surfaces, offering precise control for nanostructure fabrication. While similar to EBL, this process utilizes heavily charged ion particles with high momentum instead of electrons, enabling straighter ion travel paths both in the vacuum and in the matter and more compatibility for 3D nanostructure fabrication [34].

In the 1950s, ion sources were developed for use in particle accelerators and mass spectrometers. The 1960s saw the advent of ion implantation in semiconductor manufacturing, revolutionizing the industry. Throughout the 1980s and 1990s, ion beam technology found diverse applications in materials science, including surface analysis, thin film deposition, and micromachining. The development of gas field ion sources in the early 2000s further expanded capabilities, enabling higher resolution and new applications [43]. Over recent decades, our understanding of ion-material interactions has significantly advanced, leading to the development of reliable simulations and calculations for evaluating ion bombardment applications. Modern ion beam tools enable diverse applications spanning radiation damage studies, interface mixing, polymer transformation, defect engineering, surface sputtering, patterning, and reactive species implantation [44]. This comprehensive approach enables researchers and engineers to predict and analyze ion bombardment effects across various scales and applications, enhancing the design and optimization of ion-based processes. Consequently, ion beam technologies have become more sophisticated and tailored, driving innovations in materials science, semiconductor manufacturing, and nanotechnology and opening new possibilities for material modification and analysis at atomic and molecular levels [44]. The field of ion beam lithography has evolved into three distinct three techniques: focused ion beam (FIB) lithography, proton beam lithography, and ion projection lithography.

FIB lithography is a sophisticated nanofabrication technique that employs a finely focused beam of ions, typically gallium, to create precise nanoscale patterns and structures on a substrate. This direct-write, maskless method offers high resolution down to a few nanometers and remarkable versatility, making it capable of deposition and subtractive milling processes. FIB can modify material properties at the nanoscale, create complex three-dimensional structures, and perform site-specific alterations. The first introduction of FIB lithography dates back to the 1970s, with the first commercial FIB systems introduced in the early 1980s. The integration of FIB with SEM in the late 1990s and early 2000s marked a major milestone, creating dual-beam systems that revolutionized nanoscale imaging and fabrication. Compared to parallel lithography techniques, low-throughput limits FIB use in high-volume manufacturing. Nevertheless, its unparalleled precision and flexibility in manipulating materials at the nanoscale continue to make FIB lithography an essential tool in advanced research and specialized industrial applications.

Proton beam lithography, commonly referred to as proton beam writing (PBW), distinguishes itself through the use of focused high-energy protons—typically ranging from 1 to 3 MeV—to either pattern resist materials or directly modify substrate properties. The substantial mass of protons enables deep material penetration along straight trajectories, facilitating the fabrication of three-dimensional structures with exceptional characteristics: high aspect ratios, vertical sidewalls, and minimal line-edge roughness. This precision is further enhanced by the limited proximity effects in proton beam writing since proton-electron collisions generate only low-energy secondary electrons, as confirmed by theoretical models. These characteristics make proton beam writing particularly effective for fabricating complex, high-precision nanostructures that are challenging to achieve with other lithographic techniques, especially in applications requiring 3D structuring and high aspect ratios [45]. The proton beam lithography technique has not yet been commercially developed, and challenges remain when MeV ions are focused at the 100 nm scale. Note that the prototype for a proton beam has been developed by CIBA (Centre for Ion Beam Applications). Recently, they reported minimal proximity exposure and straight proton trajectories using focused MeV proton beam writing, with the ultimate electrode gap reaching a few nm levels [46].

Ion projection lithography uses light ions (H^+, H_2^+, or He^+) beamed through a patterned mask (stencil) to create nanoscale structures. This technique leverages the properties of ions, which are much heavier than electrons and travel in straighter paths with higher penetration power. The system employs electrostatic lenses to project a demagnified image of the mask pattern onto the substrate. While this method can achieve resolutions up to 300 nm, it requires higher energies for ion acceleration compared to electron beam lithography. The limitations include process complexity, mask stability issues, and high energy requirements. Despite these challenges, ion projection lithography offers unique advantages in creating specific types of nanopatterns [47].

2.5 Soft Lithography

Soft lithography is a set of microfabrication techniques that employ pliable, elastomeric materials as stamps or molds to generate structures at both micro- and nanoscales. Developed in the 1990s, this technique offers advantages in versatility and cost-effectiveness over traditional lithography. It employs materials like polydimethylsiloxane (PDMS) to construct different nanostructures and pattern diverse substrates, including non-planar and flexible surfaces and is compatible with a wide range of materials [48, 49]. This versatility has made soft lithography valuable in fields like microfluidics, biosensors, and flexible electronics. The elastomeric stamp is coated or inked with target molecules. When pressed onto a substrate, it directly transfers these molecules into a pattern that mirrors the stamp's surface topography, which enables precise molecular patterning at micro- and nanoscales [50]. The schematic of four major steps in soft lithography is provided in Fig. 6a, together with three soft lithography techniques [49]. Based on this concept, a large number of soft lithography techniques have been developed, and microcontact printing (μCP) is the most established one amongst them, developed in the early 1990s by George M. Whitesides and colleagues [48]. In the μCP process, an elastomeric stamp, typically made of PDMS, is "inked" with molecules capable of forming self-assembled monolayer (SAM) films, this is because PDMS has a low Young's modulus, allowing conformal contact with the substrate under mild pressure [51]. When the stamp contacts the substrate, it transfers these molecules into a pattern defined by the stamp's topography, creating a patterned SAM. One example was tailoring the properties of physical objects at the molecular level using alkanethiol SAMs on substrates coated with metal films such as gold, silver, palladium, and platinum [52]. Jun et al. developed an innovative micropatterning approach that combines μCP with vacuum-assisted, degas-driven flow patterning using PDMS stamps. Their method achieved precise surface architectures, demonstrated through two key applications: creating bi-composite patterns of FITC-BSA/PEG-silane for biomolecule arrays and developing APTES/PEG-silane patterns for self-assembled gold nanoparticle monolayers [53]. Figure 6b and c depicts the optical image for a PDMS stamp containing a circular dot array and a PDMS stamp containing line patterns. In addition, μCP is particularly effective for creating patterns of diverse materials, especially at scales of cm^2 or larger. A critical factor in successful μCP is the aspect ratio of the stamp's features. This ratio (height to width) should be between 0.2 and 2 for optimal performance, which ensures proper conformal contact without risking the structural collapse of the stamp, which could lead to issues like adhesion between neighbouring features or deformation under applied pressure [54].

Microtransfer molding (μTM) is an innovative technique derived from μCP that enables the creation of precise microstructures. The prepolymer-coated stamp is pressed against the substrate, transferring the desired pattern. Once in place, the prepolymer conforms to the stamp's design and undergoes a curing process to solidify its shape. This curing can be achieved through various methods, including irradiation, heat treatment, or application of specific gelling agents. After the curing is

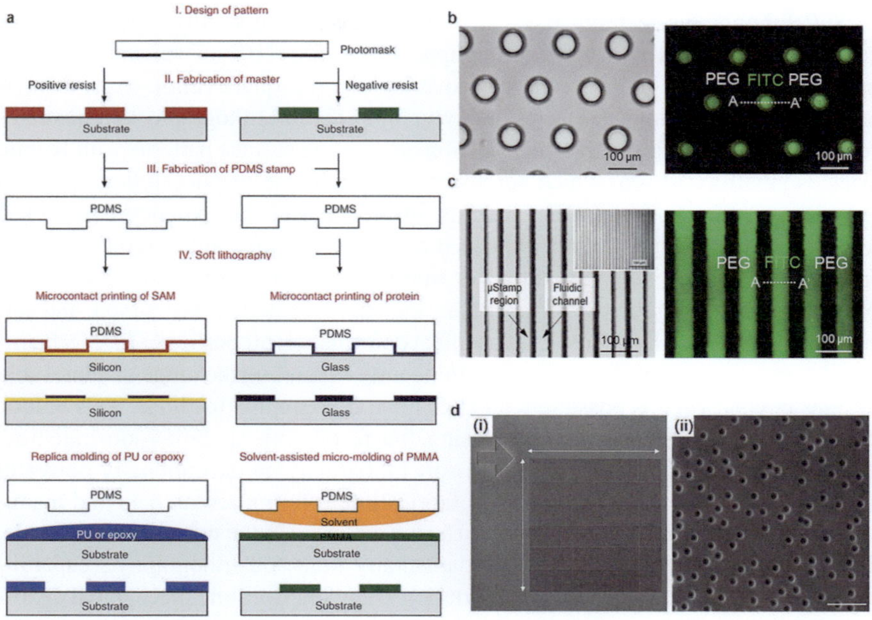

Fig. 6 **a** The diagram outlines the four key stages of soft lithography and highlights three primary soft lithographic methods. Copyright 2010 Nature Publishing Group. **b** The PDMS stamp with a circular dot pattern (left side) and a fluorescent view of FITC-BSA dots surrounded by PEG-silane (right side). **c** PDMS stamp featuring line patterns (left side) and its corresponding fluorescent image showing alternating FITC-BSA and PEG-silane lines (right side). Copyright 2018 Nature Publishing Group. **d** (i) Overview of the nanopatterned array with 1,361,700 holes on the area 540 × 540 μm^2 silicon mold. (ii) The SEM image of the array of big holes (scale bar 1 μm). Copyright 2022 Nature Publishing Group

complete, the elastomer stamp is carefully lifted off, revealing the newly formed microstructures on the substrate's surface. Replica molding (REM), another key soft lithography technique, employs the elastomeric stamp as a mold to precisely duplicate master design features using thermoset or photocurable polymers [55]. REM has shown its importance in the field of microfluidics, where it is the primary method for fabricating devices from PDMS [56–58]. In another study, poly(2-hydroxyethyl methacrylate) hydrogels were used to fabricate customizable nanopillar patterns using REM, producing uniform, defect-free structures with adjustable number, shape, and arrangement, as confirmed by atomic force microscopy while emphasizing the importance of proper hydrogel composition in imprint lithography [59]. A rapid, defect-free replication of the master's features is a prerequisite for the successful REM. Successful replica molding depends on two essential requirements: the prepolymer material, whether thermoset or photocurable, must effectively wet the prefabricated mold surface, and the kinetics of filling the master mold should be favourable.

Soft lithographic techniques excel in efficiency and versatility. These methods can be implemented with high throughput, processing (10^2 – 10^3 cm^2 h^{-1}), by using large, reusable molds or stamps covering several square centimeters. Process automation further enhances its productivity [60, 61]. Soft lithography demonstrates remarkable capabilities in microfabrication, generating surface patterns with feature sizes as small as 30 nm, which surpasses the limitations of photolithography but is constrained by optical factors such as diffraction and transparency issues. The technique's versatility is further enhanced by the use of soft stamps, typically made from PDMS, which not only allow for multiple pattern transfers but also enable patterning on a diverse range of substrates, including rigid, flexible, curved, and soft surfaces. Moreover, the isotropic mechanical deformation properties of PDMS facilitate the creation of complex patterns, significantly expanding the range of achievable designs beyond what is possible with traditional lithographic methods. This unique combination of high throughput processing, fine resolution capability, substrate flexibility, and pattern complexity renders soft lithography an exceptionally powerful and adaptable tool in the field of microfabrication, offering researchers and manufacturers unprecedented control over surface structuring at the micro and nanoscale. The inherent properties of PDMS fundamentally limit soft lithography's capabilities, particularly due to the risk of patterns experiencing isotropic mechanical distortion. To address this, more rigid stamps, such as stiffer poly(methyl methacrylate) (PMMA), are demonstrated [62, 63]. A significant constraint lies in the production process of the elastomeric stamp itself. Although creating stamps doesn't require specialized equipment or costly materials, it still depends on a silicon "master" mold. This master mold, crucial for producing the stamp, is typically fabricated via EBL. Consequently, the expenses and time associated with creating this master mold must be factored into the overall soft lithography process. The reliance on these advanced lithographic techniques for master mold creation introduces an element of complexity and cost that partially offsets the otherwise accessible nature of soft lithography.

3 Conclusion

The field of nanolithography has witnessed remarkable advancements in recent decades, with various techniques evolving to meet the ever-increasing demands of nanotechnology. From the well-established photolithography, which remains the workhorse of the semiconductor industry, to cutting-edge methods like EUVL, EBL, ion beam lithography, and soft lithography, each approach offers unique advantages and faces distinct challenges, as listed in Table 1. While photolithography has progressed to achieve resolutions down to tens of nanometers with high throughput, EUVL promises even finer features but grapples with issues such as line edge roughness and the need for high-vacuum environments. Though offering the highest practical resolution, EBL is hindered by low throughput, limiting its application in mass production. Ion beam lithography techniques provide specialized capabilities for creating specific nanopatterns but struggle with process complexity. Soft lithography has emerged as a versatile and cost-effective method, particularly valuable in fields like microfluidics and biosensors. Looking to the future, the field of nanolithography faces several critical challenges that will shape its development. The persistent trade-off between resolution and throughput remains a crucial issue, driving research into novel approaches that can deliver both high precision and high-speed fabrication. The need for broader material compatibility will likely spur innovations in resist chemistry and substrate engineering. Cost-effectiveness will continue to be a crucial factor, potentially leading to the development of more accessible high-resolution techniques. The growing demand for complex 3D nanostructures will push lithographic methods beyond traditional 2D patterning, possibly through the integration of multiple techniques or the development of entirely new approaches. Looking ahead, we can anticipate several exciting developments in the field. The integration of artificial intelligence and machine learning algorithms with lithographic processes could revolutionize pattern optimization and defect detection, leading to unprecedented levels of precision and efficiency. As nanotechnology finds applications in fields like quantum computing, neuromorphic engineering, and personalized medicine, nanolithography will need to adapt to fabricate increasingly complex and specialized nanostructures. The development of hybrid lithographic systems, combining the strengths of multiple techniques, could provide the versatility needed to meet these diverse requirements. In conclusion, while challenges remain, the future of nanolithography is bright, with ongoing research and cross-disciplinary collaborations poised to unlock new possibilities in nanostructure fabrication, driving innovations across a wide spectrum of technological fields.

Table 1 Advantages and disadvantages of the five nanolithographic techniques

Nanolithographic techniques	Advantages	Disadvantages
Photolithography	• High throughput • Well-established industrial standard • Cost-effective for mass production • Good for large-scale manufacturing • High yield and reliability	• Resolution limited by UV light wavelength • Complex alignment requirements • Requires photomasks • Expensive equipment for advanced systems • Multiple processing steps needed
Extreme Ultraviolet lithography	• Multiple processing steps needed • High throughput • Process compatibility	• Requires complex vacuum chamber operation • Sophisticated mirror systems • Line edge roughness issues • Resist outgassing problems • Very expensive equipment
Electron beam lithography	• Highest practical resolution • Maskless direct writing capability • Extremely precise pattern control • Excellent for custom designs • High accuracy	• Very low throughput • Multiple scattering phenomena affect resolution • Proximity effects issue • Sensitive to electronic/magnetic interference • Cost-prohibitive for mass production
Ion beam lithography	• Direct substrate modification • Enables 3D nanostructure creation • Multiple technique options • Deep material penetration • Minimal proximity effects in PBW	• Requires complex equipment • Lower throughput than photolithography • High energy requirements • Process complexity • Limited commercial development
Soft lithography	• Cost-effective approach • High throughput • Versatile material compatibility • Works on non-planar surfaces • Reusable stamps/molds	• Limited by PDMS properties • Mechanical deformation issues • Requires master mold • Lower resolution than advanced techniques • Pattern quality depends on stamp integrity • Potential defects during pattern transfer

References

1. Pan DZ, Yu B, Gao J-R (2013) Design for manufacturing with emerging nanolithography. IEEE Trans Comput Aided Des Integr Circuits Syst 32(10):1453–1472
2. Sharma E, Rathi R, Misharwal J, Sinhmar B, Kumari S, Dalal J, Kumar A (2022) Evolution in lithography techniques: microlithography to nanolithography. Nanomaterials 12(16):2754

3. Pimpin A, Srituravanich W (2012) Review on micro-and nanolithography techniques and their applications. Eng J 16(1):37–56
4. Traub MC, Longsine W, Truskett VN (2016) Advances in nanoimprint lithography. Annu Rev Chem Biomol Eng 7(1):583–604
5. Chen Y, Shu Z, Zhang S, Zeng P, Liang H, Zheng M, Duan H (2021) Sub-10 nm fabrication: methods and applications. Int J Extreme Manuf 3(3):032002
6. Jung WB, Jang S, Cho SY, Jeon HJ, Jung HT (2020) Recent progress in simple and cost-effective top-down lithography for $\approx$ 10 nm scale nanopatterns: from edge lithography to secondary sputtering lithography. Adv Mater 32(35):1907101
7. Balram KC, Westly DA, Davanco M, Grutter KE, Li Q, Michels T, Ray CH, Yu L, Kasica RJ, Wallin CB (2016) The nanolithography toolbox. J Res Nat Inst Stand Technol 121:464
8. Fu N, Liu Y, Ma X, Chen Z (2019) EUV lithography: state-of-the-art review. J Microelectron Manuf 2(2):1–6
9. Harinarayana V, Shin YC (2021) Two-photon lithography for three-dimensional fabrication in micro/nanoscale regime: a comprehensive review. Opt Laser Technol 142:107180
10. del Barrio J, Sánchez-Somolinos C (2019) Light to shape the future: from photolithography to 4D printing. Adv Opt Mater 7(16):1900598
11. Traub MC, Longsine W, Truskett VN (2016) Advances in nanoimprint lithography. Annu Rev Chem Biomol Eng 7:583–604
12. Grzelczak M, Vermant J, Furst EM, Liz-Marzán LM (2010) Directed self-assembly of nanoparticles. ACS Nano 4(7):3591–3605
13. Fujita M (1998) Metal-directed self-assembly of two-and three-dimensional synthetic receptors. Chem Soc Rev 27(6):417–425
14. Jeong S-J, Kim JY, Kim BH, Moon H-S, Kim SO (2013) Directed self-assembly of block copolymers for next generation nanolithography. Mater Today 16(12):468–476
15. Cox LM, Martinez AM, Blevins AK, Sowan N, Ding Y, Bowman CN (2020) Nanoimprint lithography: emergent materials and methods of actuation. Nano Today 31:100838
16. Bhagoria P, Mathew Sebastian E, Kumar Jain S, Purohit J, Purohit R (2020) Nanolithography and its alternate techniques. Mater Today: Proc 26:3048–3053
17. Fruncillo S, Su X, Liu H, Wong LS (2021) Lithographic processes for the scalable fabrication of micro- and nanostructures for biochips and biosensors. ACS Sens 6(6):2002–2024
18. Sebastian EM, Jain SK, Purohit R, Dhakad SK, Rana RS (2020) Nanolithography and its current advancements. Mater Today: Proc 26:2351–2356
19. Bruning JH (2007) Optical lithography: 40 years and holding. In: Optical microlithography XX, SPIE, pp 62–74
20. Thompson LF (1983) An introduction to lithography. ACS Publications
21. Okazaki S (2015) High resolution optical lithography or high throughput electron beam lithography: the technical struggle from the micro to the nano-fabrication evolution. Microelectron Eng 133:23–35
22. Ito T, Okazaki S (2000) Pushing the limits of lithography. Nature 406(6799):1027–1031
23. Feldman M (2014) Nanolithography: the art of fabricating nanoelectronic and nanophotonic devices and systems. Woodhead Publishing
24. Wagner C, Harned N (2010) Lithography gets extreme. Nat Photonics 4(1):24–26
25. van de Kerkhof M, Jasper H, Levasier L, Peeters R, van Es R, Bosker J-W, Zdravkov A, Lenderink E, Evangelista F, Broman P (2017) Enabling sub-10nm node lithography: presenting the NXE: 3400B EUV scanner. In: Extreme ultraviolet (EUV) lithography VIII, SPIE, pp 34–47
26. Dong Y, Xu Z, Li N, Tong J, Fu YH, Zhou Y, Hu T, Zhong Q, Bliznetsov V, Zhu S, Lin Q, Zhang DH, Gu Y, Singh N (2020) Si metasurface half-wave plates demonstrated on a 12-inch CMOS platform. Nanophotonics 9(1):149–157
27. Park J-S, Zhang S, She A, Chen WT, Lin P, Yousef KMA, Cheng J-X, Capasso F (2019) All-glass, large metalens at visible wavelength using deep-ultraviolet projection lithography. Nano Lett 19(12):8673–8682
28. Rothschild M (2005) Projection optical lithography. Mater Today 8(2):18–24

29. Karim W, Tschupp SA, Oezaslan M, Schmidt TJ, Gobrecht J, van Bokhoven JA, Ekinci Y (2015) High-resolution and large-area nanoparticle arrays using EUV interference lithography. Nanoscale 7(16):7386–7393
30. Ronse K (2006) Optical lithography—a historical perspective. C R Phys 7(8):844–857
31. Bencher C, Chen Y, Dai H, Montgomery W, Huli L (2008) 22nm half-pitch patterning by CVD spacer self alignment double patterning (SADP). In: Optical microlithography XXI, SPIE, pp 1440–1446
32. Kim S-S, Chalykh R, Kim H, Lee S, Park C, Hwang M, Park J-O, Park J, Kim H, Jeon J (2017) Progress in EUV lithography toward manufacturing. In: Extreme ultraviolet (EUV) lithography VIII, SPIE, p 1014306
33. Smeets C, Benders N, Bornebroek F, Carbone J, van Es R, Minnaert A, Salmaso G, Young S (2023) 0.33 NA EUV systems for high volume manufacturing. In: Optical and EUV nanolithography XXXVI, SPIE, pp 34–42
34. Sharma E, Rathi R, Misharwal J, Sinhmar B, Kumari S, Dalal J, Kumar A (2022) Evolution in lithography techniques: microlithography to nanolithography nanomaterials [online]
35. Pfeiffer H, Langner G (1981) Advanced deflection concept for large area, high resolution e-beam lithography. J Vac Sci Technol 19(4):1058–1063
36. Manfrinato VR, Stein A, Zhang L, Nam C-Y, Yager KG, Stach EA, Black CT (2017) Aberration-corrected electron beam lithography at the one nanometer length scale. Nano Lett 17(8):4562–4567
37. Williams C, Gordon GSD, Wilkinson TD, Bohndiek SE (2019) Grayscale-to-color: scalable fabrication of custom multispectral filter arrays. ACS Photonics 6(12):3132–3141
38. Göktaş O, Işık Göktaş N (2017) Hough transform as a quality test tool for electron beam lithography. Microelectron Eng 177:6–8
39. Valenta J, Juhasz R, Linnros J (2002) Photoluminescence from single silicon quantum dots at room temperature. J Lumin 98(1):15–22
40. Putnam WP, Hobbs RG, Keathley PD, Berggren KK, Kärtner FX (2017) Optical-field-controlled photoemission from plasmonic nanoparticles. Nat Phys 13(4):335–339
41. Randall JN, Owen JHG, Lake J, Fuchs E (2019) Next generation of extreme-resolution electron beam lithography. J Vac Sci Technol B 37(6):061605
42. Manfrinato VR, Wen J, Zhang L, Yang Y, Hobbs RG, Baker B, Su D, Zakharov D, Zaluzec NJ, Miller DJ, Stach EA, Berggren KK (2014) Determining the resolution limits of electron-beam lithography: direct measurement of the point-spread function. Nano Lett 14(8):4406–4412
43. Watt F, Bettiol A, Van Kan J, Teo EJ, Breese M (2005) Ion beam lithography and nanofabrication: a review. Int J Nanosci 4(03):269–286
44. Baglin J (2012) Ion beam nanoscale fabrication and lithography—a review. Appl Surf Sci 258(9):4103–4111
45. Watt F, Breese MBH, Bettiol AA, van Kan JA (2007) Proton beam writing. Mater Today 10(6):20–29
46. Whitlow HJ, Ng ML, Auželyté V, Maximov I, Montelius L, van Kan JA, Bettiol AA, Watt F (2004) Lithography of high spatial density biosensor structures with sub-100 nm spacing by MeV proton beam writing with minimal proximity effect. Nanotechnology 15(1):223
47. Melngailis J, Mondelli A, Berry IL III, Mohondro R (1998) A review of ion projection lithography. J Vacuum Sci Technol B: Microelectron Nanometer Struct Process Meas Phenom 16(3):927–957
48. Xia Y, Whitesides GM (1998) Soft Lithography. Angew Chem Int Ed 37(5):550–575
49. Qin D, Xia Y, Whitesides GM (2010) Soft lithography for micro-and nanoscale patterning. Nat Protoc 5(3):491
50. Tran KT, Nguyen TD (2017) Lithography-based methods to manufacture biomaterials at small scales. J Sci: Adv Mater Devices 2(1):1–14
51. Rose MA, Bowen JJ, Morin SA (2019) Emergent soft lithographic tools for the fabrication of functional polymeric microstructures. ChemPhysChem 20(7):909–925
52. Love JC, Estroff LA, Kriebel JK, Nuzzo RG, Whitesides GM (2005) Self-assembled monolayers of thiolates on metals as a form of nanotechnology. Chem Rev 105(4):1103–1170

53. Lee SH, Rho W-Y, Park SJ, Kim J, Kwon OS, Jun B-H (2018) Multifunctional self-assembled monolayers via microcontact printing and degas-driven flow guided patterning. Sci Rep 8(1):16763
54. Delamarche E, Schmid H, Michel B, Biebuyck H (1997) Stability of molded polydimethylsiloxane microstructures. Adv Mater 9(9):741–746
55. Xia Y, McClelland JJ, Gupta R, Qin D, Zhao XM, Sohn LL, Celotta RJ, Whitesides GM (1997) Replica molding using polymeric materials: a practical step toward nanomanufacturing. Adv Mater 9(2):147–149
56. Sticker D, Rothbauer M, Lechner S, Hehenberger M-T, Ertl P (2015) Multi-layered, membrane-integrated microfluidics based on replica molding of a thiol–ene epoxy thermoset for organ-on-a-chip applications. Lab Chip 15(24):4542–4554
57. McDonald JC, Duffy DC, Anderson JR, Chiu DT, Wu H, Schueller OJ, Whitesides GM (2000) Fabrication of microfluidic systems in poly (dimethylsiloxane). ELECTROPHORESIS: Int J 21(1):27–40
58. Chan HN, Chen Y, Shu Y, Chen Y, Tian Q, Wu H (2015) Direct, one-step molding of 3D-printed structures for convenient fabrication of truly 3D PDMS microfluidic chips. Microfluid Nanofluid 19:9–18
59. Krajňák T, Černá E, Šuráňová M, Šamořil T, Zicha D, Vojtová L, Čechal J (2022) Replica-mold nanopatterned PHEMA hydrogel surfaces for ophthalmic applications. Sci Rep 12(1):14497
60. Zhou X, Xu H, Cheng J, Zhao N, Chen S-C (2015) Flexure-based roll-to-roll platform: a practical solution for realizing large-area microcontact printing. Sci Rep 5(1):10402
61. Kang H, Lee D, Yang Y, Kyo Oh D, Seong J, Kim J, Jeon N, Kang D, Rho J (2023) Emerging low-cost, large-scale photonic platforms with soft lithography and self-assembly. Photonics Insights 2(2):R04
62. Mukherjee R, Sharma A, Patil G, Faruqui D, Sarathi P, Pattader G (2008) Soft lithography meets self-organization: some new developments in meso-patterning. Bull Mater Sci 31:249–261
63. Zhang J, Yang B (2010) Patterning colloidal crystals and nanostructure arrays by soft lithography. Adv Funct Mater 20(20):3411–3424

Microbial Factories for Scaled-up Production of Nanoparticles

Bavatharny Thevarajah, Wanni Arachchige Jalitha Wasath, P. H. V. Nimarshana, Nadun H. Madanayake, and Thilini U. Ariyadasa

Abstract The latest trend in the field of nanotechnology is the adoption of novel green technologies by utilizing microorganisms as "microbial cell factories" in the production of nanoparticles. In recent years, microorganisms, including bacteria, fungi, and microalgae, have been widely explored for the intracellular/extracellular synthesis of nanoparticles. Remarkably, scaled-up production of nanoparticles via biogenic synthesis is an eco-friendly approach resulting in less/non-toxic nanomaterials, without any necessity of toxic/hazardous chemicals and high pressure/temperature process conditions used in conventional methods. Hence, microbial nanoparticles are mainly preferred in the fields of biomedical and pharmaceutical for diverse applications including but not limited to, drug delivery systems, antimicrobials, tissue regenerative biomaterials, biosensors, and bioimaging. In accordance, the chapter holistically focuses on the various types of microorganisms used for nanoparticle production and their potential in real-world applications. Also, the challenges and limitations associated with the scaled-up production of biogenic nanoparticles, and potential solutions are discussed, to distinguish the scope of future research enhancing the economic feasibility and sustainability.

B. Thevarajah · W. A. J. Wasath · T. U. Ariyadasa (✉)
Department of Chemical and Process Engineering, Faculty of Engineering, University of Moratuwa, Moratuwa, Sri Lanka
e-mail: thilini@uom.lk

B. Thevarajah
e-mail: bavatharny.t@gmail.com

W. A. J. Wasath
e-mail: jalithawasath@gmail.com

P. H. V. Nimarshana
Department of Mechanical Engineering, Faculty of Engineering, University of Moratuwa, Moratuwa, Sri Lanka
e-mail: nimarshana@mech.mrt.ac.lk

N. H. Madanayake
Department of Botany, Faculty of Science, University of Peradeniya, Peradeniya, Sri Lanka

N. M. Adassooriya et al. (eds.), *Industrial Scale Production of Nanoparticles*, Smart Nanomaterials Technology, https://doi.org/10.1007/978-981-96-9176-0_15

Keywords Microorganisms · Biogenic synthesis · Scaled-up production · Nanomaterials · Sustainability

1 Introduction

In recent years, a tremendous shift towards novel green techniques in the field of nanotechnology has been witnessed due to eco-friendliness, non-toxicity, ease of scale-up, and sustainability, in comparison to the conventional chemical and physical methods. This is mainly due to the expansion of nanoparticle (NP) applications in medical, biotechnological, and pharmaceutical sectors, due to the beneficial properties including biocompatibility, antimicrobial activities, bioavailability, and biosorption (Fig. 1) [36]. In particular, the utilization of metal NPs including gold (Au), silver (Ag), platinum (Pt), and palladium (Pd) are clinically accepted in biomedical fields, due to high cytocompatibility, minimal toxicity, and effective interaction with therapeutically active moieties [63]. Moreover, micelles, liposomes, solid lipid NPs, dendrimers, and carbon (C) nanotubes are also widely explored in biomedical applications [1, 31]. For instance, biological activities exhibited by magnetic NPs, attributed to the particle size, magnetic properties, and ease of surface modifications, alongside high biocompatibility, have been aided in several applications, including tumour detection, cancer therapy, drug delivery, and resonance imaging [26]. Therefore, "microbial nanofactories" incorporating microorganisms, namely bacteria, fungi, and microalgae, have been extensively studied as an innovative green technology in NPs production. A diversified spectrum of metal nanoparticles, including iron (Fe), lead (Pb), Ag, selenium (Se), Au, Pt, and Ag-Au alloy, among others, could be biosynthesized by microbes.

Bioproduction of NPs is classified as extracellular and intracellular based on the location of synthesis, i.e., outside or inside the cell wall [19]. During the extracellular synthesis of NPs, the metal ions are entrapped on the cell wall and undergo degradation/transformation in the presence of biomolecules, including enzymes, proteins, and peptides [19]. On the other hand, the ions are transported into the microbial cell in the intracellular synthesis, wherein the NPs are extracted from the biomass [48]. The microbial biomass would be harvested after the optimal cultivation period, and the cell walls are disrupted, during the recovery of NPs. Due to this complex downstream processing, extracellular synthesis of NPs is highly preferred. Notably, the microbial NP formation reflects a "bottom-up" approach [2]. Remarkably, the metabolic pathways of microorganisms involved in NPs synthesis minimize the metal ion toxicity through bioreduction or by aggregation with non-soluble complexes. Moreover, a tailored synthesis of NPs via optimized culture conditions for distinct morphology, i.e., shapes and sizes, is beneficial in minimizing the overall production cost, in comparison to the chemical synthesis method [34]. Furthermore, green synthesis of NPs is beneficial for humans and the environment due to the requirement of mild production conditions, including temperature, pH, and pressure. Meanwhile, the use of toxic/hazardous reagents and incorporation of external reducing, capping,

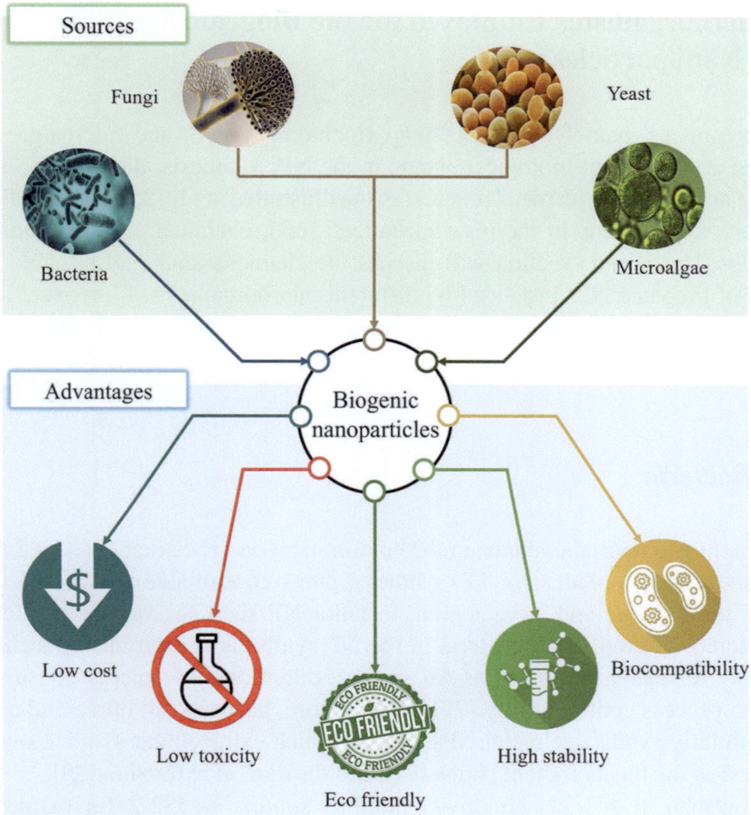

Fig. 1 Microorganisms-based synthesis of nanoparticles and their advantages

and stabilizing agents, required in the chemical and physical techniques are eliminated by the biological synthesis of NPs [34]. Hence, scaled-up production of NPs using microbes would be economically viable and sustainable, in comparison to the traditionally employed methods.

In this context, this chapter aims to holistically scrutinize the different microbial cell factories which are used for the scaled-up production of microbial NPs. Moreover, the metabolic pathways and mechanisms involved in the biosynthesis of NPs by bacteria, fungi, and microalgae are discussed. In addition, this chapter will emphasize potential challenges and solutions associated with the scaling up of microbial NP production, while highlighting the perspectives.

2 Microorganisms Employed for the Biogenic Synthesis of Nanoparticles

Microorganisms, namely bacteria, fungi (including yeast), and microalgae, have emerged as significant biological agents in the NPs synthesis, due to their diverse range of interactions with metal precursors. As illustrated in Fig. 2, biogenic NPs have numerous applications in the pharmaceutical, food, chemical, and environmental industries. Hence, this section will discuss the characteristics and possible applications of biogenic NPs produced by different microorganisms. Moreover, Table 1 summarizes the promising microbial species used in the biosynthesis of diversified NPs.

2.1 *Bacteria*

Bacteria, due to their abundance in the environment and remarkable adaptability to diverse conditions, are ideal for NPs synthesis, due to controllable culture parameters, such as temperature and oxygenation, to tailor NP sizes for various applications [17]. Moreover, employing bacteria in the NP synthesis is a promising technique, especially for metal NPs, such as Au and Ag, due to the low energy consumption and high process controllability [29]. Furthermore, bacteria can intracellularly and extracellularly synthesize metal NPs, among which extracellular synthesis is often identified as the most efficient pathway due to the ease of extraction [29].

Mathivanan et al. [27] employed *Bacillus subtilis* KMS2-2 for extracellular synthesis of Ag NPs (AgNPs) which were 18–100 nm spherical particles before

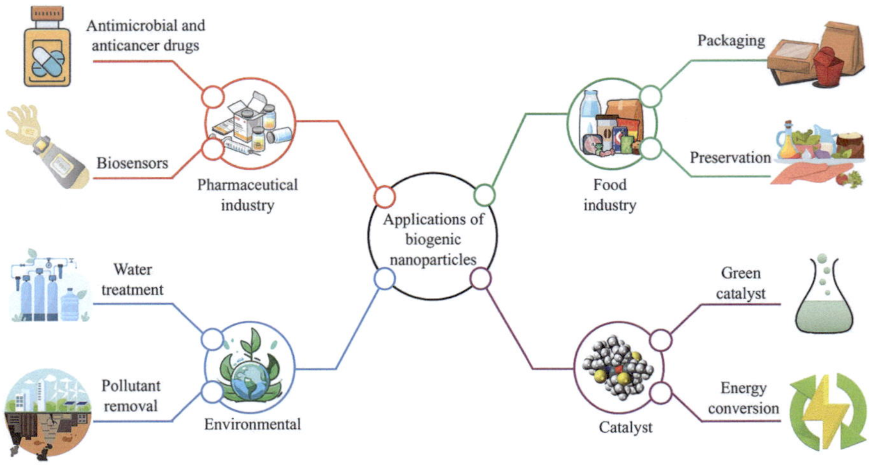

Fig. 2 Applications of biogenic nanoparticles

Table 1 Green synthesis of nanoparticles using microorganisms and their potential applications

Microorganism	Strains	Microbial-derived nanoparticles			Applications	References
		Metal/metal oxide	Particle size (nm)	Bioactive properties		
Bacteria	*Bacillus subtilis* KMS2-2	Silver	18–100 (pre-calcination) 49–153 (post-calcination)	• Significant antibacterial properties against *Pseudomonas fluorescens* and *Staphylococcus aureus*	• Food packaging • Biomedicine	Mathivanan et al. [27]
	Paracoccus haeundaensis BC74171T	Gold	20.93 ± 3.46	• Antioxidant activity • Selective cancer cell inhibition	• Drug delivery • Nanomedicine	Patil et al. [38]
	Bacillus subtilis	Zinc oxide	10–15	• Photocatalytic properties under sunlight	• Dye degradation • Photocatalytic application	Dhandapani et al. [10]
	Bacillus sp. B2	Selenium	20–50	• Crystalline properties • Antibacterial properties • Cytotoxic properties	• Therapeutic application	Bharathi et al. [4]
	Lactobacillus kimchicus	Gold	5–30	• Free radical scavenging • Non-toxic to human and murine cells	• Drug delivery • Cancer diagnostics • Photothermal therapy • Biosensing • Medical imaging	Markus et al. [25]
	Staphylococcus epidermidis	Gold	20–25	• Catalytic activity for methylene blue degradation	• Catalysts in pollutant degradation	Srinath and Ravishankar Rai [50]

(continued)

Table 1 (continued)

Microorganism	Strains	Microbial-derived nanoparticles			Applications	References
		Metal/metal oxide	Particle size (nm)	Bioactive properties		
	Staphylococcus aureus	Zinc oxide	10–50	• Antibacterial against gram-positive and gram-negative bacteria • Inhibition of MRSA and MRSE	• Treatment of bacterial infections • Antibacterial coatings	Rauf et al. [42]
Fungi	*Aspergillus niger* BSC-1	Iron oxide	20–40	• Superparamagnetic High Cr(VI) removal from aqueous solutions	• Heavy metal remediation • Wastewater treatment	Chatterjee et al. [6]
	Cladosporium cladosporioides	Silver	30–60	• Antioxidant properties • Antimicrobial properties	• Biomedical applications	Manjunath Hulikere and Joshi [24]
	Fusarium oxysporum	Silver	1–50	• Strong antibacterial activity against *Pseudomonas aeruginosa*	• Prevention of chronic infections	Srivastava et al. [51]
	Aspergillus nidulans	Cobalt oxide	20.29	• High specific capacitance	• Energy storage applications	Vijayanandan and Balakrishnan [60]

(continued)

Table 1 (continued)

Microorganism	Strains	Microbial-derived nanoparticles			Applications	References
		Metal/metal oxide	Particle size (nm)	Bioactive properties		
	Aspergillus niger strain STA9	Copper	500	• Antibacterial, Anticancer (Huh-7 cell lines), Antidiabetic	• Drug delivery • Antimicrobial agents • Anticancer agents • Antidiabetic agents	Noor et al. [32]
	Mariannaea sp. HJ	Selenium	45.19 (intracellular) 212.65 (extracellular)	• Antioxidant properties	• Nanoparticle stabilization • Biomedical applications	Zhang et al. [64]
	Cladosporium cladosporioides	Gold	< 100	• Antioxidant properties • Antimicrobial properties	• Pharmaceutical industries • Cosmetic industries • Antioxidant agents • Antimicrobial agents	Joshi et al. [16]
Yeast	*Saccharomyces cerevisiae*	Silver	13.8	• Antimicrobial properties • Long-term dispersion without precipitation	• Industrial disinfectants • Overcoming antibiotic resistance	Shu et al. [46]
	Saccharomyces cerevisiae	Selenium	4–51	• Antimicrobial activity	• Food preservation • Medical applications	Salem [43]

(continued)

Table 1 (continued)

Microorganism	Strains	Microbial-derived nanoparticles			Applications	References
		Metal/metal oxide	Particle size (nm)	Bioactive properties		
	Magnusiomyces ingens LH-F1	Selenium	70–90	• Selective antibacterial activity against gram-positive bacteria	• Targeted antimicrobial applications	Lian et al. [21]
	Novel strain *HX-YS, LPP-12Y* isolated from *Crataegus pinnatifida* and *Vitis vinifera*	Silver	50–500	• Antibacterial and antiproliferative activity against lung cancer cell lines	• Bacterial infection treatment • Cancer treatment	Liu et al. [23]
Microalgae	*Galdieria* sp.	Silver, Iron (II), Zinc	20–80	• Antimicrobial activity	• Antimicrobial therapies • Diagnostic applications	Çalışkan et al. [5]
	Botryococcus braunii	Silver	40–90	• High stability • Catalytic activity	• Pharmaceutical applications (anti-cancer, anti-inflammatory) Catalysis	Sathishkumar et al. [44]

(continued)

Table 1 (continued)

Microorganism	Strains	Microbial-derived nanoparticles			Applications	References
		Metal/metal oxide	Particle size (nm)	Bioactive properties		
	Trichodesmium erythraeum	Silver	26.5	• Antioxidant properties • Antibacterial properties • Anticancer properties	• Antioxidant therapies • Antibacterial therapies Anticancer therapies	Vanlalveni et al. [59]
	Dunaliella salina	Gold	50–60	• Antimicrobial properties against gram-positive bacteria	• Medical and pharmaceutical applications	Basiratnia et al. [3]
	Dunaliella salina	Gold	22.4	• Selective anticancer activity (MCF7) • Non-toxic to normal cells (MCF 10A)	• Anticancer agent	Singh et al. [47]
	Spirulina platensis	Silver	2.23–14.68	• Anticancer activity (A549, HCT, Hep2) • Cytotoxic to cancer cells, safe for normal cells (WISH)	• Biomedical applications • Anticancer agent	Doman et al. [11]
		Copper oxide	2.4			
	Spirulina platensis	Silver	6	• Antibacterial potency against isolates from HIV patients	• Biomedical applications	Suganya et al. [52]
	Neodesmus pupukensis	Silver	52–179	• Antimicrobial properties • Free radical scavenging properties	• Antimicrobial and antioxidant agents • Biotechnology applications	Omomowo et al. [33]
		Gold	5–34			

calcination and 49–153 nm after calcination at 200 °C. The crude AgNPs demonstrated significant antibacterial properties against pathogenic bacteria, including *Pseudomonas fluorescens* and *Staphylococcus aureus*, while the calcined AgNPs exhibited no antibacterial activity, suggesting that crude AgNPs are beneficial in food packaging and biomedical sectors [27]. Furthermore, Patil et al. [38] investigated the extracellular synthesis of Au NPs (AuNPs) in the marine bacterium *Paracoccus haeundaensis* BC74171T, which showcased 20.93 ± 3.46 nm NPs with an ultraviolet–visible (UV–Vis) spectroscopy peak at 535 nm. These AuNPs exhibited concentration-dependent antioxidant activity and selectively inhibited the growth of cancer cells, including A549 and AGS, while sparing normal human keratinocyte and kidney cells [38]. Thus, the study highlighted the potential of AuNPs for non-toxic drug delivery and nanomedicine applications. In addition, Dhandapani et al. [10] demonstrated the production of hairy-shaped zinc oxide (ZnO) nanostructures with notable photocatalytic properties under sunlight via a ureolytic *Bacillus subtilis*-mediated synthesis method. Among the ZnO NPs, 10–15 nm nanostructures with a spherical framework of 1.2–1.3 μm showcased a high surface area and effective hydroxyl radical (OH) generation. Moreover, these hairy ZnO nanostructures efficiently degrade various dyes, including methylene blue, methyl orange, and Rhodamine B, emphasizing their potential for sunlight-driven photocatalytic applications [10]. Furthermore, Bharathi et al. [4] investigated the potential of freshwater bacterium *Bacillus* sp. B2 in extracellular synthesis of Se NPs (SeNPs), which produced spherical NPs with crystalline properties and an average diameter of 20–50 nm. Besides, surface plasmon resonance was indicated by the UV–Vis analysis exhibiting a peak at 280 nm, and the Fourier-transform infrared (FTIR) spectroscopy analysis revealed that proteins from the bacterial supernatant were crucial for NP reduction and stabilization. Remarkably, the SeNPs demonstrated significant antibacterial activity against *P. aeruginosa* and notable cytotoxicity against A549 human lung cancer cells, suggesting their potential for therapeutic applications.

2.2 Fungi

NP synthesis using fungi and yeasts has gained significant attention, mainly due to scalability, efficient downstream processing, economic viability, and ecological benefits, in comparison to bacterial methods [29]. Moreover, fungi offer ease of biomass handling, high wall-binding capacity, and the ability to transform metal ions into NPs via intracellular and extracellular enzymes [20]. For instance, Chatterjee et al. [6] reported that the manglicolous fungus *Aspergillus niger* BSC-1 synthesized superparamagnetic Fe oxide NPs (IONPs) of 20–40 nm. Notably, these IONPs effectively removed hexavalent chromium (Cr(VI)) from aqueous solutions, achieving over 99% removal at 40 °C and pH 3 with a 2.5 g/L dose, and retained 79.7% of their removal capacity after 5 regeneration cycles, highlighting their effectiveness in heavy metal remediation and wastewater treatment [6]. Furthermore, Manjunath Hulikere and Joshi [24] utilized the marine endophytic fungus *Cladosporium cladosporioides* for the biosynthesis of AgNPs (30–60 nm), with UV–Vis spectroscopy confirming successful formation through a strong absorption peak at 440 nm. The AgNPs showed notable antioxidant activity, with radical scavenging ability comparable to ascorbic acid, and exhibited strong antimicrobial activity against bacterial strains, including *E. coli, S. aureus, B. subtilis, S. epidermis*, and yeast pathogen *Candida albicans*. Similarly, AgNPs synthesized by *Fusarium oxysporum* exhibited particle size of 1–50 nm, which were predominantly spherical, and demonstrated strong binding to proteins in the outer membrane of *E. coli* and *P. aeruginosa*, with greater antibacterial potency against *P. aeruginosa*. Moreover, the in-vitro validation confirmed effective inhibition of bacterial growth, with minimum inhibitory concentrations of 20 μg/mL and 10 μg/mL for *E. coli* and *P. aeruginosa*, respectively. Additionally, the AgNPs significantly reduced bacterial biofilm formation by up to 75%, highlighting their potential in preventing chronic infections [51]. Furthermore, cobalt oxide NPs (Co_3O_4-NPs) synthesis using *Aspergillus nidulans*, isolated from *Nothapodytes foetida*, resulted in spinel Co_3O_4-NPs of 20–29 nm and spherical [60]. Significantly, the NPs demonstrated 389 F/g specific capacitance, which indicated the potential for energy storage applications.

In addition, Shu et al. [46] investigated the synthesis of antimicrobial AgNPs with *S. cerevisiae* which yielded uniformly spherical NPs averaging 13.8 nm in size. Remarkably, the key biomolecules in the yeast extract, including amino acids and carbohydrates, played essential roles in NP formation and stabilization, ensuring their long-term dispersion without precipitation. Moreover, the AgNPs displayed significant antibacterial activity against *E. coli*, achieving complete growth inhibition at concentrations greater than 20.0 μg/mL. Also, ampicillin resistance was effectively oppressed by the AgNPs which highlighted their potency in industrial applications, including disinfecting multidrug-resistant bacterial strains and medical products [46]. Similarly, Salem [43] synthesized selenium NPs (SeNPs) in *S. cerevisiae*, which were 4–51 nm (diameter) spherical particles. Furthermore, the biosynthesized SeNPs using baker's yeast extract demonstrated substantial antimicrobial activity against food-borne pathogens, including *E. coli* and *S. aureus*, with minimal inhibitory concentrations of 62.5 and 125 μg/mL, respectively, suggesting their promising potential as an efficacious antimicrobial agent for food preservation and medical applications [43].

In addition, Lian et al. [21] developed a green, one-pot synthesis method for biogenic SeNPs from *Magnusiomyces ingens* LH-F1 using cell-free extracts, which were predominantly spherical or quasi-spherical with 70–90 nm diameter and an average size of 87.82 ± 2.71 nm. Moreover, the SeNPs were composed of high-purity hexagonal crystalline Se, which showcased selective antibacterial activity against gram-positive *Arthrobacter* sp. W1. However, the SeNPs were ineffective against gram-negative *E. coli* BL21, which indicated promising targeted antimicrobial applications [21]. Furthermore, Liu et al. [23] explored the biosynthesis of AgNPs with novel yeast strains HX-YS and LPP-12Y, isolated from *Crataegus pinnatifida* and *Vitis vinifera*. The study reported the production of 50–500 nm AgNPs, which exhibited significant antibacterial effects, with the largest inhibition zones reaching 21.44 mm against *Pseudomonas aeruginosa*. Additionally, the

AgNPs demonstrated strong antiproliferative effects against lung cancer cell lines *A549* and *H1975*, reducing cell viability below 20% while maintaining low toxicity in normal cells, underscoring their potential in bacterial infection and cancer treatment [23].

2.3 Microalgae

During the past decades, microalgae have been emerging as ideal candidates for NPs synthesis, mainly due to their rapid growth and ease of handling, while inheriting the ability to hyper-accumulate metals and conversion into NPs [54, 58]. Moreover, the intracellular or extracellular synthesis of NPs is species-specific and highly dependent on the microalgal cultivation conditions [30]. For instance, Çalışkan et al. [5] reported that the microalga *Galdieria* sp. are capable of synthesizing extracellular Ag, Fe (II), and zinc NPs with 20–80 nm diameters, with immense antimicrobial activities, against *E. coli* and *S. aureus*. Moreover, this study highlighted the potential of microalgae-derived NPs for antimicrobial therapies and diagnostic applications, emphasizing environmentally friendly, cost-effective production as a sustainable alternative to conventional chemical synthesis methods [5]. Similarly, *Botryococcus braunii*, a green alga, synthesized 40–90 nm particle size AgNPs, by combining the algal extract with a 1 mM silver nitrate solution [44]. Moreover, a yield of 70–85% was achieved from the one-pot synthesis of 2-aryl benzimidazoles, upon the utilization of the biogenic AgNPs as catalysts, which exhibited high stability and reusability and 90% retention of the catalytic activity more than 10 cycles. Notably, the bioproduction of benzimidazoles would be ideal in pharmaceutical applications, including anti-cancer and anti-inflammatory treatments.

Furthermore, Vanlalveni et al. [59] employed *Trichodesmium erythraeum* for AgNP synthesis, which resulted in crystalline, cubic, 26.5 nm average size particles. This study highlighted that AgNPs showcased notable antioxidant properties with radical scavenging activities of 77.01%, 67.5%, 52.77%, and 88.12% with 1-diphenyl-2-picrylhydrazyl, deoxyribose, 2,2-azino-bis-3-ethylbenzothiazoline-6-sulphonic acid, and nitric oxide radical assays, respectively. Additionally, the Ag-NPs demonstrated strong antibacterial effects against clinical isolates, namely *S. aureus* and *Proteus mirabilis*, along with drug-resistant strains, including *E. coli*, *S. aureus*, and *S. pneumoniae*. Notably, the antiproliferative effect was significant at a concentration of 50 mg, reducing cancer cell viability and showing potential for antioxidant, antibacterial, and anticancer therapies [59]. Similarly, AuNPs synthesized by the microalga *Dunaliella salina* acted as both reducing and stabilizing agents, which exhibited significant antibacterial activity against gram-positive bacteria, including *S. aureus* and *B. subtilis*, at 0.05 mg/mL. Remarkably, the enhanced antimicrobial effect is attributed to the high ratios of surface area to volume of NPs, which improves their interaction with bacterial cell walls and facilitates cellular entry, leading to bacterial cell disruption and death. Moreover, this study underscored the potential of algae-synthesized AuNPs for medical and pharmaceutical applications due to their effective antimicrobial properties and eco-friendly synthesis process [3].

In addition, microalgae cultivation integrated with waste resources and the inherent resource recovery potential is highly beneficial in establishing sustainable and low-carbon production processes of NPs [55, 57]. For instance, Chokshi et al. [8] demonstrated AgNP production by using lipid-extracted residual biomass of the thermotolerant microalga *Acutodesmus dimorphus* cultivated in dairy wastewater, which was 2–20 nm sized spherical particles. Notably, NPs synthesis exhibited antioxidant potential, highlighting the potential of phyconanotechnology for eco-friendly NP production [8].

3 Mechanisms Involved in the Microbial Biosynthesis of Nanoparticles

In microbial biosynthesis of NPs, extracellular and intracellular metabolites reduce and stabilize the metal ions into nanoscale particles, which involves several mechanisms depending on the type of microorganisms. Biomolecules including enzymes, proteins, amino acids, and carbohydrates are involved in microbial synthesis of NPs. Notably, many studies have reported that the microbial biosynthesis of NPs generally depends on 3 mechanisms involving reductase enzymes and proteins, electron shuttle quinones, and exopolysaccharides (EPS). Various enzymes, including reductases, acetyl xylan esterase, cellobiohydrolase, and glucosidase, play key roles in the microbial production of metal NPs [35]. These enzymes contribute not only to the formation of NPs but also function as stabilizing and capping agents, presenting a greener alternative to conventional chemical methods [12]. Furthermore, the defence mechanisms, namely reduction of metal ions, formation of complex compounds, precipitation, and dissimilatory oxidation, adopted by the microbes to react against metal toxicity play key roles in the NP synthesis, which occurs intracellularly and extracellularly.

Bacteria are potential cellular factories in the synthesis of metal NPs including Ag, Au, and Fe, and metal oxide NPs. Many bacteria can synthesize AgNPs via bioreduction, often as part of their defense mechanisms, as Ag^+ can cause genotoxic effects by disrupting DNA structure and function. Moreover, Ag^+ has a high affinity for thiol-containing proteins, which significantly inhibits protein functions. Notably, bacteria mainly involve specific enzyme groups, such as reductases, to reduce Ag^+ to Ag-NPs [15]. Talekar et al. [53] reported the possibility of synthesizing AgNPs using nicotinamide adenine dinucleotide (NADH)-dependent nitrate reductase cross-linked enzyme aggregates, where Ag ions were reduced to Ag at pH 7.2 using NADH as the electron source and 8-hydroxyquinoline as the electron shuttle. Furthermore, it was reported that the 8-hydroxyquinoline used in the fabrication acts as a capping agent for AgNPs (spherical; size 5–7 nm). Figure 3 schematically represents the

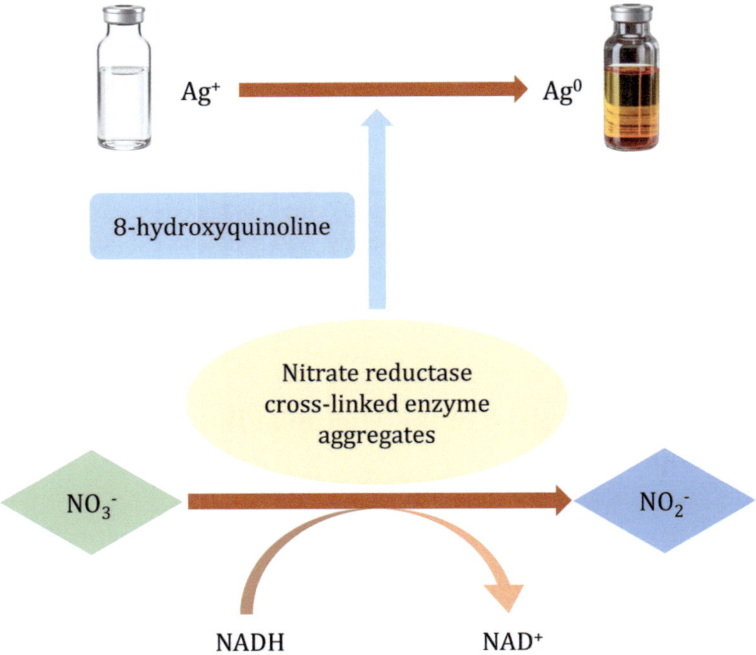

Fig. 3 A schematic representation of the mechanism for silver nanoparticle synthesis using nitrate reductase cross-linked enzyme aggregates

mechanisms involved in AgNPs synthesis via NADH-dependent nitrate reductase enzyme.

The extracellular production of NPs by fungi typically involves various biomolecules, including enzymes, proteins, polysaccharides, flavonoids, alkaloids, phenolic compounds, and organic acids (Fig. 4) [40]. Mousa et al. [28] described a two-step process employed by the endophytic fungus *Aspergillus terreus* for synthesizing Co_3O_4, CuO, Fe_3O_4, NiO, and ZnO NPs. This process involves an initial bioreduction stage followed by a capping phase [28].

Accordingly, the authors have highlighted that the cell-free extract of *A. terreus* ORG-1 contains bioactive metabolites, such as phenols, amines, and carboxyl groups function as reducing agents in the process. The FTIR analysis confirmed the formation of new absorption bands unique to the synthesized NPs which may be due to NP conjugation. These biomolecules act as reducing agents, transforming metal ions into NP form, with specific bands corresponding to each type of NP synthesized. Following reduction, the NPs are stabilized through capping by organic compounds present in the filtrate. These bioactive compounds coat the NPs, preventing aggregation and ensuring stability.

Wanarska and Maliszewska [61] studied AgNP synthesis via *Penicillium cyclopium*, using both fungal cell mass and cell-free extract, in which AgNPs synthesis occurred on the fungal cell wall where Ag^+ ions were trapped via

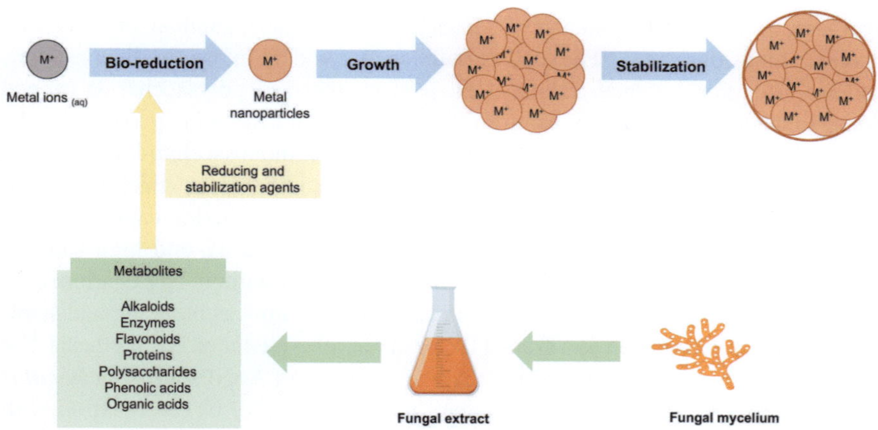

Fig. 4 Fungal-based synthesis of metallic nanoparticles and its stabilization

biosorption was hypothesised. In addition, the authors highlighted that NP synthesis occurs via non-enzyme-mediated pathways in cell-free extracts, and both synthesis methods involve saccharides and proteins in the biofabrication of AgNPs [61]. Moreover, Pomastowski et al. [39] highlighted the extracellular generation of ZnO nanocomposites utilizing *Lactobacillus paracasei* LC20, emphasizing the role of specific peptide fragments in facilitating the biosynthesis process of these nanocomposites. Notably, the functional groups associated with amino acids, such as phenylalanine, leucine, cysteine, glycine, glutamic acid, and aspartic acid, showcased interactions with Zn^{2+} ions, binding through negatively charged and polar residues, which contributed to the formation of ZnO.

Dağlıoğlu and Yılmaz Öztürk [9] reported the intracellular synthesis of AgNPs (spherical; sized between 15 and 30 nm) using *Desmodesmus* sp., where the synthesized NPs accumulated within the algal cytoplasm rather than on the cell membrane, indicating a metabolism-dependent mechanism. The shifts in chlorophyll b peaks indicated the involvement of cellular pigments, such as chlorophylls a and b, to reduce Ag ions during this metabolism-dependent mechanism. In addition, Jena et al. [14] studied AgNP synthesis by *Scenedesmus* sp. and highlighted that both intracellular and extracellular mechanisms were involved in the synthesis. For intracellular synthesis, *Scenedesmus* sp. cells accumulate Ag^+ ions in their biomass through biosorption, with subsequent reduction facilitated by negatively charged ligands on the cell surface, where enzymes and proteins catalyze the transformation [14]. In the extracellular synthesis using raw and boiled extracts of *Scenedesmus* sp., peptides and proteins present in excess amounts played a significant role in the bioreduction process while simultaneously functioning as capping and stabilizing agents.

Furthermore, EPS secreted by microorganisms, including bacteria, fungi and algae, capable of serving as natural reducing, capping, and stabilizing agents for NP synthesis. Bacterial EPS, such as alginates, levans, dextrans, xanthan, and succinoglycan, alongside fungal EPS, such as pullulan, are widely in use for diverse applications [45]. For instance, Jakhu et al. [13] synthesized polysaccharide-capped AuNPs (spherical; 30–40 nm) using EPS extracted from *Chlorella* sp. Accordingly, authors have highlighted that the EPS contained a mix of monosaccharides, including hexoses, pentoses, deoxy sugars, amino sugars, and acidic sugars, which facilitated this synthesis. Additionally, it was reported that the polysaccharide-capped AuNPs exhibited excellent stability across a broad pH range of 2–12 and high salinity, tolerating ionic concentrations up to 750 mg/mL of NaCl without aggregation.

Furthermore, the use of enzymes and chemical mediators as reducing and stabilizing agents is recognized as a crucial factor in microbe-mediated NP synthesis. For example, Singh et al. [49] mentioned that the synthesis of AuNPs utilizing *Bacillus licheniformis* cell lysate, which contained nitrate reductase, in combination with Tween 20 and dodecanethiol had produced monodispersed AuNPs (10.4 $\pm$ 0.6 nm and 12.5 $\pm$ 0.9 nm), organized in an ordered self-assembly, with nitrate reductase playing a key role in the fabrication. It was proposed that nitrate ions induce cellular enzymes such as NADH- and NADPH-dependent nitrate reductase to shuttle free electrons from NADH to Au^{3+}, thereby reducing gold ions to the Au^0 state. In addition to nitrate reductase, redox mediators, such as quinones and hydroquinones, act as electron shuttles, facilitating electron transfer from NADPH or NADH to metal ions and thereby enhancing the reduction process. Moreover, Liu et al. [22] explored the impact of quinones on graphene oxide (GO) reduction mediated by *Shewanella oneidensis* MR-1. It was highlighted that anthraquinone-2-sulfonate, anthraquinone-2,6-disulfonate, and 5-hydroxy-1,4-naphthoquinone significantly enhanced the GO reduction rate, confirmed by direct GO reduction mediated by microbially reduced anthraquinone-2-sulfonate [22]. In addition, Pat-Espadas et al. [37] stated that bioreduction of Pd(II) to Pd(0) NPs facilitated by *Geobacter sulfurreducens,* utilizingacetate as the electron donor. It was observed that anthraquinone-2,6-disulfonate significantly increased the Pd reaction rate and proposed that NP synthesis involves both direct and quinone-mediated approaches. The direct mechanism includes Pd bioreduction by outer membrane c-type cytochromes, along with amide, sulfur, and phosphoryl groups associated with bacterial proteins [37].

In addition, large-scale biosynthesis of NPs requires biotechnological approaches, especially for the design and optimization of cultivation systems namely fermenters and other bioreactors and the growth parameters, despite the limited availability of literature studies. Wang et al. [62] highlighted that Se NPs fabricated using *Bacillus licheniformis* F1 (spherical, 110–170 nm) showed that key factors required for large-scale production were seed liquid inoculum (2%), temperature (37 °C), and agitation speed (180 rpm). Moreover, substrate selection is also a critical factor during the optimization process, where corn powder (11.18 g/L), soybean meal (10.34 g/L), and NaCl (10.68 g/L) are required at optimized levels. This enables the synthesis of nearly 100.00% Se NPs from 5 mmol/L sodium selenite. Subsequently, pilot-scale verification in a 5 L fermenter using the optimized medium resulted in a shortened fermentation time, significantly reducing production costs [62]. Furthermore, scale-up studies indicate that integrating continuous bioprocessing systems with inline monitoring, such as spectroscopic methods, enhances control over NP size and stability.

4 Challenges and Potential Solutions for Scaled-up Production of Microbial Nanoparticles

Numerous challenges are faced during the scaling up of NP production using microorganisms, due to the research and development at lab-scale. Microbial species selection, optimization of process parameters during upstream processing for monodispersed NP synthesis, and selection of energy-efficient and cost-effective downstream processing, among others, are essential in the large-scale production of biogenic NPs. Hence, this section will discuss the major challenges associated with the scaled-up production of biogenic NPs and potential solutions to enhance the feasibility and sustainability.

The foremost hurdle to overcome during the scaled-up NP production is the selection of microbial species suitable for the production requirements, such as morphology, physical properties, and bioactive properties of NPs. Notably, the utilization of locally isolated species is highly preferred due to the ease of controlling environmental conditions and the conservation of biodiversity. Besides, microorganisms with lower residence time, i.e., fast growth of NPs, are highly recommended for scaled-up production to enhance the economic feasibility of the overall process. In addition, the major challenge during the scaled-up production of NPs is maintaining growth conditions, including pH, nutrient availability, temperature, oxygen levels, and light intensities, within optimal ranges, which is a key factor for biological systems. As discussed in Sect. 2, a diversified spectrum of microorganisms is employed to produce various NPs at different culture conditions and properties, which are mainly based on lab-scale production. The culture conditions are species-specific, and thus, optimization is of paramount importance.

For instance, pH is a critical cultivation parameter during NP synthesis by microorganisms, which has exhibited drastic variations in the size and morphology of the synthesized NPs. Notably, pH variations alter the overall charges of bioactive molecules and affect the binding affinity during microbial NP biosynthesis. Chitra and Annadurai [7] reported that high AgNP concentrations were obtained at pH 9 than pH 5 during *Bacillus brevis* cultivation which were spherical and 10–40 nm in size, based on the color intensities of the aqueous solution. Moreover, the study concluded that the particle size was high at acidic pH due to the low rate of nucleation process for Ag nanocrystals, which causes a reduction in NP formation. Meanwhile, the nucleation process is fast at high pH ranges due to the availability of hydroxyl ions, resulting in high quantities of small-size NPs. Similarly, temperature is another crucial parameter to be controlled during the microbial synthesis of NPs, due to the impacts on the biosynthetic pathways of microorganisms [18]. In general, high temperatures are employed due to high reaction rates and elevated process kinetics during microbial biosynthesis of NPs. For example, Ramanathan et al. [41] reported that *Morganella psychrotolerans* synthesized spherical AgNPs at 20 °C while a mixture of triangular and hexagonal nanoplates in spherical AgNPs was obtained at 25 °C. Therefore, temperature control is essential in the green synthesis of NPs to achieve desired particle properties, especially at scaled-up production.

Furthermore, microbes, especially microalgae, requiring light for the biosynthesis of NPs require optimal light irradiances for efficient production [56]. In addition, agitation and mixing of culture are essential in microbial factories to ensure homogeneity of cell suspension and nutrients, alongside reagents to promote nucleation and growth of NPs. Nevertheless, excess agitation could result in aggregation and coarsening of NPs, and hence, optimization of mixing speed is important during the scaled-up production. On the other hand, one of the important hurdles to overcome with biological systems is the risk of contamination by other microorganisms and impurities, which has negative impacts on the yield and product quality in NP production. Therefore, aseptic techniques have to be integrated with the scale-up NP production systems, along with regular monitoring, to prevent a reduction in yields and quality compromises.

In this context, the aforementioned factors, including controlling reaction parameters, culture conditions, prone to contamination and mixing rate, must be taken into consideration for the designing and fabrication of bioreactors employed in the scaled-up NP production. Remarkably, automated bioreactors aided with control systems to maintain parameters within the optimized ranges are beneficial in the real-time monitoring of NP production. Moreover, modes of production, i.e., batch, continuous and semi-continuous, need to be implemented during the scaled-up production of microbial NPs. Notably, pilot-scale studies of microbial synthesis of NPs are essential to identify critical parameters and drawbacks in scaling up, prior to the scaled-up production.

In addition, downstream processing of scaled-up production of microbial NPs demonstrates several challenges including the extraction and purification of NPs to meet the required product quality standards, especially applicable to pharmaceutical and medical sectors. Commonly employed extraction techniques, including precipitation methods, centrifugation, and filtration are associated with excess requirements of chemicals and materials for scale-up microbial NP productions, which would increase the overall production costs and energy requirements. Similarly, NP purification methods, such as chromatography and magnetic separation, are energy-intensive processes and generally have been employed in lab-scale green synthesis. Therefore, the downstream processing of microbial NP production must be streamlined to optimize the energy requirements and capital and operational costs. Moreover, comprehensive studies based on techno-economic assessment and life cycle analysis are essential in identifying critical processes and overcoming scaling up hurdles associated with microbial NP production. Furthermore, continued research and innovation of the applicability and employability of microbial NPs in biomedical applications, environmental bioremediation, and agriculture and food safety would be beneficial in the identification of novel applications and advancements in sustainable materials and technologies. In particular, toxicology, cellular, and physiological factors influencing NP-mediated drug delivery and other biomedical applications, alongside the mechanisms associated with permeability and retention of particles and pharmacokinetic profiles in humans must be further studied in detail.

5 Conclusion

Microbial cell factories, namely bacteria, fungi and microalgae offer a greener alternative to conventional methods for the NPs production. Notably, microbial-based processes are environmentally sustainable, non-toxic, and cost-effective in the production of various metals and metal oxide NPs. Moreover, the fast growth and ease of cultures are beneficial in the safer and rapid production of microbial-derived NPs with desirable particle properties, which are highly dependent on cultivation parameters such as temperature and pH. This biogenic approach has been widely studied for novel applications of microbial NPs in biomedical, environmental bioremediation, and agricultural and food-related fields. However, several drawbacks are associated with the homogeneity of NP properties namely shape, symmetry, particle size and compositions, and scaling-up of the microbial process for commercial production. Therefore, further investigations and process optimization are of paramount importance for commercial prospects of microbial-derived NPs.

References

1. Ajith S, Almomani F, Elhissi A, Husseini GA (2023) Nanoparticle-based materials in anticancer drug delivery: current and future prospects. Heliyon 9:e21227. https://doi.org/10.1016/j.heliyon.2023.e21227
2. Alsaiari NS, Alzahrani FM, Amari A, Osman H, Harharah HN, Elboughdiri N, Tahoon MA (2023) Plant and microbial approaches as green methods for the synthesis of nanomaterials: synthesis, applications, and future perspectives. Molecules 28:463. https://doi.org/10.3390/molecules28010463
3. Basiratnia E, Einali A, Azizian-Shermeh O, Mollashahi E, Ghasemi A (2021) Biological synthesis of gold nanoparticles from suspensions of green microalga *Dunaliella salina* and their antibacterial potential. Bionanoscience 11:977–988. https://doi.org/10.1007/S12668-021-00897-4
4. Bharathi S, Kumaran S, Suresh G, Ramesh M, Thangamani V, Pugazhvendan SR, Sathiyamurthy K (2020) Extracellular synthesis of nanoselenium from fresh water bacteria *Bacillus* sp., and its validation of antibacterial and cytotoxic potential. Biocatal Agric Biotechnol 27:101655. https://doi.org/10.1016/J.BCAB.2020.101655
5. Çalışkan G, Mutaf T, Öncel SŞ, Elibol M (2020) Green synthesis of metal nanoparticles using microalga *Galdieria* sp. IFMBE Proc 73:219–224. https://doi.org/10.1007/978-3-030-17971-7_34
6. Chatterjee S, Mahanty S, Das P, Chaudhuri P, Das S (2020) Biofabrication of iron oxide nanoparticles using manglicolous fungus *Aspergillus niger* BSC-1 and removal of Cr(VI) from aqueous solution. Chem Eng J 385:123790. https://doi.org/10.1016/J.CEJ.2019.123790

7. Chitra K, Annadurai G (2014) Antibacterial activity of pH-dependent biosynthesized silver nanoparticles against clinical pathogen. Biomed Res Int 2014:1–6. https://doi.org/10.1155/2014/725165
8. Chokshi K, Pancha I, Ghosh T, Paliwal C, Maurya R, Ghosh A, Mishra S (2016) Green synthesis, characterization and antioxidant potential of silver nanoparticles biosynthesized from de-oiled biomass of thermotolerant oleaginous microalgae *Acutodesmus dimorphus*. RSC Adv 6:72269–72274. https://doi.org/10.1039/C6RA15322D
9. Dağlıoğlu Y, Yılmaz Öztürk B (2019) A novel intracellular synthesis of silver nanoparticles using *Desmodesmus* sp. (Scenedesmaceae): different methods of pigment change. Rend Lincei Sci Fis Nat 30:611–621. https://doi.org/10.1007/s12210-019-00822-8
10. Dhandapani P, Prakash AA, AlSalhi MS, Maruthamuthu S, Devanesan S, Rajasekar A (2020) Ureolytic bacteria mediated synthesis of hairy ZnO nanostructure as photocatalyst for decolorization of dyes. Mater Chem Phys 243. https://doi.org/10.1016/J.MATCHEMPHYS.2020.122619
11. Doman KM, Gharieb MM, Abd El-Monem AM, Morsi HH (2024) Synthesis of silver and copper nanoparticle using Spirulina platensis and evaluation of their anticancer activity. Int J Environ Health Res 34:661–673. https://doi.org/10.1080/09603123.2022.2163987
12. Herrera Pérez GM, Castellano LE, Ramírez Valdespino CA (2024) Trichoderma and mycosynthesis of metal nanoparticles: role of their secondary metabolites. J Fungi 10:443. https://doi.org/10.3390/jof10070443
13. Jakhu S, Sharma Y, Sharma K, Vaid K, Dhar H, Kumar V, Singh RP, Shekh A, Kumar G (2021) Production and characterization of microalgal exopolysaccharide as a reducing and stabilizing agent for green synthesis of gold-nanoparticle: a case study with a *Chlorella* sp. from Himalayan high-altitude psychrophilic habitat. J Appl Phycol 33:3899–3914. https://doi.org/10.1007/s10811-021-02580-3
14. Jena J, Pradhan N, Nayak RR, Dash BP, Sukla LB, Panda PK, Mishra BK (2014) Microalga *Scenedesmus* sp.: a potential low-cost green machine for silver nanoparticle synthesis. J Microbiol Biotechnol 24:522–533. https://doi.org/10.4014/jmb.1306.06014
15. John Wilson J, Harimuralikrishnaa T, Ponmanickam P, Ponseetha Lakshmi M (2023) Bacterial silver nanoparticles: method, mechanism of synthesis and application in mosquito control. In: Mosquito research—recent advances in pathogen interactions, immunity, and vector control strategies. IntechOpen
16. Joshi CG, Danagoudar A, Poyya J, Kudva AK, Dhananjaya BL (2017) Biogenic synthesis of gold nanoparticles by marine endophytic fungus—*Cladosporium cladosporioides* isolated from seaweed and evaluation of their antioxidant and antimicrobial properties. Process Biochem 63:137–144. https://doi.org/10.1016/J.PROCBIO.2017.09.008
17. Karunakaran G, Sudha KG, Ali S, Cho EB (2023) Biosynthesis of nanoparticles from various biological sources and its biomedical applications. Molecules 28. https://doi.org/10.3390/MOLECULES28114527
18. Koul B, Poonia AK, Yadav D, Jin J-O (2021) Microbe-mediated biosynthesis of nanoparticles: applications and future prospects. Biomolecules 11:886. https://doi.org/10.3390/biom11060886
19. Kulkarni D, Sherkar R, Shirsathe C, Sonwane R, Varpe N, Shelke S, More MP, Pardeshi SR, Dhaneshwar G, Junnuthula V, Dyawanapelly S (2023) Biofabrication of nanoparticles: sources, synthesis, and biomedical applications. Front Bioeng Biotechnol 11. https://doi.org/10.3389/fbioe.2023.1159193
20. Li Q, Liu F, Li M, Chen C, Gadd GM (2022) Nanoparticle and nanomineral production by fungi. Fungal Biol Rev 41:31–44. https://doi.org/10.1016/J.FBR.2021.07.003
21. Lian S, Diko CS, Yan Y, Li Z, Zhang H, Ma Q, Qu Y (2019) Characterization of biogenic selenium nanoparticles derived from cell-free extracts of a novel yeast *Magnusiomyces ingens*. 3 Biotech 9:221. https://doi.org/10.1007/S13205-019-1748-Y
22. Liu G, Zhang X, Zhou J, Wang A, Wang J, Jin R, Lv H (2013) Quinone-mediated microbial synthesis of reduced graphene oxide with peroxidase-like activity. Bioresour Technol 149:503–508. https://doi.org/10.1016/j.biortech.2013.09.115

23. Liu X, Chen J Le, Yang WY, Qian YC, Pan JY, Zhu CN, Liu L, Ou W Bin, Zhao HX, Zhang DP (2021) Biosynthesis of silver nanoparticles with antimicrobial and anticancer properties using two novel yeasts. Sci Rep 11. https://doi.org/10.1038/S41598-021-95262-6
24. Manjunath Hulikere M, Joshi CG (2019) Characterization, antioxidant and antimicrobial activity of silver nanoparticles synthesized using marine endophytic fungus—*Cladosporium cladosporioides*. Process Biochem 82:199–204. https://doi.org/10.1016/J.PROCBIO.2019.04.011
25. Markus J, Mathiyalagan R, Kim YJ, Abbai R, Singh P, Ahn S, Perez ZEJ, Hurh J, Yang DC (2016) Intracellular synthesis of gold nanoparticles with antioxidant activity by probiotic *Lactobacillus kimchicus* DCY51T isolated from Korean kimchi. Enzyme Microb Technol 95:85–93. https://doi.org/10.1016/J.ENZMICTEC.2016.08.018
26. Materón EM, Miyazaki CM, Carr O, Joshi N, Picciani PHS, Dalmaschio CJ, Davis F, Shimizu FM (2021) Magnetic nanoparticles in biomedical applications: a review. Appl Surf Sci Adv 6:100163. https://doi.org/10.1016/j.apsadv.2021.100163
27. Mathivanan K, Selva R, Chandirika JU, Govindarajan RK, Srinivasan R, Annadurai G, Duc PA (2019) Biologically synthesized silver nanoparticles against pathogenic bacteria: synthesis, calcination and characterization. Biocatal Agric Biotechnol 22:101373. https://doi.org/10.1016/J.BCAB.2019.101373
28. Mousa SA, El-Sayed E-SR, Mohamed SS, Abo El-Seoud MA, Elmehlawy AA, Abdou DAM (2021) Novel mycosynthesis of Co_3O_4, CuO, Fe_3O_4, NiO, and ZnO nanoparticles by the endophytic *Aspergillus terreus* and evaluation of their antioxidant and antimicrobial activities. Appl Microbiol Biotechnol 105:741–753. https://doi.org/10.1007/s00253-020-11046-4
29. Mughal B, Zaidi SZJ, Zhang X, Hassan SU (2021) Biogenic nanoparticles: synthesis, characterisation and applications. Appl Sci 11:2598. https://doi.org/10.3390/APP11062598
30. Mukherjee A, Sarkar D, Sasmal S (2021) A review of green synthesis of metal nanoparticles using algae. Front Microbiol 12:693899. https://doi.org/10.3389/FMICB.2021.693899/BIBTEX
31. Nikolova MP, Chavali MS (2020) Metal oxide nanoparticles as biomedical materials. Biomimetics 5:27. https://doi.org/10.3390/biomimetics5020027
32. Noor S, Shah Z, Javed A, Ali A, Hussain SB, Zafar S, Ali H, Muhammad SA (2020) A fungal based synthesis method for copper nanoparticles with the determination of anticancer, antidiabetic and antibacterial activities. J Microbiol Methods 174. https://doi.org/10.1016/J.MIMET.2020.105966
33. Omomowo IO, Adenigba VO, Ogunsona SB, Adeyinka GC, Oluyide OO, Adedayo AA, Fatukasi BA (2020) Antimicrobial and antioxidant activities of algal-mediated silver and gold nanoparticles. IOP Conf Ser Mater Sci Eng 805:012010. https://doi.org/10.1088/1757-899X/805/1/012010
34. Osman AI, Zhang Y, Farghali M, Rashwan AK, Eltaweil AS, Abd El-Monaem EM, Mohamed IMA, Badr MM, Ihara I, Rooney DW, Yap P-S (2024) Synthesis of green nanoparticles for energy, biomedical, environmental, agricultural, and food applications: a review. Environ Chem Lett 22:841–887. https://doi.org/10.1007/s10311-023-01682-3
35. Ovais M, Khalil AT, Ayaz M, Ahmad I, Nethi SK, Mukherjee S (2018) Biosynthesis of metal nanoparticles via microbial enzymes: a mechanistic approach. Int J Mol Sci 19:4100. https://doi.org/10.3390/ijms19124100
36. Pandit C, Roy A, Ghotekar S, Khusro A, Islam MN, Bin ET, Lam SE, Khandaker MU, Bradley DA (2022) Biological agents for synthesis of nanoparticles and their applications. J King Saud Univ Sci 34:101869. https://doi.org/10.1016/j.jksus.2022.101869
37. Pat-Espadas AM, Razo-Flores E, Rangel-Mendez JR, Cervantes FJ (2014) Direct and quinone-mediated palladium reduction by geobacter sulfurreducens: mechanisms and modeling. Environ Sci Technol 48:2910–2919. https://doi.org/10.1021/es403968e
38. Patil MP, Kang Mj, Niyonizigiye I, Singh A, Kim JO, Seo YB, Kim GD (2019) Extracellular synthesis of gold nanoparticles using the marine bacterium *Paracoccus haeundaensis* BC74171T and evaluation of their antioxidant activity and antiproliferative effect on normal and cancer cell lines. Colloids Surf B Biointerfaces 183:110455. https://doi.org/10.1016/J.COLSURFB.2019.110455

39. Pomastowski P, Król-Górniak A, Railean-Plugaru V, Buszewski B (2020) Zinc oxide nanocomposites—extracellular synthesis, physicochemical characterization and antibacterial potential. Materials 13:4347. https://doi.org/10.3390/ma13194347
40. Rai M, Bonde S, Golinska P, Trzcińska-Wencel J, Gade A, Abd-Elsalam KA, Shende S, Gaikwad S, Ingle AP (2021) Fusarium as a novel fungus for the synthesis of nanoparticles: mechanism and applications. J Fungi 7:139. https://doi.org/10.3390/jof7020139
41. Ramanathan R, O'Mullane AP, Parikh RY, Smooker PM, Bhargava SK, Bansal V (2011) Bacterial kinetics-controlled shape-directed biosynthesis of silver nanoplates using *Morganella psychrotolerans*. Langmuir 27:714–719. https://doi.org/10.1021/la1036162
42. Rauf MA, Owais M, Rajpoot R, Ahmad F, Khan N, Zubair S (2017) Biomimetically synthesized ZnO nanoparticles attain potent antibacterial activity against less susceptible *S. aureus* skin infection in experimental animals. RSC Adv 7:36361–36373. https://doi.org/10.1039/C7RA05040B
43. Salem SS (2022) Bio-fabrication of selenium nanoparticles using baker's yeast extract and its antimicrobial efficacy on food borne pathogens. Appl Biochem Biotechnol 194:1898–1910. https://doi.org/10.1007/S12010-022-03809-8
44. Sathishkumar RS, Sundaramanickam A, Srinath R, Ramesh T, Saranya K, Meena M, Surya P (2019) Green synthesis of silver nanoparticles by bloom forming marine microalgae *Trichodesmium erythraeum* and its applications in antioxidant, drug-resistant bacteria, and cytotoxicity activity. J Saudi Chem Soc 23:1180–1191. https://doi.org/10.1016/J.JSCS.2019.07.008
45. Sathiyanarayanan G, Dineshkumar K, Yang Y-H (2017) Microbial exopolysaccharide-mediated synthesis and stabilization of metal nanoparticles. Crit Rev Microbiol 43:731–752. https://doi.org/10.1080/1040841X.2017.1306689
46. Shu M, He F, Li Z, Zhu X, Ma Y, Zhou Z, Yang Z, Gao F, Zeng M (2020) Biosynthesis and antibacterial activity of silver nanoparticles using yeast extract as reducing and capping agents. Nanoscale Res Lett 15. https://doi.org/10.1186/S11671-019-3244-Z
47. Singh AK, Tiwari R, Singh VK, Singh P, Khadim SR, Singh U, Laxmi SV, Hasan SH, Asthana RK (2019) Green synthesis of gold nanoparticles from *Dunaliella salina*, its characterization and in vitro anticancer activity on breast cancer cell line. J Drug Deliv Sci Technol 51:164–176. https://doi.org/10.1016/J.JDDST.2019.02.023
48. Singh NA, Narang J, Garg D, Jain V, Payasi D, Suleman S, Swami RK (2023) Nanoparticles synthesis via microorganisms and their prospective applications in agriculture. Plant Nano Biology 5:100047. https://doi.org/10.1016/j.plana.2023.100047
49. Singh S, Dev A, Gupta A, Nigam VK, Poluri KM (2019) Nitrate Reductase mediated synthesis of surface passivated nanogold as broad-spectrum antibacterial agent. Gold Bull 52:197–216. https://doi.org/10.1007/s13404-019-00264-y
50. Srinath BS, Ravishankar Rai V (2015) Rapid biosynthesis of gold nanoparticles by *Staphylococcus epidermidis*: its characterisation and catalytic activity. Mater Lett 146:23–25. https://doi.org/10.1016/J.MATLET.2015.01.151
51. Srivastava S, Bhargava A, Pathak N, Srivastava P (2019) Production, characterization and antibacterial activity of silver nanoparticles produced by *Fusarium oxysporum* and monitoring of protein-ligand interaction through in-silico approaches. Microb Pathog 129:136–145. https://doi.org/10.1016/J.MICPATH.2019.02.013
52. Suganya KSU, Govindaraju K, Kumar VG, Dhas TS, Karthick V, Singaravelu G, Elanchezhiyan M (2015) Size controlled biogenic silver nanoparticles as antibacterial agent against isolates from HIV infected patients. Spectrochim Acta A Mol Biomol Spectrosc 144:266–272. https://doi.org/10.1016/J.SAA.2015.02.074
53. Talekar S, Joshi A, Chougle R, Nakhe A, Bhojwani R (2016) Immobilized enzyme mediated synthesis of silver nanoparticles using cross-linked enzyme aggregates (CLEAs) of NADH-dependent nitrate reductase. Nano-Struct Nano-Objects 6:23–33. https://doi.org/10.1016/j.nanoso.2016.03.002
54. Thevarajah B, Nishshanka GKSH, Premaratne M, Wasath WAJ, Nimarshana PHV, Malik A, Ariyadasa TU (2023) Cyanobacterial pigment production in wastewaters treated for heavy

metal removal: current status and perspectives. J Environ Chem Eng 11:108999. https://doi.org/10.1016/J.JECE.2022.108999
55. Thevarajah B, Piyatilleke S, Nimarshana PHV, Gunawardena S, Ariyadasa TU (2024) Harnessing algae: advancements in wastewater treatment and resource recovery. In: Yadav S, Ghangrekar MM, Yadava RN (eds) Emerging technologies in biological and hybrid wastewater treatment: lessons from developed to enhancing practices in developing countries. Earth and Environmental Sciences Library. Springer Nature Switzerland Cham, pp 145–175
56. Thevarajah B, Piyatilleke S, Nimarshana PHV, Koushalya S, Malik A, Ariyadasa TU (2024) Exploring effective light spectral conversiontechniques for enhanced production of Spirulina-derived blue pigment protein c-phycocyanin. Bioresour Technol 399:130612. https://doi.org/10.1016/j.biortech.2024.130612
57. Thevarajah B, Piyathilleke S, Sahu A, Nimarshana PHV, Malik A, Ariyadasa TU (2024) Microalgae-based bioproducts and biomaterials towards a sustainable circular bioeconomy. In: Bioeconomy for sustainability. Springer Nature, Singapore, pp 125–162
58. Thevarajah B, Nimarshana PHV, Sandani WP, Wasath WAJ, Ariyadasa TU (2025) Bottlenecks in applications of algal-based technologies toward industrial commercialization. In: Algal biorefinery. Elsevier, pp 307–324
59. Vanlalveni C, Ralte V, Zohmingliana H, Das S, Anal JMH, Lallianrawna S, Rokhum SL (2024) A review of microbes mediated biosynthesis of silver nanoparticles and their enhanced antimicrobial activities. Heliyon 10. https://doi.org/10.1016/J.HELIYON.2024.E32333
60. Vijayanandan AS, Balakrishnan RM (2018) Biosynthesis of cobalt oxide nanoparticles using endophytic fungus *Aspergillus nidulans*. J Environ Manage 218:442–450. https://doi.org/10.1016/J.JENVMAN.2018.04.032
61. Wanarska E, Maliszewska I (2019) The possible mechanism of the formation of silver nanoparticles by *Penicillium cyclopium*. Bioorg Chem 93:102803. https://doi.org/10.1016/j.bioorg.2019.02.028
62. Wang Z, Li N, Zhou X, Wei S, Zhu Y, Li M, Gong J, He J, Dong X, Gao C, Cheng S (2024) Optimization of fermentation parameters toimprove the biosynthesis of selenium nanoparticles by Bacillus licheniformis F1 and its comprehensive application. BMC Microbiol 24:271. https://doi.org/10.1186/s12866-024-03410-5
63. Zambonino MC, Quizhpe EM, Mouheb L, Rahman A, Agathos SN, Dahoumane SA (2023) Biogenic selenium nanoparticles in biomedical sciences: properties, current trends, novel opportunities and emerging challenges in theranostic nanomedicine. Nanomaterials 13:424. https://doi.org/10.3390/nano13030424
64. Zhang H, Zhou H, Bai J, Li Y, Yang J, Ma Q, Qu Y (2019) Biosynthesis of selenium nanoparticles mediated by fungus *Mariannaea* sp. HJ and their characterization. Colloids Surf A Physicochem Eng Asp 571:9–16. https://doi.org/10.1016/J.COLSURFA.2019.02.070

Product Separation, Purification, Storage and Packaging of Nanoparticles

K. M. M. Bimsara Chandrasekera and S. M. W. T. P. K. Ariyarathna

Abstract Nanoparticles have revolutionized a variety of industries, from medicine to energy, due to their unique properties. However, their widespread application depends significantly on the efficiency of their separation, purification, storage, and packaging processes. These critical steps in the production of nanoparticles directly impact their quality, stability, and performance in practical applications. **Separation** of nanoparticles often involves techniques such as centrifugation, filtration, and magnetic separation, depending on the type of nanoparticle and the medium they are synthesized in. The goal is to isolate the nanoparticles from unwanted byproducts, residual solvents, or aggregation products that may have formed during synthesis. **Purification** aims to remove any contaminants or byproducts from the nanoparticle suspension. Common methods include dialysis, size-exclusion chromatography, and solvent exchange techniques. Purification ensures that the nanoparticles retain their desired properties, such as size, shape, and surface charge, which are crucial for their performance in applications like drug delivery, catalysis, and electronics. **Storage** of nanoparticles is another critical factor, as they are often prone to agglomeration, oxidation, or chemical degradation over time. Factors such as temperature, pH, solvent type, and exposure to air must be carefully controlled to ensure long-term stability. Methods such as lyophilization (freeze-drying), suspension in stabilizing agents, or storage under inert atmospheres are employed to prevent degradation and maintain the functionality of the nanoparticles. **Packaging** considerations include ensuring that nanoparticles are protected from environmental factors while minimizing any health risks during handling. Nanoparticles are often packaged in sterile, air-tight, or light-resistant containers. For larger-scale production, specialized packaging materials that minimize contamination and degradation while being easy to handle in transportation and storage are essential. In conclusion, the separation, purification, storage, and packaging of nanoparticles are integral to ensuring their high-quality production and safe, effective application. Advances in these areas are critical to realizing the full potential of nanotechnology in various sectors, including healthcare, electronics, and environmental science.

K. M. M. Bimsara Chandrasekera · S. M. W. T. P. K. Ariyarathna (✉)
Department of Chemical and Process Engineering, Faculty of Engineering, University of Peradeniya, Peradeniya, Sri Lanka
e-mail: thusharia@eng.pdn.ac.lk

N. M. Adassooriya et al. (eds.), *Industrial Scale Production of Nanoparticles*, Smart Nanomaterials Technology, https://doi.org/10.1007/978-981-96-9176-0_16

Keywords Nanoparticles · Separation · Purification · Storage · Packaging

1 Introduction

Nano particles are ultra-small particles that range in size from 1 to 100 nm. They are classified into different classes such as organic, inorganic, ceramic and carbon based Nano particles. The inorganic Nano particles are further classified into metal and metal oxide particles. Nano particles have unique physical and chemical properties due to their small size and large surface area relative to the volume, which make them in a wide range of applications across various fields. They have been broadly engaged for industrial applications covering from photovoltaic, super capacitors, cosmetics, food and drug delivery, medical diagnostics and therapy [43].

In order to assure the quality and functionality of the finished product, careful procedures for product separation and purification are required during the industrial manufacturing of nanoparticles. Nanoparticles are often sensitive to environmental factors like temperature, light, and air exposure, which can lead to agglomeration or degradation of properties and therefore storage of nanoparticles requires careful control to maintain their stability. Isolating the targeted nanoparticles from contaminants and by-products can be extremely difficult because the synthesis of nanoparticles frequently produces a complex mixture of polydispersed size. And, all the physical and chemical properties of Nano particles depend on their size [43].

Separation of nanoparticles involves techniques designed to isolate nanoparticles of different sizes, shapes, surface charges, or compositions. There are several conventional techniques including filtration, ultrafiltration, electrophoresis, chromatography, size selective precipitation, solvent addition and advance methods as micro fluidic technologies. Each method offers unique advantages and may be selected based on the specific properties of the nanoparticles being separated.

Once the nanoparticles are separated, they often require purification to remove unwanted by-products, residual chemicals, or solvents. They include ultrafiltration, chromatography and dialysis. To purify nanoparticles in a single step, novel techniques including diafiltration and reversed-micelle production have been used [3]. These methods have the advantage of requiring less energy and removing the need for extra organic solvents [17, 49]. Additionally, systems for continuous-flow purification have been developed, which combine membrane separation techniques to remove excess ligands and contaminants and preserve nanoparticle form [23].

Robust downstream processes are essential for large scale production. In order to ensure consistency and scalability in the manufacturing process, techniques including tangential flow filtration (TFF) and lyophilization have been developed for the washing, concentration, and storage of polymeric nanomedicines [37].

The storage of nanoparticles requires careful control to maintain their stability and properties. They can be suspended in solvents like water or ethanol, depending on their chemical properties. Stabilizers or surfactants can be often added to prevent agglomeration. In cases where long-term stability is needed, nanoparticles are dried into a powder form. Freeze-drying (lyophilization) or spray drying is commonly used to preserve the structure and prevent aggregation. Low temperatures (refrigeration) and inert gas environments (e.g., argon or nitrogen) are sometimes used to minimize oxidation and other degradation processes.

Proper packaging is critical for the safe handling, transportation, and distribution of nanoparticles. Sterility, material compatibility, protection from environmental factors and safety and labeling are the factors consider during Nano particle packaging.

Each step in the production chain, from synthesis to packaging is critical to maintaining the quality and functionality of the nanoparticles. Additionally, regulatory standards and safety protocols must be followed, especially when nanoparticles are used in medical, environmental, or industrial applications. Here in this chapter, those are comprehensively evaluated including advantages, disadvantages and current trends.

2 Separation of Nano Particles

The separation of nanoparticles is a critical process in various fields like materials science, biotechnology, and environmental engineering. Several methods are employed for the separation of nanoparticles, depending on the desired outcome, the nature of the particles, and the matrix they are suspended in. They are categorized as size based and shape based in liquid and gaseous phase. Table 1 provides an overview of various nanomaterial separation techniques, highlighting the associated advantages and disadvantages of each method.

2.1 Filtration

Filtration is a common separation technique, but separating nanoparticles specifically presents unique challenges due to their extremely small size. Nano porous membranes can be used to filter the Nano scale particles and there are several types of membranes such as polymeric, ceramic and graphene oxide. Nano particle filtration depends on the factors as per size of Nano particles, Shape and charge, pressure and the flow rate. Membrane fouling, cost and the selectivity are the main challenges of using membrane filtration.

In dead-end filtration nanoparticles tend to accumulate on the surface of the filter, forming a cake layer that clogs the pores and reduces filtration efficiency. This buildup leads to increased resistance to flow and eventually causes the filtration process to fail or become inefficient.

In contrast, **cross-flow filtration**, as explained by Robertson et al. [42], is designed to mitigate this issue by introducing a tangential flow of fluid along the surface of the filter. Instead of the fluid flowing directly through the filter as in dead-end

Table 1 Overview of nanomaterial separation techniques with associated advantages and disadvantages

Category	Nanomaterial	Separation technique	Advantages	Disadvantages	Key references
Inorganic	Graphene (2D)	Ultra-centrifugation	Allows for precise control over size and thickness of nanomaterials	Can be time-consuming and requires specialized equipment	Kang et al. [20]
Inorganic	AgNPs AuNPs	Ultrafiltration Ultrafiltration through porous polymeric membranes	Greater control over size, concentration and aggregation state, high efficiency, high scalability	Loss of gold particles with inappropriate membranes	Alele et al. [3], Trefry et al. [50]
Inorganic	Magnetic nanoparticles (MNP)	Magnetic separation	Rapid and efficient separation due to magnetic properties	Limited to magnetic materials and potential for agglomeration	Ahmadi et al. [2]
Inorganic	Silica nanoparticles	Filtration	Cost-effective and simple	Potential for clogging and limited to certain sizes	Minakshi et al. [35]
Organic and Inorganic	Metal–organic frameworks	Magnetic separation	High selectivity and reusability	Limited to specific applications and materials	Minakshi et al. [35]

filtration, cross-flow filtration keeps particles suspended by sweeping them away from the pores. This significantly reduces the buildup of particles on the filter surface, allowing continuous filtration without rapid clogging. Additionally, the application of high pressure across the membrane aids in pushing the filtrate (permeate) through the filter while keeping the particles in the spongy flow.

The tangential flow essentially helps prevent the formation of the filter cake, maintaining a steady filtration rate over a longer period. Cross-flow filtration is widely used in industries for processes like water purification, protein separation, and nanoparticle filtration for these reasons.

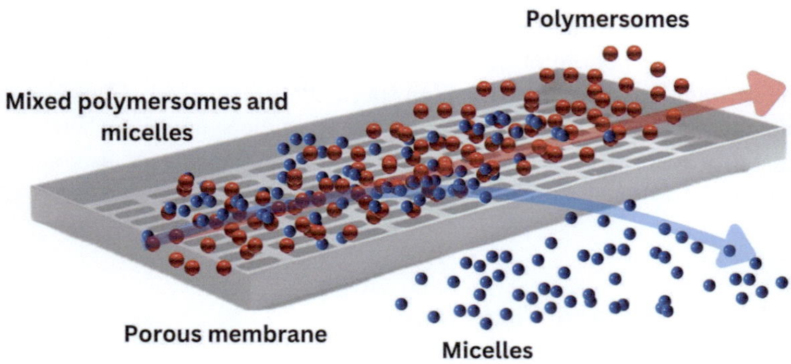

Fig. 1 Purification of polymersomes from micelles using cross flow filtration [42]

This method was further elaborated with mixed polymersomes and micelles in [42], where Fig. 1 (in their study) illustrates how the tangential flow in cross-flow filtration mitigates pore blockage, ensuring more effective nanoparticle separation. By maintaining a steady flow of particles along the filter, cross-flow filtration enhances the longevity and efficiency of the filtration process compared to dead-end filtration [42].

Ultrafiltration is a very effective and precisely controllable technology that is perfect for generating high purity dispersions of nanoparticles [50]. It is a useful technique in the creation of nanoparticles on an industrial scale because of its capacity to manage high volumes while preserving the integrity of the particles [3, 11].

2.2 *Capillary Electrophoresis*

A method for separating products throughout the nanoparticle production process, capillary electrophoresis is especially appreciated for its capacity to separate particles according to size and charge. In this procedure, an electric field is used to push nanoparticles down a capillary tube, where their charge to size ratio determines how quickly they migrate. High resolution separation is provided by this technique, which makes it perfect for differentiating between nanoparticles with minute variations in characteristics. Capillary electrophoresis is an essential technique in the refinement of nanoparticle products because it works particularly well in applications where precise control over the purity and composition of nanoparticles is required [18, 30].

2.3 *Ultra Centrifugation*

In the product separation stage of the creation of nanoparticles, centrifugation is a commonly employed process that is crucial for effectively separating nanoparticles according to their size, shape, and density. This procedure allows concentrated particles to gather at the bottom of the centrifuge tube by separating the nanoparticles from the surrounding medium using high speed centrifugal force. Centrifugation is a fundamental approach in industrial-scale nanoparticle synthesis, where accuracy and scalability are critical. It is very effective at processing enormous volumes and ensuring consistent separation [3, 9, 31].

2.4 Chromatographic Techniques

Size exclusion chromatography (SEC), which was used in a study on the shape separation of nanometer sized gold particles, was shown to be a successful technique for differentiating between gold nanoparticles of various sizes and shapes. The researchers controlled the adsorption characteristics of spherical and rod-shaped gold nanoparticles onto column packing materials using a mixed-surfactant mixture containing polyoxyethylene dodecanol (Brij-35) and sodium dodecyl sulfate (SDS). The scientists increased separation efficiency by varying the SDS and Brij-35 concentrations. The work also demonstrated the possibility of SEC in conjunction with diode-array detection for gold nanoparticle size and shape identification. The significance of chromatographic methods in the separation and characterization of nanoparticles is highlighted by this study [51].

Hydrodynamic Chromatography (HDC) is a separation technique that occurs within the inter-particle channels of tightly packed narrow capillaries, or within wider capillaries containing non-porous packing materials. These materials form capillary like pathways, facilitating separation. In HDC, larger molecules elute earlier than smaller ones, as separation is driven by the differential eluent velocities experienced by particles of varying sizes. This velocity gradient arises from the flow dynamics in the interstitial spaces of the packed bed, where the effective capillaries enable the differentiation of sample components [53].

One method used to speciate designed silver containing nanoparticles (Ag-NPs) is reverse-phase high performance liquid chromatography (RP-HPLC). To enable the accurate and simultaneous assessment of both nanoparticle-bound silver and ionic silver species, a method utilizing RP-HPLC has been developed [45]. HPLC is a strait forward and a robust technique. Depending on the compound to be separated sensitive detectors can be selected and both the stationary and mobile phases can be selected according to the necessary separation of the components [38].

2.5 Magnetic Separation

Magnetic separation techniques, such as High Gradient Magnetic Separation (HGMS), are promising for separating coated magnetic nanoparticles. HGMS uses magnetic fields and gradients to separate magnetic materials from non magnetic liquids. It is used to extract target solutes like polymer coated nanoparticles for organic contaminant extraction and phospholipid coated particles for protein separation. However, the capture efficiency of polymer coated nanoparticles may differ from phospholipid coated particles. The minimum diameter required for successful particle capture in HGMS is approximately 20 nm for individually dispersed nanoparticles and 40 -70 nm for aggregates, depending on the coating and properties of the particles. The study emphasizes the importance of considering particle size and coated nanoparticle properties for effective separation using HGMS. Further research is needed to optimize magnetic filtration of magnetic fluids for nanoparticle separation [36].

3 Purification of Nanoparticles

A crucial step in the creation of nanoparticles is product purification, which focuses on getting rid of contaminants and byproducts that might affect the final product's functionality and quality. After the first separation, the nanoparticles are refined to the required level of purity using purification techniques such solvent exchange, dialysis, and chromatography. This process is crucial to improving the performance, safety, and consistency of the nanoparticles in use by making sure they meet the strict quality criteria needed for their intended industrial applications. Table 2 provides an overview of various nanomaterial purification techniques, highlighting their associated advantages and disadvantages.

3.1 Exchange of Solvents

This purification method involves moving nanoparticles from one solvent to another in order to increase their stability, dispersibility, or to get rid of undesired contaminants or leftover solvents. In this procedure, the old solvent is progressively removed from the nanoparticle dispersion while a new solvent is gradually added, frequently using techniques like dialysis or evaporation. To ensure that the nanoparticles stay stable and disseminated in the new environment, the new solvent must be carefully chosen, taking into account the chemical characteristics of the particles. When the original solvent is incompatible with further processing stages or the final use of the nanoparticles, solvent exchange is frequently utilized to customize the surface chemistry of the particles [7, 26, 29, 44].

Table 2 Overview of nanomaterial purification techniques with associated advantages and disadvantages

Category	Nanomaterial	Purification technique	Advantages	Disadvantages	Key references
Organic and inorganic	Aminopropyl-functionalized silica	Dialysis	Maintains stability; effective separation of unreacted agents	Dialysis is slow	Pálmai et al. [39]
Inorganic	Gold nanoparticles	Liquid–liquid extraction	Fast, simple, inexpensive, and efficient; complete purification in a few hours	May not be suitable for all nanoparticle types	Kulu et al. [27]
Inorganic	Silver nanoparticles	Continuous-flow purification	Faster and more efficient than batch methods Scalability for industrial applications	Reduces intermediate handling during synthesis and purification processes High initial setup and equipment costs Partial removal of Impurities	Khositanon et al. [23]
Organic and inorganic	Alkyl-thiol-stabilized gold nanoparticles and cadmium selenide	Chromatographic techniques	High precision in separating nanoparticles based on size and shape Scalable and widely accessible	Techniques may require optimization for different nanoparticle types Shape-selective reactivity may involve complex chemistries	Kowalczyk et al. [25]

3.2 *Dialysis*

A mild yet efficient purification method that distinguishes between nanoparticles and tiny molecular contaminants such byproducts, salts, and unreacted materials. A semipermeable membrane containing a nanoparticle dispersion is submerged in a solvent or buffer solution to complete the process. The larger nanoparticles are trapped by the membrane, which only permits small molecules to flow through. A solution of cleansed nanoparticles is produced as the contaminants eventually permeate through the barrier and into the surrounding solvent. Since dialysis functions in a moderate environment and shields tiny nanoparticles from strong chemicals or high energy inputs, it is especially beneficial for preserving their integrity. When creating nanoparticles for biomedical applications, where purity and biocompatibility are crucial, this method is frequently employed [4, 6, 27, 34].

3.3 *Chromatography*

Depending on a nanoparticle's size, charge, or chemical affinity, it can be separated using this incredibly accurate and adaptable purification method. Using this technique, a mobile phase (solvent) passes through a column that is filled with a stationary phase material as a combination of nanoparticles is transported through it. Separation results from the nanoparticles' movement through the column at various rates due to their interaction with the stationary phase. Size-exclusion chromatography is used to separate nanoparticles by size, ion-exchange chromatography is used to separate by charge, and affinity chromatography is used to target particular surface functions, among other types of chromatography that are used based on given demands [15, 41]. Chromatography is an indispensable technique for both research and industrial-scale nanoparticle production since it is very effective at producing high-purity nanoparticles and is frequently utilized when a great degree of control over the end product's composition is required [48].

4 Storage Methods for Nanoparticles in Industry

Product storage is a crucial step in the creation of nanoparticles because it helps preserve the stability, utility, and integrity of the particles over time [32]. Adequate storage conditions are essential to prevent unwanted changes like aggregation, oxidation, or degradation. These characteristics consist of controlled humidity, light exposure, and temperature. Use of stabilizing chemicals and sealed containers further ensures that the nanoparticles remain in the right state and are ready for use in industrial applications. Effective storage methods are crucial to maintaining the quality and shelf life of nanoparticles and ensuring that they retain their intended properties until needed [1, 32].

Nanoparticles (NPs) have emerged as pivotal materials in various industrial applications due to their unique properties, which differ significantly from their bulk counterparts [46]. These properties include increased surface area, enhanced reactivity, and specific optical characteristics, making nanoparticles valuable in fields such as pharmaceuticals, electronics, food technology, and energy storage [22]. However, the effective storage of nanoparticles is crucial for maintaining their stability, functionality, and safety during handling and application. This section research into the various methods of nanoparticle storage in industrial settings, emphasizing the importance of appropriate container specifications, environmental conditions, safety protocols, and the challenges faced in this domain.

4.1 Container Specifications

4.1.1 Material Selection

A key factor in guaranteeing stable nanoparticle storage is the choice of container materials, which need to be non-reactive to avoid chemical reactions that could change the structure, characteristics, or functioning of the nanoparticles. Because of its superior chemical resistance, glass is one of the most extensively used materials and is ideal for a wide variety of nanoparticles [55]. Because of its inert nature, which stops any reaction with the item being held, and its transparency, which makes it possible to visually analyze the contents, it is a preferred option in laboratory settings where accurate monitoring is crucial [55].

Another material that is frequently utilized is High-Density Polyethylene (HDPE), especially in industrial or bulk storage applications [54]. It is preferred over more delicate materials like glass because of its low weight and sturdy resistance to a wide range of pollutants. It is also more affordable. In order to reduce the possibility of contamination from outside sources and maintain the integrity of the nanoparticles throughout storage and transportation, HDPE containers can also be firmly sealed [10].

Applications where temperature fluctuations may occur during storage frequently choose polypropylene. Polypropylene, which is well-known for having strong thermal resistance, is a great option for nanoparticles that need particular temperature-controlled settings since it can tolerate mild heating or cooling [52]. Its chemical inertness further guarantees that it stays stable when various kinds of nanoparticles are present, lowering the possibility of any negative reactions [16].

Metal containers can be utilized in specific situations, but choosing the right one needs thought. To prevent reactivity with the nanoparticles, the metal must be selected very carefully because any interaction between the two could cause the properties of the nanoparticles to degrade or change. As a result, using metal containers becomes less typical and increasingly reliant on the particular qualities of the nanoparticles being kept. To guarantee the long-term stability and efficacy of the nanoparticles, the selection of container material should be based on the chemical makeup of the particles, the storage environment, and the duration of the intended storage [8].

4.1.2 Labeling and Documentation

For the safe handling, storage, and management of nanoparticles, especially in industrial and research settings where safety and accuracy are critical, proper and thorough labeling is essential. To reduce the possibility of improper handling or exposure to potentially dangerous products, each container should have a clear label that provides comprehensive information about its contents. The particular kind of nanoparticle and its concentration should be identified on the label so that staff members may promptly and precisely determine the contents of the container. In order to track the age of the nanoparticles and ensure their efficacy and safety, it is also necessary to indicate the date of synthesis or purchase. This is because the stability and attributes of the nanoparticles can alter with time.

The applicable Safety Data Sheets (SDS), which include vital details regarding the possible risks connected to the nanoparticles as well as suggested safety precautions, should also be cited on the label. SDS should be easily accessible to ensure that anyone handling the materials are aware of any potential risks and have the tools necessary to handle them. Furthermore, to guarantee that the nanoparticles are handled safely throughout their storage and use, handling instructions should be clearly indicated on the label [47]. These instructions should include any unique precautions, such as temperature control, avoiding particular environmental conditions, or requiring personal protective equipment (PPE).

For efficient tracking and monitoring of nanoparticle inventories, upkeep of a centralized, well organized database or inventory management system is necessary, in addition to labeling. These systems help avoid overstocking or the depletion of essential commodities by providing real-time data on inventory levels, storage conditions, and expiration dates. Furthermore, as it offers a clear record of the nanoparticles on hand, their amounts, and their particular handling and disposal requirements, a well managed inventory system helps with regulatory compliance [13]. All employees involved in the usage and research of nanoparticles will operate in a safer environment thanks to this all encompassing approach to inventory management and labeling, which also greatly lowers the chance of mishaps or incorrect handling [14].

5 Method of Packaging Nanoparticles in Industry

In the last phases of producing nanoparticles, product packaging is essential because it guarantees that the particles are securely confined and safeguarded throughout handling, storage, and transportation. The packaging needs to be carefully thought out in order to protect the nanoparticles from light, oxygen, and moisture, all of which can lower their quality. Furthermore, packaging needs to be sensible so that the nanoparticles can be handled safely and easily without compromising their integrity. An essential aspect of the entire production process, good packaging not only maintains the qualities of the nanoparticles but also makes it easier for them to be used effectively in industrial applications.

5.1 Facilitating Safe Handling

Since efficient packaging is essential for reducing contamination and exposure hazards to hazardous materials, it is imperative that researchers and industry people place a high priority on safety and convenience of use when designing nanoparticle packaging. Achieving this objective requires an ergonomic design; packaging should be made to be easy to handle while maintaining the integrity of the nanoparticles [5]. An ergonomic design is essential. This means that packaging should be easy to open and reseal and should be made with as little chance of spills or exposure during handling as possible.

Tamper-evident features are another crucial safety precaution that should be included in the packing. These characteristics offer outward cues, like labels or seals, that show whether the object has been tampered with before usage. This improves the nanoparticles' safety while also giving users more assurance about the caliber and dependability of the materials they are working with. Particularly in industrial and research settings where material integrity is essential to experimental results and product efficacy, tamper-evident packaging is crucial.

Additionally essential to the efficient use of nanoparticle packaging is thorough and unambiguous labeling. Important details on proper handling practices, storage conditions, and precautions should be included on labels. Employees are better equipped to handle nanoparticles securely thanks to this information, which guarantees that they are aware of the particular risks connected to the materials and what steps to take to reduce those risks. For instance, labeling ought to specify whether the nanoparticles are susceptible to changes in temperature, light, or moisture. It should also include advice on the ideal storage settings needed to preserve the stability and functionality of the nanoparticles.

Relevant safety data sheets (SDS) or summaries on the packaging or supporting paperwork are also advantageous. These resources can offer comprehensive details about the chemical characteristics of the nanoparticles, possible risks, and how to treat an accidental exposure. Organizations can improve staff readiness to handle safety concerns linked to nanoparticles by providing easy access to this knowledge.

Environmental effect and sustainability should also be taken into account when designing packaging. Using recyclable or biodegradable materials reduces the environmental impact of handling and disposing of nanoparticles while also being in line with modern sustainability norms. The significance of this design element is growing among scholars and institutions dedicated to ecological accountability.

To effectively support researchers and industry staff, safety, usability, and sustainability must be given top priority in the design of nanoparticle packaging. Organizations can develop packaging solutions that improve safety and enable the responsible handling of nanoparticles throughout their lifecycle by emphasizing ergonomic design, including tamper-evident elements, guaranteeing clear labeling, and taking environmental effect into account. This all-encompassing strategy promotes the efficient use of nanoparticles in numerous scientific and industrial domains while also making the workplace safer.

6 Protection from Environmental Factors

6.1 Temperature Control

Since temperature variations or unfavorable thermal conditions can significantly affect the physical and chemical properties of nanoparticles, temperature is a crucial component in their stability and preservation. For example, high temperatures can hasten the processes of agglomeration, in which nanoparticles group together, and degradation, in which they lose some of their special qualities and functions [56]. These modifications not only lessen the nanoparticles' efficiency but also have the potential to affect their behavior in ways that could endanger human safety or result in unfavorable interactions. Lower temperatures, on the other hand, help to reduce these dangers by offering a more regulated setting where the structure and properties of the nanoparticles are preserved for longer.

The ideal temperature for storing nanoparticles is mostly determined by their unique composition and sensitivity. Temperatures for storage generally vary from 4 °C to room temperature (20–25 °C) [28]. Room temperature storage may be sufficient for many types of nanoparticles, especially those utilized in more durable applications, as it provides a compromise between practicality and maintaining their fundamental qualities. Stricter temperature controls are required for nanoparticles that are more susceptible to temperature changes, such as those utilized in highly reactive, biological, or medicinal applications [40].

For specialized nanoparticles, such as biological nanoparticles, which are frequently vulnerable to temperature deterioration, refrigerated storage is especially crucial. In these situations, keeping the nanoparticles in storage at a temperature between 2 and 8 °C can greatly lower the chance of degradation and increase their shelf life. For nanoparticles that could become unstable or lose their functionality in the presence of even slight temperature variations, refrigeration is an important storage option. Their reactive qualities can be better preserved by maintaining them at continuously low temperatures, guaranteeing that they will continue to be viable for their intended purpose over time [40].

Maintaining the stability of nanoparticles in locations where they are often handled and stored requires the use of robust refrigeration systems. This guarantees that the nanoparticles are maintained at ideal temperatures and shielded from the damaging effects of heat, which can jeopardize their functionality or security. Maintaining the quality of nanoparticles and making sure they can be used safely and effectively in a variety of applications requires careful observation and control of storage conditions, whether they are kept at room temperature or under refrigeration.

6.2 Humidity Control

A crucial environmental component that can have a big impact on the performance and stability of nanoparticles is humidity, especially for those that are hydrophilic or extremely sensitive to moisture. High humidity levels can cause unintended clumping or aggregation of nanoparticles, in which individual particles stick to one another to create bigger clusters [21]. Since of this aggregation, the nanoparticles become less effective since their uniform dispersion is disrupted, which is a crucial quality for numerous applications in areas including medication delivery, catalysis, and material science. Moisture induced agglomeration of nanoparticles reduces their surface area, reactivity, and capacity to interact with other substances, which may cause them to be less effective or less efficient in their intended uses [12].

The usage of desiccants is one of the best ways to regulate humidity and shield nanoparticles from moisture. Desiccants are substances that actively draw moisture from the air inside a storage container. Examples of these substances are silica gel and activated alumina. These desiccants assist in preventing the buildup of water vapor surrounding the nanoparticles, lowering the risk of aggregation and preserving the dispersion stability of the particles by establishing a dry, moisture-free environment. Because it is widely available and has a great capacity to absorb moisture, silica gel is particularly useful in low-humidity environments. Activated alumina has comparable advantages but concentrates on eliminating particular moisture levels [33].

Furthermore enhancing the protection of nanoparticles against moisture exposure are appropriate storage practices, such as making sure containers are well sealed and stored in low-humidity conditions, in addition to the use of desiccants. To guarantee that ideal conditions are maintained throughout the storage period, it is also necessary to regularly monitor the humidity levels in storage rooms or containers. Desiccants and other moisture-control methods can be used to proactively manage humidity, which can prolong the shelf life of nanoparticles, preserve their functional qualities, and keep them effective in a variety of applications.

6.3 Light Exposure

Light sensitivity is an important factor to take into account while storing nanoparticles, especially since exposure to light specifically ultraviolet (UV) radiation can induce substantial alterations in the chemical composition and physical characteristics of the particles through a process called photodegradation. Nanoparticles may disintegrate or undergo structural modifications as a result of this deterioration, which reduces their usefulness in a variety of applications, especially in areas like biology, electronics, and optics where even little structural changes can have a significant impact. It is crucial to keep light-sensitive nanoparticles under conditions that reduce their exposure to light in order to reduce the danger of photodegradation [19, 24].

One of the best ways to shield nanoparticles from light exposure is to store them in opaque containers that block out visible and UV light to create a safe environment. Light induced degradation of nanoparticles can be further minimized by storing them in dark environments like closed cabinets or dimly lit rooms. UV filtering films or coatings can be used to windows and other possible light sources in storage spaces to improve protection. This allows for controlled ambient lighting while limiting UV exposure. Organizations can greatly increase the shelf life and stability of light sensitive nanoparticles and guarantee their efficacy in a variety of scientific and industrial applications by combining these preventive measures with clear labeling of containers with information regarding light sensitivity.

7 Support Framework for Nanomaterials (NMs) Type-Based Selection of Separation, Purification, and Storage Methods

The effective selection of separation, purification, and storage techniques for nanomaterials (NMs) is critical for ensuring the quality, reproducibility, and stability of nanomaterial products. A decision tree or support framework based on the specific type of nanomaterial (NM) can significantly aid practitioners in making informed choices that align with the material's unique properties, thus optimizing the overall process efficiency. This approach is essential as NMs, due to their diverse physicochemical properties, exhibit varying behaviors that necessitate tailored processing strategies. Figure 2 illustrates a support framework for selecting appropriate methods of separation, purification, and storage based on the type of nanomaterial (NM).

The first step in the decision-making process involves identifying the type of NM, whether it is organic, inorganic, or hybrid. Organic NMs, such as dendrimers, biopolymers, or liposomes, typically require methods that preserve their structural integrity and avoid degradation. In contrast, inorganic NMs, such as silica, gold, or silver nanoparticles, often require distinct handling due to their different surface characteristics, stability, and potential for aggregation. Hybrid NMs, combining both organic and inorganic components, necessitate a deeper understanding of both constituent materials to determine the most appropriate approach for separation and purification.

Once the NM type is established, a framework for selecting suitable methods should be based on key material properties such as size, surface charge, solubility,

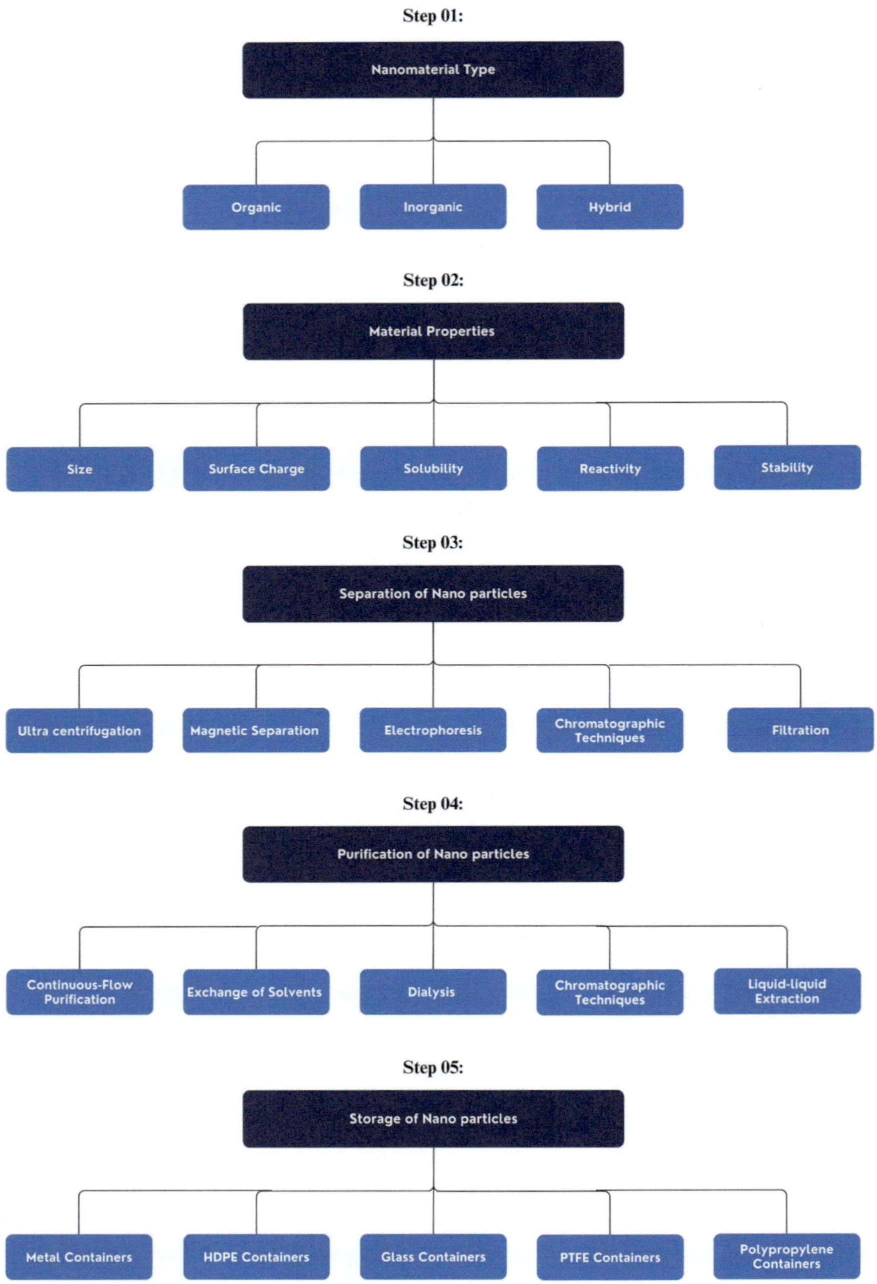

Fig. 2 Support framework for nanomaterials (NMs) type-based selection of separation, purification, and storage methods

reactivity, and stability. The size of NMs plays a pivotal role in determining appropriate separation techniques. For instance, ultrafiltration is highly effective for separating particles smaller than 50 nm, while centrifugation or size-exclusion chromatography is better suited for separating nanoparticles based on size distribution. Surface functionalization, including the presence of hydrophobic or hydrophilic groups, must also be considered when selecting methods like electrophoresis or ion-exchange chromatography, which exploit surface charge and chemical interactions. Additionally, the reactivity of the NMs influences method selection, highly reactive materials may require inert processing environments, such as nitrogen atmospheres, to prevent unwanted reactions.

For purification, a tiered approach can be adopted. Techniques like dialysis and size-exclusion chromatography are standard for purifying NMs by removing contaminants, small molecules, or by achieving homogeneity in particle size. The choice of purification method should take into account the scale of production, with laboratory-scale methods such as filtration being suitable for small volumes, while larger-scale operations may benefit from continuous-flow systems or liquid–liquid extraction. A combination of techniques may be required to achieve the desired purity levels, particularly when dealing with complex NMs that exhibit a broad range of surface modifications.

The storage of NMs is another crucial aspect, with environmental stability being a key consideration. Temperature, moisture, and light sensitivity must be factored into the decision framework. Temperature-sensitive NMs should be stored at controlled temperatures, often within the range of 4–8 °C, to maintain stability. Materials that are sensitive to moisture require controlled humidity conditions to prevent aggregation or degradation, while light-sensitive NMs may need to be stored in opaque containers or shielded from UV radiation. Moreover, the material of the storage container plays a critical role, containers made of inert materials, such as glass, HDPE, or PTFE, are preferred for reactive or sensitive NMs due to their non-reactive nature.

Incorporating a decision tree or support framework based on these considerations allows practitioners to systematically evaluate the specific requirements of the NM in question. The framework begins with identifying the NM type, followed by an assessment of key material characteristics such as size, surface charge, reactivity, and environmental sensitivity. From there, the appropriate separation, purification, and storage techniques can be selected based on the material's unique properties. This structured approach ensures that the most effective and efficient methods are chosen, resulting in the highest quality nanomaterial products that meet the desired specifications for both laboratory and industrial applications.

8 Conclusion

The separation, purification, storage, and packaging of nanoparticles are critical processes that ensure their stability, functionality, and safety for diverse applications. Effective separation techniques, such as filtration, chromatography, and ultracentrifugation, enable precise control over nanoparticle properties, while advanced purification methods enhance their quality by removing contaminants. Proper storage and packaging strategies safeguard nanoparticles from environmental factors like temperature, humidity, and light, maintaining their integrity over time.

The choice of storage materials, labeling, and ergonomic packaging design not only ensures safe handling but also aligns with regulatory and sustainability standards, paving the way for responsible industrial and research practices. By addressing these key aspects, this chapter underscores the importance of meticulous handling throughout the lifecycle of nanoparticles, supporting their successful integration into technological advancements across various fields.

References

1. Abdelwahed W, Degobert G, Stainmesse S, Fessi H (2006) Freeze-drying of nanoparticles: formulation, process and storage considerations. Adv Drug Deliv Rev 58(15):1688–1713. https://doi.org/10.1016/j.addr.2006.09.017
2. Ahmadi M, Elmongy H, Madrakian T, Abdel-Rehim M (2017) Nanomaterials as sorbents for sample preparation in bioanalysis: a review. Analytica Chimica Acta 958:1–21. https://doi.org/10.1016/j.aca.2016.11.062
3. Alele N, Streubel R, Gamrad L, Barcikowski S, Ulbricht M (2016) Ultrafiltration membrane-based purification of bioconjugated gold nanoparticle dispersions. Sep Purif Technol 157:120–130. https://doi.org/10.1016/j.seppur.2015.11.033
4. Al-kattan A, Dufour P, Drouet C (2011) Purification of biomimetic apatite-based hybrid colloids intended for biomedical applications: a dialysis study. Colloids Surf, B 82(2):378–384. https://doi.org/10.1016/j.colsurfb.2010.09.022
5. Amoabediny H, Naderi A, Malakootikhah J, Koohi M, Mortazavi A, Naderi M, Rashedi H (2009) Guidelines for safe handling, use and disposal of nanoparticles. J Phys Conf Ser 170. https://doi.org/10.1088/1742-6596/170/1/012037
6. Atmatzidis K, Alimohammadi F, Aich N, Tehrani R (2020) Nano-enhanced dialytic fluid purification: CFD modeling of Pb(II) removal by manganese oxide. ACS Omega 5(50):32697–32705. https://doi.org/10.1021/acsomega.0c05069
7. Bai S, Nguyen TL, Mulvaney P, Wang D (2010) Using hydrogels to accommodate hydrophobic nanoparticles in aqueous media via solvent exchange. Adv Mater 22(30):3247–3250. https://doi.org/10.1002/adma.201000336
8. Bendall JS, Ilie A, Welland ME, Sloan J, Green MLH (2006) Thermal stability and reactivity of metal halide filled single-walled carbon nanotubes. J Phys Chem B 110(13):6569–6573. https://doi.org/10.1021/jp056405t
9. Bootz A, Vogel V, Schubert D, Kreuter J (2004) Comparison of scanning electron microscopy, dynamic light scattering and analytical ultracentrifugation for the sizing of poly(butyl cyanoacrylate) nanoparticles. Eur J Pharm Biopharm 57(2):369–375. https://doi.org/10.1016/S0939-6411(03)00193-0
10. Chrissafis K, Paraskevopoulos KM, Tsiaoussis I, Bikiaris D (2009) Comparative study of the effect of different nanoparticles on the mechanical properties, permeability, and thermal

degradation mechanism of HDPE. J Appl Polym Sci 114(3):1606–1618. https://doi.org/10.1002/app.30750
11. Dalwadi G, Sunderland VB (2007) Purification of PEGylated nanoparticles using tangential flow filtration (TFF). Drug Dev Ind Pharm 33(9):1030–1039. https://doi.org/10.1080/03639040601180143
12. Endres SC, Ciacchi LC, Mädler L (2021) A review of contact force models between nanoparticles in agglomerates, aggregates, and films. J Aerosol Sci 153. https://doi.org/10.1016/j.jaerosci.2020.105719
13. Glisovic S, Pesic D, Stojiljkovic E, Golubovic T, Krstic D, Prascevic M, Jankovic Z (2017) Emerging technologies and safety concerns: a condensed review of environmental life cycle risks in the nano-world. Int J Environ Sci Technol 14(10):2301–2320. https://doi.org/10.1007/s13762-017-1367-2
14. Groso A, Petri-Fink A, Magrez A, Riediker M, Meyer T (2010) Management of nanomaterials safety in research environment. Particle Fibre Toxicol 7. https://doi.org/10.1186/1743-8977-7-40
15. Guerreiro AR, Chianella I, Piletska E, Whitcombe MJ, Piletsky SA (2009) Selection of imprinted nanoparticles by affinity chromatography. Biosens Bioelectron 24(8):2740–2743. https://doi.org/10.1016/j.bios.2009.01.013
16. Hambir S, Bulakh N, Jog JP (n.d.) Polypropylene/clay nanocomposites: effect of compatibilizer on the thermal, crystallization and dynamic mechanical behavior
17. Hollamby MJ, Eastoe J, Chemelli A, Glatter O, Rogers S, Heenan RK, Grillo I (2010) Separation and purification of nanoparticles in a single step. Langmuir 26(10):6989–6994. https://doi.org/10.1021/la904225k
18. Hwang M, Lee C-Y, Boo DW, Choi J-G (2003) Separation of nanoparticles in different sizes and compositions by capillary electrophoresis. Bull Korean Chem Soc 24(5)
19. Izak-Nau E, Huk A, Reidy B, Uggerud H, Vadset M, Eiden S, Voetz M, Himly M, Duschl A, Dusinska M, Lynch I (2015) Impact of storage conditions and storage time on silver nanoparticles' physicochemical properties and implications for their biological effects. RSC Adv 5(102):84172–84185. https://doi.org/10.1039/c5ra10187e
20. Kang J, Sangwan VK, Wood JD, Hersam MC (2017) Solution-based processing of monodisperse two-dimensional nanomaterials. Acc Chem Res 50(4):943–951. https://doi.org/10.1021/acs.accounts.6b00643
21. Kelesidis GA, Furrer FM, Wegner K, Pratsinis SE (2018) Impact of humidity on silica nanoparticle agglomerate morphology and size distribution. Langmuir 34(29):8532–8541. https://doi.org/10.1021/acs.langmuir.8b00576
22. Khan I, Saeed K, Khan I (2019) Nanoparticles: properties, applications and toxicities. Arab J Chem 12(7):908–931. https://doi.org/10.1016/j.arabjc.2017.05.011
23. Khositanon C, Adpakpang K, Bureekaew S, Weeranoppanant N (2020) Continuous-flow purification of silver nanoparticles and its integration with flow synthesis. J Flow Chem 10(1):353–362. https://doi.org/10.1007/s41981-020-00084-8
24. Korshed P, Li L, Ngo DT, Wang T (2018) Effect of storage conditions on the long-term stability of bactericidal effects for laser generated silver nanoparticles. Nanomaterials 8(4). https://doi.org/10.3390/nano8040218
25. Kowalczyk B, Lagzi I, Grzybowski BA (2011) Nanoseparations: strategies for size and/or shape-selective purification of nanoparticles. Curr Opin Colloid Interface Sci 16(2):135–148. https://doi.org/10.1016/j.cocis.2011.01.004
26. Krivtsov I, Ilkaeva M, Avdin V, Amghouz Z, Khainakov SA, García JR, Díaz E, Ordóñez S (2015) Exceptional thermal stability of undoped anatase TiO_2 photocatalysts prepared by a solvent-exchange method. RSC Adv 5(46):36634–36641. https://doi.org/10.1039/c5ra01114k
27. Kulu I, Huang R, Kalyanaraman B, Rotello VM (2020) A modified and simplified method for purification of gold nanoparticles. MethodsX 7. https://doi.org/10.1016/j.mex.2020.100896
28. Lemoine D, Francois C, Kedzierewicz F, Preat V, Hoffman M, Maincent P (1996) Stability study of nanoparticles of poly(e-caprolactone), poly(D,L-lactide) and poly(D,L-lactide-co-glycolide). Biomaterials 17

29. Li N, Nikoubashman A, Panagiotopoulos AZ (2018) Multi-scale simulations of polymeric nanoparticle aggregation during rapid solvent exchange. J Chem Phys 149(8). https://doi.org/10.1063/1.5046159
30. Liu FK (2011) Extremely highly efficient on-line concentration and separation of gold nanoparticles using the reversed electrode polarity stacking mode and surfactant-modified capillary electrophoresis. Anal Chim Acta 694(1–2):167–173. https://doi.org/10.1016/j.aca.2011.03.056
31. Luykx DMAM, Peters RJB, Van Ruth SM, Bouwmeester H (2008) A review of analytical methods for the identification and characterization of nano delivery systems in food. J Agric Food Chem 56(18):8231–8247. https://doi.org/10.1021/jf8013926
32. Masoudipour E, Kashanian S, Azandaryani AH, Omidfar K, Bazyar E (2017) Surfactant effects on the particle size, zeta potential, and stability of starch nanoparticles and their use in a pH-responsive manner. Cellulose 24(10):4217–4234. https://doi.org/10.1007/s10570-017-1426-3
33. Md Zulkiflie FA, Muhamad Sarih N, Hashim NA, Mohd Zubir MN, Abdullah S, Mohd Amin AS (2023) Moisture adsorption–desorption behaviour in nanocomposite copolymer films. Polymers 15(14). https://doi.org/10.3390/polym15142998
34. Mihaila R, Chang S, Wei AT, Hu ZY, Ruhela D, Shadel TR, Duenwald S, Payson E, Cunningham JJ, Kuklin N, Mathre DJ (2011) Lipid nanoparticle purification by Spin Centrifugation-Dialysis (SCD): a facile and high-throughput approach for small scale preparation of siRNA-lipid complexes. Int J Pharm 420(1):118–121. https://doi.org/10.1016/j.ijpharm.2011.08.017
35. Minakshi P, Ghosh M, Brar B, Ranjan K, Patki HS, Kumar R (2019) Separation techniques with nanomaterials. In: Handbook of nanomaterials in analytical chemistry: modern trends in analysis. Elsevier, pp 99–158. https://doi.org/10.1016/B978-0-12-816699-4.00006-2
36. Moeser GD, Roach KA, Green WH, Hatton TA, Laibinis PE (2004) High-gradient magnetic separation of coated magnetic nanoparticles. AIChE J 50(11):2835–2848. https://doi.org/10.1002/aic.10270
37. Operti MC, Bernhardt A, Sincari V, Jager E, Grimm S, Engel A, Hruby M, Figdor CG, Tagit O (2022) Industrial scale manufacturing and downstream processing of PLGA-based nanomedicines suitable for fully continuous operation. Pharmaceutics 14(2). https://doi.org/10.3390/pharmaceutics14020276
38. Ouwmeester HANSB (2015) A review of analytical methods for the identification and characterization of nano delivery systems in food, pp 8231–8247
39. Pálmai M, Nagy LN, Mihály J, Varga Z, Tárkányi G, Mizsei R, Szigyártó IC, Kiss T, Kremmer T, Bóta A (2013) Preparation, purification, and characterization of aminopropyl-functionalized silica sol. J Colloid Interface Sci 390(1):34–40. https://doi.org/10.1016/j.jcis.2012.09.025
40. Parakhonskiy BV, Parak WJ, Volodkin D, Skirtach AG (2019) Hybrids of polymeric capsules, lipids, and nanoparticles: thermodynamics and temperature rise at the nanoscale and emerging applications. Langmuir 35(26):8574–8583. https://doi.org/10.1021/acs.langmuir.8b04331
41. Pirok BWJ, Abdulhussain N, Aalbers T, Wouters B, Peters RAH, Schoenmakers PJ (2017) Nanoparticle analysis by online comprehensive two-dimensional liquid chromatography combining hydrodynamic chromatography and size-exclusion chromatography with intermediate sample transformation. Anal Chem 89(17):9167–9174. https://doi.org/10.1021/acs.analchem.7b01906
42. Robertson JD, Rizzello L, Avila-Olias M, Gaitzsch J, Contini C, MagoÅ MS, Renshaw SA, Battaglia G (2016) Purification of nanoparticles by size and shape. Sci Rep 6:1–9. https://doi.org/10.1038/srep27494
43. Salafi T, Zeming KK, Zhang Y (2017) Advancements in microfluidics for nanoparticle separation. Lab Chip 17(1):11–33. https://doi.org/10.1039/C6LC01045H
44. Seo D, Yoon W, Park S, Kim J, Kim J (2008) The preparation of hydrophobic silver nanoparticles via solvent exchange method. Colloids Surf, A 313–314:158–161. https://doi.org/10.1016/j.colsurfa.2007.04.086
45. Soto-Alvaredo J, Montes-Bayón M, Bettmer J (2013) Speciation of silver nanoparticles and silver(I) by reversed-phase liquid chromatography coupled to ICPMS. Anal Chem 85(3):1316–1321. https://doi.org/10.1021/ac302851d

46. Stark WJ, Stoessel PR, Wohlleben W, Hafner A (2015) Industrial applications of nanoparticles. Chem Soc Rev 44(16):5793–5805. https://doi.org/10.1039/c4cs00362d
47. Stuttgen V, Giffney HE, Anandan A, Alabdali A, Twarog C, Belhout SA, O'Loughlin M, Podhorska L, Delaney C, Geoghegan N, Mc-Fadden J, Alhadhrami NA, Fleming A, Phadke S, Yadav R, Fattah S, McCartney F, Alsharif SA, McCaul J., Singh K, Erikandath S, O'Meara F, Wychowaniec JK, Cutrona MB, MacMaster G, Reynolds AL, Gaines S, Hogg B, Farrelly N, d'Alton M, Coulahan P, Bhattacharjee S (2019) The UCD nanosafety workshop (03 December 2018): towards developing a consensus on safe handling of nanomaterials within the Irish university labs and beyond—a report. Nanotoxicology 13(6):717–732. https://doi.org/10.1080/17435390.2019.1621402
48. Süß S, Metzger C, Damm C, Segets D, Peukert W (2018) Quantitative evaluation of nanoparticle classification by size-exclusion chromatography. Powder Technol 339:264–272. https://doi.org/10.1016/j.powtec.2018.08.008
49. Sweeney SF, Woehrle GH, Hutchison JE (2006) Rapid purification and size separation of gold nanoparticles via diafiltration. J Am Chem Soc 128(10):3190–3197. https://doi.org/10.1021/ja0558241
50. Trefry JC, Monahan JL, Weaver KM, Meyerhoefer AJ, Markopolous MM, Arnold ZS, Wooley DP, Pavel IE (2010) Size selection and concentration of silver nanoparticles by tangential flow ultrafiltration for SERS-based biosensors. J Am Chem Soc 132(32):10970–10972. https://doi.org/10.1021/ja103809c
51. Wei GT (1999) Shape separation of nanometer gold particles by size-exclusion chromatography. Anal Chem 71(11):2085–2091. https://doi.org/10.1021/ac990044u
52. Wen X, Wang Y, Gong J, Liu J, Tian N, Wang Y, Jiang Z, Qiu J, Tang T (2012) Thermal and flammability properties of polypropylene/carbon black nanocomposites. Polym Degrad Stab 97(5):793–801. https://doi.org/10.1016/j.polymdegradstab.2012.01.031
53. Williams A, Varela E, Meehan E, Tribe K (2002) Characterisation of nanoparticulate systems by hydrodynamic chromatography. Int J Pharmaceutics 242. www.elsevier.com/locate/ijpharm
54. Yao YL, De Guzman MR, Duan H, Gao C, Lin X, Wen YH, Du J, Lin L, Chen JC, Wu CS, Suen MC, Sun YL, Hung WS, Tsou CH (2020) Infusing high-density polyethylene with graphene-zinc oxide to produce antibacterial nanocomposites with improved properties. Chin J Polym Sci (Engl Ed) 38(8):898–907. https://doi.org/10.1007/s10118-020-2392-z
55. Zhang Y, Huang L, Shi Y (2021) Towards damage resistant Al_2O_3–SiO_2 glasses with structural and chemical heterogeneities through consolidation of glassy nanoparticles. Acta Materialia 215. https://doi.org/10.1016/j.actamat.2021.117016
56. Zyuzin MV, Honold T, Carregal-Romero S, Kantner K, Karg M, Parak WJ (2016) Influence of temperature on the colloidal stability of polymer-coated gold nanoparticles in cell culture media. Small 12(13):1723–1731. https://doi.org/10.1002/smll.201503232

Quality Assurance in Nanoparticles Production

Asanka Wijerathne and Kasun Thennakoon

Abstract Nanomaterials attract high research interest due to their favorable properties, such as high exposed surface area and higher bioavailability. Nanomaterial synthesis using chemical, physical, and biological methods includes process variables such as pressure, temperature, crystallization time, pH, solvents and surfactants, and mixing patterns in the synthesis media. These process variables and post-processing steps determine the uniformity of the synthesized nanomaterials and the degree of reproducibility of the nanomaterial synthesis protocol. Quality assurance aims to minimize the variability of the targeted properties of synthesized nanoparticles characterized using various techniques, including UV–vis, Fourier transform infrared spectroscopy, transmission electron microscopy, scanning electron microscopy, X-ray diffraction, and dynamic light scattering. Once the desired level of control of a particular synthesis is achieved, nanomaterials can be incorporated into products in various sectors including catalysis (oxide-supported Pt, Pd, Cu), agri-food (mesoporous silica, fullerene, Ag, ZnO), cosmetics (solid lipid nanoparticles, fullerene, TiO_2, ZnO, Cu), lubricants and fuel additives (CeO_2, MoS_3), and nanomedicine and nanocarriers (Au, Ag, Fe, magnetic materials). This chapter highlights the current advances in quality assurance in nanomaterial synthesis, with special emphasis on metal NPs, process variables affecting quality control, statistical analysis of the nanomaterials to ensure their uniform properties and the prospects of nanomaterial quality assurance.

Keywords Metal nanoparticles · Particle size distribution · Particle morphology · Process control · Quality assurance

A. Wijerathne (✉)
Department of Chemical Engineering, University of Virginia, Charlottesville, USA
e-mail: akw7uj@virginia.edu

K. Thennakoon
Cleantech Pvt Ltd., Colombo, Sri Lanka

N. M. Adassooriya et al. (eds.), *Industrial Scale Production of Nanoparticles*, Smart Nanomaterials Technology, https://doi.org/10.1007/978-981-96-9176-0_17

1 Introduction

Nanoscale materials and particles, typically less than 100 nm in size, have gained high research and industrial interest due to their unique properties that differ significantly from their bulk counterparts. Most of these unique properties, such as enhanced reactivity and adsorption capacity of nanosized materials, are governed by their high surface area to volume ratio and the quantum effects that emerge at the nanoscale. Those properties are highly desirable in applications such as catalysis, drug delivery, and sensing technologies [40, 83]. Quantum mechanical effects lead to altered optical and electronic behaviors, as seen in semiconductor nanoparticles known as quantum dots that exhibit size-dependent optical properties. Those unique properties make them suitable for bioimaging and photovoltaic applications [2, 49]. Additionally, nanoscale materials often demonstrate superior mechanical strength and thermal conductivity, exemplified by carbon nanotubes and graphene, which have applications in advanced composites and thermal management [5]. Furthermore, metallic nanoparticles display localized surface plasmon resonance, enhancing their optical properties for use in sensors and medical diagnostics [30].

In the last two decades, nanomaterials have found applications in diverse fields such as medicine, electronics, energy storage, and environmental science due to nanomaterials' unique characteristics, such as high surface area-to-volume ratio, quantum effects, and superior mechanical properties. However, highly uniform nanomaterials with tightly controlled size, shape, morphology, and distribution are critical factors for achieving the desired properties in these applications. Nanomaterial's uniformity directly influences product functionality, including reaction rates, electrical conductivity, drug delivery efficiency, and catalytic performance.

1.1 Key Applications of Nanomaterials

1.1.1 Medicine and Drug Delivery

Targeted drug delivery systems have evolved using nanoparticles, such as liposomes, chitosan, and dendrimers, to deliver drugs specifically to diseased cells [24, 66, 77]. The uniform size and distribution of nanoparticles enhance cell uptake and reduce toxicity, improving treatment efficacy [35]. Furthermore, the solubility characteristics of drugs critically depend on particle size, and therefore, even small deviations in particle size in the nanoscale can alter the product bioavailability [61]. The bioavailability of NPs also depends on the absorption barriers of the targeted cells to the NPs that depend on nanoparticle surface characteristics such as hydrophobicity and surface ligands [78]. Therefore, consistent nanoparticle surface functionalization is highly desirable to achieve therapeutic efficacy.

1.1.2 Electronics

In semiconductors and transistors, uniform nanomaterials like quantum dots or nanowires offer consistent electrical properties. For example, variability of size and nanowire–superconductor interface in superconductor-nanowire hybrids can disrupt the flow of electrons, impacting device performance. Therefore, techniques such as selectively depositing the superconductor on the nanowire have been extensively studied to minimize the variability of the nanowire–superconductor interface [3]. Furthermore, uniform nanoparticles have been crucial for the miniaturization and enhancement of electronic components [60], and bottom-up synthesis approaches, such as atomic layer deposition, are widely used in semiconductor research due to the precise control they offer in manufacturing [69].

1.1.3 Energy Storage

Nanomaterials are used to develop advanced lithium-ion batteries and supercapacitors. Highly uniform materials in the electrodes, such as nanostructured carbon or metal oxides, lead to stable energy storage cycles and improved charge–discharge rates, contributing to better battery performance and longevity [41] as long as the conditions in the cells kept in a regime where reactive nanoparticles do not coarsen in time [67].

1.1.4 Environmental Science

In water purification and pollution control, uniform nanoparticles like TiO_2 or silver have been employed for their antimicrobial and photocatalytic properties. Uniformity ensures consistent interaction with contaminants, leading to efficient degradation or removal of pollutants [22]. Furthermore, enhanced environmental imaging in applications such as subsurface hydrocarbon detection in oil-field rocks or sensing in contaminated systems requires monodispersed iron oxide (magnetite) based nanoparticles [39] to increase the stability of the particles.

1.1.5 Catalysis

Supported metal or metal oxide nanoparticles are key in various catalysis applications ranging from pollution abatement in vehicle's catalytic converter to industrial-scale chemical processes such as hydrocarbon reforming. The active sites [76, 75] for structure-sensitive catalytic reactions are typically specific ensembles of surface atoms [81] or particle-support interface atoms [53] although all exposed metal atoms are active for structure insensitive catalytic reactions [8, 70]. Therefore, for structure-sensitive catalytic reactions where specific atomic ensembles are active sites, it is paramount to control the size of supported metal nanoparticles to maximize the fraction of active sites per given metal loading [47, 18, 23]. Moreover, the smaller nanoparticles have a higher fraction of exposed surface atoms, making them attractive for even structure-insensitive catalytic reactions because all those exposed metal atoms can be utilized for catalysis. However, small nanoparticles typically have higher chemical potential [12, 13] due to the presence of more under-coordinated surface sites compared to larger size particles and the bulk material and, therefore, susceptible to sintering and agglomeration during their catalytic lifetime [21, 52]. To minimize the sintering of the supported nanoparticles, it is desired to have narrower particle size distributions because the driving force for Ostwald ripening, a dominant sintering mechanism in most supported metal catalysts, is lower in narrower size distributions [37, 80].

In the above applications where nanoparticles are being used, the heterogeneities in particle size, shape, surface functionalization, and dispersion can occur at any stage, from synthesizing NPs to integrating into products [71]. These heterogeneities can significantly affect product performance and safety. Therefore, quality assurance (QA) in nanoparticle production involves developing and maintaining processes that prevent defects, minimize variability, and ensure that the final product meets predetermined specifications covering the entire production life span (Fig. 1), from raw material selection and synthesis methods to product testing and post-production handling. The QA process in nanoparticle production is built on several foundational components, including standard operating procedures (SOPs), regulatory guidelines, risk management, and quality control testing. In brief, SOPs ensure consistency across batches and reduce human error by outlining detailed protocols for the steps to be followed in nanoparticle synthesis, characterization, and handling. Nanoparticle production must also adhere to national and international regulatory frameworks, such as good manufacturing practice (GMP) guidelines for pharmaceuticals or ISO standards for nanomaterials, to ensure product safety and process efficiency [64]. Identifying and mitigating risks throughout the production process includes evaluating potential hazards such as contamination, particle agglomeration, and environmental impact, as well as implementing corrective actions when necessary. Predefined quality specifications of nanoparticles, such as size distribution, surface charge, purity, and stability, are maintained at different stages of production. This chapter will explore key QA principles, methods, and tools used to maintain the desired uniformity, stability, and safety of nanoparticles during production.

2 QA Methods and Tools to Maintain Desired Characteristics of Nanoparticles

The desired NP characteristics are quantified during nanoparticle production and after integrating into products using well-established advanced characterization techniques. These characterization methods include imaging and non-imaging methods. Transmission electron microscopy (TEM and scanning electron microscopy (SEM)

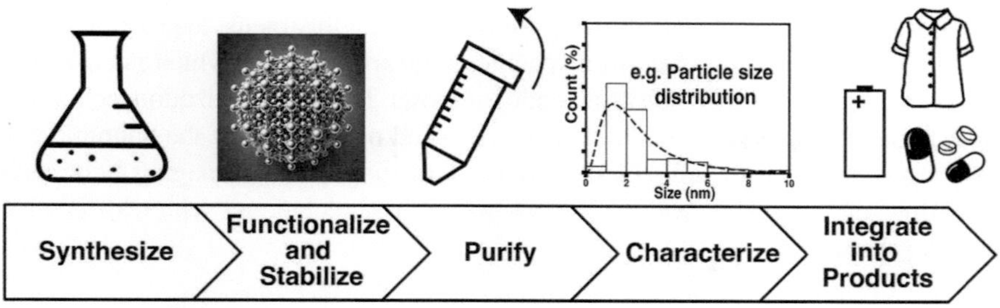

Fig. 1 Various processing stages of nanoparticle synthesis that require quality control

can directly visualize the size and shape of individual nanoparticles. High-resolution imaging can reveal details about morphology, particle distribution, and structural integrity. Furthermore, the energy-dispersive X-ray spectroscopy (EDX) can be done in conjunction with TEM or SEM to obtain the elemental distribution of the NPs. Those elemental mappings are useful in assessing the elemental heterogeneity of the nanoparticles. For example, EDX can identify special nanoparticles such as core–shell structures and Janus particles where a clear separation between elements in bi-metallic nanoparticles exists [19]. Real-time monitoring of nanoparticle synthesis is highly desirable to improve process control and consistency. However, in their current form, imaging techniques such as SEM, TEM, and EDX are labor-intensive and require manual intervention to acquire and interpret data. We will discuss recent improvements to imaging techniques in later sections of the chapter.

The surface properties of nanomaterials can be identified using techniques such as zeta potential measurement, Fourier-transform infrared spectroscopy (FTIR), and high-performance liquid chromatography (HPLC). Zeta potential measures the surface charge of nanoparticles in a colloidal solution and provides information about their stability in suspension. In general, higher zeta potentials indicate a stable nanoparticle suspension, while a lower value suggests a higher tendency of NP aggregation [34]. FTIR characterizes surface functional groups (e.g., hydroxyl, methyl) on nanoparticles by detecting specific chemical bonds and, therefore, commonly used to confirm surface modifications, such as coating with ligands or polymers [29, 84]. HPLC can be used to analyze and quantify the attachment of ligands or drugs, which is particularly important in pharmaceutical applications, where the accurate dosing and the range of ligand to particle ratio of a therapeutic agent is crucial [48].

In contrast to the imaging techniques, the average properties of NPs can be quantified using techniques such as Dynamic Light Scattering (DLS), chemisorption, and X-ray absorption near edge spectroscopy (XANES). DLS is a widely used technique to measure nanoparticle size distribution in a liquid suspension (hydrodynamic size) by analyzing fluctuations in light scattering caused by the Brownian motion of nanoparticles. Chemisorption uses a titrant such as CO, H_2, or O_2 to count the surface metal atoms of nanoparticles to assess the size, whereas in XANES, the coordination number of metal atoms can suggest the average size of particles. Recently, in-line spectroscopy [6, 45] or automated particle sizing systems gained interest as they can provide continuous feedback, allowing for immediate adjustments to process parameters if deviations from the desired specifications are detected, which is critical in process variable control in industrial-scale synthesis. The characterization techniques for nanomaterials (NMs) should be selected based on the type of the nanomaterial and property of interest. Table 1 lists most used imaging and non-imaging techniques for various nanomaterials and the importance of these techniques with respective to the QA parameters.

3 Controlling the Process Variables of NP Synthesis

Batch-to-batch reproducibility while combating particle aggregation and stability issues are the central focus of the process variable control in nanoparticle-based production. Minor variations in synthesis conditions, such as temperature, pH, or reagent purity, could result in product heterogeneity and propagate through the manufacturing process. If not properly stabilized, high surface energy per volume present in nanoparticles can lead to aggregation and alter the functional properties of the nanoparticles, which may complicate their performance in applications. Stability studies, including long-term storage and environmental impact, are critical to ensuring the viability of nanoparticle-based products. Process variables can significantly influence the quality of nanoparticles (NPs) at each stage of their life cycle [71]. These variables can lead to uniform or non-uniform distributions in terms of nanoparticle phase purity, size, shape, and surface properties. Below, we discuss the key process variables and the potential consequences of their improper control at different processing stages, as shown in Fig. 1.

3.1 Synthesis Stage

Key process variables in the synthesis stage are the concentration of reactants, temperature, reaction time, and pH of the synthesis media. For example, nucleation and growth rate of nanoparticles depend primarily on the precursor concentrations (degree of supersaturation) and the temperature in synthesis media [72], and improper controlling of these variables can cause uneven growth rates, resulting in size variability [42]. Short or excessively long reaction times can also affect the completeness of nanoparticle formation. Furthermore, pH influences the ionic environment and can alter the nanoparticles' stability, surface charge distribution, and morphology [4, 62].

Table 1 Characterization techniques for various NMs and their importance in assessing QA parameters

Characterization technique	Nanomaterials (NMs) characterized	Importance of technique	QA parameters assessed
Transmission electron microscopy (TEM)	Nanoparticles, nanorods, nanosheets	Provides high-resolution imaging to visualize the size, shape, and structure of individual NPs at the atomic level	Particle size and shape, morphology, structural integrity, defects, particle distributions
Scanning electron microscopy (SEM)	Nanoparticles, nanostructures, nanowires	Allows direct visualization of NPs' surface morphology, size, shape, and distribution	Surface morphology, particle distribution, surface integrity, particle size, uniformity
Energy-dispersive X-ray spectroscopy (EDX)	Nanoparticles, core–shell, Janus particles	Provides elemental analysis, revealing the composition and distribution of elements in individual NPs or specific regions	Elemental distribution, spatial heterogeneity, core–shell structure, composition mapping
Zeta potential measurement	Colloidal nanoparticles, suspensions	Measures the surface charge of NPs in suspension, providing insights into stability and aggregation tendencies	Surface charge, suspension stability, aggregation tendency, dispersion quality
Fourier-transform infrared spectroscopy (FTIR)	Functionalized NPs, drug-loaded NPs, coatings	Identifies surface functional groups and confirms surface modifications (e.g., coatings or ligand attachments)	Surface functionalization, surface chemistry, ligand/polymer attachment, coating integrity
High-performance liquid chromatography (HPLC)	Drug-loaded NPs, conjugated NPs	Quantifies and analyzes the attachment of drugs or ligands to NPs, especially in pharmaceutical applications	Drug/ligand attachment, quantification, ligand-to-particle ratio, dosage accuracy
Dynamic light scattering (DLS)	Nanoparticles in suspension, colloidal NPs	Measures the hydrodynamic size and distribution of NPs in solution based on light scattering due to Brownian motion	Particle size distribution, polydispersity index (PDI), stability in suspension, size consistency

(continued)

Table 1 (continued)

Characterization technique	Nanomaterials (NMs) characterized	Importance of technique	QA parameters assessed
Chemisorption	Metal NPs, catalysts	Assesses the number of surface metal atoms by using specific titrant gases, providing size-related data and surface characteristics	Surface area, surface metal atom density, size estimation, active site density
X-ray absorption near edge spectroscopy (XANES)	Metal NPs, nano catalysts	Provides information on the coordination environment of metal atoms, helping to estimate particle size and surface properties	Coordination number, particle size estimation, surface structure, oxidation state
In-line spectroscopy	Nanoparticles in suspension, in-process NPs	Provides real-time, continuous monitoring of particle properties during production, enabling process adjustments on the fly	Real-time size monitoring, production consistency, process optimization, deviation detection
Automated particle sizing systems	NPs in suspension, colloidal NPs	Offers continuous feedback and real-time monitoring of particle size distribution, especially in industrial-scale synthesis	Particle size consistency, process control, deviation detection, automated feedback for adjustments

3.2 Functionalization and Stabilization Stage

The choice and concentration of stabilizing agents, the solvent used, the type, pattern, and the speed of mixing can affect the uniformity of surface coatings, leading to either homogeneous or heterogeneous surface functionalization [11]. For example, the solvent can impact how evenly the functional groups are attached to the nanoparticle surface and the overall functionality and behavior of the nanoparticles [31].

3.3 Purification Stage

Centrifugation speed and duration are critical for the near-complete separation of impurities and minimizing aggregation of particles. In filtration methods, adequate pore sizes and consistent filtering pressure allow only particles of desired sizes to pass, resulting in uniform size distributions. Moreover, proper washing procedures should be followed to remove impurities that can cause clumping or inconsistent behavior in further applications.

3.4 Characterization Stage

Inaccurate or inconsistent measurement methods may fail to detect size variations or distribution uniformity, leading to misleading results. Furthermore, the sample sizes for imaging techniques should be sufficient to extract representative statistical quantities. Sample preparation for characterization techniques should follow proper protocols to avoid agglomeration or particle breakage during the measurement. Inline, high-throughput characterization techniques desirable in process control are discussed in Sect. 4 of this chapter.

3.5 Integration into the Products Stage

Concentration, dispersion, and factors like temperature, humidity, or pH during integration can affect performance consistency. Dosage or concentration levels when integrating nanoparticles into products such as pharmaceuticals are critical. Inefficient dispersion techniques can cause aggregation, resulting in non-uniform distribution within the product matrix.

By carefully controlling these variables at each stage, the distribution of nanoparticles can be made more uniform, which is essential for maintaining consistent quality and performance in their applications.

4 Statistical Analysis to Ensure Proper Characterization of NPs

Nanoparticles after synthesis media isolating show a distribution of properties such as size, morphology, and aspect ratio. Therefore, statistical analysis is paramount in quantifying the quality deviations of nanomaterials, especially in techniques such as TEM, where only a small fraction of NPs is being imaged.

Particle size distributions obtained from TEM are widely used in NP characterization. However, challenges for successful TEM are also reported. For example, it is difficult to detect and distinguish all the small particles in the presence of bigger size particles or agglomerates [79], and systematic rules should be applied when counting particles in TEM for better reproducibility [10]. Likewise, particles with two to three times larger than the average may present as only a small fraction of the population owing to the characteristics of typically observed log-normal distributions of NPs and, therefore, pose challenges to detect by counting a small number of nanoparticles. Therefore, choosing an appropriate number of particles to sample and measure is critical when using microscopic technique such as TEM to determine the size distribution of NP samples. Although, the imaged nanoparticles will be a small fraction of the actual NP population with an appropriate number of particles chosen for imaging, the statistics of the population and sample could be closely matched. The ISO standard 13322-1:2004 states that the size distribution of the sample and the degree of confidence required determines the number of particles to be imaged. Assuming that the particles sizes follow a known distribution with mean size (μ), the number of particles (n) that need to be measured for a given confidence limit is given by [9]:

$$\ln(n) = -2\ln(\mu) + K \tag{1}$$

where K depends on the required confidence limit and the particle distribution function. For example, in a log-normal distribution $K = \ln\left[\sigma^2\left(\frac{\sigma^2}{2} + 1\right)\right]$, where σ^2 is the variance of the nanoparticle size. Therefore, Eq. 1 can be rewritten for a log-normally distributed particle population as follows [9].

$$\ln(n) = -2\ln(\mu) + \ln\left[\sigma^2\left(\frac{\sigma^2}{2} + 1\right)\right] \tag{2}$$

We should note that the population mean (μ), and the variance (σ^2) are unknown as we only measure sample properties. However, if we constrain the minimum number of particles imaged to a sufficiently large number ($n > 30$), the difference between the population mean and sample mean will be minimal. Figure 2 plots Eq. 2 and reveals that for most nearly spherical log-normally distributed particle size distributions, $n =$ 30–200 should be sufficient for extracting statistics of the size distribution. However, Fig. 2 also shows that for samples with smaller mean and large variance (e.g., lower right corner), a higher number of particles should be imaged. Furthermore, if the aspect ratios are different in major and minor axis of particles, then more particles need to be imaged for accurate quantification.

It is desirable to have a scalable, low-cost, and high-throughput statistical analysis of particle properties for industrial-scale nanoparticle production. Such characterization procedures enable the process variables to be adjusted using a process-control feedback loop to meet the required properties, thereby ensuring the quality of the product. However, currently widely used imaging techniques in NP characterization such as TEM (discussed above), despite developments of software such as ImageJ [1] and Pebbles [46], are still time and labor-intensive, making them hard to utilize in real time process control. Nevertheless, there have been attempts to modify the data acquisition and analysis of such characterization methods such that rapid automated readings are possible [6, 38, 45]. These methods include Spatially Resolved Dynamic Light Scattering (SR-DLS) where an illuminated volume of nanoparticle

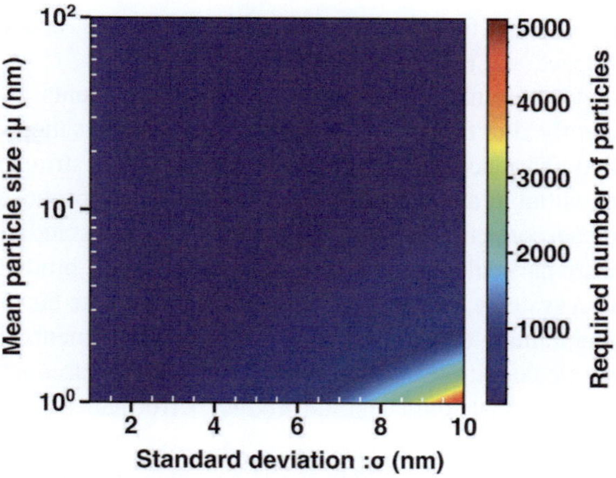

Fig. 2 Minimum number of particles (spherical) required for imaging to extract accurate statistical properties of nanoparticles based on ISO standard 13322-1:2004 for nanoparticles that follow a log-normal distribution with mean (μ) and standard deviation (σ)

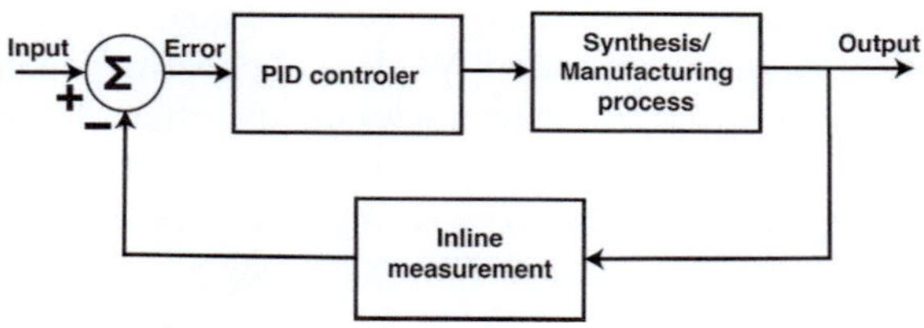

Fig. 3 Feedback loop for nanoparticle production process control to assure product quality

suspension's scattered light signal is measured as function of sample depth [6]. At timescales required for manufacturing control where NPs in flow and at high turbidity solutions, continuous measurement has been integrated in PID controls (Fig. 3) [20]. Quality assurance of nanomaterial production at industrial scale demands feedback loops in each unit operations. Therefore, Fig. 3 represents a general QA framework applicable for each process unit to guarantee the synthesized NMs are within required specifications. Moreover, some high-throughput NP characterization methods such as combining image processing, machine learning techniques such as Gaussian mixture model based on expectation maximization (EM) algorithms [82] and Markov chain Monte Carlo (MCMC) to deconvolute the complex morphological features of nanoparticle ensembles [38] could be utilized in such QA frameworks.

5 Risk Management, Regulatory Considerations and Compliance in Nanoparticle Production

The same unique physical, chemical, and biological properties, making them highly valuable across industries such as pharmaceuticals, electronics, and environmental sciences, can introduce risks related to human health, workplace safety, and environmental sustainability. Quality Assurance (QA) protocols are essential in managing these risks, focusing on contamination control, occupational safety, environmental impact, and risk management.

Contaminants, including dust, biological agents, or chemical residues, can significantly alter the properties and functionality of nanoparticles, particularly in medical and pharmaceutical applications. Maintaining stringent cleanroom environments is essential to avoid contamination that can compromise product integrity. Therefore, cleanroom environments must adhere to strict standards such as ISO 14644 to minimize particulate contamination in nanoparticle production [36]. Advanced air filtration systems, such as high-efficiency particulate air (HEPA) filters, are essential for maintaining the sterility of production environments [63]. For example, maintaining sterile conditions is critical to preventing the introduction of unwanted pathogens or particles that could degrade product performance is essential in medical applications [7].

Nanoparticles could pose significant health risks to workers due to inhalation and dermal exposure. Therefore, risk assessments and protective measures are necessary to minimize exposure and prevent potential health hazards, including respiratory diseases or skin irritation. Workers exposed to nanoparticles may face respiratory and skin issues, as nanoparticles can penetrate deeper into the respiratory system than larger particles [14]. In addition, dermal exposure can result in nanoparticle absorption through the skin, posing long-term health risks [68]. QA protocols should enforce protective measures such as using personal protective equipment (PPE) such as respirators, gloves, and protective clothing. Engineering controls like ventilation systems and fume hoods are essential for minimizing airborne nanoparticles in production areas [44].

Nanoparticle production generates chemical solvents, hazardous waste, and byproducts, which must be properly managed to minimize environmental risks. Improper disposal or accidental release of nanoparticles into the environment can contaminate soil, water, and air. Safe disposal and treatment of waste materials generated during nanoparticle production are critical for environmental protection. Recycling or neutralizing chemical solvents and safely disposing of hazardous waste are key strategies for reducing the ecological footprint of nanoparticle manufacturing [73]. The potential for nanoparticles to be released into the environment poses long-term risks, as these particles may persist in ecosystems and bioaccumulate, leading to toxic effects on wildlife and humans [32].

A comprehensive risk management approach is necessary to address nanoparticle production's health, safety, and environmental risks. Effective QA protocols should include real-time monitoring, hazard assessments, and compliance with international standards for nanomaterials production. Risk assessment is essential for identifying potential hazards throughout the nanoparticle production process. Those risk assessments should incorporate into risk management strategies including continuously monitoring production environments and safety audits to prevent exposure incidents and ensure adherence to best practices [55].

Regulatory compliance is critical for ensuring the safety and quality of nanoparticle-based products. In the European Union, the REACH (Registration,

Evaluation, Authorization, and Restriction of Chemicals) regulation requires manufacturers to assess the environmental and health risks of substances, including nanoparticles, throughout their lifecycle. This ensures that potential hazards associated with nanoscale materials are thoroughly evaluated [65]. In addition, the European commission has released specific guidance for nanomaterials, which focuses on areas such as toxicity, bioaccumulation, and exposure assessment [54].

In the United States, the Food and Drug Administration (FDA) has issued guidelines to evaluate nanoparticle products, particularly in pharmaceuticals and medical devices, to ensure safety, efficacy, and appropriate labeling [74]. Furthermore, the U.S. Environmental Protection Agency (EPA), through the Toxic Substances Control Act (TSCA), mandates manufacturers to report the production, use, and health impacts of engineered nanomaterials [17]. Alongside these regulatory frameworks, international standards like ISO 9001 [26] and ISO 13485 [27] are essential in maintaining quality assurance. ISO 9001 ensures consistent quality management practices, focusing on continuous improvement and risk management in nanoparticle production [26]. For medical devices incorporating nanomaterials, ISO 13485 provides stricter guidelines, ensuring that processes are auditable, traceable, and meet rigorous safety standards [27]. Compliance with these frameworks enables manufacturers to meet regulatory requirements while fostering consumer trust and minimizing risks during production scale-up [58].

The industrial production of nanomaterials (NMs) requires stringent practices to ensure both product quality and operational safety. As nanotechnology advances, several industry-level practices and guidelines have been established to address these concerns. Below are some of the key practices:

5.1 Product Quality Management

- **Standardization of Materials**: To maintain consistency, industries rely on established standards like ISO 9001 (Quality Management Systems) [26] and ISO/TS 80004 (Terminology and Definitions for Nanotechnologies) [28]. These standards ensure that the properties, size, shape, and surface characteristics of nanomaterials meet the required specifications.
- **Characterization Techniques**: Advanced characterization techniques such as Scanning Electron Microscopy (SEM), Atomic Force Microscopy (AFM), X-ray Diffraction (XRD), and Dynamic Light Scattering (DLS) are routinely used to analyze particle size, morphology, and surface properties of nanomaterials. This ensures uniformity and reproducibility in the industrial scale manufacturing processes of NM incorporated products, such as cement-based composites [59] and catalyst layer designs in membrane electrode assemblies [43].
- **Control of Nanomaterial Properties**: The control over the synthesis process (e.g., sol–gel, chemical vapor deposition, and ball milling) allows for fine-tuning of the material properties such as surface area, particle distribution, and functionality, which are crucial for product performance [26].

5.2 *Operational Safety Practices*

- **Risk Assessment and Management**: The potential risks of nanomaterials, such as toxicity or environmental impact, are assessed using methods like quantitative structure–activity relationship (QSAR) modeling, which helps predict the hazard associated with nanomaterials before their widespread use. Risk assessment includes consideration of exposure routes, physicochemical properties, and the intended application [25].
- **Occupational Safety**: The handling of nanomaterials in industrial settings often requires specialized safety protocols to mitigate inhalation, ingestion, or skin contact risks. These measures include:
  - **Engineering Controls**: Ventilation systems (e.g., fume hoods, local exhaust systems) and containment strategies (e.g., closed production systems) to minimize exposure to airborne nanoparticles.
  - **Workplace Monitoring**: Continuous monitoring of air quality and nanoparticle concentration in the workplace using specialized equipment helps maintain safe exposure levels [57].
  - **Personal Protective Equipment (PPE)**: Employees at NM processing facilities must be equipped with appropriate PPE such as gloves, face shields, respirators, and protective clothing to prevent direct contact with nanoparticles [25].
- **Hazardous Material Handling and Disposal**: Nanomaterials may require special handling and disposal protocols due to their potential toxicity. Industries follow guidelines provided by OSHA [56], NIOSH [50, 51] and, local environmental agencies for safe disposal and containment of waste [16].

5.3 *Environmental, Health, and Safety (EHS) Regulations*

- **Regulatory Compliance**: Compliance with national and international regulatory frameworks such as REACH, TSCA in the U.S., or the European Union's regulation on the classification, labeling, and packaging of substances and mixtures ensures that nanomaterials are evaluated for safety, environmental impact, and human health risks [15].
- **Toxicology and Biocompatibility Studies**: The impact of nanomaterials on human health and the environment is a highly active area of research. Toxicology studies, including in vitro and in vivo testing are conducted to assess the potential hazards of NMs and establish safe exposure levels [15].
- **Eco-design Principles**: Sustainable practices are integrated into nanomaterial production processes, emphasizing minimizing waste, energy consumption, and environmental harm [15].

5.4 Supply Chain and Traceability

- **Traceability Systems**: Effective traceability is implemented to track the nanomaterials from raw material sourcing to final product delivery. This includes labeling and documentation that inform about the nature of the nanomaterials, their production processes, and safety measures [33].
- **Sourcing of Raw Materials**: Ethical and quality standards for sourcing raw materials ensure the responsible procurement of materials, reducing risks related to contamination or the use of hazardous substances in the production process.

5.5 Innovation in Production Processes

- **Continuous Improvement**: Lean manufacturing principles and Six Sigma methodologies are applied to continuously improve product quality and operational efficiency, reduce waste, and minimize defects in nanomaterial production
- **Automation and Advanced Manufacturing**: Automated systems in nanomaterial synthesis, packaging, and quality testing can improve precision and reduce human error, thereby enhancing both product quality and safety.

6 Conclusion

While novel synthesis strategies provide different ways of synthesizing nanomaterials of interest, controlling the nanomaterial properties is vital to achieve predictable and well-behaved composites in both laboratory and industrial applications. Quality assurance in nanoparticle production ensures that these advanced nanomaterials can be safely and effectively integrated into various applications. Manufacturers can reduce variability, mitigate risks, and maintain compliance with regulatory requirements by implementing rigorous QA practices, such as statistical analysis and inline process monitoring. The complexity of nanoparticle systems and propagation of quality deviations through processing steps demands a comprehensive approach to quality, incorporating advanced characterization tools, real-time monitoring, and robust risk management strategies. Therefore, QA practices will play a crucial role in realizing the full potential of nanoparticles in applications.

References

1. Abràmoff MD, Magalhães PJ, Ram SJ (2004) Image processing with Image. J Biophotonics Int 11:36–42. https://imagescience.org/meijering/publications/download/bio2004.pdf
2. Alivisatos AP (1996) Perspectives on the physical chemistry of semiconductor nanocrystals. J Phys Chem 100:13226–13239. https://doi.org/10.1021/jp9535506

3. Badawy G, Bakkers EPAM (2024) Electronic transport and quantum phenomena in nanowires. Chem Rev 124:2419–2440. https://doi.org/10.1021/acs.chemrev.3c00656
4. Badawy AME, Luxton TP, Silva RG, Scheckel KG, Suidan MT, Tolaymat TM (2010) Impact of environmental conditions (pH, ionic strength, and electrolyte type) on the surface charge and aggregation of silver nanoparticles suspensions. Environ Sci Technol 44:1260–1266. https://doi.org/10.1021/es902240k
5. Baughman RH, Zakhidov AA, de Heer WA (2002) Carbon nanotubes—the route toward applications. Science 297:787–792. https://doi.org/10.1126/science.1060928
6. Besseling R, Damen M, Wijgergangs J, Hermes M, Wynia G, Gerich A (2019) New unique PAT method and instrument for real-time inline size characterization of concentrated, flowing nanosuspensions. Eur J Pharm Sci 133:205–213. https://doi.org/10.1016/j.ejps.2019.03.024
7. Borm PJ, Robbins D, Haubold S, Kuhlbusch T, Fissan H, Donaldson K, Schins R, Stone V, Kreyling W, Lademann J, Krutmann J, Warheit D, Oberdorster E (2006) The potential risks of nanomaterials: a review carried out for ECETOC. Part Fibre Toxicol 3:11. https://doi.org/10.1186/1743-8977-3-11
8. Boudart M (1985) Heterogeneous catalysis by metals. J Mol Catal 30:27–38. https://doi.org/10.1016/0304-5102(85)80014-6
9. Boyd R, Cuenat A, Meli F, Klein T, Frase CG, Gleber G, Krumrey M, Duta A, Duta S, Hogstrom R, Prieto E (2011) Good practice guide for the determination of the size and size distribution of spherical nanoparticle samples, measurement good practice guide. National Physical Laboratory. http://eprintspublications.npl.co.uk/id/eprint/5444
10. Bresch H, Hodoroaba V-D, Schmidt A, Rasmussen K, Rauscher H (2022) Counting small particles in electron microscopy images—proposal for rules and their application in practice. Nanomaterials 12:2238. https://doi.org/10.3390/nano12132238
11. Buyukgoz GG, Castro JN, Atalla AE, Pentangelo JG, Tripathi S, Davé RN (2021) Impact of mixing on content uniformity of thin polymer films containing drug micro-doses. Pharmaceutics 13:812. https://doi.org/10.3390/pharmaceutics13060812
12. Campbell CT (2013) The energetics of supported metal nanoparticles: relationships to sintering rates and catalytic activity. Acc Chem Res 46:1712–1719. https://doi.org/10.1021/ar3003514
13. Campbell CT, Mao Z (2017) Chemical potential of metal atoms in supported nanoparticles: dependence upon particle size and support. ACS Catal 7:8460–8466. https://doi.org/10.1021/acscatal.7b03090
14. Donaldson K, Seaton A (2012) A short history of the toxicology of inhaled particles. Part Fibre Toxicol 9:13. https://doi.org/10.1186/1743-8977-9-13
15. ECHA (European Chemicals) (2017) Guidance on the biocidal products regulation volume IV environment-assessment and evaluation (Parts B+ C).
16. Eastlake A, Hodson L, Geraci C, Crawford C (2012) A critical evaluation of material safety data sheets (MSDSs) for engineered nanomaterials. J Chem Health Saf 19:1–8. https://doi.org/10.1016/j.jchas.2012.02.002
17. Environmental Protection Agency (2017) Chemical substances when manufactured or processed as nanoscale materials; TSCA reporting and recordkeeping requirements. Fed Regist 82(8):3641–3655. https://www.federalregister.gov/d/2017-00052
18. van Etten MPC, Zijlstra B, Hensen EJM, Filot IAW (2021) Enumerating active sites on metal nanoparticles: understanding the size dependence of cobalt particles for CO dissociation. ACS Catal 11:8484–8492. https://doi.org/10.1021/acscatal.1c00651
19. Ferrando R, Jellinek J, Johnston RL (2008) Nanoalloys: from theory to applications of alloy clusters and nanoparticles. Chem Rev 108:845–910. https://doi.org/10.1021/cr040090g
20. Grau-Carbonell A, Wang Y, Verbeek M, van Tuijn R, Schuurmans C, Besseling R, Geirich A (2023) Real-time process control in nanodispersion manufacturing via continuous nanoparticle size monitoring [WWW Document]. AZoNano. https://www.azonano.com/article.aspx?ArticleID=6591. Accessed 10 July 2024
21. Hansen TW, Delariva AT, Challa SR, Datye AK (2013) Sintering of catalytic nanoparticles: particle migration or ostwald ripening? Acc Chem Res 46:1720–1730. https://doi.org/10.1021/ar3002427

22. ul Haq A, Saeed M, Khan SG, Ibrahim M (2021) Photocatalytic applications of titanium dioxide (TiO_2). In: Titanium dioxide—advances and applications. IntechOpen. https://doi.org/10.5772/intechopen.99598
23. Van Hardeveld R, Hartog F (1969) The statistics of surface atoms and surface sites on metal crystals. Surf Sci 15:189–230. https://doi.org/10.1016/0039-6028(69)90148-4
24. Hasan M, Belhaj N, Benachour H, Barberi-Heyob M, Kahn CJF, Jabbari E, Linder M, Arab-Tehrany E (2014) Liposome encapsulation of curcumin: physico-chemical characterizations and effects on MCF7 cancer cell proliferation. Int J Pharm 461:519–528. https://doi.org/10.1016/j.ijpharm.2013.12.007
25. Hoet P, Legiest B, Geys J, Nemery B (2009) Do nanomedicines require novel safety assessments to ensure their safety for long-term human use? Drug-Safety 32:625–636. https://doi.org/10.2165/00002018-200932080-00002
26. International Organization for Standardization (2015) ISO 9001:2015 – Quality Management Systems – Requirements. ISO, Geneva. https://www.iso.org/standard/62085.html
27. International Organization for Standardization (2016) ISO 13485:2016 – Medical Devices – Quality Management Systems – Requirements for Regulatory Purposes. ISO, Geneva. https://www.iso.org/standard/59752.html
28. International Organization for Standardization (2015) ISO/TS 80004-1:2015 – Nanotechnologies — Vocabulary — Part 1: Core Terms. ISO, Geneva. https://www.iso.org/standard/66642.html
29. Jadhav SA, Brunella V, Scalarone D (2015) Polymerizable ligands as stabilizers for nanoparticles. Part Part Syst Charact 32:417–428. https://doi.org/10.1002/ppsc.201400074
30. Kelly KL, Coronado E, Zhao LL, Schatz GC (2003) The optical properties of metal nanoparticles: the influence of size, shape, and dielectric environment. J Phys Chem B 107:668–677. https://doi.org/10.1021/jp026731y
31. Kim Y-G, Thérien-Aubin H (2020) Impact of the solvent quality on the local dynamics of soft and swollen polymer nanoparticles functionalized with polymer chains. Macromolecules 53:7561–7569. https://doi.org/10.1021/acs.macromol.0c00346
32. Klaine SJ, Alvarez PJJ, Batley GE, Fernandes TF, Handy RD, Lyon DY, Mahendra S, McLaughlin MJ, Lead JR (2008) Nanomaterials in the environment: behavior, fate, bioavailability, and effects. Environ Toxicol Chem 27:1825–1851. https://doi.org/10.1897/08-090.1
33. Kuhl K, Stepa R, Schmitz-Felten E, Klueser R, Winski T, Graveling R (2012) OSH WIKI: nanomaterials. Nanomaterials. https://oshwiki.osha.europa.eu/en/themes/nanomaterials. Accessed 1 July 2025
34. Kumar A, Dixit CK (2017) 3—Methods for characterization of nanoparticles. In: Nimesh S, Chandra R, Gupta N (eds) Advances in nanomedicine for the delivery of therapeutic nucleic acids. Woodhead Publishing, pp 43–58. https://doi.org/10.1016/B978-0-08-100557-6.00003-1
35. Kumar A, Sadhu SD, Singh R (2018) Nanotechnology: key for sustainable future. In: Nanotechnology in environmental science. Wiley, pp 759–804. https://doi.org/10.1002/9783527808854.ch24
36. Lakshmana Prabu S, Suriyaprakash TNK, Ruckmani K, Thirumurugan R (2017) Chapter 3—GMP in pharma manufacturing—description of GMP as related to air-handling units and prevention of contamination and implementation of GMP regulatory requirements. In: Kohli R, Mittal KL (eds) Developments in surface contamination and cleaning: types of contamination and contamination resources. William Andrew Publishing, pp 85–123. https://doi.org/10.1016/B978-0-323-43158-3.00005-8
37. Lardinois TM, Mandal K, Yadav V, Wijerathne A, Bolton BK, Lippie H, Li CW, Paolucci C, Gounder R (2022) Kinetic and thermodynamic factors influencing palladium nanoparticle redispersion into mononuclear Pd(II) cations in zeolite supports. J Phys Chem C. https://doi.org/10.1021/acs.jpcc.2c01613
38. Lee B, Yoon S, Lee JW, Kim Y, Chang J, Yun J, Ro JC, Lee J-S, Lee JH (2020) Statistical characterization of the morphologies of nanoparticles through machine learning based electron microscopy image analysis. ACS Nano 14:17125–17133. https://doi.org/10.1021/acsnano.0c06809

39. Li W, Liu D, Wu J, Kim C, Fortner JD (2014) Aqueous aggregation and surface deposition processes of engineered superparamagnetic iron oxide nanoparticles for environmental applications. Environ Sci Technol 48:11892–11900. https://doi.org/10.1021/es502174p
40. Love GMW, Christopher J (2001) The art of building small [WWW Document]. Scientific American. https://www.scientificamerican.com/article/the-art-of-building-small/. Accessed 10 Aug 2024
41. Mahmud S, Rahman M, Kamruzzaman M, Ali MO, Emon MSA, Khatun H, Ali MR (2022) Recent advances in lithium-ion battery materials for improved electrochemical performance: a review. Results Eng 15:100472. https://doi.org/10.1016/j.rineng.2022.100472
42. Manta P, Chandra Singh S, Deep A, Kapoor DN (2021) Temperature-regulated gold nanoparticle sensors for immune chromatographic rapid test kits with reproducible sensitivity: a study. IET Nanobiotechnol 15:338–346. https://doi.org/10.1049/nbt2.12024
43. Maric R, Ayers K, Wagner A, Bliznakov S, Bonville L (2022) Catalyst layer design, manufacturing and in-line quality control. University of Connecticut, Storrs, CT, p 06269
44. Maynard AD, Kuempel ED (2005) Airborne nanostructured particles and occupational health. J Nanopart Res 7:587–614. https://doi.org/10.1007/s11051-005-6770-9
45. McLeod E, Dincer TU, Veli M, Ertas YN, Nguyen C, Luo W, Greenbaum A, Feizi A, Ozcan A (2015) High-throughput and label-free single nanoparticle sizing based on time-resolved on-chip microscopy. ACS Nano 9:3265–3273. https://doi.org/10.1021/acsnano.5b00388
46. Mondini S, Ferretti AM, Puglisi A, Ponti A (2012) PEBBLES and PEBBLEJUGGLER: software for accurate, unbiased, and fast measurement and analysis of nanoparticle morphology from transmission electron microscopy (TEM) micrographs. Nanoscale 4:5356–5372. https://doi.org/10.1039/C2NR31276J
47. Montejano-Carrizales JM, Aguilera-Granja F, Morán-López JL (1997) Direct enumeration of the geometrical characteristics of clusters. Nanostruct Mater 8:269–287. https://doi.org/10.1016/S0965-9773(97)00168-2
48. Mullen DG, Banaszak Holl MM (2011) Heterogeneous ligand-nanoparticle distributions: a major obstacle to scientific understanding and commercial translation. Acc Chem Res 44:1135–1145. https://doi.org/10.1021/ar1001389
49. Murray CB, Norris DJ, Bawendi MG (1993) Synthesis and characterization of nearly monodisperse CdE (E = sulfur, selenium, tellurium) semiconductor nanocrystallites. J Am Chem Soc 115:8706–8715. https://doi.org/10.1021/ja00072a025
50. NIOSH (2019) NIOSH nanotechnology research plan for 2018–2025. https://doi.org/10.26616/NIOSHPUB2019116
51. National Institute for Occupational Safety and Health (2016) Building a safety program to protect the nanotechnology workforce: a guide for small to medium-sized enterprises. DHHS (NIOSH) Publication No. 2016-102. U.S. Department of Health and Human Services, Centers for Disease Control and Prevention. https://www.cdc.gov/niosh/docs/2016-102/default.html
52. Ndolomingo MJ, Bingwa N, Meijboom R (2020) Review of supported metal nanoparticles: synthesis methodologies, advantages and application as catalysts. J Mater Sci 55:6195–6241. https://doi.org/10.1007/s10853-020-04415-x
53. Neyman KM, Kozlov SM (2022) Quantifying interactions on interfaces between metal particles and oxide supports in catalytic nanomaterials. NPG Asia Mater 14:1–8. https://doi.org/10.1038/s41427-022-00405-4
54. Nielsen MB, Skjolding L, Baun A, Hansen SF (2023) European nanomaterial legislation in the past 20 years—closing the final gaps. NanoImpact 32:100487. https://doi.org/10.1016/j.impact.2023.100487
55. Oberdörster G, Oberdörster E, Oberdörster J (2005) Nanotoxicology: an emerging discipline evolving from studies of ultrafine particles. Environ Health Perspect 113:823–839. https://doi.org/10.1289/ehp.7339
56. Occupational Safety and Health Administration (2013) Working safely with nanomaterials. OSHA Fact Sheet No. 3634. U.S. Department of Labor. https://www.osha.gov/sites/default/files/publications/OSHA_FS-3634.pdf

57. Occupational Safety and Health Administration (2013) Working safely with nanomaterials. OSHA Fact Sheet No. 3634. U.S. Department of Labor. https://www.osha.gov/sites/default/files/publications/OSHA_FS-3634.pdf
58. Onyeaka H, Passaretti P, Miri T, Al-Sharify ZT (2022) The safety of nanomaterials in food production and packaging. Curr Res Food Sci 5:763–774. https://doi.org/10.1016/j.crfs.2022.04.005
59. Paul SC, van Rooyen AS, van Zijl GPAG, Petrik LF (2018) Properties of cement-based composites using nanoparticles: a comprehensive review. Constr Build Mater 189:1019–1034. https://doi.org/10.1016/j.conbuildmat.2018.09.062
60. Payal, Pandey P (2022) Role of nanotechnology in electronics: a review of recent developments and patents. Recent Patents Nanotechnol 16:45–66. https://doi.org/10.2174/1872210515666210120114504
61. Peltonen L (2018) Practical guidelines for the characterization and quality control of pure drug nanoparticles and nano-cocrystals in the pharmaceutical industry. Adv Drug Deliv Rev 131:101–115. https://doi.org/10.1016/j.addr.2018.06.009
62. Peng C, Zhang W, Gao H, Li Y, Tong X, Li K, Zhu X, Wang Y, Chen Y (2017) Behavior and potential impacts of metal-based engineered nanoparticles in aquatic environments. Nanomaterials 7:21. https://doi.org/10.3390/nano7010021
63. Rauscher H, Roebben G, Boix SA, Emons H, Gibson P, Koeber R, Linsinger T, Rasmussen K, Riego SJ, Sokull-Kluettgen B, Stamm H (2015) Towards a review of the EC Recommendation for a definition of the term "nanomaterial": Part 3: Scientific-technical evaluation of options to clarify the definition and to facilitate its implementation [WWW Document]. JRC Publications Repository. https://doi.org/10.2788/770401
64. Roubert F, Beuzelin-Ollivier M-G, Hofmann-Amtenbrink M, Hofmann H, Hool A (2016) "Nanostandardization" in action: implementing standardization processes in a multidisciplinary nanoparticle-based research and development project. NanoEthics 10:41–62. https://doi.org/10.1007/s11569-015-0248-8
65. Rudén C, Hansson SO (2010) Registration, evaluation, and authorization of chemicals (REACH) is but the first step-how far will it take us? Six further steps to improve the European chemicals legislation. Environ Health Perspect 118:6–10. https://doi.org/10.1289/ehp.0901157
66. Sanyakamdhorn S, Agudelo D, Tajmir-Riahi H-A (2013) Encapsulation of antitumor drug doxorubicin and its analogue by chitosan nanoparticles. Biomacromol 14:557–563. https://doi.org/10.1021/bm3018577
67. Schoonman J (2000) Nanostructured materials in solid state ionics. Solid State Ionics 135:5–19. https://doi.org/10.1016/S0167-2738(00)00324-6
68. Schulte P, Geraci C, Zumwalde R, Hoover M, Kuempel E (2008) Occupational risk management of engineered nanoparticles. J Occup Environ Hyg 5:239–249. https://doi.org/10.1080/15459620801907840
69. Sheng J, Lee J-H, Choi W-H, Hong T, Kim M, Park J-S (2018) Review Article: Atomic layer deposition for oxide semiconductor thin film transistors: advances in research and development. J Vac Sci Technol, A 36:060801. https://doi.org/10.1116/1.5047237
70. Somorjai GA, Carrazza J (1986) Structure sensitivity of catalytic reactions. Ind Eng Chem Fund 25:63–69. https://doi.org/10.1021/i100021a009
71. Stavis SM, Fagan JA, Stopa M, Liddle JA (2018) Nanoparticle manufacturing—heterogeneity through processes to products. ACS Appl. Nano Mater. 1:4358–4385. https://doi.org/10.1021/acsanm.8b01239
72. Thanh NTK, Maclean N, Mahiddine S (2014) Mechanisms of nucleation and growth of nanoparticles in solution. Chem Rev 114:7610–7630. https://doi.org/10.1021/cr400544s
73. Tyagi P, Singh R, Vats S, Kumar D, Tyagi S (2012) Nanomaterials use in wastewater treatment. In: Nanomaterials use in wastewater treatment. Presented at the International conference on nanotechnology and chemical engineering, Bangkok
74. U.S. Food and Drug Administration (2022) Drug products, including biological products, that contain nanomaterials: guidance for industry. U.S. Department of Health and Human Services. https://www.fda.gov/media/157812/downloadfda.gov

75. Vogt C, Weckhuysen BM (2022) The concept of active site in heterogeneous catalysis. Nat Rev Chem 6:89–111. https://doi.org/10.1038/s41570-021-00340-y
76. Védrine JC (2014) Revisiting active sites in heterogeneous catalysis: their structure and their dynamic behaviour. Appl Catal A General B Delmon Honor Special Issue 474:40–50. https://doi.org/10.1016/j.apcata.2013.05.029
77. Wang X, Cai X, Hu J, Shao N, Wang F, Zhang Q, Xiao J, Cheng Y (2013) Glutathione-triggered "off–on" release of anticancer drugs from dendrimer-encapsulated gold nanoparticles. J Am Chem Soc 135:9805–9810. https://doi.org/10.1021/ja402903h
78. Wang Y, Pi C, Feng X, Hou Y, Zhao L, Wei Y (2020) The influence of nanoparticle properties on oral bioavailability of drugs. Int J Nanomed 15:6295–6310. https://doi.org/10.2147/IJN.S257269
79. Weibel A, Bouchet R, Boulc' F, Knauth P (2005) The big problem of small particles: a comparison of methods for determination of particle size in nanocrystalline anatase powders. Chem Mater 17:2378–2385. https://doi.org/10.1021/cm0403762
80. Wettergren K, Schweinberger FF, Deiana D, Ridge CJ, Crampton AS, Rötzer MD, Hansen TW, Zhdanov VP, Heiz U, Langhammer C (2014) High sintering resistance of size-selected platinum cluster catalysts by suppressed Ostwald ripening. Nano Lett 14:5803–5809. https://doi.org/10.1021/nl502686u
81. Yang A-C, Choksi T, Streibel V, Aljama H, Wrasman CJ, Roling LT, Goodman ED, Thomas D, Bare SR, Sánchez-Carrera RS, Schäfer A, Li Y, Abild-Pedersen F, Cargnello M (2020) Revealing the structure of a catalytic combustion active-site ensemble combining uniform nanocrystal catalysts and theory insights. Proc Natl Acad Sci 117:14721–14729. https://doi.org/10.1073/pnas.2002342117
82. Zelenty J, Dahl A, Hyde J, Smith GDW, Moody MP (2017) Detecting clusters in atom probe data with Gaussian mixture models. Microsc Microanal 23:269–278. https://doi.org/10.1017/S1431927617000320
83. Zhang Y, Kohler N, Zhang M (2002) Surface modification of superparamagnetic magnetite nanoparticles and their intracellular uptake. Biomaterials 23:1553–1561. https://doi.org/10.1016/S0142-9612(01)00267-8
84. Zhang F, Lees E, Amin F, Rivera_Gil P, Yang F, Mulvaney P, Parak WJ (2011) Polymer-coated nanoparticles: a universal tool for biolabelling experiments. Small 7:3113–3127. https://doi.org/10.1002/smll.201100608

Process Economics of Nanomaterials Production—A Decision Analysis Framework for Assessing Technical, Business and Environmental Viability

Aruna Manipura

Abstract Demand for nanomaterials is increasing with emerging applications in electronics, biomedical, chemical catalysis, paints and coating, fertilizer, construction, composites, energy harnessing and storage industries. However, compared to bulk chemicals, only smaller quantities of nanomaterials are required in formulations or integrations into novel products arising from various industrial sectors. Markets are not well defined and there are many barriers for entering to those industries due to evolving nature of nanomaterial applications, inadequate analysis of long-term impact, fate and transport in the environment, and high cost of production etc. Despite these challenges, the estimated global market for nanomaterials will be over USD 27.4 billion by 2030 with a compound annual growth rate (CAGR) greater than 15%. This situation encourages both established and novel players to enter the nanomaterial market as these have shown significant improvements in performance of formulated products. Therefore, a general decision analysis framework for assessing technical, business, and environmental viability of industrial scale nanomaterials production is proposed through a literature review. The proposed framework consists of decision analysis for market and business viability, process chemistry R&D focusing on nanomaterial synthesis, process design and equipment selection, and criteria for overall optimization of the whole manufacturing process. Then, a case study is presented by applying the proposed decision analysis framework for evaluating process economics of graphene oxide production at industrial scale. Further improvement opportunities and limitations of the proposed framework were also identified.

Keywords Nanomaterials process economics · Scale up

A. Manipura (✉)
Formerly Department of Chemical and Process Engineering, Faculty of Engineering, University of Peradeniya, Peradeniya, Sri Lanka
e-mail: amanipura@yahoo.com

N. M. Adassooriya et al. (eds.), *Industrial Scale Production of Nanoparticles*, Smart Nanomaterials Technology, https://doi.org/10.1007/978-981-96-9176-0_18

1 Introduction

Nanomaterials (NMs) are increasingly used for manufacturing various products such as consumer, energy, electronics, medical, textiles, paints and coatings, and adhesives etc. to improve product performances [39]. The global market for nanomaterials was valued at USD 12.42 billion in 2023 and USD 14.34 billion in 2024, assuming a steady growth rate (Grand View Research, 2024). However, there are varying estimations about the net global demand and type of nanomaterials by various market research organizations due to uncertainties associated with emerging applications, inability to access internal market information of various companies, and different estimation techniques used by different market research organizations.

Generally, NMs are considered with the size less than 100 nm. Shape, dimension, composition, and high specific surface area lead to improve their optical, mechanical, physical, and chemical properties [21]. Further Khan et al. [21] stated that electrons energy levels, affinities, transitions, magnetic characteristics, phase transition temperature, melting point, and affinities to polymeric, biological, and organic substances are only some of the many properties that are influenced by changes in the surface area. These NMs are produced by various methods, which can be broadly categorized as bottom-up and top-down approaches [23]. While top-down reduces bulk materials using physical processes like milling and grinding, bottom-up constructs materials from atoms or molecules, allowing uniformity in shape and size. Statvis et al. (2018) identified six generic steps in processes of making NMs into products, namely synthesize, stabilize, functionalize, purify, characterize, and integrate. This system level analysis can immensely help in developing a general framework for analyzing process economics associated with industrial scale NMs production by identifying manufacturing process (top-down *vs.* bottom-up) and associated unit operations, selection of operation mode (batch *vs.* continuous or semi batch), production facility configuration, estimations of capital and operating costs. However, lack of historical data on scale up of NMs production similar to Lang factors and other cost indices [34] used in standard chemical manufacturing practices for emerging technologies hinder developing such generic process economics models with an order of magnitude accuracy for estimating the capital and operational costs.

Some characteristics of the NMs production are demand for smaller quantities in a very specific area, very high value per unit mass (*cf.* kg of cement *vs.* graphene), rapid emergence of new applications, use of both bottom-up and top-down approaches to meet the desired characteristics such as shape and size of given type of NM, use of emerging technologies such as microfluidics, laser and additive manufacturing methods, adaptability of batch manufacturing practices due to low production quantities and etc. Generally, NMs are produced at using standard chemical engineering unit operations such as ball milling (e.g. top-down) and CSTRs followed by separation and purification (e.g. bottom-up) synthesis methods [33]. However, it is hard to achieve economy of scale benefits due to low volumes of products, which normally achieved in standard chemical manufacturing practices [2]. The situation can be changed with increasing demand in future. Therefore, manufacturing of NMs in industrial scale can also be categorized as high value low volume and low value high volume products according to product price, production quantities and their market demand. This enables us to estimate process economics of the NMs production using strategies adopted in fine chemical, specialty, biochemical and bulk chemical manufacturing practices.

Thus, this chapter focuses on critically reviewing challenges and opportunities for analyzing process economics of industrial scale NMs production, propose a general framework for assessing process economics at industrial scale NMs production and test the proposed general framework using a case study of graphene and related materials production.

2 Challenges and Opportunities for Analyzing Process Economics of Industrial Scale Nanomaterials (NMs) Production

Analyzing process economics of industrial-scale NMs production presents both challenges and opportunities in terms of estimating capital and operational costs, return on investment (ROI), and payback period. These aspects are critical for assessing viability of NMs production at industrial scale. Thus, this section reviews primary challenges and opportunities of industrial-scale NMs production identified in literature, emerging NMs markets, their growth rate, prices and various applications, and comparison of industrial scale NMs production under four categories namely, business analysis, process chemistry R&D and syntheses methods, large scale manufacturing process design and manufacturing process optimization of overall business, technical and environmental sustainability using multi-criteria decision analysis [31].

2.1 Business Analysis

Business analysis puts the foundation for NMs production in industrial scale similar to any manufacturing process. Market needs, product specifications, overall demand, and expected percentage of market penetration (intended production capacity) are required for conducting process chemistry research and development to assess technical feasibility of targeted NMs [13]. Neither individual nor corporate entity will invest in manufacturing unless there is a significant return on investment (ROI) compared to other low-risk options (e.g. bank deposits, sovereign bonds, treasury bills etc.). Therefore, clear business and technical risk analysis prior to investment is not an option, but a necessity. Emerging nature of NMs applications in various industrial products for improved product performances should justify the additional investments required for such large-scale production or usage of NMs. Generally, NMs are used in smaller quantities for product formulation (e.g. paint and coating, cement) or use as catalysts in chemical industry [39]. This exerts additional constraints on business feasibility of NMs production as scale of economy is difficult to achieve. Hence, high unit cost will discourage industries to use NMs unless it shows the additional benefits of using them. Table 1 lists some of the challenges and opportunities in industrial scale NMs production.

Commercial information is not publicly available in easy to access manner and various sources report different estimations. Some market research companies present estimates publicly with limited information such as shown in Table 2. It clearly shows the wide variations of their estimates irrespective of the base year used. This may be due to the word "nanomaterials" used loosely by various entities and confusion arisen from interchanging the words such as nanotechnology, nanoparticles, and nanomaterials. Further market estimation methods, sample selected (e.g. industrial sectors, say electronics, energy storage etc.), data refining technique, information sources and forecasting models could have contributed to these variations.

Due to uncertainty of estimates of nanomaterials market prediction by various market research organizations, a scenario (sensitivity) analysis with varying compound annual growth rate (CAGR) for an initial market of USD 12.4 billion (the lowest reported in Table 2) for 10 year period can be estimated using Eq. 1 below and plotted as shown in Fig. 1.

$$EV = BV * \left\{1 + \frac{CAGR}{100}\right\}^{n} \tag{1}$$

where

EV	End Value;
BV	Beginning Value;
N	number of years, and;
CAGR	Compound Annual Growth Rate.

Considering all these variations, a CAGR of 15% can lead to double the NM market size in every 5 years approximately showing an exponential growth of global demand as shown in Fig. 1. There can be multiplier impact with gradual establishments of NMs supply chains and emerging fields of NM applications due to global trends in climate change and green approaches for manufacturing (demand for energy storage materials, catalysis and carbon capture).

Status of NMs market, quantities, application area, are needed to assess the potential demand as shown in Fig. 2 and Table 3 as high value low volume and low value high volume. High-value, low-volume NMs with specialized applications and smaller production scales often serve niche purposes that justify their high value despite lower volumes [13, 39].

According to Table 3, even the higher quantities of materials such as $CaCO_3$, SiO_2 and TiO_2 are required less than 200,000 tons per annum showing the need of comparatively smaller manufacturing plants (*cf.* 330 days continuous manufacturing

Table 1 Business, technical and environment risks, opportunities and challenges in industrial scale NMs production

No.	Risk aspect	Challenges	Opportunities	Remarks
1	Market demand and price	• Unproven products in commercial scale • Long term impact on product function during prolong usage • Value proposition and market expectations	• Emerging markets in health, energy, agriculture, sensors, consumer products etc. • Early adopters • Value gain (e.g. product performance) for price	• Early-stage business to business (B2B) partnerships
2	Process chemistry knowledge	• Product and impurity profiles • Consistent product quality as per user specifications • Synthesis conditions	• Learning from fine and specialty chemicals industry • Decision support for selecting NM synthesis method	• High value low volume versus low value high volume
3	Process behavior at large scale	• Process uncertainties at large scale operation • Shape, size and formation rates • Particle aggregation	• Numbering up • Additive manufacturing • Subtractive manufacturing (Mechanical exfoliation)	• Emerging AI-driven technologies
4	Equipment and their functional fitness to required chemistry	• Impact of mass and heat transfer limitations on nanoparticle formation	• Existing equipment with well-characterized operation	• Process benchmarking
5	Reliability of novel equipment	• Scale up factors for novel equipment	• Numbering up • Additive manufacturing	• Microfluidic systems • Process benchmarking
6	Capital and operational cost of novel equipment	• Lack of historical data for novel equipment for NP production	• Modifying existing cost models for capital and operating expenditure estimations	• Modular structured equipment with standardized cost
7	EHS of NM production, usage and end of life disposal	• Regulatory requirements, standards • Life cycle costing	• Learning from fine and specialty chemicals industry	• Detailed discussion in Chap. 20

Table 2 Global nanomaterial market analysis by various market research organizations

No.	Organization*	Base year	Projected period	Base year value (USD billion)	Forecast end year Value (USD billion)	CAGR (%)
1	Grand view research	2023	2024–2030	12.4	32.9	15.0
2	Market research future	2023	2024–2032	21.8	93.9	17.3
3	Allied market research	2021	2021–2031	16.3	62.8	14.6
4	IMARC group	2024	2025–2033	36.7	136.5	14.9
5	Business wire	2020	2020–2029	10.3	38.17	17.8
6	Future market insights	2023	2023–2033	12.6	51.5	15.1

*See given web links of above market research companies under the references. Values could have been changed

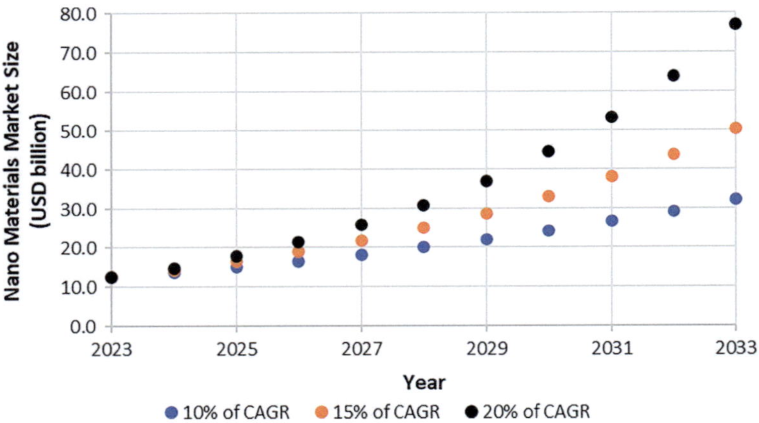

Fig. 1 Various CAGR scenarios for base year nanomaterials market value of USD 10 billion

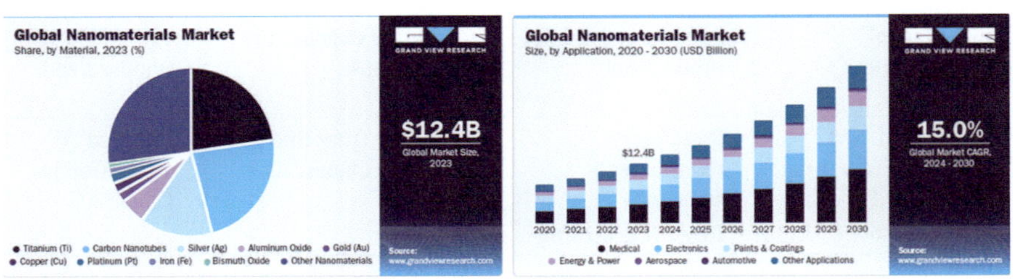

Fig. 2 Global nanomaterial market size and trends. *Source* Grandview Research (2024)

Table 3 Various types of NPs, their annual demand, CAGR, and their typical applications

No.	Nanoparticle type	Annual demand (tons)	CAGR (%)	Application areas	Sources
1	Gold nanoparticles	~ 5	N/A	Electronics, diagnostics, catalysis	Mordor intelligence, allied market research
2	Silver nanoparticles	~ 10	N/A	Antibacterial coatings, electronics, textiles	Grand view research, allied market research
3	Platinum nanoparticles	~ 2 to 3	N/A	Catalysts in automotive and chemical industries	Mordor intelligence, grand view research
4	Carbon nanotubes (single-wall)	~ 10 to 15	18.1	Electronics, energy storage, aerospace	Allied market research, grand view research
5	Zinc Oxide nanoparticles	~ 15 to 20	12.5	UV protection in cosmetics, coatings	Allied market research, grand view research
6	Silica nanoparticles	200,000+	7.8	Cement additives, coatings, electronics	ChemView consulting
7	Calcium Carbonate nanoparticles	150,000+	12.5	Paper, plastics, paints, and adhesives	Grandview research
8	Titanium Dioxide nanoparticles	130,000+	12.5	Sunscreens, paints, and coatings	BCC research
9	Zinc Oxide nanoparticles	60,000+	12.5	UV-protective coatings, textiles	Allied market research
10	Iron Oxide nanoparticles	40,000+	12.5	Magnetic storage, pigments	ChemView consulting

facility of NMs will yield about 600 tons per day *vs.* 5000 tons per day cement manufacturing plant). Therefore, ability to reduce cost of production is a critical requirement for NMs to be competitive in the market (scale of economy). On the other hand, gold and silver nanoparticles are vital in electronic components and medical diagnostics due to their unique properties, while platinum nanoparticles play essential roles in catalysis, especially for automotive applications and chemical industry [5, 40]. However, current global demand is under 1000 tons per year leading to very small throughput leading to distributed manufacturing in batch processes [17, 36].

Unless technical know-how, superior product performance improvement, required specifications with correct quality and correct price are available, not many players can run such businesses without business-to-business (B2B) partnerships due to difficulties in accessing markets and closely knitted nature of such business partnerships. Increasing electrical vehicle market may require high energy and power densities electrical energy storage systems while chemical industry may require larger quantities of better catalysts for greener manufacturing processes to reduce energy consumption and obtain better yields. Change of market behaviour, type and specifications of NMs, opening new opportunities in energy storage, electronics and composites use in automobile industry will open up new opportunities for various scale NM manufacturers involved in the global supply chains. Carbon nanotubes (CNTs) and silica nanoparticles are more diverse, appearing in electronics, aerospace, and composite materials [28]. Low-value, high-volume nanoparticles are widely used in manufacturing sectors such as construction, automotive, and polymer, paints and coating industries due to their accessibility and cost-effectiveness.

However, meeting different product specifications (size and shape) and quantities as demand by specific markets, based on specific applications and achieving commercial success in competitive manner will play a crucial role in scale up of NMs production. Lack of benefits in scale of economy type manufacturing, avoiding plant equipment idling, modular manufacturing plants to cater demand-driven manufacturing capacity planning (seasonal demand variations, multiple products with various specifications—shape and size) are some of the characteristics to consider for planning NMs production in industrial scale. Modular multi-purpose manufacturing plants will assist to scale up or down and product switching based on market demand similar to contract chemical manufacturing of intermediates for pharmaceutical, fine and specialty chemicals. Large-scale production can reduce the unit cost (scale of economy), making NMs more affordable and competitive in the market. Further, innovations in production methods are required to improve efficiency, yield, and quality, reducing costs [32]. Expanding the use of NMs across various industries (e.g., electronics, healthcare, energy storage) can increase the demand and profitability [39]. Further, implementing green manufacturing processes can reduce costs and enhance marketability in an eco-conscious market. Collaborations with research institutions and industry players can lead to shared resources, knowledge, and reduced costs. Therefore, growing demand for advanced materials and composites in high-tech industries (e.g. electronics, aviation and automobiles etc.) presents significant opportunities for profitability. Exploring new business models and ability to penetrate mass-scale applications of NMs will enhance reducing high unit cost of production [26].

2.2 Process Chemistry R&D and NP Synthesis Methods

Once business feasibility is established for a targeted NM market sector, technical feasibility is needed to explore at depth as there can be many syntheses approaches for a given NM synthesis. Therefore, systematic approach to process chemistry R&D should be carried out to screen the most viable synthesis method for a given NM specification (shape, size, quantity) and its derivatives (e.g. graphene, graphene oxide or reduced graphene oxides as needed). As discussed earlier, lessons from pharmaceutical, fine and specialty chemical industries can assist to reduce the time to market a given NM product. Safety aspects during synthesis, formulation and usage will be required to establish prior to entry to market (See Chap. 20 for further details). As particular NM can be synthesized numerous ways, cost of raw materials (starting materials/reactants), various reagents, equipment for reaction and separation will be required for an economically viable synthetic route. Thus, process chemistry research can improve conversion, yield, and selectivity to reach overall process efficiency in terms of mass and heat transfer with desired shape and size of the NM to obtain the intended functionality in the target industrial formulation (e.g. paint, cement, chemical energy storage, optical properties of materials etc.). This has an impact on process equipment selection and reducing dead zones inside equipment so that maximum space utilization is achieved for the incurred capital investment. Lower efficiencies could lead to high effort needed for workup, purification, and waste treatment (Jones, 2011 in Houson, I.). Further, ability and flexibility of controlling synthesis route in the event of exothermic reactions, which can lead to thermal runaway or product breakdown, is required to establish via modelling and experiments. Due to competitive and evolving nature of NM industry, freedom to operate a particular synthesis method may not be available, hindering the business unless licensing agreements are reached by relevant parties. Apart from those physicochemical conditions (heat and mass transfer) required for synthesizing the expected shape and size of NMs, rate limiting steps of synthesis reactions, nucleation and growth of NMs will impact on overall manufacturing efficiency. It is reported that achieving consistent and high yields are challenging, impacting the overall cost-effectiveness of NM synthesis [15]. Further, meeting industry standards and regulations can add to the cost and complexity of production of a given NM [33].

A major uncertainty in NMs production is estimation of the yield as it depends on processing conditions maintained in the system (temperature, pressure, mixing rate etc., in the case of bottom-up approach and rate of exploitation related parameters in top-down approaches) and avoidance of agglomeration of particles after the reaction as shown in Table 4. Therefore, understanding the NMs formation mechanism is highly important as simple stoichiometric relationships does not yield the final amount of NMs formed [11].

As an example during MgO nanoparticles synthesis, based on the ideal calculation, 260.5 g of $Mg(NO_3)_2.6H_2O$ reacts with 40.1 g of NaOH. This reaction, could create 116 g of MgO. However, the true yield was 85.3 g of MgO nanoparticles, which is only about 74% of the stoichiometric estimation [1]. Therefore, it is hard to generalize the prediction of yield of NM formation as they are depended on the process followed and specific conditions maintained during the synthesis. An experienced synthesis scientist may develop the gut feeling to optimize the process to reach the theoretical maximum, but it will be hard to predict based on physiochemical models, which is important in estimating overall process design and related economics. Thus, both top-down and bottom-up approaches of NM production are important factors to consider in developing correlations of theoretical maximum and practical yields via experimental models. Table 3 presents a various NM types, synthesis methods, yield, and specific applications. Clearly high yield is an important factor, but other parameters such as conversion, mass and heat transfer characteristics, chemical kinetics should be properly documented to determine the scale up strategy [7].

Table 4 Various NP types, synthesis method, yield and specific applications

No.	Nanoparticle Type	Synthesis Method	Yield (%)	Remarks	References
1	Gold (Au)	Bottom-up (e.g., chemical reduction)	80–90	High stability and biocompatibility; commonly used in medical imaging and drug delivery	Khan et al. [21], Harish et al. [18]
2	Silver (Ag)	Bottom-up (e.g., chemical, biosynthesis)	70–85	Antimicrobial properties; used in coatings and wound care	Khan et al. [21], Harish et al. [18]
3	Iron Oxide (Fe_3O_4)	Bottom-up (sol–gel, co-precipitation)	85–95	Magnetic properties ideal for biomedical and environmental applications	Harish et al. [18]
4	Silica (SiO_2)	Bottom-up (sol–gel)	80–90	Used in drug delivery and as a support material in catalysis	Harish et al. [18]
5	Titanium Dioxide (TiO_2)	Top-down (milling), Bottom-up (hydrothermal)	70–85	Commonly used in photocatalysis and as a pigment in cosmetics	Khan et al.[21]
6	Zinc Oxide (ZnO)	Bottom-up (precipitation, hydrothermal)	80–90	Exhibits photocatalytic and antibacterial properties; used in UV-blocking applications	Khan et al. [21]
7	Carbon Nanotubes (CNTs)	Top-down (arc discharge, laser ablation), Bottom-up (chemical vapor deposition	60–80	High strength and conductivity, used in electronics and composite materials	Khan et al.[21]
8	Graphene Oxide (GO)	Top-down (mechanical exfoliation); Bottom-up (modified Hummer's method)	40–95	Electronics, photonics, optoelectronics, sensors, and drug / gene delivery systems	Farjadian et al [40]

The synthesis approach impacts particle yield, cost, and scalability. Bottom-up methods, like chemical reduction and sol–gel, are often used for high purity and smaller particle sizes, while top-down methods, like milling and lithography, are more common for larger particles or structural applications. Yield percentages are typically process-dependent, ranging from around 60–95%, but often approach the higher end when using optimized chemical routes [18, 21].

2.3 Large Scale Manufacturing Process Design

Transitioning from lab-scale to industrial-scale production of NM often results in variations in particle size, shape, and distribution, which can directly affect the quality and performance of the produced NMs [15]. This is mainly due to scaling effects of unit operations in which fully utilization of available equipment space is hampered. Thus, setting up industrial-scale production facilities for NMs requires significant investment in specialized equipment and technology [9]. Generally, NM demand is in the range of 100–1000 tons per annum leading to smaller plant capacity requirements [16]. This leads to decide the appropriate production mode (batch vs. continuous) and potential use of modular smaller plants.

High-value low-volume NMs production shown in Table 2, can be carried out similar to fine and specialty chemicals production practices followed in chemical processing industry as previously stated. This can be multipurpose batch plants for contract manufacturing as per the need of various applications such as cements, composites, electronics, automobile, medical diagnosis etc. Learning from pharmaceutical industry on product quality management, health and safety aspects, and regulatory requirements can readily be applied [7]. Therefore, process equipment selection, sizing, capital and operational expenditure estimations can be done using existing knowledge. Instead of scaling up, numbering up may have higher potential in NMs production due to smaller demand for specific nanomaterials. Emerging technologies such as 3D printing, laser ablation, microfluidic systems, and biomanufacturing may also provide opportunities for direct capacity increase via numbering up reducing time and effort required for traditional scale up approaches [34]. These emerging technologies not only reduce capital cost, but also it promotes modular manufacturing leading to scale up or scale down based on market dynamics. Emerging technologies such as CNC machining has brought down cost of manufacturing of micro-structured equipment used in microchemical processing such as microreactors [3]. Further, ability to access these emerging technologies enable even small-scale fabricators to design, build and test such microchemical processing plants [24]. Further, operating expenditure can be significantly improved and this was specified to the different portions such as labor, plant costs, QA/QC, waste treatment, transport and logistics, and change over and cleaning [22].

As shown in Fig. 3a, scale of economy in chemical manufacturing is obtained from increasing the production capacity, which reduces the unit cost of production. However, majority of NMs have lower demand as showed in Table 3. Therefore, based on market demand, plant size should be deduced by estimating the target production capacity and accordingly capital (CapEx) and operating (OpEx) expenditures. Thus, product price can be reduced by optimizing the whole process to penetrate various markets of NMs applications. Current high prices hinder accessing highly functionalized NMs for product performance improvements.

For example, graphene oxide (GO) is used in cement industry for improving concrete properties. However, amount of graphene being used is less than 0.02% of the cement mortar (Li et al. 2016). Thus, 5000 tons per day clinker manufacturing

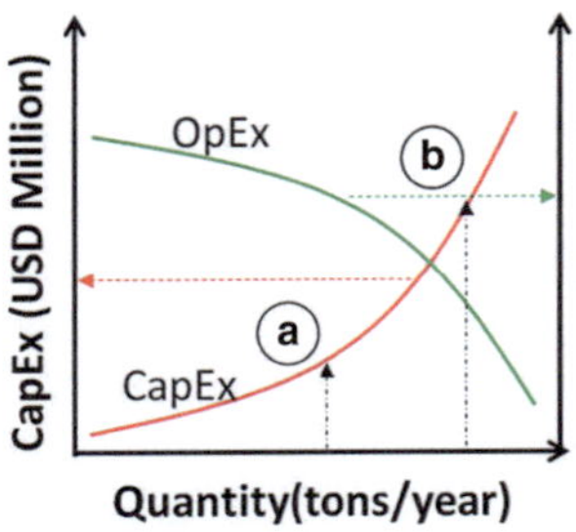

(a) Scale of economy

(b) Modular manufacturing: F^3 Factory, Source: European Community's 7th Framework Programme (FP7/2007-2013)

Fig. 3 NP production in industrial scale for achieving scale of economy

plant only needed about 1 ton of GO. This leads to GO demand about 330 tons per annum (assuming plant is operated for 330 days per year) provided that all clinker produced is mixed with GO for improving concrete properties. However, not all cement mortar is needed to blend with GO as the type of cement mortar required for a particular application may vary with the specifications needed by the end-user [12]. Thus, cement manufacturer may tend to reduce the annual GO requirement. Therefore, small modular manufacturing plants, which can be numbered up or down, could be used onsite as shown in Fig. 3b, tradition skid-mounted facilities or multi-product plants should be explored to cater various specifications demand by the various application developers to increase the production capacity and the price to be competitive.

Fuentes et al. [16] discussed evaluation of four case studies associated with the large-scale production of magnetite nanoparticles, specifically focused on environmental, economic, and exergy aspects. Isaacs et al. [19] compared three methods (arc ablation—arc, chemical vapor deposition—CVD, and high-pressure carbon monoxide—HiPco) of single-walled carbon nanotube processes and evaluated additional scenarios on the economics of environmental, health, and safety best manufacturing practices. Further, Isaacs et al. [19] stated that unlike other cost systems prevalent in industry where the manufacturing overhead is allocated based on direct labor or machine rent, process-based cost models break down manufacturing overhead into individual process steps, providing a more detailed cost assessment and an ability to explore how costs change as input parameters change. Typical cost elements include materials, labor, energy, capital equipment, tooling, building space, and maintenance. Cost model development consists of four steps namely, definition of the process steps, construction of the model logic and framework, collection of information, and validation of the model.

2.4 Optimization of Overall Business, Technical and Environmental Sustainability

Businesses cannot be run purely based on financial benefits, but overall sustainability should be looked at in wholistic manner (See Chapt. 19 for details). Short history of application of NM in formulated products, long-term health, environmental and safety aspects, fate and transport of man-made materials in living systems and natural environment are needed to critically review before large scale manufacturing of NM. Various aspects used by fine and specialty chemical industries for reducing business, technical and environmental risks such as SELECT criteria [7] can be applied for business analysis to manufacturing. Option screening for NP synthesis, scale up, process and equipment selection are usually carried out using multicriteria decision analysis (MCDA). Some of the challenges in scaling up of production and commercializing emerging materials are associated with uncertainties related to market behaviour, regulatory requirements, freedom to operate, lack of scaling effect related data, obtaining consistent product specification as per formulation industry requirements, which should be addressed systematically. Keppner-Tregoe decision analysis [31] is useful for developing such decision criteria in the absence of reliable data, yet using word models to screen options. However, multidisciplinary teams are required to capture the whole spectrum of business development from process chemistry to facility design and operation, manufacturing, sales and marketing. Nandiyanto et al. [29] carried out an economic evaluation of producing magnesium oxide nanoparticles via liquid-phase route focusing on gross profit margin (GPM), internal return rate (IRR), payback period (PBP), cumulative net present value (CNPV), breakeven point (BEP), and profit to investment (PI).

3 General Framework for Assessing Process Economics of Industrial Scale Nanomaterials Production

A general framework for assessing the process economics of industrial scale nanomaterials production is proposed by considering market demand and capacity planning, process chemistry (physicochemical requirements such as temperature, pressure, phases involved, specific conditions such as corrosiveness and toxic gas emissions), process options selection, process block diagram, equipment selection, and plant installation for capital and operation expenditure estimations.

3.1 Main Assumptions

Following assumptions were made to limit the scope of the proposed framework for decision analysis of NM production in industrial scale.

a. Framework is limited to NM production. Subsequent functionalization and formulations are not considered.
b. Standard chemical processing equipment and practices will be used for estimating the capital and operational costs.
c. Manufacturing is done as batch, semi-batch, or continuous operation according to the scale of production.
d. Emerging technologies such as flexible modular plants are considered as repetitive operations similar to batch or continuous operations in standard chemical manufacturing practices, yet scale up or scale down and numbering up or down operations are carried out to meet the varying capacity needs and product switching (multiproduct) if required in the same plant.

3.2 Decision Analysis for Process Economics of NM Production

Main aspects analyzed are shown in Fig. 4 under the categories of business analysis, process chemistry R&D, NM synthesis methods, large scale process design and whole process optimization to achieve the overall technical and business sustainability of industrial-scale NM production.

Figure 4 shows the need of multidisciplinary approach for industrial scale NM production starting from market research and business case development to overall manufacturing process optimization focusing process chemistry R&D; synthesis method selection; manufacturing process design; equipment selection and sizing; and plant erection (procurement, installation, testing) etc., which are typical in any chemical manufacturing facility development. However, standard chemical plant design approach should be carried out focusing on specific requirements needed for NM production according to the target application area, specifications needed and quantities to be produced.

One of the main challenges is how to access the markets for the synthesized NMs with various functional properties as there are no direct business opportunities for the new comers to NM manufacturing due to various concerns in supply chain reliability required by established industries. Uncertainties associated with various small-scale producers arise from supply chain reliability factors such as end user confidence on continued NM supply, product quality and consistency, right price and quantity etc. Thus, it is important to develop business-to-business (B2B) partnerships for those small-scale NM manufacturers to address market entry barriers. On the other hand, established companies will leverage their already established business partnerships and product range to expand the new opportunities arising from NMs. Even though process intensified (PI) technologies (Hakka et al 2020), modular manufacturing and emerging technologies can encourage small scale production of NMs, the above challenges will greatly impact newcomers as an entry barrier to the markets.

Ability to demonstrate value proposition created by NM for various application should be clearly spelled out to develop a business case for a given NM. Thus, end

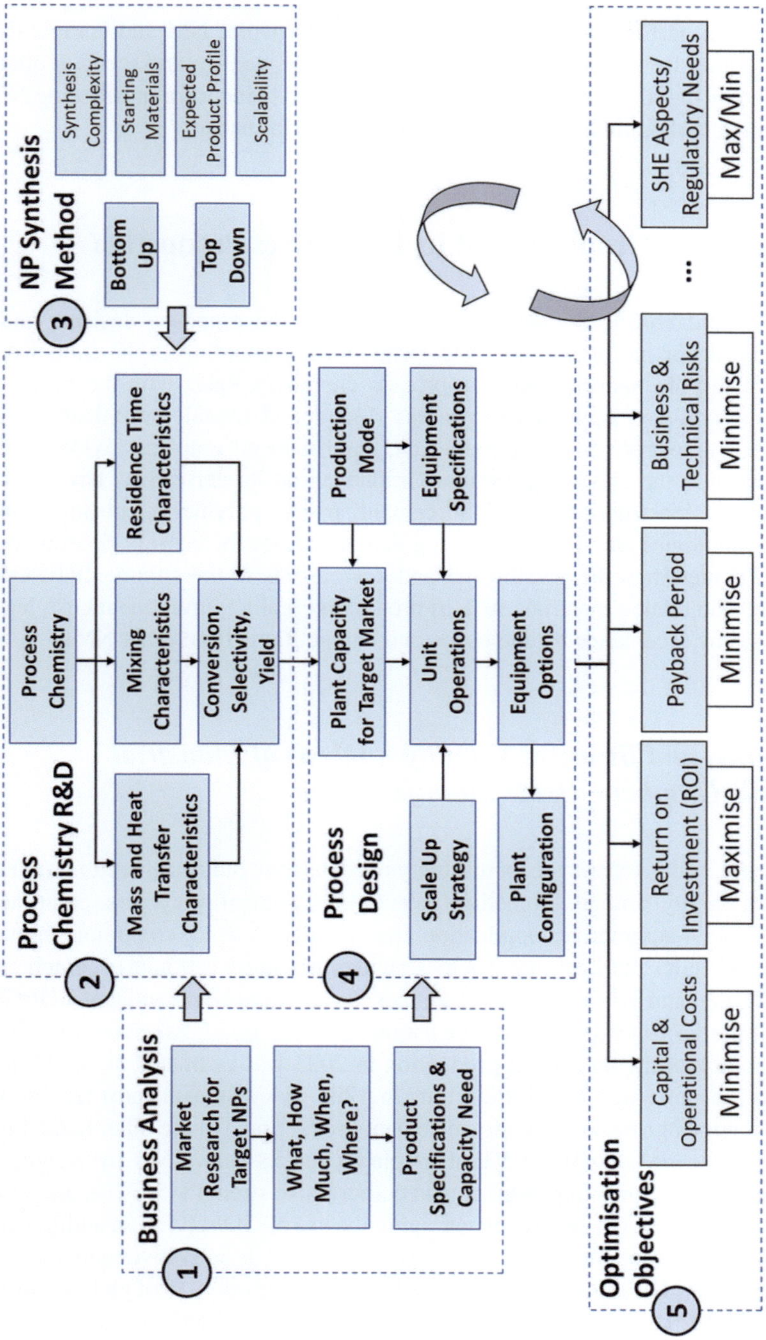

Fig. 4 Nanoparticle (NP) production from bench to industrial scale

user (formulator) will look at both comparative and competitive advantage arising from NM inclusions in their product formulations. Comparative advantage comes from additional performance improvements due a particular NM and competitive advantage comes from various NM manufacturers price, quality, and speed of quantity supplied by them. Thus, additional expenses needed to incur for introducing NPs in their products should be justified by the product performance.

4 Case Study: Graphene and Its Derivatives Production

Graphene manufacturing is considered to exemplify the proposed decision analysis framework for industrial scale NMs production. According to the framework provided in Fig. 4, business analysis, process chemistry R&D with the focus of nanoparticle synthesis approaches, process design and overall optimization are presented for graphene, graphene oxide (GO) and reduced graphene oxide (rGO) manufacturing in the following section. Graphene and its derivatives have wider applications in construction materials (cement, paint), polymer, electronics, and medical technologies etc. Thus, a growing demand exists for them [27]. However, Matyjas [27] identified the possibilities and key barriers to the commercialization of graphene and related materials such as production scaling, investment cost, legal regulations, low acceptance by end users, and competitions from other NMs.

4.1 Market and Business Viability Analysis of Industrial Scale Graphene Oxide Manufacturing

Graphene and its derivatives are increasingly used in many industrial sectors such as composites, energy storage, biomedical, electronics, cement, polymers, paint, and coatings. Table 4 summarizes application area, annual demand, compound annual growth rate (CAGR), market value. There is a wide variation of estimates of graphene, GO and rGO demand by various market sources. As an example, potential of GO/rGO for cement mortar as an additive can be estimated based on global cement production. The total worldwide cement production in 2023 was estimated as 4.1 billion tons while the same was 1.39 billion tons in 1995 (www.statista.com) leading to over 9% of annual growth rate of global cement production. On the other hand, [44] stated that inclusion of 0.01–0.05% of graphene and its derivatives by weight in cement mortar has shown improvements in compressive strength (5–77%) and flexural strength (10–90%) in 28 days encouraging the use of GO/rGO as an additive for performance improvement. Then, estimated graphene and its derivatives demand will be 4800 tons provided assuming only 0.01% by weight and only 1% of global cement production in 2026 (4.8 billion tons) will use graphene and its derivatives for performance improvements. However, estimated graphene and its derivatives demand in

2026 is 2700 tons for cement sector according to Table 4 (www.graphene-info.com). This could have been due to conservative estimates done despite the high potential of graphene and its derivatives for performance improvements and early stages of product testing in cement industry. Thus, this shows high potential of graphene and its derivatives market in cement sector if it is widely adopted (increasing the percentage usage of minimum of 10%), which can lead to graphene and its derivatives demand of 48,000 tons per year by 2026 cement sector alone. This has number of implications on reducing cost of production through scale of economy. Similarly, graphene and its derivatives demand for other sectors can be estimated.

Table 5 shows 8 application areas of GO and rGO with total annual demand in 2020 and 2026 as 14,300 tons per year and 25,100 tons per year, respectively. This shows cumulative annual growth rate over 12.6% compared to the demand in 2020. Different unit prices may have caused by various specifications needed by different application areas and their demand. Further, reported market is not clear whether it is the value of GO or rGO price paid for them or the market value of application area. Therefore, it is unable to estimate the GO and rGO price offered for the application area.

4.1.1 Capacity and Production Mode Planning

Based on above analysis, if GO/rGO demand is a small compared to traditional chemical manufacturing. Further, there are many GO/rGO manufacturers for various target markets. If 200 tons per annum capacity plant is built and operated for 300 days per year by 2026, daily production of GO/rGO will be about 670 kg. The income received from this should meet both allocations for capital equipment depreciation, total operating expenditure (e.g. salaries and incentives, consumables, utilities, insurance, wastewater treatment, sales and marketing, and other miscellaneous expenses etc.), and expected return on investment after tax (See Fig. 1a). This scenario leads to number of decisions to make in terms of manufacturing mode (continuous *vs.* batch), production scheduling based on customer demand, plant configuration etc. to meet the expected return on investment (RoI). Product mix in terms of GO/rGO specifications required by different application areas and downstream processing requirements may also play a significant role on business viability of this scale production plant. Table 6 summarizes some critical aspects to choose the operating mode.

Depending on the number of shifts, equipment capacities will be changed significantly as shown in Table 6. This has implications on capital investment required. In the event of batch manufacturing, larger batch size can lead to higher inconsistencies and the total loss will be high in the event of such incident. The emissions will also be high at a given time by a large equipment in the event of any accidental release. Thus, smaller equipment will have a better containment ability with smaller capital investment. However, flexibility to use the same plant equipment for different products will not be available when continuous manufacturing is done though product quality and process controllability are high. Further, continuous manufacturing requires smaller equipment compared to batch manufacturing of same throughput. If the plant operates

Table 5 GO/rGO application area, annual demand, CAGR, market value in 2020 and 2026

No.	GO/ rGO	Application area	Annual demand (metric tons)	CAGR (%)	Market value (USD million)	Reference
1	GO	Composites	3000 (2020); 4500 (2026)	10	$35 M (2020) $65 M (2026)	https://www.graphene-info.com/
2	GO	Energy Storage	2500 (2020); 4000 (2026)	9	$25 M (2020); $50 M (2026)	https://www.grandviewresearch.com/industry-analysis/graphene-industry
3	rGO	Biomedical	1200 (2020); 2400 (2026)	12	$20 M (2020); $45 M (2026)	https://www.gmiresearch.com/report/global-graphene-market-by-type/
4	rGO	Electronics and IoT	1500 (2020); 3000 (2026)	11	$30 M (2020); $60 M (2026	https://www.grandviewresearch.com/industry-analysis/graphene-industry
5	GO	Coatings and Additives	2000 (2020); 3800 (2026)	10	$28 M (2020); $52 M (2026)	https://www.gmiresearch.com/report/global-graphene-market-by-type/
6	GO	Cement	1500 (2020); 2700 (2026)	9	$18 M (2020); $36 M (2026)	https://www.graphene-info.com/services/market-reports/graphene-oxide-market-report
7	GO	Polymers	1,800 (2020); 3,200 (2026)	10	$22 M (2020); $44 M (2026)	https://www.grandviewresearch.com/industry-analysis/graphene-industry
8	GO	Water Treatment	800 (2020); 1500 (2026)	8	$10 M (2020); $22 M (2026)	https://www.graphene-info.com/services/market-reports/graphene-oxide-market-report

more than one shift, the salaries and incentives can be high as generally night-shift is paid higher than the day-shift. In case of day-shift only production mode, the selected process should be able to complete reaction and other related process steps under 8 h, which has implications on synthetic route selection and subsequent process design.

Table 6 Production scenarios, continuous versus batch operations for 200 tons per annum

Operational aspect	Shift category			Remarks
1. No. of hours per shift	8 h shift	16 h shift	24 h shift	
2. Throughput rate/kg/shift	670	335	225	
3. Equipment size	Large	Medium	Small	Quantity depended equipment sizing
4. Hazards	High	Medium	Low	Quantity depended hazards per unit
5. Operational cost (salaries)	Low	Medium	High	Additional cost for night shifts
6. Operation mode	Batch	Batch	Batch/Continuous	Product QA/QC

4.2 *Process Chemistry R&D of Graphene and Derived Products (GO and rGO)*

Process chemistry R&D plays an important role for identifying a viable commercial scale graphene, GO or rGO manufacturing process. Numbers of methods have been developed to synthesize graphene, GO and rGO over the years due to intense interest by scientific communities and commercial establishments attributing many applications and superior properties of graphene and related products. In order to demonstrate the second step of overall framework, few commonly used synthesis methods are presented in Table 7.

According to Table 7, majority of methods have been evolved from Hummer's method and all modifications provide various yields as low as 30% to as high as 95%. In terms of large-scale manufacturing, properties of reactants (precursors), operating temperature and pressure, and reaction time play a critical role for deciding the equipment chain, and utility requirements (cooling/heating, electricity etc.) leading to capital expenditure (CapEx) and operating expenditure (OpEx) of the facility. As strong acids and oxidants are used (e.g. H_2SO_4 and $KMnO_4$), special attention should be given to hazards and materials of construction of equipment to be employed. Further, in the case of batch manufacturing, longer reaction time will lead to longer batch time and consequently careful control of the whole process to avoid any batch inconsistencies. Lower yield can lead to high downstream processing cost to reach the expected impurity profile depending on the end user requirements. Additional cost will be incurred for waste and wastewater treatment.

For Hummers' and its variants, temperatures are generally controlled to avoid overreaction or explosion risks due to oxidants like $KMnO_4$. Electrochemical and mechanical exfoliation methods operate at ambient conditions, making them energy-efficient. Exact temperature ranges are often provided in synthesis protocols and may vary slightly based on experimental conditions. In order to select synthetic route options from Table 7 for producing GO and rGO, a set of criteria based on the business and technical interest can be developed as given below Table 8.

Table 7 Commonly use GO and rGO synthesis methods

No.	Method	Starting materials	Yield (%)	Reaction time	Temperature	Details	Reference
1	Brodie method	Graphite, $KClO_3$, HNO_3	~ 40–60%	3–4 days	60 °C	Early method; involves chlorate oxidation, producing less oxidized GO	Botas et al. [6]
2	Staudenmaier method	Graphite, H_2SO_4, HNO_3, $KClO_3$	~ 50–70%	2–4 days	Ice bath	Improved upon Brodie method; suitable for smaller-scale synthesis	Poh et al. [35]
3	Hummers' method	Graphite, $KMnO_4$, H_2SO_4, $NaNO_3$	~ 75–90%	6–12 h	~ 0–45°C	Widely used; involves strong oxidation of graphite	Botas et al. [6], Benzait et al. [4]
4	Modified hummers' method	Graphite, $KMnO_4$, H_2SO_4, H_2O_2	~ 80–95%	2–8 h	~ 0–50°C	Adds modifications for higher oxidation and safety; reduced reaction time	Botas et al. [6], Benzait et al. [4]
5	Chemical reduction	GO, hydrazine, ascorbic acid, or other reductants	Vary	Vary	Vary	Converts GO to rGO with tunable properties based on reductant	Chua and Pumera [10]
6	Thermal reduction	GO	Vary	Minutes to hours	~ 35 to 70°C	High-temperature reduction under inert atmosphere; produces less oxygenated rGO	Poh et al. [35]
7	Electrochemical exfoliation	Graphite, electrolyte solution (e.g., H_2SO_4, Na_2SO_4)	~ 50–70%	~ 2–4 h	Ambient temperature	Uses electrical potential to exfoliate graphite into graphene oxide sheets; scalable method	Zhang and Zu [43] 25g/h

(continued)

Table 7 (continued)

No.	Method	Starting materials	Yield (%)	Reaction time	Temperature	Details	Reference
8	Mechanical exfoliation	Graphite flakes, adhesive tape, ultrasonic bath	~ 40–50%	~ 1–2 h	Ambient temperature	Produces high-quality graphene layers; low yield and labor-intensive; useful for prototyping	Yi and Shen [41]
9	Mechanical exfoliation	Graphite flakes	> 30%	Vary	Below 40°C	Silk proteins were used for graphene exfoliation	Zhang et al. [42], 6g/h of FLG
10	Plasma spraying	Synthetic graphite powder	~ 85%	Milliseconds	Up to 4000 K	Chemical- and solvent-free graphite exfoliation	Islam et al. [20], 48 g/h USD 1.12/g material cost

Table 8 Screening synthetic route for GO/rGO Production

No.	Criterion	Option 1	Option 2	Option 3
		Modified hummer	Electrochemical	Mechanical
1	Operating temperature	5	5	5
2	Reaction time	4	4	5
3	Yield	4	3	2
4	No. of reactants	3	5	5
5	NFPA health hazard rank	2	4	4
6	NFPA fire hazard rank	3	4	3
7	NFPA specific hazard rank	3	4	4
8	NFPA instability hazard rank	2	5	4
9	Environmental emissions	1	4	3
Total Score		27	38	35

For simplicity, only 4 options were selected for further analysis. First 3 options were omitted due to long reaction time, which has impact on operational aspects of the plant such as reaction exceeds more than 8 hour working shift and need of process monitoring for safety reasons. The 5th and 6th options were excluded as it is limited to producing rGO from GO. Though high yield and shorter reaction time at laboratory scale were showed in some of the synthetic routes in Table 6, scale up strategy should be clearly defined to reach such results considering dimensionless behaviour of synthesis conditions.

Table 9 shows the scoring mechanism for screening GO synthetic route options. National Fire Protection Association (NFPA), USA hazard analysis approach is used for health, fire, specific and instability analysis of chemicals and subsequent reactions as an example. Other technical and business drivers can be integrated to scoring system based on strategic and legal requirements.

Electrochemical methods for synthesizing GO/rGO are more environmentally friendly compared to traditional chemical oxidation methods, as they avoid the use of hazardous and explosive chemicals like those used in the modified Hummers' method as presented in Table 8. Further, the electrochemical approach can produce GO with high quality, minimal defects, and tunable oxidation levels by controlling relevant parameters. This results in reduced graphene oxide (rGO) with high conductivity, which is beneficial for applications in electronics and energy storage [43].

4.3 Process Design for Graphene Oxide Manufacturing

Considering the business, technical and environmental criteria as described in Table 8, electrochemical methods is selected for production scale up of graphene oxide.

Table 9 Scoring system for screening synthetic route options

1. Operating Temperature	**Scale**	**2. Reaction Time**	**Scale**	**3. Yield**	**Scale**
• Operating temperature is less than 50 °C	5	• Reaction time less than 3 h	5	• Yield is greater than 90%	5
• Operating temperature is between 50 and 90 °C	3	• Reaction time less than 6 h	4	• Yield is between 75 and 90%	4
• Operating temperature is above °C	1	• Reaction time less than 8 h	3	• Yield is between 75 and 60%	3
		• Reaction time more than 8 h	2	• Yield is between 45 and 60%	2
		• Reaction time more than 1 day	1	• Yield is less than 40%	1
4. No. of Reactants	**Scale**	**5. NFPA Health Hazards**	**Scale**	**6. NFPA Fire Hazards**	**Scale**
• Less than 3	5	• NFPA health hazard category 0	5	• NFPA flammability hazard category 0	5
• Between 4 and 6	3	• NFPA health hazard category 1	4	• NFPA flammability hazard category 1	4
• More than 6	1	• NFPA health hazard category 2	3	• NFPA flammability hazard category 2	3
		• NFPA health hazard category 3	2	• NFPA flammability hazard category 3	2
		• NFPA health hazard category 4	1	• NFPA flammability hazard category 4	1
7. NFPA Specific Hazards		**8. NFPA Instability Hazards**		**9. Environmental Emissions**	
• NFPA specific hazard category 0	5	• NFPA instability hazard category 0	5	• Gas, liquid and solid emission < 0.005 t/tGO	5
• NFPA specific hazard category 1	4	• NFPA instability hazard category 1	4	• Gas, liquid and solid emission < 0.015 t/tGO	4
• NFPA specific hazard category 2	3	• NFPA instability hazard category 2	3	• Gas, liquid and solid emission < 0.03 t/tGO	3

(continued)

Table 9 (continued)

• NFPA specific hazard category 3	2	• NFPA instability hazard category 3	2	• Gas, liquid and solid emission < 0.05 t/tGO	2
• NFPA specific hazard category 4	1	• NFPA instability hazard category 4	1	• Gas, liquid and solid emission > 0.05 t/tGO	1

NFPA National Fire Protection Association, USA

However, similar to synthetic route selection, process design also can be done using either single step or two steps [4].

4.3.1 Single-Step Electrochemical Method

In a single-step electrochemical method, a two-electrode cell is typically used. The graphite intercalation compound (GIC) serves as the anode, while a platinum electrode is used as the cathode. This setup facilitates the oxidation and exfoliation of graphite to produce graphene oxide (GO). A 0.1 M ammonium sulfate ($(NH_4)_2SO_4$) aqueous solution is commonly used as the electrolyte. The method allows for the fast production of GO, typically in less than 30 min [43]. This efficiency is due to the optimized electrochemical conditions that promote rapid exfoliation and oxidation. This choice is crucial as it supports the formation of stage 1 GICs, which are essential for achieving high yields of GO. The electrochemical process is typically conducted at room temperature. This is advantageous as it simplifies the setup and reduces energy consumption compared to thermal methods that require high temperatures. The single-step electrochemical method can achieve a high yield of GO, approximately 71 wt%. This is facilitated by the efficient intercalation and oxidation processes enabled by the chosen electrode and electrolyte configuration [25]. The GO produced has a reasonable oxygen content and a high degree of monolayer flakes, which are desirable for various applications. The process also results in reduced graphene oxide (rGO) with high electrical conductivity, making it suitable for electronic applications. The process can be scaled up for mass production, making it suitable for industrial applications. This is achieved by using larger graphite foils and optimizing the cell voltage and electrolyte concentration.

4.3.2 Two-Step Electrochemical Method

In the two-step electrochemical method, a graphite electrode is used as the anode, and a platinum or graphite rod serves as the cathode. This setup is designed to facilitate the intercalation and subsequent oxidation of graphite. The electrolyte typically consists of a sulfuric acid (H_2SO_4) solution, which aids in the formation of graphite intercalation compounds (GICs) during the first step [8]. This choice is crucial for effective intercalation and subsequent oxidation. In the second step, the intercalated graphite is oxidized and exfoliated to produce GO [8]. This step is optimized to ensure complete oxidation and efficient exfoliation, resulting in high-quality GO with desirable properties. The process is generally conducted at room temperature, which simplifies the setup and reduces energy requirements compared to methods that require elevated temperatures. The two-step method can achieve a high yield of graphene oxide (GO), as the separation of intercalation and oxidation processes allows for better control over each step, enhancing the overall efficiency. The GO produced is of high quality, with a high degree of oxidation and minimal defects, making it suitable for various applications, including electronics and energy storage.

4.4 Limitation and Further Improvement Opportunities for the Proposed Framework

The proposed framework can be used to organize the information and knowledge required for analyzing business, technical and environmental viability of nanomaterials production in industrial scale. Process economics that normally an investor interested, cannot be achieved, unless capital and operational expenditures, return on investment and payback period are demonstrated. Therefore, a detailed analysis of equipment options for various NMs production processes, capital and operational expenditures of those, a set of rules of thumbs for equipment selection for NMs production will be useful for practitioners. However, to develop such rules of thumbs, significant amount of manufacturing related cost data is required. Due to scale of economy related production challenges in NMs, unit cost of production seems to be very high. Further, closely connected B2B product and market development hinder deriving models for assessing cost and capacity relationships.

The exploration of nanomaterials presents unique challenges and opportunities, necessitating a careful balance between innovation and regulatory frameworks to ensure their safe application in various sectors [30]. This balancing act involves collaboration among scientists, policymakers, and industry leaders to establish guidelines that protect public health while fostering technological advancement. Such collaborative efforts are crucial in addressing the potential risks associated with nanomaterials, as well as maximizing their benefits for sustainable development across multiple industries [37]. As the dialogue between these stakeholders evolves, it becomes essential to prioritize transparency and public engagement in the decision-making process, ensuring that societal concerns are adequately addressed while promoting responsible innovation. By fostering an inclusive environment where diverse perspectives are valued, stakeholders can work together to create comprehensive strategies that not only mitigate risks but also enhance the positive impact of nanomaterials on society and the economy [38].

5 Conclusion

A general decision analysis framework for assessing technical, business, and environmental viability of industrial scale NMs production was proposed and applied to a case study of graphene oxide manufacturing targeting various applications. The proposed framework consists of decision points to consider in market and business analysis, process chemistry R&D with the focus of NMs synthesis approaches, process design and overall optimization of the manufacturing process. Such systematic approach not only helps to organize the knowledge gathered during overall business development, but also it helps to expedite development/retrofitting of new additions to the existing plant in rapidly.

Due to poor scale of economy of NM production, emerging technologies can play an important role to cater specific NMs needs by various industries by onsite manufacturing for reducing logistics, improving quality and availability on demand through co-design of process. Business to business product development will enhance customized specifications of NMs for specific industries and reduce the cost of production. Significant performance improvements in formulated products and emerging applications of NMs will increase the demand in future though it is difficult to estimate accurate estimations. Industries such as cement, paint, rubber and polymer sectors will have a higher demand leading to lower unit cost due to scale of economy of NMs production with the time.

References

1. Abinya S, Kavitha HP (2023) Magnesium oxide nanoparticles: effective antilarvicidal and antibacterial agents. ACS Omega 8(6):5225–5233. https://doi.org/10.1021/acsomega.2c01450
2. Augusto PA, Castelo-Grande T, Vargas D, Pascual A, Hernández L, Estevez AM, Barbosa D (2020) Upscale design, process development, and economic analysis of industrial plants for nanomagnetic particle production for environmental and biomedical use. Materials 13:2477. https://doi.org/10.3390/ma13112477
3. Bagga K, McCann R, Wang M, Stalcup A, Vázquez M, Brabazon D (2015) Laser assisted synthesis of carbon nanoparticles with controlled viscosities for printing applications. J Colloid Interface Sci 447:263–268
4. Benzait Z, Chen P, Trabzon L (2021) Enhanced synthesis method of graphene oxide. Nanoscale Adv 3:223–230. https://doi.org/10.1039/D0NA00706D
5. Biehler E, Quach Q, Abdel-Fattah TM (2024) Application of platinum nanoparticles decorating mesoporous carbon derived from sustainable source for hydrogen evolution reaction. Catalysts 14:423. https://doi.org/10.3390/catal14070423
6. Botas C, Álvarez P, Blanco P, Granda M, Blanco C, Santamaría R, Romasanta LJ, Verdejo R, López-Manchado MA, Menéndez R (2013) Graphene materials with different structures prepared from the same graphite by the Hummers and Brodie methods. Carbon 65:156–164. https://doi.org/10.1016/j.carbon.2013.08.009
7. Butters M, Catterick D, Craign A, Curzons A, Dale D, Gillmore A et al (2006) Critical assessment of pharmaceutical processes—a rationale for changing the synthetic route. Chem Rev 106:3002–3027

8. Cao J, He P, Mohammed MA, Zhao X, Young RJ, Derby B, Kinloch IA, Dryfe RAW (2017) Two-step electrochemical intercalation and oxidation of graphite for the mass production of graphene oxide. J Am Chem Soc 139(48):17446–17456. https://doi.org/10.1021/jacs.7b08515
9. Charitidis CA, Georgiou P, Koklioti MA, Trompeta A-F, Markakis V (2014) Manufacturing nanomaterials: from research to industry. Manuf Rev 1:11
10. Chua CK, Pumera M (2014) Chemical reduction of graphene oxide: a synthetic chemistry viewpoint. Chem Soc Rev 43:291–312. https://doi.org/10.1039/C3CS60303B
11. Dimiev AM, Tour JM (2014) Mechanism of graphene oxide formation. ACS Nano. https://doi.org/10.1021/nn500606a
12. Dimov D, Amit I, Gorrie O, Barnes MD, Townsend NJ (2018) Ultrahigh performance nanoengineered graphene-concrete composites for multifunctional applications. Adv Funct Mater 28:1705183. https://doi.org/10.1002/adfm.201705183
13. Falsini S, Bradi U, Abou-Hassan A, Ristori S (2018) Sustainable strategies for large-scale nanotechnology manufacturing in the biomedical field. Green Chem 20:3897–3907. https://doi.org/10.1039/C8GC01248B
14. Farjadian F, Roointan A, Mohammadi-Samani S, Hosseini M (2019) Mesoporous silica nanoparticles: synthesis, pharmaceutical applications, biodistribution, and biosafety assessment. Chem Eng J 359:684–705 https://doi.org/10.1016/j.cej.2018.11.156
15. Fernandes C, Jathar M, Sawant BKS, Warde T (2023) Scale-up of nanoparticle manufacturing process. In: Jindal AB (eds) Pharmaceutical process engineering and scale-up principles. AAPS introductions in the pharmaceutical sciences, vol 13. Springer, Cham. https://doi.org/10.1007/978-3-031-31380-6_12
16. Fuentes OP, Trujillo DM, Sánchez ME, Abrego-Perez AL, Osma JF, Cruz JC (2024) Embracing sustainability in the industry: a study of environmental, economic, and exergetic performances in large-scale production of magnetite nanoparticles. ACS Sustain Chem Eng 12(2):760–772. https://doi.org/10.1021/acssuschemeng.3c05000
17. Grimaldi F, Pucciarelli M, Gavriilidis A, Dobson P, Lettieri P (2020) Anticipatory life cycle assessment of gold nanoparticles production: comparison of milli-continuous flow and batch synthesis. J Clean Prod 269:122335. https://doi.org/10.1016/j.jclepro.2020.122335
18. Harish V, Ansari MM, Tewari D, Gaur M, Yadav AB, García-Betancourt M-L, Abdel-Haleem FM, Bechelany M, Barhoum A (2022) Nanoparticle and nanostructure synthesis and controlled growth methods. Nanomaterials 12:3226. https://doi.org/10.3390/nano12183226
19. Isaacs JA, Tanwani A, Healy ML et al (2010) Economic assessment of single-walled carbon nanotube processes. J Nanopart Res 12:551–562. https://doi.org/10.1007/s11051-009-9673-3
20. Islam A, Mukherjee B, Pandey KK, Keshri AK (2021) Ultra-fast, chemical-free, mass production of high quality exfoliated graphene. ACS Nano 15(1):1775–1784. https://doi.org/10.1021/acsnano.0c09451
21. Khan Y, Sadia H, Ali Shah SZ, Khan MN, Shah AA, Ullah N, Ullah MF, Bibi H, Bafakeeh OT, Khedher NB et al (2022) Classification, synthetic, and characterization approaches to nanoparticles, and their applications in various fields of nanotechnology: a review. Catalysts 12:1386. https://doi.org/10.3390/catal12111386
22. Krtschil U, Hessel V, Kralisch D, Kreisel G, Küpper M, Schenk R (2006) Cost analysis of a commercial manufacturing process of a fine chemical compound using micro process engineering. Chimia 60:611–617. https://doi.org/10.2533/chimia.2006.611
23. Kumari S, Raturi S, Kulshrestha S, Chauhan K, Dhingra S, András K, Thu K, Khargotra R, Singh T (2023) A comprehensive review on various techniques used for synthesizing nanoparticles. J Market Res 27:1739–1763. https://doi.org/10.1016/j.jmrt.2023.09.291
24. Li L-L, Li X, Wang H (2017) Microfluidic synthesis of nanomaterials for biomedical applications. Small Methods 1700140. https://doi.org/10.1002/smtd.201700140
25. Low CTJ, Walsh FC, Chakrabarti MH, Hashim MA, Hussain MA (2013) Electrochemical approaches to the production of graphene flakes and their potential applications. Carbon 54:1–21. https://doi.org/10.1016/j.carbon.2012.11.030
26. Lubik S, Garnsey E (2016) Early business model evolution in science-based ventures: the case of advanced materials. Long Range Plan 49(3):393–408

27. Matyjas Z (2018) Graphene flakes—prospects for commercialization. Entrepreneurship Manag XIX(6):299–308
28. Modekwe HU, Ayeleru OO, Onu MA, Tobias NT, Mamo MA, Moothi K, Daramola MO, Olubambi PA (2022) The current market for carbon nanotube materials and products. In: Abraham J et al (eds) Handbook of carbon nanotubes. https://doi.org/10.1007/978-3-319-70614-6_73-2
29. Nandiyanto ABD, Fariansyah R, Ramadhan MF, Abdullah AG, Widiaty I (2018) Economic evaluation of the production magnesium oxide nanoparticles via liquid-phase route. IOP Conf Ser Mater Sci Eng 306:012011. https://doi.org/10.1088/1757-899X/306/1/012011
30. Ngarize S, Makuch KE, Pereira R (2013) The case for regulating nanotechnologies: international, European and national perspectives. Rev Eur Comp Int Environ Law. https://doi.org/10.1111/REEL.12029
31. Parker JS, Moseley JD (2008) Kepner–Tregoe decision analysis as a tool to aid route selection. Part 1. Org Process Res Dev 12(6):1041–1043
32. Patel, Pathak (2021) Emerging technologies for nanoparticle manufacturing. Springer Nature Switzerland AG. https://doi.org/10.1007/978-3-030-50703-9
33. Patel DM, Patel NN, Patel JK (2021) Nanomedicine scale-up technologies: feasibilities and challenges. In: Patel JK, Pathak YV (eds) Emerging technologies for nanoparticle manufacturing. Springer, Cham. https://doi.org/10.1007/978-3-030-50703-9_24
34. Peters MS, Timmerhaus K, West RE (2003) Plant design and economics for chemical engineers. McGraw Hill, USA
35. Poh HL, Šaněk F, Ambrosi A, Zhao G, Sofer Z, Pumera M (2012) Graphenes prepared by Staudenmaier, Hofmann and Hummers methods with consequent thermal exfoliation exhibit very different electrochemical properties. Nanoscale 4:3515–3522. https://doi.org/10.1039/C2NR30490B
36. Salmiati AS, Salim AR, Kueh ABH, Hadibarata T, Nur H (2017) A review of silver nanoparticles: research trends, global consumption, synthesis, properties, and future challenges. J Chin Chem Soc 7:732–756
37. Sass J, Musu T, Burns K, Illuminato I (2008) Nanomaterials: brief review of policy frameworks in the US and Europe and recommendations from an occupational and environmental perspective. Eur J Oncol 13(4):211–218
38. Schroeder D, Dalton-Brown S, Schrempf B, Kaplan D (2016) Responsible inclusive innovation and the nano-divide. Nanoethics 9(10):177–188. https://doi.org/10.1007/S11569-016-0265-2
39. Stark WJ, Stoessel PR, Wohlleben W, Hafner A (2015) Industrial applications of nanoparticles. Chem Soc Rev 44:5793–5805. https://doi.org/10.1039/C4CS00362D
40. Yaqoob AA, Ahmad H, Parveen T, Ahmad A, Oves M, Ismail IMI, Qari HA, Umar K, Mohamad Ibrahim MN (2020) Recent advances in metal decorated nanomaterials and their various biological applications: a review. Front Chem 8:341. https://doi.org/10.3389/fchem.2020.00341. PMID: 32509720; PMCID: PMC7248377
41. Yi M, Shen Z (2013) A review on mechanical exfoliation for scalable production of graphene. J Mater Chem A 3:11700–11715. https://doi.org/10.1039/C5TA00252D
42. Zhang X, Wang L, Lu Q, Kaplan DL (2018) Mass production of biocompatible graphene using silk nanofibers. ACS Appl Mater Interfaces 10(27):22924–22931. https://doi.org/10.1021/acsami.8b04777
43. Zhang Y, Xu Y (2019) Simultaneous electrochemical dual-electrode exfoliation of graphite toward scalable production of high-quality graphene. Adv Funct Mater 1902171. https://doi.org/10.1002/adfm.201902171
44. Zhao L, Guo X, Song L, Song Y, Dai G, Liu J (2020) An intensive review on the role of graphene oxide in cement-based materials. Constr Build Mater 241:1–17. https://doi.org/10.1016/j.conbuildmat.2019.117939

Web References

45. https://www.futuremarketinsights.com/. Accessed on 7 Nov 2024
46. https://www.alliedmarketresearch.com/nano-materials-market. Accessed on 7 Nov 2024
47. https://www.marketresearchfuture.com/reports/nanomaterials-market-7814. Accessed on 7 Nov 2024
48. https://www.grandviewresearch.com/industry-analysis/nanotechnology-and-nanomaterials-market. Accessed on 7 Nov 2024
49. https://www.fortunebusinessinsights.com/nanotechnology-market-108466. Accessed on 7 Nov 2024
50. https://www.businesswire.com/. Accessed on 7 Nov 2024
51. https://www.imarcgroup.com/nanomaterials-market. Accessed on 7 Nov 2024
52. https://www.thegraphenecouncil.org/. Accessed on 7 Nov 2024
53. https://euon.echa.europa.eu/nanodata/sectors/manufacturing/overview/market. Accessed on 7 Nov 2024
54. https://www.persistencemarketresearch.com/market-research/nanomaterials-market.asp?form=MG0AV3. Accessed on 7 Nov 2024

Life Cycle Assessment of Nanomaterials Production

Punethri Hewage, Sandeepanie Bandara, and Mahinsasa Rathnayake

Abstract Life cycle assessment (LCA) can be utilized to assess the environmental, social, and economic impacts and energy sustainability of a certain product or process. This chapter elaborates the current knowledge on LCA of nanomaterials and nano-enabled products. It is important to perform LCAs for identifying the product life cycle stages and operations with significant impacts on the environment and thereby find solutions to mitigate those impacts. The objective of this chapter is to familiarize the reader with the LCA methodology in the context of nanomaterials. The chapter includes a brief introduction to the LCA methodology followed by highlights from the recent peer-reviewed LCA studies on nanomaterials. Finally, it discusses the challenges of conducting LCAs for nanomaterials and future work to be continued.

Keywords Life cycle assessment · Nanomaterials · Nano-enabled products · Environmental impacts · Sustainability

1 Introduction

Nanomaterial production, being a vast study area evolving rapidly, has now been identified by researchers to evaluate the environmental impacts of the nanomaterials despite claiming without any scientifically proven methodology. Several environmental and human health impacts have been identified in the production process of nanomaterial production even though few of the applications claim to provide improvements in environmental protection, such as wastewater treatment advancements and the human health sector with modernized diagnostic and surgical equipment [5]. In addition to the conventional methods of production, green synthesis and production methods have been currently identified. The need for an early-stage approach to potential environmental impacts has now been considered with the rapid change and implementations of the production methods applied by organizations,

P. Hewage · S. Bandara · M. Rathnayake (✉)
Department of Chemical and Process Engineering, University of Moratuwa, Moratuwa, Sri Lanka
e-mail: mratnayake@uom.lk

N. M. Adassooriya et al. (eds.), *Industrial Scale Production of Nanoparticles*, Smart Nanomaterials Technology, https://doi.org/10.1007/978-981-96-9176-0_19

moving towards sustainability. A systematic Life Cycle Assessment (LCA) approach can analyze and evaluate the environmental impacts of nanomaterial production throughout its life cycle. Several studies have been performed conducting life cycle assessments. However, the unavailability of an acceptable database, variety, and product-specific inventory has affected the quality of the assessment [22]. In this chapter, with the understanding of the life cycle assessment and how it can be implemented in nanomaterials production, the recent LCA studies conducted for nanomaterials are compared. Finally, the limitations and issues identified in the recent life cycle assessment studies and the prospects of applying life cycle assessment in nanomaterial production are discussed in this chapter.

2 Life Cycle Assessment Methodology in Brief

Life Cycle Assessment (LCA) is a powerful tool for analyzing and evaluating the environmental impacts of a product or a process throughout its entire life cycle, from natural resource acquisition to end-of-life waste management, with a holistic understanding. This enables organizations and industries to identify the critical stages of significant environmental emissions and material and energy consumption through the lifecycle, encouraging continuous process improvements and sustainable decision-making. This fosters a transition to a circular economy by promoting environmental emission reduction, resource efficiency, waste reduction, and recyclability [10].

Figure 1 illustrates the ISO 14040/44 standard framework that provides the systematic methodology for conducting life cycle assessments. The methodology contains four main phases: Goal and scope definition and system boundary, Life cycle inventory, Life cycle impact assessment, and Interpretation [21, 23]. The first phase, goal and scope definition provide a comprehensive overview of the objectives of conducting the life cycle assessment and defines the system boundary.

The first stage, goal and scope definition are included of a comprehensive overview of the objectives of conducting a life cycle assessment and defining the system boundary, based on the specified objectives. The system boundary may contain the raw material acquisition stage to the disposal stage (cradle-to-grave), or to a stage where the product is recycled back to the process (cradle-to-cradle), or it is acceptable to analyze the impacts until the end of the product manufacturing stage (cradle-to-gate). Depending on the product's functionality or application, the functional unit is assigned. The major constraints and assumptions are also identified in this phase.

Developing the life cycle inventory for 1 functional unit with the material, energy inputs calculations and the outputs and emissions calculations are performed in the second stage, the life cycle inventory analysis. The environmental impacts of the emissions are evaluated in the third stage, impact assessment. This can be performed in either the midpoint approach or the end-point approach. If the objective is to specifically evaluate the environmental impacts, then the mid-point method is followed. If the focus is to assess the impact on the areas of protection such as human health,

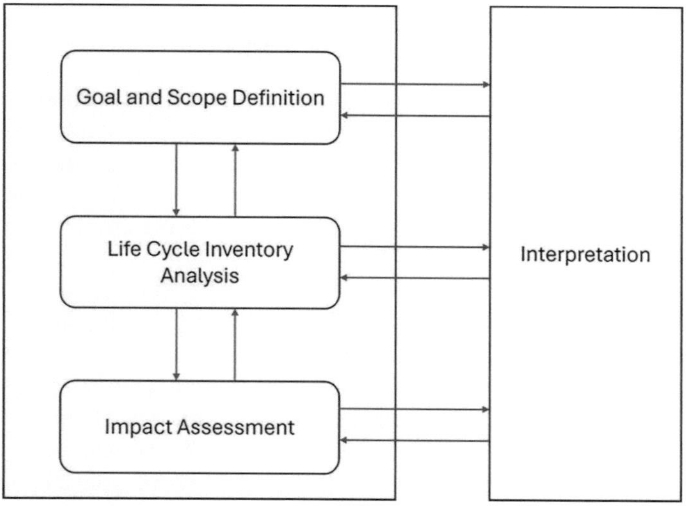

Fig. 1 ISO 14040/44 standard LCA framework

ecosystem quality, etc., and then the end-point approach is followed. Several other factors such as energy intensity and resource efficiency are also being analyzed during this phase. The interpretation of results is generated simultaneously with the three phases to indicate the results and evaluations more systematically [33].

3 Life Cycle Assessment Methodology Applied for Nanomaterial Production

Nanomaterial production is currently widely evolving from early-stage conventional methods to green and sustainable pathways. Although the applications of several nanomaterials are claimed environmentally sustainable and aid in overcoming several human health challenges, the potential environmental and human health impacts across the entire life cycle of these nanomaterials are not fully comprehended. The availability of a wide range of production pathways of the various nanomaterials evaluates the impacts of the production of nanomaterials challenging, regarding data quality, acceptance, and availability. However, the early-stage approach of conducting a life cycle assessment of the numerous pathways of nanomaterials production is essential to developing the modifications environmentally sustainable [5].

3.1 Goal and Scope Definition

The scope of the life cycle assessment can be identified starting from the raw material acquisition and depending on the goal of the study, the system boundary can be limited to the production stage as a cradle-to-gate assessment, or it can be further evaluated with the usage and the end-of-life stage as a cradle-to-grave assessment. Figure 2 shows the detailed and generalized cradle-to-grave system boundary of nanomaterial production.

The raw materials synthesis and purification are key operations in the material processing and manufacturing stages. Pre-treatment for preparing the materials for processing, synthesis of nanomaterials by top-down or bottom-up approaches, and post-treatment after the synthesis increase the stability and functionality of the materials. Several conventional and green approaches are available in the industry therefore the inventory can be varied accordingly. The airborne emissions and other discharges from chemicals are vital when evaluating the environmental impacts. The

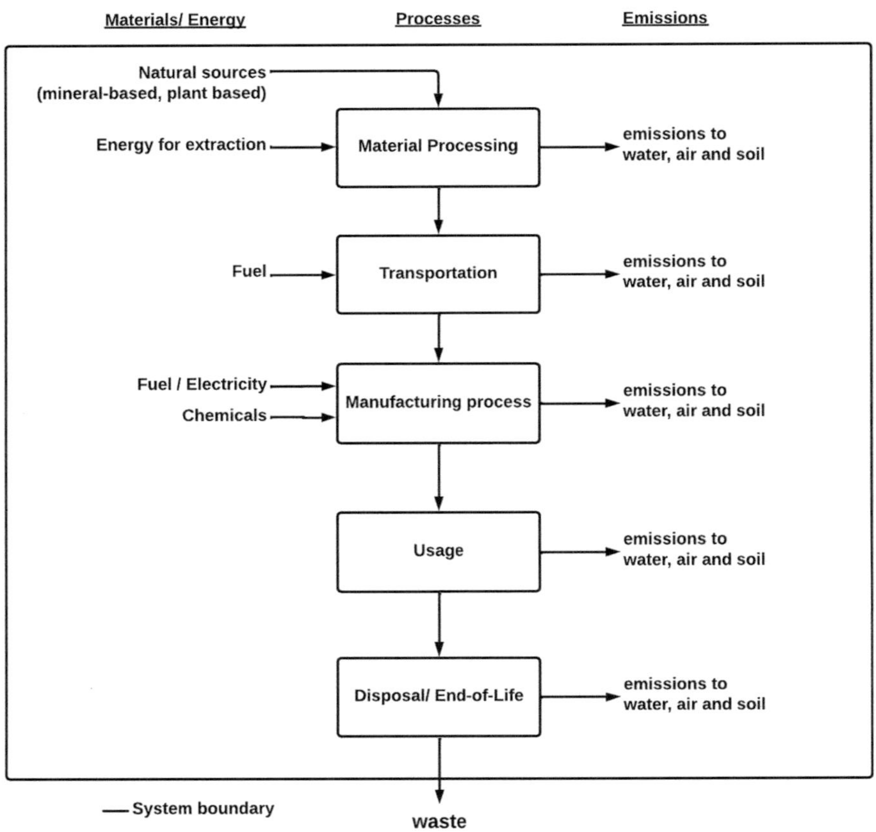

Fig. 2 System boundary of a general nanomaterial production cradle-to-grave LCA

transportation of processed materials finished goods, and other required materials are indicated as the stage "Transportation". The purified products are either directly used or used after blending with other polymeric materials. The use stage of nanomaterials is also crucial when evaluating the environmental impacts due to the release of nanoparticles into the atmosphere as airborne emissions or fugitive dust emissions. The available options for the disposal of used nanomaterials are landfilling, incineration, reuse, or recycling [34]. Researchers are interested in improving the recyclability of nanomaterials to reduce the impact of landfilling and other available disposal methods.

Selection of an appropriate functional unit is critical when conducting a life cycle assessment. In most of the studies, a unit mass (1 kg, 1 g, etc.) has been considered as the functional unit. However, when deciding the functional unit, the unit basis should be functionally comparable with other products. If the functional unit is inadequate for the study, it might lead to higher uncertainty in the study. Therefore, based on the objective, selecting a unit basis aligning with the functionality is crucial during a life cycle assessment [22].

3.2 *Inventory Analysis*

Life cycle inventory analysis is the crucial stage of the assessment when it comes to nanomaterial production. Due to the variation of the production processes and the implementations happening day by day, the availability of acceptable data is compromised, since many of the products are patented and product-specific [22]. The material efficiency including water, solvents, and chemicals is quite low and the waste generation is higher since many of the materials are required to produce at a low yield. The energy intensity of the production stage is higher compared to the other stages; therefore, this stage is excessively energy and material intensive. The energy intensity of several nanomaterial productions such as Carbon Nanotubes (minimum 2200 MJ/kg) stated higher compared to other metal manufacturing energy consumption such as steel (30 MJ/kg) or aluminum (218 MJ/kg) [34]. High purity is essential when manufacturing 2-dimensional and 3-dimensional nanomaterials, therefore when the purification process is repeated, the energy requirement is increased rapidly, when the number of purification and pre-treatment processes increase [5]. Exposure during production can impact human health causing acute and chronic diseases.

Compared to other materials of similar applications, nanomaterials may have stage emissions to be considered such as paints and coatings, packaging materials, electronic items, etc. Therefore, proper evaluation is necessary for the lifespan of the product. This factor is also product-specific; thus, it is convenient to assess the effect in a case study approach [35].

The unavailability of end-of-life disposal or disposal regulations on-site landfilling emissions evaluation, and challenges can arise for recycling the nanomaterials after the life cycle due to the changes in physical and chemical properties after usage. Therefore, the disposal of nanomaterials is evaluated similarly to many conventional products. This might cause the uncertainty of the evaluation to be increased, with the various potential consequences available, without strong confirmation [22].

However, it is advised by the researchers that the industries should not wait until the adequate database is released and should conduct life cycle assessments in a case study-based approach [15].

3.3 Impact Assessment

The environmental impact of the production of nanomaterials can be evaluated in two analysis methods: mid-point analysis and end-point analysis. Mid-point analysis focuses on the specific impact whereas the end-point impact focuses on the point of protection. Conducting an impact assessment will help the user/organization to understand the potential hotspots to be treated or modified towards the path of sustainable conversion of the process.

The major midpoint impacts considered in the life cycle assessment are climate change, human toxicity, ecotoxicity, eutrophication, acidification ozone depletion, and photo-oxidant formation. Among the studies conducted so far, the average global warming potential is provided as 0.7–3.0 kg CO_2 eq/functional unit and acidification ranges from 1.0 to 20 kg SO_2 eq/functional unit. The ecotoxicity ranges between 3.0 and 8.0 CTUeq/functional unit where the functional unit is mostly 1 kg of product [22].

Several factors have been identified when evaluating the environmental impacts of mid-point and end-point approaches.

The fate factor is used to determine the behavior of the nanomaterials once they are emitted into the environment. This means that the fate of nanomaterials encompasses sources, distribution, transformations, and how long they persist or adapt in different environments. As of now, the models with a view of predicting fate lose ground from accuracy because of the combination of the materials and processes models are developed [22]. Studies use models like probabilistic material flow analysis (MFA) to assess their environmental behavior. However, fate factor analysis remains an area least explored because of its complexities.

Exposure factor in nanomaterials studies estimates the degree of exposure nanomaterials are likely to have in the environment, especially in the water systems. Only a few such studies have been conducted in the past, and such studies assumed the worst cases of the most likely exposure. Those that inclusive identification is developed by data lack on nanomaterials concentration, exposure route, and use or manufacture of nanomaterials guarantee unused. Existing models such as MFA aim at estimating concentrations in the environment of these materials, but they have great uncertainties and validation challenges. Further, issues in the transport and disposal of nanomaterials remain elusive, fueling barriers to exposure measurement [29].

Effect factor is a measure used to evaluate the potential impact of a substance on human health or ecosystems per unit of exposure or release. It links the dose (or concentration) of a substance to its potential effect, typically within categories such as human toxicity or ecotoxicity. Several studies have conducted the effect factor analysis on human toxicity, environmental eco-toxicity, etc. However, the unit measurements vary with effective concentration to be affected [22].

Modelling tools can be used for the evaluation of the impact of the life cycle such as SimaPro, GaBi, OpenLCA software, etc. However, limitations have arisen with the unavailability of acceptable and adaptable data in the databases, such as the eco-invent and think-step databases.

4 Life Cycle Assessment Overview of Nanomaterials and Nanoproducts

Even though it is difficult to gather life cycle inventory data, many studies on life cycle assessment of nanoparticle production have been carried out in past years. It can be seen from Table 1, which includes the studies on the life cycle assessment of nanomaterials in the past 10 years (from 2014 to 2024). The studies carried out by [1, 2, 6, 13, 16–18, 24, 28] are related to the life cycle assessments carried out for the nanomaterials synthesizing processes while there are some studies related to the life cycle assessments conducted for the development of nano-enabled products such as batteries, gas sensors, composites, coatings, medical bandages, etc. [11, 12, 14, 25–27, 30]. Using nanomaterials for different purposes such as wastewater treatment is also categorized under the second one [3, 9].

According to the data presented in Table 1, carbon nanoparticles (e.g.: carbon nanotubes, graphene) are the most studied nanoparticles [1, 12, 25, 26, 28]. While life cycle assessments have been carried out for the other types such as silver nanoparticles [14, 27], gold nanoparticles [24], titanium dioxide nanoparticles [14], cellulose nanoparticles [2, 13], iron nanoparticles [3, 9], silica nanoparticles [11, 30] etc. Table 1 summarizes those studies including the impacts they assessed, system boundaries, functional unit, methods, software used, and the impact assessment method (midpoint and endpoint).

Almost all the studies have used cradle-to-gate system boundaries since it is difficult to collect data when it goes for a further system boundary such as cradle-to-grave or cradle-to-cradle. However, there are few studies that have used cradle-to-grave system boundaries [26, 27]. In most of the studies they have taken mass-based functional units as 1 g or 1 kg of the particle or the product. In the case of carbon nanotubes, sometimes they use 1 nm and for some products, 1 piece of the product is used as the functional unit [25, 28].

Energy consumption, greenhouse gas emissions, eutrophication, acidification, human health, ozone depletion, and freshwater ecotoxicity are the often-assessed impact categories in conventional processes. In some of the studies they have addressed nano-specific impacts but most of the studies have limited assessing those impacts due to the unavailability of accessing data. Health and toxicity issues for living beings related to nanoparticles, environment and fate issues, and energy

Table 1 Summary of existing LCA studies on nanomaterials and nano-enable products

Reference	Type of nano particle	The focus of the study in which LCA was conducted	System boundary	Impact categories	Method/ Software	Functional unit	Impact assessment		Assessment conclusion
							Mid point	End point	
Ingrao et al. [16]	Nano hydroxy apatite	Using nano hydroxy apatite as a biomaterial for improving leather fireproofing properties	Cradle to gate	GHG emissions, nonrenewable energy, respiratory inorganics, human health, climate change, resources, ecosystem quality	IMPACT/ SimaPro	10 g	Yes	Yes	Major production hotspots are electricity and production of phosphoric acid and in the transportation of chemicals
Lee et al. [17]	Nano calcium carbonate	For the synthesis process using steel slag	Cradle to gate	GHG emissions, CO_2 emissions, (sensitivity analysis included)	Not stated/ SimaPro Ecoinvent database used	1 g	Yes	No	Using steel slag instead of limestone to produce calcium carbonate nanoparticles reduce CO_2 emissions by 8% The unit cost of production is 484 UDS for the new process
Martins et al. [18]	Iron nanoparticles	For the synthesis process	Cradle to gate	Energy consumption, human health, atmospheric emissions	IMPACT 2002+/ SimaPro Ecoinvent database used	1 kg	Yes	No	The green synthesis method presents lower environmental impacts than the traditional method that uses sodium borohydride to produce iron nanoparticles In the green synthesis method, the critical stage is the extraction process which has high electricity consumption, while in the traditional method reactant use is the critical stage Economic evaluation shows that the traditional method is 8 times more expensive than the green method

(continued)

Table 1 (continued)

Reference	Type of nano particle	The focus of the study in which LCA was conducted	System boundary	Impact categories	Method/ Software	Functional unit	Impact assessment		Assessment conclusion
							Mid point	End point	
Pizza et al. [26]	Graphite nanoplatelets (GnP)	For the manufacturing process of epoxy-based nanocomposite	Cradle to grave	Energy consumption	ReCiPe, USEtox, EDIP, CML/ SimaPro Ecoinvent database used	1 kg of epoxy composite loaded with 0.058 kg of GnP	Yes	Yes	1 kg of GnP filler requires 1879 MJ of primary energy and epoxy resin requires 187 MJ. The preparation of 1 kg of epoxy composite loaded with 0.058 kg of GnP is 303 MJ. It shows environmental impact from composite production is negligible compared to the raw material extraction and filler and resin production
Hischier et al. [14]	Titanium dioxide nanoparticles Silicon-dioxide nanoparticles Silver nanoparticles	Using titanium dioxide nanoparticles in the facade coatings/ paints	Cradle to gate	GHG emissions, freshwater eutrophication, fossil fuel depletion, acidification, ecotoxicity, human toxicity, human health, resource availability	USEtox and ReCiPe /Open LCA tool, The ecoinvent database is used	1 square meter of (indoor or outdoor) wall for 80 years	Yes	Yes	Environmental performance of paints containing manufactured nanomaterials (MNMs) and paints without MNMs were conducted. It depends on: 01. The MNMs must substitute an (active) ingredient of the initial paint composition and not simply be an additional ingredient 02. The new composition must extend the lifetime of the paints 03. Releases of MNMs must be reduced to the lowest level possible

(continued)

Table 1 (continued)

Reference	Type of nano particle	The focus of the study in which LCA was conducted	System boundary	Impact categories	Method/ Software	Functional unit	Impact assessment		Assessment conclusion
							Mid point	End point	
Rodriguez-Garcia et al. [28]	Single-walled and multi walled carbon nanotubes (CNTs)	Using CNTs in batteries	Cradle to gate	Human toxicity, freshwater ecotoxicity	USEtox model/ Microsoft Excel	1 nm	Yes	No	CNTs cause less environmental harm than alternative materials but their potential effects on human health might discourage its implementation
Pourzahedi and Eckelman [27]	Silver nanoparticles	Using silver nanoparticles in medical bandages	Cradle to grave	Ozone depletion, GHG emissions, smog formation, respiratory effects, water and soil quality impacts, acidification, eutrophication, human health, eco-toxicity	TRACI 2.1 Impact assessment method/ SimaPro eco-invent database used	1 g	Yes	Yes	01. Dominating environmental impact is upstream electricity production, except life cycle ecotoxicity where the largest contributor is mining waste 02. AgNPs are the largest contributor to the impacts of the bandage for all impact categories, considered despite low AgNP loading 03. Impacts of bandage production are several times those of bandage incineration, including nano silver releases to the environment

(continued)

Table 1 (continued)

Reference	Type of nano particle	The focus of the study in which LCA was conducted	System boundary	Impact categories	Method/ Software	Functional unit	Impact assessment		Assessment conclusion
							Mid point	End point	
Petrakli et al. [25]	Carbon nanotubes	Using carbon nano tubes in prototypes of sailing boat and handbrake lever made from carbon fiber reinforced polymer (CFRP)	Cradle to gate	Human toxicity, respiratory effects, ionizing radiation, photochemical oxidation, climate change, ozone depletion, GHG emissions, human health, ecotoxicity, acidification, eutrophication, land occupation, water consumption, mineral extraction	ILCD midpoint/ SimaPro Ecoinvent database used	1 product piece	Yes	No	Potential benefits to recycling CFRP waste are: 01. Materials contribution dominates in all 14 impact categories 02. Recycling via pyrolysis can lead to significant impact savings due to material recovery 03. Incineration with energy recovery also leads to environmental savings, but much lower than recycling 04. Incineration without energy recovery contributes negatively by approximately 5 and 2% to climate change for the boat and lever, respectively
Gu et al. [13]	Cellulose nanocrystals	For the synthesis process from wood pulp	Cradle to gate	Energy consumption, GHG emissions, Ozone depletion, acidification, eutrophication, human health, ecotoxicity, fossil fuel depletion	TRACI and primary data of pilot scale production/ SimaPro Ecoinvent database used	1 kg of cellulose nanocrystals	Yes	No	For all the impacts, sodium hydroxide is the dominant contributor in the cradle-to-gate LCA. In this process, sodium hydroxide is used for neutralizing purposes

(continued)

Table 1 (continued)

Reference	Type of nano particle	The focus of the study in which LCA was conducted	System boundary	Impact categories	Method/ Software	Functional unit	Impact assessment		Assessment conclusion
							Mid point	End point	
Gilbertson et al. [12]	Carbon nanotubes	Using carbon nanotubes in a chemical gas sensor	Cradle to gate	GHG emissions, acidification, eutrophication, ozone depletion, smog formation, human health impacts from carcinogenic and non-carcinogenic, respiratory diseases, ecotoxicity	ReCipe and TRACI 2/ SimaPro Ecoinvent database used	1 g per chip	Yes	Yes	The midpoint impacts associated with producing carbon nanotubes (CNTs) are marginal compared with other manufacturing stages Unit of disability-adjusted life years (DALYs) is used to normalize the impact data because there is a tangible health benefit associated with product implementation The equivalent DALY impact from the manufacture of the single-walled carbon nanotubes (SWNT) enabled H_2S sensor is negligible compared with the potential DALY gain from the prevention of occupational fatalities
Pati et al. [24]	Gold nanoparticles	For the synthesis process	Cradle to gate	Energy consumption, climate change, metal depletion, agricultural land occupation, freshwater ecotoxicity	Not stated/ SimaPro Ecoinvent database used	1 mg of gold nanoparticle	Yes	Yes	A substantial part of the overall impact of a given synthesis can be associated with heating, stirring, and sonication rather than replacing toxic reducing agents in conventional processes with renewable reagents in green synthesis methods Reaction yield is a key parameter in figuring out the cumulative energy demand (CED) of a synthesis process, and it is especially important in the case of nanotechnologies based on resource-limited raw materials (e.g., gold)

(continued)

Table 1 (continued)

Reference	Type of nano particle	The focus of the study in which LCA was conducted	System boundary	Impact categories	Method/ Software	Functional unit	Impact assessment		Assessment conclusion
							Mid point	End point	
Bai et al. [3]	Al_2O_3-supported Fe–Mn binary oxides nanoparticles	For the process of Oxidation and adsorption of Antimony from water	Cradle to gate	Ozone depletion, GHG emissions, smog, acidification, eutrophication, carcinogenic and non-carcinogenic, respiratory effects, ecotoxicity, fossil fuel depletion	TRACI 2.1/ SimaPro, Ecoinvent and US life cycle inventory used	1 kg	Yes	No	Al_2O_3-supported Fe–Mn binary oxide nanoparticles were derived using green technologies to remove antimony from surface water. LCA indicates that the synthesis of this material met the green chemistry principles
Feijoo et al. [9]	Magnetic nano particles made of magnetite-Fe_3O_4	Using nanoparticles in Fenton or Photo-Fenton catalysts used in wastewater treatment	Cradle to gate	Climate change, ozone depletion, acidification, eutrophication, toxicity, fossil depletion	IMPACT and ReCiPe/ SimaPro Ecoinvent database used	1 kg	Yes	No	Synthesized Fenton and Photo-Fenton catalysts for wastewater treatment using magnetic nanoparticles. 03 different stabilizing agents; Polyacrylic acid (PAA), polyethyleneimine (PEI), and Silica (SiO_2) were used to increase catalytic activity. Magnetic nanoparticles coated with PAA or PEI are the most suitable choice for their applications which have the lowest environmental impacts compared to other types

(continued)

Table 1 (continued)

Reference	Type of nano particle	The focus of the study in which LCA was conducted	System boundary	Impact categories	Method/ Software	Functional unit	Impact assessment		Assessment conclusion
							Mid point	End point	
Arvidsson et al. [2]	Cellulose Nanofibrils	For the synthesis process of cellulose nanofibrils from wood pulp	Cradle to gate	Energy use, climate change, acidification, water use (sensitivity analysis included)	ReCiPe, Ecoinvent database used	1 kg	Yes	No	Compared 3 synthesis routes of the production of cellulose nanofibrils (CNF) from wood pulp: 01. Enzymatic production route which includes an enzymatic pre-treatment 02. Carboxymethylation route which includes carboxymethylation pretreatment and 03. Route without pretreatment CNF made using the 2nd route has the highest environmental impacts due to the large use of solvents made from crude oil compared to the other 2 routes
Arvidsson et al. [1]	Graphene	For the synthesis process	Cradle to gate	Energy use, water use, human toxicity, ecotoxicity, (sensitivity analysis included)	USEtox model/ not stated	1 kg of graphene in solution	Yes	No	LCA was conducted for 2 graphene synthesizing routes, namely ultrasonication and chemical reduction Ultrasonication route has lower water and energy use but, higher human and ecotoxicity compared to the other method A sensitivity analysis showed that the solvent recovery in the ultrasonication process gives lower impacts for all the impacts Solvent recovery in the chemical reduction process is important to reduce the blue water footprint

(continued)

Table 1 (continued)

Reference	Type of nano particle	The focus of the study in which LCA was conducted	System boundary	Impact categories	Method/Software	Functional unit	Impact assessment		Assessment conclusion
							Mid point	End point	
Sackey et al. [30]	Nano silica	For nano-silica-modified asphalt mixtures	Cradle to gate	Ecotoxicity, carcinogens, GHG emissions, ozone depletion, acidification, eutrophication, respiratory effects	TRACI/Open LCA tool, Ecoinvent database used	1000 kg	Yes	No	Compared the environmental impacts of conventional asphalt mixture production and nano-silica modified asphalt mixture (NMAM) production NMAM had a global warming potential of 7.44563 * 10^3 kg/CO_2-eq per functional unit (FU) compared to 7.41900 * 10^3 kg/CO_2-eq per functional unit of conventional asphalt mixture production According to the calculated normalized impact scores, NMAM performs better in five categories; global warming (0.3102 pts/FU), Ozone depletion (0.2323 pts/FU), eutrophication (0.6779 pts/FU), photochemical oxidation (0.0217 pts/FU) and ecotoxicity (0.0634 pts/FU) and performs worst in 4 categories; respiratory effects (36.9556 pts/FU), human health carcinogenic (5.8258 pts/FU), human health non-carcinogenic (182 pts/FU) and acidification (41.0101 pts/FU)

(continued)

Table 1 (continued)

Reference	Type of nano particle	The focus of the study in which LCA was conducted	System boundary	Impact categories	Method/ Software	Functional unit	Impact assessment		Assessment conclusion
							Mid point	End point	
Gao et al. [11]	Hollow silica nanospheres	Using hollow silica nanoparticles in nano insulators	Cradle to gate	Energy consumption	Not stated/ used primary data in laboratory scale study	1 g	Yes	No	Synthesis of hollow silica nanospheres nano insulation materials have high environmental impact
Bobba et al. [6]	Tungsten di-sulphide nanoparticles	For the synthesis route	Cradle to gate	Energy resources, GHG emissions, acidification, eutrophication, human toxicity	ReCiPe, CML SimaPro Ecoinvent database used	1 g	Yes	No	LCA was conducted for large-scale production of tungsten di-sulfide nanoparticles The synthesis of nano tungsten di-sulfide is highly energy-consuming and the raw material with the higher contribution is the tungsten trioxide in terms of weight

consumption are the main nano-specific issues that need to be addressed. When considering health and toxicity issues related to nanoparticles, nanomaterials like carbon nanotubes and nano-silver show significant toxicity even at low concentrations due to their high surface area, reactivity, and potential for bioaccumulation. This can lead to health issues such as inflammation, oxidative stress, and organ damage when inhaled or ingested [7, 32]. Due to the small particle size and the high reactivity, mobility of them in the environment can lead to unintended distribution within air, water, and soil affecting organisms and the environment and causing long-term environmental risks [32]. When comparing the energy consumption of nanoparticle synthesizing processes such as the top-down milling process or the bottom-up chemical synthesis process with conventional processes, they are highly energy intensive. For example, the synthesis of nano calcium carbonate using steel slag has higher energy consumption than the conventional method which was studied by Lee et al. [17].

For the life cycle assessment of every process, several types of software have been used. SimaPro is the most used software while few studies used Open LCA [14] as well as Microsoft Excel software [28]. The most used database is an eco-invent database, while the US life cycle inventory [3] and primary lab scale inventory data [11] were also used. These software tools include various impact assessment methods such as ReCiPe, USEtox, IMPACT, etc. ReCiPe is a very comprehensive impact assessment method that offers both midpoint and endpoint assessment facilities, while USEtox is specially focused on human toxicity and ecotoxicity.

Midpoint assessments focus on specific environmental issues or impact categories at an early stage of the cause-effect chain while the endpoint method focuses on the final consequences or damage categories resulting from the midpoint impacts. It quantifies the overall damage to human health, ecosystem quality, and resource availability. Examples of midpoint indicators are global warming potential, acidification potential, eutrophication potential, ozone depletion, and human toxicity potential. Human health, ecosystem quality, and resource depletion are some end-point indicators [4]. Table 1 shows the type of impact assessment methods used in each of the studies.

5 Challenges in Conducting LCA for Nanomaterials and Future Recommendations

When performing an LCA for nanomaterials, certain limitations are there. Some of them are summarized below [20, 22, 31].

- In most of the studies scope is taken as cradle to gate, due to the difficulty in assessing the whole life cycle with the scopes cradle to grave or cradle to cradle. That is because of the lack of inventory data since their manufacturing processes are new and often subject to changes. The lack of data also limits the accuracy of the environmental impacts calculated.

- Due to the lack of knowledge on data such as release potential pathways, characterization factors, models of calculating impacts, etc. to calculate the impacts related to nanomaterials, most of the authors assume that nanomaterials are handled similarly to conventional products. However, conventional processes often use mass as a baseline to assess impact. For nanomaterials, when assessing their impacts, it is important to consider their surface area, particle number, or specific chemical reactivity. It can lead to underestimating or misrepresenting the impacts of relying on the mass-based cutoffs since the unique characteristics of nanomaterials do not scale linearly with mass.
- The development of characterization factors for nanomaterials is still inadequate.
- Nanomaterial synthesizing processes are not yet standardized, but they are changing constantly. Therefore, it is another issue arising when developing inventory data and so doing the impact assessments.

To address these issues and to reduce the negative impacts arising from the manufacturing of nanomaterials and products following are some of the solutions that can be implemented. [8, 19, 20, 22, 31].

- To address the gap in lack of inventory data, it is needed to develop country-based or region-based databases. In addition to that, laboratory and modeling-based studies can be coupled to advance information generation. Furthermore, standardized methods for nanomaterial release should be developed per product category [20].
- It is also recommended to collaborate with industries to gather LCI data related to the release and exposure of nanomaterials.
- It is needed to develop suitable characterization factors based on transparent characterization models such as fate modeling, exposure modeling, and effect modeling.
- As a solution for the inadequate LCI data, it is possible to combine LCA with risk assessment (RA). To evaluate the environmental and human health consequences of nanomaterials, this LCA-RA approach would be immensely useful.

6 Conclusion

Considering the high use of nanomaterials in different fields, it is needed to have proper life cycle assessments for the nanomaterial synthesizing processes and the development of nano-enabled products. LCA has been conducted for numerous studies related to nanomaterials used in a wide range of applications. However, most of the studies have conducted LCA, assuming the process as a conventional process and neglecting the output flows related to releases of nanomaterials into the environment. Therefore, most of these studies have limited their scope to cradle-to-gate without going for cradle-to-grave or cradle-to-cradle scopes. That is because of the lack of knowledge and inventory data and due to uncertainties about the end-of-life management of waste products. Also, when considering the impact assessment phase, toxicity-related impact categories have been weakly discussed due to the absence of characterization factors. Therefore, it is recommended to fill these gaps by developing proper inventory databases by collaborating with industries and developing models to find characterization factors related to toxicity-related impacts to have comprehensive LCAs for nanomaterials and nanomaterials-based products.

References

1. Arvidsson R, Kushnir D, Sandén BA, Molander S (2014) Prospective life cycle assessment of graphene production by ultrasonication and chemical reduction. Environ Sci Technol 48(8):4529–4536. https://doi.org/10.1021/es405338k
2. Arvidsson R, Nguyen D, Svanström M (2015) Life cycle assessment of cellulose nanofibrils production by mechanical treatment and two different pretreatment processes. Environ Sci Technol 49(11):6881–6890. https://doi.org/10.1021/acs.est.5b00888
3. Bai Y, Wu F, Gong Y (2020) Oxidation and adsorption of antimony(iii) from surface water using novel Al_2O_3-supported Fe–Mn binary oxide nanoparticles: effectiveness, dynamic quantitative mechanisms, and life cycle analysis. Environ Sci Nano 7(10):3047–3061. https://doi.org/10.1039/d0en00609b
4. Bare JC, Hofstetter P, Pennington DW, Udo de Haes HA (2000) Life cycle impact assessment workshop summary. Midpoints versus endpoints: the sacrifices and benefits. Int J Life Cycle Assess 5(6):319–326
5. Basiuk VA, Basiuk EV (2015) Green processes for nanotechnology: from inorganic to bioinspired nanomaterials. Green Process Nanotechnol Inorg Bioinspired Nanomater 1–446. https://doi.org/10.1007/978-3-319-15461-9/COVER
6. Bobba S, Deorsola FA, Blengini GA, Fino D (2016) LCA of tungsten disulphide (WS2) nanoparticles synthesis: state of art and from-cradle-to-gate LCA. J Clean Prod 139:1478–1484. https://doi.org/10.1016/j.jclepro.2016.07.091
7. Boverhof DR, David RM (2010) Nanomaterial characterization: considerations and needs for hazard assessment and safety evaluation. Anal Bioanal Chem 396(3):953–961. https://doi.org/10.1007/s00216-009-3103-3
8. Dhingra R, Naidu S, Upreti G, Sawhney R (2010) Sustainable nanotechnology: through green methods and life cycle thinking. Sustainability 2(10):3323–3338. https://doi.org/10.3390/su2103323
9. Feijoo S, González-Rodríguez J, Fernández L, Vázquez-Vázquez C, Feijoo G, Moreira MT (2020) Fenton and photo-fenton nanocatalysts revisited from the perspective of life cycle assessment. Catalysts 10(1):1–11. https://doi.org/10.3390/catal10010023
10. Finnveden G, Potting J (2014) Life cycle assessment. Encycl Toxicol Third Ed 74–77. https://doi.org/10.1016/B978-0-12-386454-3.00627-8
11. Gao T, Sandberg LIC, Jelle BP (2014) Nano insulation materials: synthesis and life cycle assessment. Procedia CIRP 15:490–495. https://doi.org/10.1016/j.procir.2014.06.041
12. Gilbertson LM, Busnaina AA, Isaacs JA, Zimmerman JB, Eckelman MJ (2014) Life cycle impacts and benefits of a carbon nanotube-enabled chemical gas sensor. Environ Sci Technol 48(19):11360–11368. https://doi.org/10.1021/es5006576
13. Gu H, Reiner R, Bergman R, Rudie A (2015) LCA study for pilot scale production of cellulose nano crystals (CNC) from wood pulp. In: LCA XV paper proceedings—a "bright green" future, April, pp 33–42
14. Hischier R, Nowack B, Gottschalk F, Hincapie I, Steinfeldt M, Som C (2015) Life cycle assessment of façade coating systems containing manufactured nanomaterials. J Nanopart Res 17(2). https://doi.org/10.1007/s11051-015-2881-0

15. Hischier R, Walser T (2012) Life cycle assessment of engineered nanomaterials: state of the art and strategies to overcome existing gaps. Sci Total Environ 425:271–282. https://doi.org/10.1016/J.SCITOTENV.2012.03.001
16. Ingrao C, Vesce E, Evola RS, Rebba E, Arcidiacono C, Martra G, Beltramo R (2021) Chemistry behind leather: life cycle assessment of nano-hydroxyapatite preparation on the lab-scale for fireproofing applications. J Clean Prod 279:123837. https://doi.org/10.1016/j.jclepro.2020.123837
17. Lee J, Ryu KH, Ha HY, Jung KD, Lee JH (2020) Techno-economic and environmental evaluation of nano calcium carbonate production utilizing the steel slag. J CO_2 Utilization 37:113–121. https://doi.org/10.1016/j.jcou.2019.12.005
18. Martins F, Machado S, Albergaria T, Delerue-Matos C (2017) LCA applied to nano scale zero valent iron synthesis. Int J Life Cycle Assess 22(5):707–714. https://doi.org/10.1007/s11367-016-1258-7
19. Miseljic M, Olsen SI (2014) Life-cycle assessment of engineered nanomaterials: a literature review of assessment status. J Nanopart Res 16(6). https://doi.org/10.1007/s11051-014-2427-x
20. Moloi MS, Lehutso RF, Erasmus M, Oberholster PJ, Thwala M (2021) Aquatic Environment Exposure and Toxicity of Engineered Nanomaterials Released from Nano-Enabled Products: Current Status and Data Needs. Nanomaterials 11(11):2868
21. Muralikrishna IV, Manickam V (2017) Life cycle assessment. Environ Manag 57–75. https://doi.org/10.1016/B978-0-12-811989-1.00005-1
22. Nizam NUM, Hanafiah MM, Woon KS (2021) A content review of life cycle assessment of nanomaterials: current practices, challenges, and future prospects. Nanomaterials 11(12):1–27. https://doi.org/10.3390/nano11123324
23. Pajula T, Behm K, Vatanen S, Saarivuori E (2017) Managing the life cycle to reduce environmental impacts. In: Dynamics of long-life assets: from technology adaptation to upgrading the business model, pp 93–113. https://doi.org/10.1007/978-3-319-45438-2_6
24. Pati P, McGinnis S, Vikesland PJ (2014) Life cycle assessment of "green" nanoparticle synthesis methods. Environ Eng Sci 31(7):410–420. https://doi.org/10.1089/ees.2013.0444
25. Petrakli F, Gkika A, Bonou A, Karayannis P, Koumoulos EP, Semitekolos D, Trompeta AF, Rocha N, Santos RM, Simmonds G, Monaghan G, Valota G, Gong G, Charitidis CA (2020) End-of-life recycling options of (nano)enhanced CFRP composite prototypes waste-a life cycle perspective. Polymers 12(9). https://doi.org/10.3390/POLYM12092129
26. Pizza A, Metz R, Hassanzadeh M, Bantignies JL (2014) Life cycle assessment of nanocomposites made of thermally conductive graphite nanoplatelets. Int J Life Cycle Assess 19(6):1226–1237. https://doi.org/10.1007/s11367-014-0733-2
27. Pourzahedi L, Eckelman MJ (2015) Environmental life cycle assessment of nanosilver-enabled bandages. Environ Sci Techno; 49(1):361–368
28. Rodriguez-Garcia G, Zimmermann B, Weil M (2014) Nanotoxicity and life cycle assessment: first attempt towards the determination of characterization factors for carbon nanotubes. IOP Conf Ser Mater Sci Eng 64(1). https://doi.org/10.1088/1757-899X/64/1/012029
29. Romeo D, Hischier R, Nowack B, Jolliet O, Fantke P, Wick P (2022) In vitro-based human toxicity effect factors: challenges and opportunities for nanomaterial impact assessment. Environ Sci Nano 9(6):1913–1925. https://doi.org/10.1039/D1EN01014J
30. Sackey S, Lee DE, Kim BS (2019) Life cycle assessment for the production phase of nano-silica-modified asphalt mixtures. Appl Sci (Switzerland) 9(7):1–13. https://doi.org/10.3390/app9071315
31. Salieri B, Turner DA, Nowack B, Hischier R (2018) Life cycle assessment of manufactured nanomaterials: where are we? NanoImpact 10:108–120. https://doi.org/10.1016/j.impact.2017.12.003
32. Tran TK, Nguyen MK, Lin C, Hoang TD, Nguyen TC, Lone AM, Khedulkar AP, Gaballah MS, Singh J, Chung WJ, Nguyen D (2024) Review on fate, transport, toxicity and health risk of nanoparticles in natural ecosystems: emerging challenges in the modern age and solutions toward a sustainable environment. Sci Total Environ 912:169331. https://doi.org/10.1016/j.scitotenv.2023.169331

33. UNEP/SETAC (2007) Life cycle management: a business guide to sustainability acknowledgements. United Nations Environment Programme, 51. http://www.unep.fr/shared/publications/pdf/DTIx0889xPA-LifeCycleManagement.pdf
34. Upadhyayula VKK, Meyer DE, Curran MA, Gonzalez MA (2012) Life cycle assessment as a tool to enhance the environmental performance of carbon nanotube products: a review. J Clean Prod 26:37–47. https://doi.org/10.1016/J.JCLEPRO.2011.12.018
35. Vejerano EP, Leon EC, Holder AL, Marr LC (2014) Characterization of particle emissions and fate of nanomaterials during incineration. Environ Sci Nano 1(2):133–143. https://doi.org/10.1039/C3EN00080J

Health and Safety Aspects of Nanoparticle Production at Industrial Scale: An Overview

Kalinda Dasanayaka

Abstract The industrial-scale production of nanomaterials (NMs) presents significant health and safety challenges due to their unique properties, which can lead to adverse biological interactions. Engineered NMs, designed with specific characteristics, pose occupational hazards primarily through inhalation, dermal contact, and ingestion routes. Inhalation is specifically concerning as it may cause respiratory issues such as inflammation and fibrosis. The lack of comprehensive data on human exposure effects necessitates rigorous risk assessment and management strategies. To mitigate these risks, assessment of the process lifecycle of NMs is crucial to define the most relevant hierarchical method for hazard control, consistent with regulatory standards: elimination or substitution with safer materials when possible, engineering controls like ventilation systems equipped with HEPA filters, administrative controls such as good hygiene practices, and personal protective equipment as a last resort. Occupational exposure limits (OELs) have been established for some NMs but remain largely undefined for many others. Control banding strategies offer an alternative risk management tool when OELs are unavailable. Regulatory frameworks are evolving but face challenges due to the rapid advancement in nanotechnology. The prevention through design (PtD) approach emphasizes integrating safety considerations early in the product design phase to reduce risks proactively. This proactive strategy not only enhances the safety of stakeholders but also contributes to more efficient production processes. Continuous monitoring through occupational health surveillance programs further supports maintaining safe working conditions. As nanoparticle production expands globally, prioritizing health and safety measures will be essential for sustainable growth in the sector. In conclusion, ensuring safe nanoparticle production involves ongoing research into toxicity mechanisms, developing robust regulatory standards, and adhering to stringent safety protocols. This chapter provides an overview of current health and safety management principles

K. Dasanayaka (✉)
Department of Civil and Environmental Engineering, Faculty of Engineering, University of Ruhuna, Hapugala, Galle, Sri Lanka
e-mail: kalinda.dasanayaka@siamcitycement.com

Siam City Cement (Lanka) Ltd, Colombo 02, Sri Lanka

N. M. Adassooriya et al. (eds.), *Industrial Scale Production of Nanoparticles*, Smart Nanomaterials Technology, https://doi.org/10.1007/978-981-96-9176-0_20

associated with NM production at an industrial scale while highlighting regulatory directions for safer manufacturing practices.

Keywords Risk assessment · Exposure risk · Prevention through design (PtD) · Occupational exposure limits (OELs) · Hierarchy of controls · Regulations

1 Introduction

Significant nanomaterials applications have been developed in nanotechnology over the past few years. Nanomaterials, defined as materials with dimensions between 1 and 100 nm, include nanoparticles, nanofibers, nanotubes, and nanosheets. Due to their unique physicochemical properties, they are utilized in several industries, including electronics, energy, medicine, cosmetics, food engineering, telecommunications, construction, and agriculture, nanotechnology is advancing quickly. Below are some examples to elaborate on their value additions in such industries [1–5].

- Medical: Drug delivery, diagnostics, and regenerative medicine.
- Construction: Enhanced durability and strength of materials, thermal insulation.
- Electronics: Components in computers, mobile devices, and sensors.
- Energy: Solar cells, batteries, and supercapacitors for improved energy storage and efficiency.

Due to their small size, nanomaterials have special optical, magnetic, electrical, physical, reactivity, strength, surface area, sensitivity, and stability features [1, 2, 5–9]. Surprisingly, the property change occurs when bulk materials are converted into nanomaterials, meaning that previously non-magnetic materials become magnetic at the nanoscale. Because of its unique features, nanoscale matter is a separate form of matter from the solid, liquid, gaseous, and plasma states. Nanomaterials' characteristics are mostly determined by their shapes and sizes [9, 10]. However, the production and use of NMs also present new challenges in HSE management. Nanoparticles, due to their size, unique characteristics, and interactions, can set unknown risks to health, safety, and the environment. This chapter aims to provide an overview of the key HSE aspects related to general nanoparticle production and explore strategies for managing these risks in the industry.

2 Understanding Health and Safety Related Risk Perception on Nanomaterials

Effective workplace protection requires knowledge within the industry that nanoparticles are being handled. Understanding the Risk Perception Theory [11] examines how individuals or groups perceive and respond to risks, emphasizing that these perceptions often differ significantly from expert assessments of risk. It explores the psychological, cultural, and social factors influencing how people evaluate potential dangers and make decisions about them. Therefore, understanding perceptions related to nanomaterials is critically important to regulate risks for three main reasons as highlighted by the studies conducted by the European Union [12–14].

i. Studying risk perception is a mandate to grasp necessary approaches and attitudes to adapt and promote HSE measures focusing scientific context of public uncertainty about almost every aspect of nanomaterials (definition, characterization, toxicity, exposure levels)
ii. The sensitivity of the new areas in nanotechnology such as biotechnology is affected by how the public reacts to its development and marketing.
iii. The lack of knowledge, lack of transparency, involuntary nature of exposure, or delayed or uncontrollable effects could impact the delaying of appropriate prevention measures.

Considering the above critical factors, the understanding of the total process stream from raw materials supply to dispatch of finished products is crucial to breaking down into individual risks to identify their impacts and adequate control methods in a scientific background.

The study below was conducted by the European Commission [12] related to the main US and EU stakeholders' attitudes on nanotechnology was as follows in Table 1.

The risk perception of nanomaterials is a critical consideration for key industry stakeholders, as highlighted in Table 1. This perception stems from a combination of application-specific risk–benefit analyses and the integration of existing value frameworks into novel, hypothetical scenarios posed by emerging technologies. However, the current level of risk awareness and perception is dynamic and subject to change over time. Future developments, technological advancements, and unexpected events could significantly influence how risks associated with nanomaterials are understood and addressed.

Table 1 Perception and actions of US and EU stakeholders on nanotechnology [12]

	What they say on nano	What they do on nano
Industry	Risk assessment is adequate	Setting guidelines
Start-ups	Assessing risk is expensive	Reluctant to raise safety issues
NGOs	Focus on risk	Ask for more testing/regulations
Regulatory bodies	Learning curve	Enhancing current regulations
Insurers	Dialogue on risk	Worrying publicly
Researchers	Funding is needed	Studying public attitudes
Media	Enthusiasm, suspicions, sci-friction	Wondrous prospects/haunting stories
Consumers	Magic, out of control	Disorientation

3 Health and Safety Hazards Related to the Nanomaterial Industry

Current and potential applications in nanotechnology generally represent incremental advancements rather than revolutionary breakthroughs. However, over the long term, it is anticipated that nanotechnologies will exert a transformative influence on manufacturing processes, like the extensive impact of ICT today.

A wide range of techniques are available for fabricating nanostructures, each varying in quality, speed, and cost. These manufacturing methods can be broadly classified into two categories: 'bottom-up' and 'top-down' approaches, as discussed in earlier chapters. Workplace exposure to nanoparticles largely depends on the production techniques employed. Understanding the manufacturing processes involved is crucial for identifying potential hazards, assessing risks, and developing effective risk mitigation strategies. Defining appropriate control measures depends on the type of nanomaterial and its associated health risks to ensure exposure remains within acceptable limits.

Exposure limits for engineered nanomaterials remain largely undefined due to limited toxicological data. However, as outlined in Table 2, various agencies and research organizations have proposed recommended exposure limits (RELs) for several commonly used nanomaterials.

A comprehensive analysis of the current state of knowledge on the health effects of engineered nanomaterials on workers was conducted by Schulte et al. [27], while NIOSH [17] carried out studies on developing occupational exposure limits (OELs) and exposure bands for engineered nanomaterials, compiling findings from various human epidemiology studies that can be used for further studies.

Referring to Table 2, it is evident that nanomaterial-related hazard identification and control depend on the entire process and the specific form of the NMs. As an example, Fig. 1 illustrates a generic example of metallic NM production through chemical synthesis (Sol–Gel technique), a commonly used bottom-up approach. This process highlights the key steps necessary in nanoparticle manufacturing.

If the example in Fig. 1 is adapted to the Nickel (Ni) nanomaterial production process [28, 29], most unit operations involve liquid phases, with the final product obtained in the solid phase. However, if the process shifts to a top-down thermal decomposition technique, most operations occur in the gas phase [28, 29] increasing the risk of airborne Ni NM inhalation compared to the Sol–Gel method. Referring to Table 2, the recommended exposure limit (REL) for soluble Ni NMs is 1 mg/m^3 (8 h TWA), whereas for insoluble Ni NMs, it is significantly lower at 0.015 mg/m^3 (10 h TWA). Not only the material's form, but also its particle size must be considered in risk assessment. For example, the REL for TiO_2 nanomaterials varies based on particle size, with limits of 0.3 mg/m^3 for ultrafine NMs and 2.4 mg/m^3 for fine NMs and even a lower value of 0.8 $\mu g/m^3$ to prevent lung inflammation [16, 18].

Therefore, the potential exposure of the workforce to nanoparticles is highest when the materials are processed in a gaseous environment, necessitating stringent monitoring of worker exposure. In contrast, when the NMs are manufactured in a liquid

Table 2 Key Recommended Exposure Limits (RELs) for Nanomaterials

Nanomaterial	Agency/ Source	Recommended exposure limit (REL) for 8 h' time-weighted average (TWA), respirable fraction
Carbon nanotubes and carbon nanofibers	NIOSH [15]	1.0 µg/m^3
Titanium dioxide (TiO_2) ultrafine particles	NIOSH [16]	0.3 mg/m^3
Titanium dioxide (TiO_2) fine particles	NIOSH [16, 17]	2.4 mg/m^3
Titanium dioxide (TiO_2)	ANSES [18]	0.80 µg/m^3
Silica nanoparticles	OSHA and NIOSH [19]	50 µg/m^3 (respirable crystalline silica including Silica nanoparticles) 25 µg/m^3 as an action level (AL)
Silver nanoparticles	NIOSH [20, 21]	0.9 µg/m^3 (< 100 nm primary particle size) 10 µg/m^3 for silver dust and soluble silver compounds
Nickel	OSHA [22]	1 mg/m^3 for soluble nickel compounds (studies suggest REL for respirable nickel nanoparticles as low as 79 µg/m^3 [23, 24])
Nickel	NIOSH [22]	0.015 mg/m^3 (TWA 10 h)
Graphene and graphene oxide	No endorsement	Graphene: up to 18 µg/m^3 Graphene Oxide: 0.33 µg/m (Proposed by research studies [25, 26])

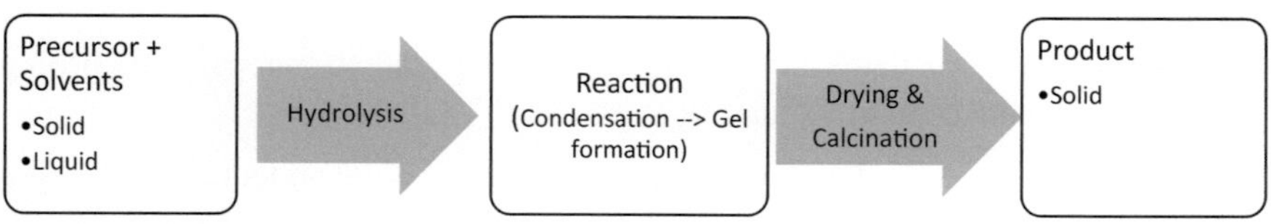

Fig. 1 The generic process involved in the production of nanoparticles using the chemical synthesis (Sol–Gel technique) method

phase it significantly reduces the potential exposure risk for workers. Producing and storing nanoparticles in the liquid phase can reduce the risks associated with airborne exposure. However, fugitive emissions in the form of vapors or wet aerosols may still occur, necessitating careful management. Therefore, this difference in hazard based on the material's form and size must be carefully considered when designing processes and implementing safety control mechanisms.

Despite these precautions, the health impacts of nanoparticles remain a critical concern due to their unique physicochemical properties. Nanoparticles can penetrate biological barriers, such as the skin, respiratory system, and even the blood–brain

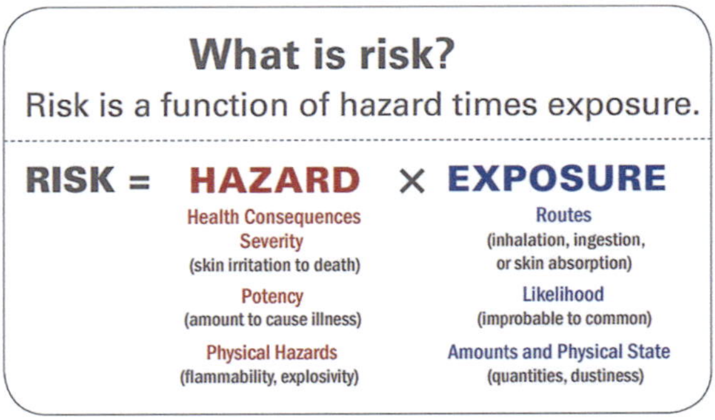

Fig. 2 Risk is a function of hazard times exposure [9]

barrier, potentially leading to toxicological effects. These include respiratory inflammation, oxidative stress, and long-term risks such as carcinogenicity or neurotoxicity, depending on the nanoparticle type and exposure level [10, 24, 27, 30–35].

The general approach to assessing and controlling risk involves two key steps: identifying the hazard (the potential of a substance to cause harm) and systematically determining the likelihood of exposure to the hazard along with its associated consequences. This is explained in Fig. 2.

Therefore, conducting thorough, process-specific hazard identification is essential to implementing effective workplace regulations by identifying the hazard and the likelihood of exposure to safeguard workers' health and ensure that the rapid advancement of nanotechnologies aligns with robust safety and regulatory practices.

Below are critical health-related risks/consequences identified for managing the health and safety of industrial exposures to nanomaterials [1, 10, 17, 24, 25, 27, 36–39]:

- Inhalation Risks: Inhalation is a primary route of exposure to nanoparticles, posing significant health concerns such as,
  - Respiratory Issues: Nanoparticles can penetrate deep into the lungs, reaching the alveoli and potentially triggering inflammatory responses, fibrosis, and, in severe cases, carcinogenesis.
  - Systemic Health Effects: Once in the bloodstream, circulating nanoparticles can affect various organs, including the liver, kidneys, and heart, potentially leading to systemic toxicity.
- Dermal and Ingestion Risks: While inhalation poses the most significant risk, exposure through skin contact or ingestion should not be overlooked.
  - Dermal Absorption: Certain nanoparticles can penetrate the skin barrier, potentially causing localized irritation or systemic toxicity through absorption into the bloodstream.
  - Ingestion Risks: Accidental ingestion of nanoparticles, often through hand-to-mouth transfer, can result in gastrointestinal irritation and systemic absorption, posing further health risks.

Table 3 LD_{50} and LC_{50} Values of Ti, Cu and Ag based nanomaterials

Nanomaterial	LD_{50} (mg/kg body weight)	LC_{50} (mg/L)	Remarks	Reference
Titanium dioxide (TiO2)	Oral LD_{50}: > 5000	Inhalation LC50: > 3.43 for male rats (particle size < 3.5 μm) > 5.09 for female rats (20% particle size < 3.5 μm)	Rats and mice	[38]
Silver	250 350	3	LD_{50}—mice with two types of Si NMs LC_{50}—fish	[37, 40]
Copper		3	Fish	[40]

- Long-Term Health Risks: Chronic exposure to nanomaterials raises concerns about long-term health implications such as,
  - Carcinogenic Potential: Some nanomaterials, such as carbon nanotubes, have structural similarities to asbestos fibers, raising concerns about their potential to cause cancers after prolonged exposure.
  - Neurotoxic Effects: Emerging research indicates that nanoparticles can cross the blood–brain barrier, potentially impacting the central nervous system and raising concerns about long-term neurotoxicity.

Addressing these risks requires robust workplace safety practices, continuous monitoring, and proactive research into the toxicological effects of nanomaterials to safeguard workers' health and safety. Table 3 provides insights into the toxicological behavior of a select few NMs. Referring to the toxicity classification based on LD_{50} and LC_{50} values, silver-based nanomaterials (NMs) are classified as highly toxic, while TiO_2 and Cu NMs also pose significant risks. Therefore, further toxicity assessment is a critical priority to ensure the safe application of NMs without compromising human health and environmental well-being.

4 Occupational Health and Safety Risk Management Methodologies

Effective risk management centers on three critical elements: hazard identification, exposure assessment, and exposure control. A straightforward way to initiate this process is by considering the following key questions:

1. Hazard Identification:

   a. Do any of the chemicals, materials, or processes involved possess hazardous properties?

   b. What are the specific risks associated with these properties, such as toxicity, flammability, or reactivity?

2. Exposure Assessment:
   a. Are any of my employees potentially exposed to these hazards?
   b. How can I detect or measure such exposure? This includes identifying exposure pathways (inhalation, dermal contact, ingestion) and using appropriate monitoring techniques.

3. Exposure Control:
   a. If exposure is identified, what measures can I implement to control or eliminate it?
   b. Can I eliminate hazardous materials or substitute hazardous materials with safer alternatives, use engineering/administrative controls (e.g., ventilation or containment), or provide personal protective equipment (PPE)?

By systematically addressing these questions, organizations can establish a robust risk management framework that not only minimizes hazards but also ensures a safe working environment for employees.

Understanding the nature of the hazard, such as its toxicity or the likelihood of incidents like explosions, is crucial to determining the extent to which exposure must be controlled. Risk control measures often focus on limiting the release of the material into air or water and/or disrupting the pathways through which the substance could reach receptors, such as human organs or environmental systems, where it might cause harm. Consequently, a comprehensive understanding of exposure pathways and the quantities involved is essential for effective risk management. The risk management model illustrated in Fig. 3, adapted by NIOSH, has been widely implemented for effectively managing risks associated with the utilization of nanomaterials.

Several risk management concepts have been developed by various authorities and agencies, with most guidelines following a similar approach to that outlined by NIOSH. For example, the model proposed by the Finnish Occupational Health and Safety Administration [42] emphasizes integrating safety protocols into workplace practices, incorporating specific improvements to enhance context-specific implementation and long-term sustainability with integrated feedback mechanism. Similarly, the U.S. Environmental Protection Agency (EPA) [43] has developed a framework closely aligned with NIOSH, offering a structured approach to managing nanomaterial-related risks effectively.

The model proposed by Schulte et al. [44] offers a comprehensive approach that closely aligns with the NIOSH model, while incorporating enhancements to improve safety in nanomaterial-related industries at a cultural level. This model begins with societal understanding and emphasizes risk management through effective communication. It integrates behavioral safety aspects into risk assessment and mitigation while also identifying potential failure modes and defining OELs within operations.

Fig. 3 Risk management process from hazard identification [41]

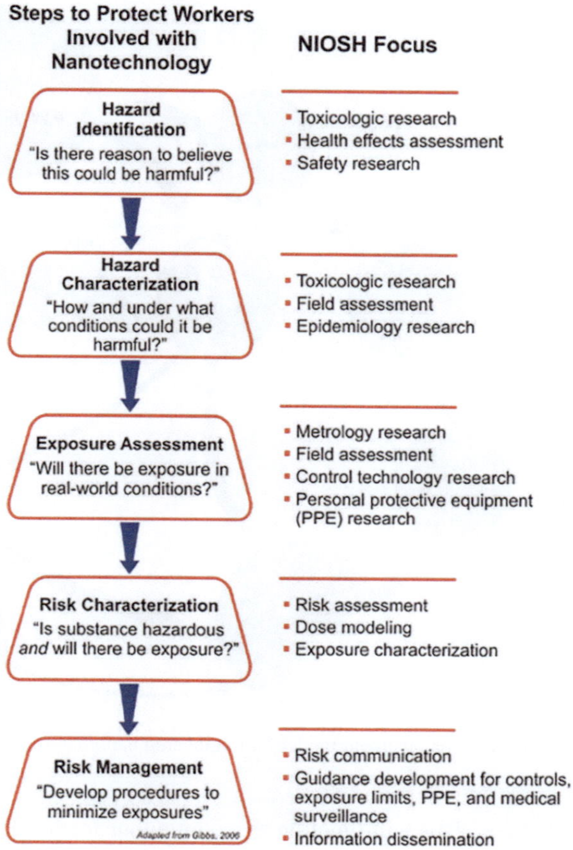

Therefore, this model serves as a valuable framework for developing a more industry-focused approach, ensuring proactive and scientific safety integration throughout the nanomaterial production process. The model represents Fig. 4.

After a thorough analysis of four previously highlighted safety management models related to the NMs industry, three key pillars emerge as critical areas for establishing a strong safety management culture: **m**easurement, **m**anagement, and **p**olicy. These pillars are interdependent, functioning as a continuous feedback loop that requires ongoing evaluation and revision. This dynamic interaction ensures that the system remains agile, adaptable, and robust, effectively mitigating potential nanomaterial-related risks and enhancing overall workplace safety. These three key pillars (2MP), along with their respective focus areas, are illustrated in Fig. 5. Policies should be established based on quantitative and qualitative risk assessments, incorporating necessary measurements and validations to determine appropriate OELs. Using this benchmarking approach, a comprehensive risk management framework can be developed, integrating key aspects such as evaluation, adaptation, mitigation, decision-making, and communication to ensure effective and sustainable safety

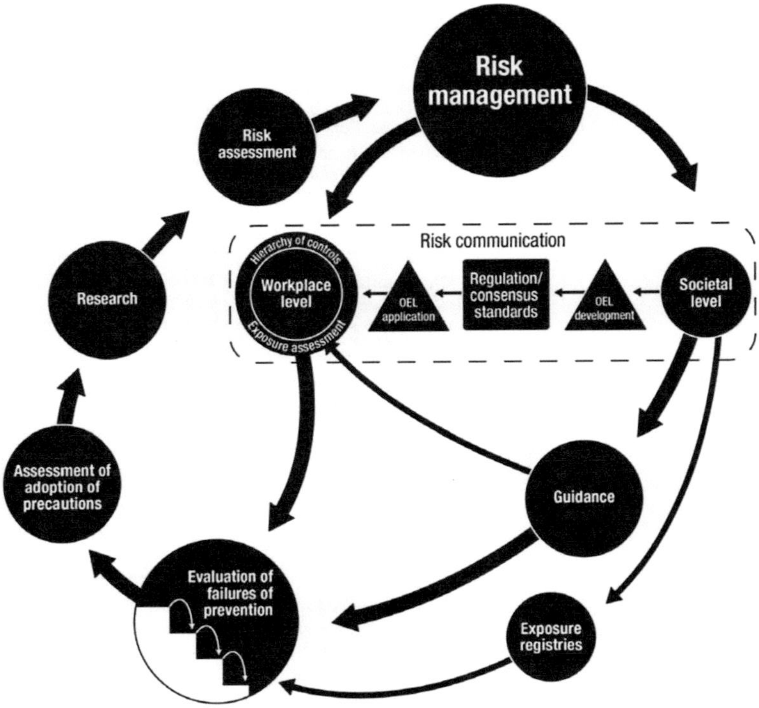

Fig. 4 Risk management model for engineered nanomaterials proposed by Schulte et al. [44]

practices. This will function as a continuous feedback loop, ensuring that the safety culture evolves over time with necessary improvements and adaptations to maintain effectiveness and resilience.

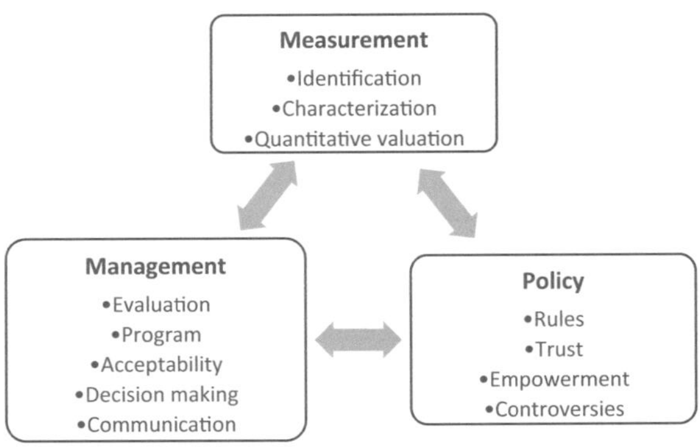

Fig. 5 Essential pillars (2MP) in a risk management system related to a nanomaterial industry

5 Occupational Health and Safety Risk Assessment and Quantification

All the methodologies highlighted in the previous section share a common foundation about the anticipation of potential risks through a comprehensive consideration of the material's life cycle. This approach incorporates:

- **Manufacturing Processes and Inputs**: Evaluating the processes and materials involved in production to identify potential hazards at the source.
- **Interactions During Use**: Assessing how the product interacts with individuals and the environment throughout its active life, including unintended exposures.
- **End-of-Life Disposal**: Understanding disposal methods and potential environmental or health impacts associated with waste management.

By adopting a life cycle perspective, risks associated with emerging technologies can be identified and mitigated proactively, ensuring that safety is maintained from production to disposal [45].

A typical life cycle stage of a nanomaterial consists of key processes from its initial development to its eventual disposal. Each stage presents unique considerations for risk assessment and management:

- Material Design and Synthesis: Development of nanomaterials through bottom-up or top-down approaches, including fabrication techniques, and quality control measures.
  - Risk considerations: Toxicity potential, handling hazards, and exposure during production.
- Manufacturing and Processing: Scaling up nanomaterial production for commercial use, involving processes like material formulation, processing, and incorporation into products.
  - Risk considerations: Release of nanoparticles into the air, liquid, or during manipulation; potential for worker exposure.
- Product Use and Application: Integration of nanomaterials into products or systems, ranging from electronics and coatings to biomedical devices and pharmaceuticals.
  - Risk considerations: Exposure pathways through contact, inhalation, or ingestion during the product's active life.
- End-of-Life Management: Disposal, recycling, or repurposing of products containing nanomaterials after their useful life has ended.
  - Risk considerations: Environmental release, degradation, or leaching of nanoparticles, as well as potential impacts on ecosystems.

- Post-Use and Environmental Impact: Monitoring and assessment of any long-term environmental or health impacts caused by the nanomaterials once they have been released into the environment.
  - Risk considerations: Persistence in ecosystems, bioaccumulation, and potential toxicity to humans and wildlife.

This life cycle framework provides a comprehensive view of the potential hazards and exposure risks associated with nanomaterials and underlines the importance of considering safety at each stage of the material's existence.

After identifying critical points of potential exposure to nanomaterials through life cycle analysis, individual exposure groups can be evaluated for their specific exposure risks. This assessment helps define the risks associated with nanomaterials. Table 4 illustrates the potential exposure groups and their corresponding risks in the event of system or process failures.

Table 4 Steps in the life cycle of a nanomaterial, from research-scale product development to end-user applications, with potential exposure levels [46]

Locations	Quantities	Exposure groups	Potential exposure
Research laboratory	mg to grams	Researchers, technicians, maintenance and waste-handling	Inhalation, dermal contact, or accidental spills
Laboratory sale up	kg		
Process development	kg to 100 s kg	Researchers, pilot personnel, technicians, engineers, process and facility maintenance, waste handling and transport, testing staff	Inhalation, dermal contact from airborne particles and residues during process adjustments
Product development	> 100 s kg		
Operate and maintain production	> 100 s kg	R&D staff, production personnel, engineers, process and facility maintenance, waste handling and transport, testing staff, storage and transport staff	Inhalation, dermal contact from bulk handling, processing, packaging, and potential fugitive emissions
Primary consumer (end user direct application)	mg to 100 s kg	Direct user, ancillary personnel, maintenance and waste management	Inhalation, dermal contact based on product type, with risks arising from wear, degradation, misuse, environmental release or worker exposure during reprocessing, disposal and recycling
Secondary use consumer (create on incorporate into nano enabled product)	kg to 100 s kg		

After the identification of possibilities for risks involved with the manufactured nanomaterials. Conducting a proper risk assessment and then a proper quantification is necessary to define the correct risk management method. There are different methods developed to identify the risk. Risk assessment and risk quantification are two critical components in managing the safety of manufactured nanomaterials.

5.1 Risk Assessment

Risk assessment is a systematic process used to identify, evaluate, and characterize potential hazards associated with manufactured NMs. It involves understanding the nature and magnitude of risks that may arise from exposure to these materials as described in Tables 2, 3 and 4. The primary goal of risk assessment is to provide a comprehensive understanding of potential risks to inform decision-making and risk management strategies. It helps stakeholders identify which risks need to be addressed and prioritize actions accordingly.

Three key components are involved in risk assessment:

- Hazard Identification: Determining whether a nanomaterial poses a potential health or environmental risk based on its properties and potential exposure scenarios.
- Exposure Assessment: Estimating the extent of exposure to NM among workers, consumers, or the environment. This includes assessing routes of exposure (inhalation, dermal contact, ingestion).
- Risk Characterization: Integrating data from hazard identification and exposure assessment to describe the overall risk, often using qualitative or semi-quantitative methods.

5.2 Risk Quantification

Risk quantification involves assigning numerical values to the likelihood and severity of identified risks. This process translates qualitative assessments into quantitative metrics that can be used for comparison and decision-making. The goal of risk quantification is to provide a clear, numerical basis for comparing different risks, facilitating more informed decisions regarding resource allocation for risk management efforts. It allows for a more objective evaluation of risks across various scenarios or materials.

Below are the three key components in risk quantification [6, 25, 46–52].

- Probability Estimation: Calculating or modeling the likelihood of specific adverse effects occurring because of exposure to nanomaterials.
- Consequence Analysis: Evaluating the potential severity of health impacts or environmental damage if exposure occurs.

- Risk Priority Number (RPN): Often used in methodologies like Failure Mode and Effects Analysis (FMEA), RPN combines probability and consequence scores to prioritize risks.

The below methods are common in identifying risks involved with manufactured nanomaterials and most of them are equipped with the facility to assess and quantify the risk involved with engineered nanomaterials.

- Screening Risk Assessment Methods:

  Various screening methods have been developed to qualitatively assess risks associated with nanomaterials. These methods typically use ordinal scales to rank risks based on available data about the materials' properties and potential hazards. Examples include the ANSES method, NanoRiskCat, and NanoSafer among others, which categorize nanomaterials based on toxicity, exposure potential, and other relevant parameters.
- Control Banding Tools:

  Control banding is a widely recommended approach for managing occupational exposure to engineered nanomaterials (ENMs). This method involves assigning hazard and exposure bands based on the characteristics of the nanomaterials and the processes involved in their handling. Tools such as OECD, ISO/TS 12901-1, CB Nanotool and Stoffenmanager-Nano help practitioners determine appropriate control measures by evaluating factors like dust generation, toxicity, and physical form of the materials.
- Case-by-Case Risk Assessment:

  Currently, risk assessments for manufactured nanomaterials are often conducted on a case-by-case basis due to the unique properties of each material. This involves characterizing potential hazards under controlled conditions and measuring exposure levels to quantify risk accurately. The complexity of this process is compounded by the need for specific testing protocols tailored to the unique characteristics of nanomaterials. Below exposure assessment techniques could be involved in the process.

  1. Airborne Exposure Assessment: Techniques such as personal monitoring, area sampling, and computational fluid dynamics (CFD) modeling can quantify worker exposure to airborne nanomaterials.
  2. Biological Monitoring: Analyzing biological samples (e.g., blood, urine) can help quantify internal exposure to nanomaterials and assess potential health impacts.

- Grouping and Categorization:

  Given the diverse properties of nanomaterials, grouping them based on intrinsic properties or activities is a practical strategy for risk assessment. Projects like NanoToxClass focus on establishing categorization strategies that facilitate risk assessment by allowing for reading across well-characterized materials to similar nanomaterials.

- Biological Risk Assessment Methodologies:

  For specific applications, such as medical devices containing nanomaterials, methodologies based on frameworks like ISO 10993-1 guide biological risk assessments. These involve evaluating particle release, distribution, persistence, and toxicological assessments to determine potential health risks associated with nanomaterials in medical applications.

These methods collectively support a comprehensive approach to identifying risks associated with manufactured nanomaterials, enabling informed decision-making regarding their safe use and management in various applications. A systematic approach not only enhances workplace safety but also fosters innovation in the safe use of nanotechnology across various industries.

The OECD [53] provides a comprehensive benchmarking of available models/tools subjected to scope analysis, user-friendliness examination, sensitivity analysis, and performance testing to identify the most suitable method for the assessment of risks incorporated into the individual manufacturing system. Further benchmarking of individual models with their applications was highlighted by the OECD [54] publications. Insights on measurement and testing methodologies can be drawn from OECD and NIOSH publications to enhance the risk assessment process [7, 10, 53–60].

Comprehensive case studies on risk assessments related to nanomaterials (NMs) have been provided by OECD, NIOSH, Employment and Social Development Canada, the Canadian Centre for Occupational Health and Safety, and other regulatory bodies [4, 7, 8, 10, 50, 53, 54, 58–71].

Figure 6 is used as an example to assess the risk related to the NM industry. The algorithm was adapted from the Canada-United States Regulatory Cooperation Council [10]. The method has been developed to assess human exposure to novel nanoparticles and determine whether additional testing is required for proper risk evaluation. As for the given decision-making process,

- If no exposure to nanoparticles is identified, no further testing is needed. However, if exposure is likely or possible, additional assessments may be necessary to establish proper control measures and prevent employee exposure.
- Routes of Entry and Testing Considerations: Nanoparticles can enter the body through three major routes: inhalation, ingestion (oral), and dermal absorption.

1. Oral Exposure:

- If oral exposure is identified, the solubility of nanoparticles in biological fluids and low pH dissolution must be considered.
- Low solubility: Additional testing is needed to assess particle size and surface chemistry, as smaller particles have higher uptake and surface properties influence absorption.
- High solubility: If sufficient information is available on molecular or ionic dissociation/toxicity, no further testing is needed. Otherwise, additional testing is required.

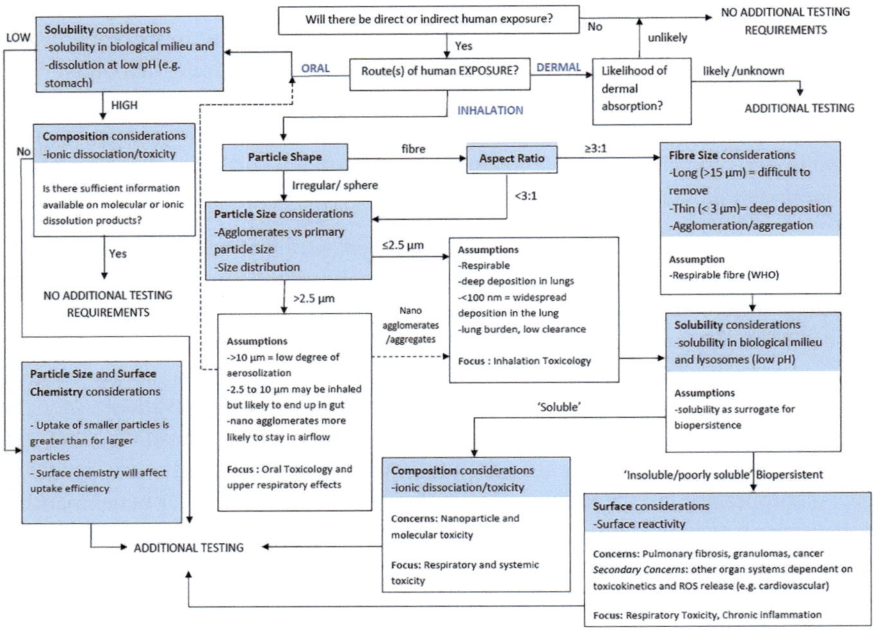

Fig. 6 Algorithm for determining concerns with exposure to novel nanoparticles developed by the Canada-United States Regulatory Cooperation Council, 2015

2. Inhalation Exposure:
- Particle shape must be evaluated. If the particle is fibrous, the aspect ratio (length-to-width ratio) must be determined:
  - Aspect ratio $\geq$ 3:1 $\rightarrow$ Assess fiber size and solubility:

    If soluble, further testing is needed to evaluate respiratory and systemic toxicity.
    If poorly soluble or insoluble (bio persistent), additional testing is required due to risks of pulmonary fibrosis, granulomas, and cancer.

  - Aspect ratio < 3:1 $\rightarrow$ Determine particle size:

    $> 2.5\ \mu m \rightarrow$ Follow the oral exposure pathway to assess additional testing needs.
    $\leq 2.5\ \mu m \rightarrow$ Follow the inhalation pathway and evaluate solubility and surface properties to determine further testing requirements.

3. Dermal Exposure:
- If dermal absorption is likely or unknown, additional testing is required.
- If dermal absorption is unlikely, no further testing is needed.

This structured approach ensures a comprehensive risk assessment, guiding the need for further testing based on exposure likelihood, nanoparticle properties, and potential health risks. Based on this evaluation, the necessary data for risk assessment and risk quantification can be identified. Once these assessments are completed, the overall risk associated with the specific nanomaterial (NM) can be determined, ensuring a comprehensive understanding of potential hazards and appropriate risk management strategies.

6 Occupational Health and Safety Risk Management

As highlighted in the previous section on risk identification and risk assessment, it is crucial to collect the following key criteria before defining a risk management plan for NMs.

1. Material-Specific Factors:

   - Characteristics of the material (e.g., composition, size, surface area, reactivity)
   - Amount of material being used or handled
   - Physical state of the material (e.g., dry powder, liquid suspension, embedded in a solid).

2. Exposure-Related Factors:

   - Degree of containment (e.g., presence of engineering controls such as fume hoods, glove boxes, or sealed systems)
   - Number of workers exposed
   - Duration of exposure (e.g., frequency and length of time workers are in contact with the material).

3. Risk Uncertainties and Additional Considerations: Given the limited toxicological data on nanomaterials, exposure should be minimized as much as possible. Additional information may be required, such as:

   - Toxicological properties of the specific nanomaterial or its elemental components (based on human, animal, or laboratory studies)
   - Potential explosion hazards or other safety concerns.

4. Hazard Mapping for Nanotechnology Processes: To enhance workplace safety, consider creating a hazard map using a facility floor plan. Mark locations where hazards exist, focusing on:

   - Physical hazards (e.g., risk of spills, airborne exposure, or fire hazards)
   - Frequency of exposure (e.g., daily handling vs. occasional contact)
   - Level of exposure (e.g., high-concentration areas, potential for inhalation or dermal contact)
   - Specific chemicals or agents involved
   - Workers or job roles most at risk.

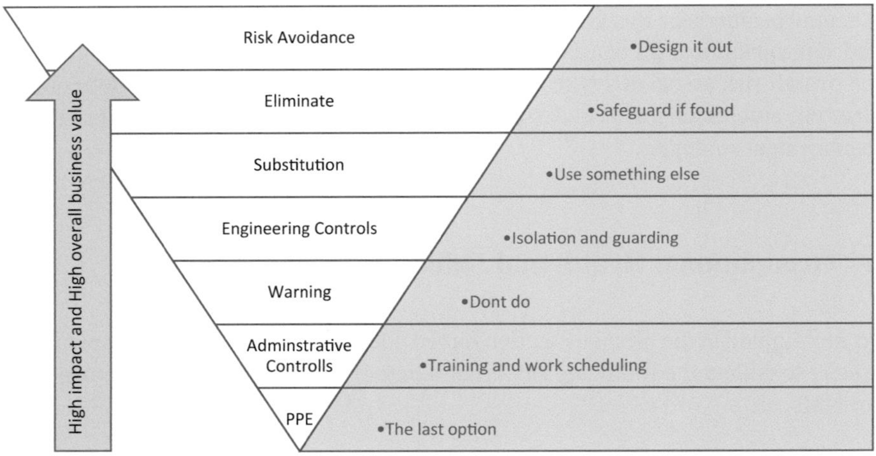

Fig. 7 Hierarchy of controls principle to the risk management of engineered NMS

Implementing a comprehensive risk assessment and hazard mapping strategy will help identify and address critical areas of concern, enabling proactive risk mitigation and ensuring safer nanotechnology practices.

After completing the hazard and risk mapping for the nanomaterials (NMs) production industry, the hierarchy of control principles can be applied to effectively manage and mitigate identified risks. Applying the Hierarchy of Controls principle to the risk management of engineered nanomaterials involves systematically implementing strategies to minimize or eliminate exposure to these materials across various industrial settings. The Hierarchy of Controls is a widely recognized framework in occupational health and safety that prioritizes control measures based on their effectiveness. The following framework shown in Fig. 7 has been developed by prevention through design (PtD) guidelines established by the American Society of Safety Professionals and approved by the American National Standards Institute [3, 9].

6.1 *Risk Avoidance, Risk Elimination and Risk Substitution*

As companies implement hazard control measures that are higher in the hierarchy, they enhance their overall business value. These improvements in business value are associated with:

- Accelerated Time to Market
- Enhanced Operational Efficiency
- Superior Product Quality
- Increased Market Share.

This strategic approach not only promotes safety but also drives competitive advantage and profitability.

The **Prevention through Design (PtD)** concept is a proactive approach that can be effectively applied to avoid risks associated with engineered nanomaterials. This strategy emphasizes the integration of safety considerations into the design process of products, processes, and systems involving nanomaterials. Here's how PtD can be implemented in this context:

1. Design for Safety

   The design for safety approach can be applied to the entire manufacturing process by evaluating individual unit operations and identifying potential risks at each stage. With a completed risk/hazard map, the most effective control measures can be integrated during the design phase itself, ensuring proactive risk management.

   - **Material Selection:** Choose nanomaterials that have lower toxicity profiles or reduced potential for harmful exposure. For example,
     - Not use the nanomaterial at all.
     - Selecting biocompatible nanomaterials for medical applications can minimize health risks.
     - Replace the nanomaterial with a "normal" sized particle that has known information about its hazards and risks.
     - Select fewer toxic materials for the same application. For example, consider nano ceramics which may have a lower toxicity than nano metals due to a lower dissolution rate in water.
   - Modifying Nanomaterial Characteristics:
     - Size: Increase the size of the nanomaterial to reduce its ability to penetrate cells.
     - Shape: Alter the shape of the nanomaterial to reduce its toxicity. For example, rod-shaped nanoparticles may cause more severe toxic effects than other shapes [72].
     - Surface Charge: Modify the surface charge to reduce electrostatic interactions with cell membranes. Nanoparticles with positive zeta potential are generally more toxic [72].
     - Surface Functionalization: Altering the surface functionalization can also alter biological activity and reduce toxicity [72].
     - Coatings: Remediate (coat) pristine NM to reduce emission and hazard rates [73].

Implement an iterative design process where feedback from risk assessments informs modifications to material designs, ensuring that safety is prioritized throughout development as explained in Fig. 8.

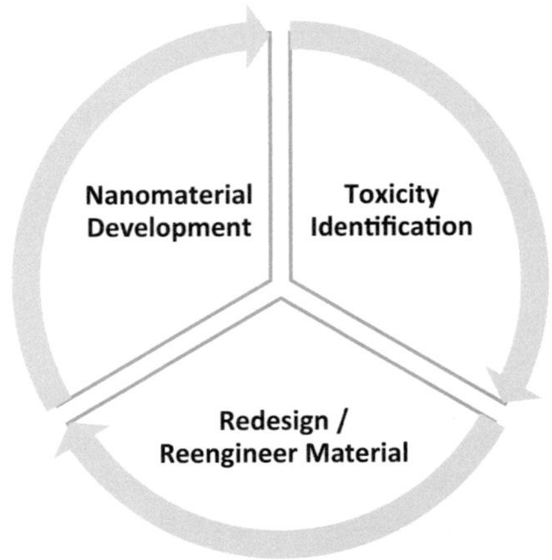

Fig. 8 Iterative nanomaterial designing process to avoid/eliminate the associated risk

- **Process Engineering**:
  - Design manufacturing processes that limit the generation of airborne nanoparticles, such as using enclosed systems or automated handling techniques to reduce human exposure.
  - Select manufacturing processes with lower toxic phase risks. For example, in Nickel NM production, the Sol–Gel method can be preferred over Thermal Decomposition to minimize the use of gaseous phases, thereby reducing the risk of airborne exposure and inhalation hazards [28, 29].

2. Regulatory Compliance and Best Practices

- **Adherence to Standards**: Ensure that designs comply with existing regulations and guidelines related to nanomaterials, such as those set forth by organizations like OSHA and NIOSH.
- **Continuous Improvement**: Foster a culture of continuous improvement by adding the PDCA approach by regularly reviewing and updating designs based on new research findings and technological advancements related to engineered nanomaterials.

By applying the Prevention through Design concept, industries can proactively address risks associated with engineered nanomaterials at the earliest stages of product development. This approach not only enhances worker safety but also contributes to the responsible use of nanotechnology, ultimately leading to safer products and processes in various applications.

Integrating Hazard and Operability Analysis (HAZOP) and Failure Mode and Effects Analysis (FMEA) into the Prevention through Design (PtD) framework can significantly enhance the risk management processes related to engineered nanomaterials. As an example, FMEA is a systematic tool that can be effectively applied in the manufactured nanomaterials industry to identify potential failure modes, assess their impact, and implement preventive measures. Here's how FMEA can be utilized in this context integrating with the PtD framework:

1. **Identification of Failure Modes**
   - **Process Analysis**: FMEA begins with a thorough analysis of the manufacturing processes involving nanomaterials. This includes identifying all steps in production, handling, and application where failures could occur, such as during synthesis, formulation, or processing of nanomaterials.
   - **Material Characteristics**: Assess the unique properties of nanomaterials (e.g., size, shape, surface area) that may lead to specific failure modes, such as agglomeration or unintended chemical reactions.
2. **Assessment of Effects and Causes**
   - **Impact Evaluation**: For each identified failure mode, evaluate the potential effects on product quality, worker safety, and environmental impact. This includes considering health risks associated with exposure to nanomaterials.
   - **Root Cause Analysis**: Determine the underlying causes of each failure mode. This may involve examining factors like equipment malfunction, human error, or inadequate safety protocols.
3. **Risk Prioritization**
   - **Risk Priority Number (RPN)**: Assign an RPN to each failure mode based on its severity, occurrence likelihood, and detection difficulty. This quantitative measure helps prioritize which risks require immediate attention.
   - **Focus on High-Risk Areas**: Concentrate efforts on failure modes with high RPNs that could lead to significant health risks or operational disruptions.
4. **Development of Mitigation Strategies**
   - **Control Measures**: For prioritized risks, develop and implement control measures to mitigate potential failures. This could include engineering controls (e.g., ventilation systems), administrative controls (e.g., training), and personal protective equipment (PPE).
   - **Continuous Monitoring**: Establish monitoring protocols to ensure that control measures are effective and to detect any deviations from expected performance.
5. **Documentation and Review**
   - **Comprehensive Documentation**: Maintain detailed records of the FMEA process, including identified failure modes, assessments, RPN calculations, and implemented control measures.
   - **Periodic Review**: Regularly review and update the FMEA as new information becomes available or as processes change. This ensures that the analysis remains relevant and effective in managing risks associated with nanomaterials.

6.2 *Risk Control, Warnings, and Personal Protective Equipment (PPE)*

The integration of risk control strategies is vital for managing the unique challenges posed by engineered nanomaterials in the industry. By prioritizing and implementing effective control measures, organizations can enhance workplace safety while continuing to innovate in the field of nanotechnology.

Risk control incorporates a range of strategies designed to minimize exposure to hazards associated with engineered nanomaterials. Key control measures include Engineering controls and administrative controls.

Engineering Controls: Engineering controls are essential methods to eliminate hazards or create a barrier between workers and potential risks. As an example, inhalation is the most common route of exposure to airborne particles. To minimize exposure, ventilation is an effective measure to reduce the concentration of nanoparticles released during various work processes. Whenever feasible, design work processes to be incorporate appropriate safety Features.

- **Built-in Controls**: Integrate engineering controls directly into the design of products or processes. For example, incorporating ventilation systems or filtration mechanisms within the equipment that handles nanomaterials can help mitigate exposure risks [9, 10, 31, 32, 34, 74]. As examples,
  - **Enclosures:**

    Small-scale enclosures for equipment like weighing scales.
    Large-scale enclosures to house entire nanomaterial reactors.
  - **Local Exhaust Ventilation:**

    Directly captures nanoparticles at their source to prevent them from becoming airborne.
  - **Isolation Devices:**

    Use of glove boxes or other sealed systems to handle nanomaterials safely.
  - **Air Locks and Sealed Containers:**

    Air locks prevent contamination and maintain controlled environments.
    Sealed containers reduce the risk of nanoparticle release during transport or storage.

These measures not only protect workers and the environment but also ensure compliance with workplace safety standards and regulations. As shown in Fig. 9, Hodson et al. explained the suitable Engineering control selection considering the critical risk factors like physical form. Exposure duration, exposure quantity and the potential health hazard due to the exposure [9, 44].

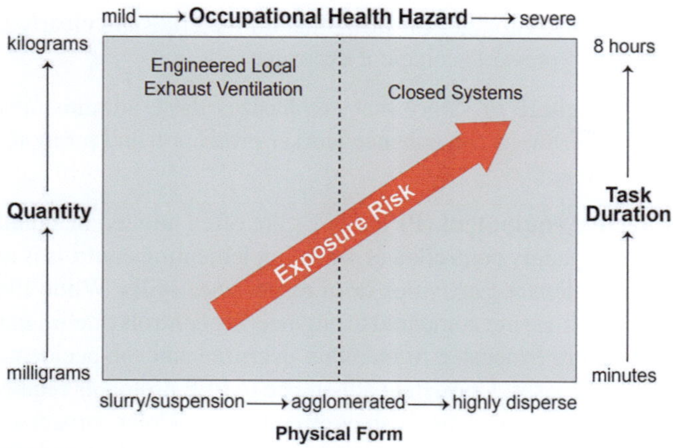

Fig. 9 Selection of engineering control mechanism for engineered nanomaterials [9]

Detailed examples of engineering controls for handling nanomaterials can be found in the NIOSH and OECD guidance documents including "Current Strategies for Engineering Controls in Nanomaterial Production and Downstream Handling Processes, General Safe Practices for Working with Engineered Nanomaterials in Research Laboratories" [8, 32, 50, 60, 63, 67, 68, 70, 75–77] provides valuable insights into designing and implementing effective engineering controls for research and working environments to ensure safe handling and containment of engineered nanomaterials.

Administrative Controls: Work practices and administrative controls are most effective when integrated into a broader safety and health culture within an organization. Administrative controls, including good housekeeping practices such as wet-wiping surfaces and using HEPA-filtered vacuums, can significantly reduce airborne concentrations of workplace contaminants, thereby decreasing workers' potential exposure. These administrative controls cover various strategies, including:

- Employee Training: Providing comprehensive training to ensure employees are aware of hazards and understand safe handling procedures.
- Standard Operating Procedures (SOPs): Developing SOPs that outline safe practices for working with nanomaterials ensures consistency and adherence to safety protocols. Additionally, workers must wash their hands after completing any tasks. They should also wash their hands before eating, drinking, smoking, or leaving the workplace. Consumption of food and beverages should be strictly prohibited in areas where nanomaterials or other hazardous substances are handled.
- Limiting Exposure Time: Reducing the duration that workers handle hazardous materials to minimize risk.
- Good Housekeeping Practices: Implementing regular cleaning protocols to maintain a safe work environment.

- Proper Labeling and Storage: Ensuring that all materials are clearly labeled and stored correctly to prevent accidental exposure.

By fostering a culture of safety that emphasizes these administrative controls and practices, organizations can enhance worker protection and promote a healthier workplace environment.

Personal protective equipment (PPE): PPEs are often utilized in situations where hazards are not adequately controlled or when implementing controls is not feasible, such as during maintenance activities or in response to spills. While PPE can be a cost-effective initial measure compared to engineering controls or elimination, it may ultimately prove more expensive to maintain overtime and can accidentally lead to unexpected exposure risks. Consistent adherence to PPE protocols requires ongoing effort from workers, which can diminish its effectiveness. Unless processes have been thoroughly evaluated and appropriate containment or control devices installed, operators should wear PPE as a precautionary measure. Wearing PPE is strongly recommended during maintenance tasks or when opening sealed enclosures. To minimize dermal and respiratory exposure, the NIOSH [33, 34, 74, 76, 78, 79]: and OECD [68–70, 80, 81], and AIHA [82] recommends the following PPE.

- **Long Pants and Long-Sleeved Shirts**: Workers should wear long pants without cuffs and long-sleeved shirts to protect the skin.
- **Laboratory Coats or Coveralls**: Use laboratory coats or coveralls, preferably with cuffs and Tyvek-type wrist covers, to shield exposed skin between the lab coat and gloves. Note that cotton-woven laboratory coats are not advisable for protection against nanoparticle exposure due to their tendency to trap particles.
- **Chemically Impervious Gloves**: Nitrile gloves or other chemically resistant gloves should be used when handling nanomaterial powders and liquids. Latex gloves are not recommended, as they do not provide adequate protection against most chemical solvents and may pose allergy risks.
- **Closed-Toe Shoes**: Workers should wear closed-toe shoes made from low-permeability materials, such as leather. Disposable over-the-shoe booties may also be utilized.
- **Eye Protection**: Safety glasses, goggles, and/or face shields should be worn based on the risk assessment of potential liquid splashes. It is important to note that a face shield alone does not provide sufficient protection against unbound dry materials.
- **Respirators**: Appropriate respiratory protection should be used as necessary based on exposure assessments.[81, 83]

By implementing these PPE recommendations, organizations can better safeguard workers against potential hazards associated with engineered nanomaterials while fostering a culture of safety and compliance within the workplace. The NIOSH provides a summary stating "Questions to Ask Before You Strat" that serves as a guideline for workers to assess personal risks associated with the handling and engagement of engineered nanomaterials [33].

After establishing the risk-controlling hierarchy in the workplace it is essential to verify the effectiveness of established safety mitigation measures and emergency preparedness related to engineered nanomaterials to ensure they are functioning as intended.

This can be achieved through traditional industrial hygiene sampling methods, such as area and personal sampling, along with real-time measurements of airborne concentrations of nanomaterials. Regular audits and assessments, including emergency preparedness drills and evaluations of procedures for managing accidental emissions, are essential for identifying areas of improvement in safety protocols. These activities should align with the guidelines established by relevant authorities to ensure compliance and enhance workplace safety. By systematically reviewing and testing these protocols, organizations can strengthen their response capabilities and better protect workers from potential hazards.

The NIOSH and OECD have developed comprehensive guidelines aimed at controlling exposure to engineered nanomaterials. These include recommended exposure limits (RELs) and various guide documents that address workplace practices, medical screening, and hazard surveillance. For instance, OECD publications and NIOSH's Current Intelligence Bulletins provide specific recommendations for managing exposure to materials like titanium dioxide, silver, and carbon nanotubes, which can serve as benchmarks for evaluating safety measures in workplaces handling nanomaterials [8, 15, 16, 20, 21, 56, 68, 69, 77, 84–87].

7 Regulatory Frameworks and Standards

Regulatory oversight specific to nanomaterials remains limited, although several agencies are actively working to establish comprehensive frameworks. Many regulatory bodies recognize NMs that exhibit novel characteristics that may lead to unforeseen health and environmental risks. Therefore, there is a pressing need for mandatory, nano-specific regulations that can effectively govern their use and ensure safety.

Below is a summary of current key agencies that are more engaged in developing new regulators to establish the safety of engineered nanomaterials. Further details regarding the Risk Governance Framework can be obtained through the document titled "Risk Governance of Emerging Technologies Demonstrated in Terms of its Applicability to Nanomaterials" [88].

- **Occupational Safety and Health Administration (OSHA)**: The OSHA is actively developing guidelines for the safe handling of nanomaterials in workplaces. Recognizing the unique hazards associated with nanotechnology, OSHA has published a fact sheet titled Working Safely with Nanomaterials [2], which provides essential information for both workers and employers regarding potential risks and control measures. OSHA has established recommended exposure limits for certain nanomaterials such as carbon nanotubes and carbon nanofibers, TiO_2.

- **Environmental Protection Agency (EPA)**: EPA's focus on regulating nanomaterial emissions, waste management, and environmental impact reflects a commitment to ensuring that advancements in nanotechnology do not compromise public health or environmental safety. Through initiatives like the Nanotechnology Reporting Rule and ongoing information-gathering efforts, the agency aims to establish a robust framework for managing the risks associated with engineered nanomaterials [43].

In 2017, the EPA issued a final rule under the Toxic Substances Control Act (TSCA) that requires manufacturers and processors of certain nanoscale materials to report specific information. This includes:

- Chemical Identity: Identification of the nanoscale materials being handled.
- Production Volume: Reporting on the quantity produced or processed.
- Methods of Manufacture: Details on how nanomaterials are produced.
- Exposure and Release Information: Data on how these materials are used and any emissions that occur.

This rule specifically targets solid nanomaterials and their aggregates or agglomerates in the size range of 1–100 nanometers, establishing a reporting obligation for companies involved in their production or processing. While there are currently no specific regulations solely focused on the environmental impacts of nanoparticles, the EPA assesses potential risks associated with emissions from processes involving nanomaterials under existing environmental laws, such as the Clean Air Act. If adverse health effects are identified, the agency may classify certain nanomaterials as hazardous air pollutants, leading to stricter controls. The EPA is working to develop a comprehensive regulatory framework that addresses the unique properties and risks associated with nanomaterials. This includes adapting existing chemical regulations to better manage these substances while facilitating innovation in nanotechnology [89].

- **The International Organization for Standardization (ISO)**: ISO, through its technical committee ISO/TC 229, is actively engaged in creating standardized protocols for the safety and testing of nanomaterials. Established to address the unique challenges posed by nanotechnology, ISO/TC 229 focuses on various aspects of nanomaterials, including their characterization, safety assessment, and environmental impact. Several standards have been developed by the ISO that are directly related to the safety of nanomaterials. These standards provide essential guidelines for risk management, safety data sheets, and occupational health practices. Key standards include:
  - ISO/TS 12901-1:2012: Nanotechnologies—Occupational risk management applied to engineered nanomaterials—Part 1: Principles and approaches.
  - ISO/TS 12901-2:2014: Nanotechnologies—Occupational risk management applied to engineered nanomaterials—Part 2: Use of the control banding approach.
  - ISO/TR 13121:2011: Nanotechnologies—Nanomaterial risk evaluation.

  - ISO/TR 13329:2024: Nanomaterials—Preparation of the material safety data sheet (MSDS).
  - ISO 21363: 2020: Nanotechnologies—Measurements of particle size and shape distributions by transmission electron microscopy.

- **The European Standardization Committee (CEN)**: The CEN established a technical committee for nanotechnology, known as CEN/TC 352, in 2006. This committee is tasked with coordinating standardization efforts initiated by an EU mandate granted by the EU/EFTA. CEN/TC 352 is organized into three working groups, each addressing specific aspects of nanotechnology:

  - WG1—Measurement, characterization, and performance evaluation
  - WG2—Commercial and other stakeholder aspects
  - WG3—Health, safety and environmental aspects.

The establishment of CEN/TC 352 is significant for several reasons:

- Harmonization of Standards: By developing standardized protocols for nanotechnology, CEN/TC 352 helps harmonize regulations across EU member states, facilitating trade and ensuring that products meet consistent safety criteria.
- Support for Innovation: The committee's work supports innovation in nanotechnology by providing clear guidelines that enable companies to develop new products safely while complying with regulatory requirements.
- Public Health Protection: Through its focus on health, safety, and environmental aspects, CEN/TC 352 contributes to protecting public health by ensuring that potential risks associated with nanomaterials are adequately assessed and managed.
- Collaboration with Other Standards Organizations: CEN/TC 352 collaborates closely with other international bodies such as ISO/TC 229 (International Organization for Standardization) to align efforts in standardizing nanotechnology practices globally.

- **Registration, Evaluation, Authorization, and Restriction of Chemicals (REACH)**: The REACH regulation (Regulation No. 1907/2006) is a cornerstone of chemical legislation in the European Union. It aims to ensure the safe use of chemicals, including those in nanoform. Currently, discussions are ongoing regarding adjustments to the information requirements for chemical substances in nanoform, as well as the potential for specific assessments based on the generated data.

The European Commission has proposed changes to the information requirements under REACH to better address the unique properties of nanomaterials. These proposals have been subject to public consultation, and various EU impact assessments have evaluated their implications for both industry and society.

The European Chemicals Agency (ECHA) plays a pivotal role in implementing REACH regulations related to nanomaterials. ECHA has recognized the unique properties of nanomaterials in its guidelines and assessments and has initiated the "Better

Control of Nanomaterials" initiative to propose relevant information requirements for these substances in a regulatory context. As part of its commitment to enhancing safety measures, ECHA continues to update its guided documents and testing protocols specific to nanoforms. This includes developing standardized testing methods that align with international standards to ensure comprehensive risk assessments.

- **Organization for Economic Cooperation and Development (OECD)**: In 2006, the OECD initiated a program to ensure that nanomaterials are safe for workers and consumers on a global level. The program was called "OECD Working Party on Manufactured Nanomaterials (WPMN)". It focuses on safety aspects relating to human health and the environment over the whole lifecycle. Eight projects have been selected to be worked on in subgroups:
  - Project 1: development of an OECD database on EHS research
  - Project 2: EHS research strategies on manufactured nanomaterials
  - Project 3: safety testing of a representative set of manufactured nanomaterials
  - Project 4: manufactured nanomaterials and test guidelines
  - Project 5: co-operation on voluntary schemes and regulatory programs
  - Project 6: co-operation on risk assessment
  - Project 7: the role of alternative methods in nanotoxicology
  - Project 8: co-operation on exposure measurement and exposure mitigation.

A model based on the concept of Prevention through Design has been developed by the OECD to enhance safety and sustainability in advanced technologies, particularly regarding engineered nanomaterials. This model is known as the Safe(r) and Sustainable Innovation Approach (SSIA) [90]. Building upon the earlier Safe(r) Innovation Approach (SIA) [91], SSIA aims to empower all stakeholders to effectively address safety and sustainability assessments of innovations in a robust yet agile manner. The primary goal of SSIA is to reduce the time gap between the emergence of technological innovations and the development of appropriate risk assessment tools and frameworks.

The approach incorporates two key components:

- Safe(r) and Sustainable-by-Design (SSbD): Development on the former Safe(r) by Design (SbD) concept encourages innovators to integrate safety and sustainability considerations early in the innovation process. By doing so, potential risks can be identified and mitigated before products reach the market.
- Regulatory Preparedness (RP): This component focuses on enhancing regulators' ability to anticipate changes in technology, facilitating the development of adaptable regulations that can keep pace with advancements in nanomaterials, nano-enabled products, and other advanced materials.

Both SSbD and RP are supported by processes that promote knowledge sharing and dialogue within a Trusted Environment (TE). This environment fosters collaboration between innovators and regulators, allowing for the exchange of information, insights, and best practices. Such dialogue is crucial for ensuring that safety and sustainability issues are addressed proactively during the design phase, thereby preventing challenges later in the life cycle of nanomaterials or nano-enabled products.

- **World Health Organization (WHO)**: The WHO has established various activity areas related to occupational health, with a specific focus on emerging issues such as nanomaterials. Within this framework, projects concerning nanomaterials are organized under the subgroup NM Nanomaterials in Activity Area 6 which focuses:
  - Risk Assessment: Developing methodologies for assessing the health risks posed by exposure to nanomaterials in occupational settings.
  - Guideline Development: Creating comprehensive guidelines that inform stakeholders about safe handling practices and risk management strategies for nanomaterials.
  - Training and Capacity Building: Providing training resources for workers and employers to enhance their understanding of nanomaterial safety and promote safe practices in workplaces.
  - Research Collaboration: Collaborating with international partners, regulatory agencies, and industry stakeholders to gather data and share knowledge on nanomaterial safety.
  - Policy Development Support: Assisting in the formulation of policies that address the unique challenges posed by nanotechnology, ensuring that public health considerations are integrated into regulatory frameworks.
- **The National Institute for Occupational Safety and Health (NIOSH)**: The NIOSH has developed a comprehensive approach to address the health and safety issues associated with engineered nanomaterials through various initiatives and best practice documents. Here's an overview of NIOSH's key activities related to nanomaterials:
  - Nanotechnology Information Exchange: NIOSH has established a web-based platform known as the Nanotechnology Information Exchange, which serves as a resource for sharing knowledge, best practices, and guidelines regarding the safe handling of nanomaterials in occupational settings. NIOSH's current Intelligence Bulletin focused on working with engineered nanomaterials. This bulletin will provide updated information on hazards, exposure risks, and recommended safety practices to protect workers in industries utilizing nanotechnology. NIOSH's NANO-Comms initiative functions as a technical observatory dedicated to disseminating information about health and safety issues related to nanoparticles. It aims to bridge gaps between research findings and practical applications in the workplace.
  - Standardization Contributions: NIOSH engages with regulators, manufacturers, downstream users, and researchers to assist in developing relevant standards that meet the needs of all stakeholders involved in nanotechnology. Below are key areas which NIOSH contributes to developing comprehensive frameworks related to emerging nanotechnology.

Exposure Process Identification
Exposure Modeling Review
Health Risk Categorization:
Toxicity Assessment
Risk Assessment Framework.

The establishment of a global framework for nanomaterial safety is essential for effectively managing the risks associated with these substances. A harmonized global framework provides,

- **Consistency Across Jurisdictions**: A harmonized framework helps ensure that safety standards and regulatory practices are consistent across different countries, facilitating international trade and cooperation.
- **Adaptation to Unique Properties**: Given that nanomaterials exhibit unique properties that can affect their behavior and toxicity, a global framework allows for the adaptation of existing chemical regulations to better address these characteristics.
- **Enhanced Risk Assessment**: A comprehensive framework supports improved methodologies for risk assessment, enabling regulators to make informed decisions regarding the safety of nanomaterials in various applications.
- **Collaboration Among Stakeholders**: Establishing a global framework encourages collaboration among governments, industry stakeholders, researchers, and public health organizations, fostering knowledge sharing and best practices.

By leveraging resources like the OECD website [92] and promoting international cooperation, stakeholders can work towards harmonizing regulations that protect public health while fostering innovation in nanotechnology. Figure 10 offers an overview of the current legislative frameworks across various countries concerning the applications of engineered nanomaterials provided by the OECD. It highlights the regulatory conditions and approaches adopted internationally to manage the use and safety of these advanced materials.

8 Conclusion

Ensuring workplace health and safety is crucial for the success and growth of businesses, especially those that handle engineered nanomaterials. Research indicates that certain nanomaterials pose respiratory, cardiovascular and carcinogenic risks, as evidenced by studies showing adverse effects in laboratory animals. Employees may be exposed to these risks through inhalation, skin absorption, or ingestion, necessitating robust safety measures. Developing a proactive safety program that addresses the unique challenges posed by engineered nanomaterials is essential for safeguarding worker health and ensuring long-term success.

Several factors were identified as influencing factors that create a potential risk for employee exposure to nanomaterials:

Fig. 10 The chemical legislation indicator of the OECD [92]

- Route of Exposure: How exposure occurs—whether through inhalation, dermal contact, or ingestion—can significantly impact health outcomes.
- Concentration and Duration: The concentration of nanomaterials present in the workplace and the duration and frequency of exposure are critical determinants of risk.
- Dispersion Ability: Nanomaterials that can easily disperse into dust or aerosols pose a higher risk of inhalation and subsequent health effects.
- Available Control Measures: The effectiveness of existing control measures to limit exposure plays a vital role in protecting workers.

To mitigate these risks, businesses should establish a robust safety program aligned with a comprehensive risk assessment and risk management plan. This plan should follow the hierarchy of controls, prioritizing measures at the highest levels of the hierarchy for maximum effectiveness.

A successful safety program should align with the evolving landscape of the nanomaterial market. Key elements include the **2MP**:

- **M**anagement
- **M**easurement
- **P**olicy.

All safety programs should evolve through **Continuous Evaluation**, regularly assessing and updating the program using the **Plan-Do-Check-Act (PDCA) cycle** to maintain and enhance its effectiveness.

References

1. Gupta R, Xie H (2018) Nanoparticles in daily life: applications, toxicity and regulations. J Environ Pathol Toxicol Oncol 37(3):209–230. https://doi.org/10.1615/JEnvironPatholToxicolOncol.2018026009
2. OSHA (2013) FactSheet: working safely with nanomaterials. (Online). Available: www.cdc.gov/niosh/docs/2009-116pdfs/
3. ANSI/ASSE (2016) ANSI/ASSE Z590.3–2011 (R2016) Prevention through design guidelines for addressing occupational hazards and risks in design and redesign processes
4. OECD (2022) Important issues on risk assessment of manufactured nanomaterials series on the safety of manufactured nanomaterials. (Online). Available: www.oecd.org/chemicalsafety/
5. NIOSH (2021) Approaches to developing occupational exposure limits or bands for engineered nanomaterials: user guide and technical report
6. Arvidsson R, Furberg A, Molander S (2016) Review of screening risk assessment methods for nanomaterial
7. OECD (2023) Towards tailored safety testing methods for nanomaterials
8. OECD (2017) Strategies, techniques and sampling protocols for determining the concentrations of manufactured nanomaterials in air at the workplace
9. Hodson L, Cincinnati HM (2016) Building a safety program to protect the nanotechnology workforce: a guide for small to medium-sized enterprises
10. Employment and Social Development Canada. Engineered nanoparticles: health and safety considerations. Employment and Social Development Canada
11. Wildavsky A, Dake K (1990) Theories of risk perception: who fears what and why? Accessed:17 Nov 2024. (Online). Available: https://www.jstor.org/stable/20025337
12. EC (2010) Communicating nanotechnology. European Commission
13. Berube D, Faber B, Scheufele DA (2010) Communicating risk in the 21st century: the case of nanotechnology
14. Brun E, Gibson R, Stacey N, Drais E, Wallin H, Zatorski W (2012) Risk perception and risk communication with regard to nanomaterials in the workplace : European risk observatory, literature review. Publications Office. https://doi.org/10.2802/93075
15. NIOSH (2010) Current intelligence bulletin 65 occupational exposure to carbon nanotubes and nanofibers. (Online). Available: http://www.particleandfibretoxicology.com/content/7/1/28
16. NIOSH (2011) Current intelligence bulletin 63: occupational exposure to titanium dioxide. (Online). Available: www.cdc.gov/niosh
17. NIOSH—CDC (2021) Approaches to developing occupational exposure limits or bands for engineered nanomaterials: user guide and technical report
18. ANSES. Recommended occupational exposure limits for titanium dioxide nanoparticles. ANSES
19. OSHA. Occupational exposure to respirable crystalline silica 29 C.F.R. § 1910.1053. OSHA
20. Kuempel E et al (2021) Current intelligence bulletin 70: health effects of occupational exposure to silver nanomaterials. https://doi.org/10.26616/NIOSHPUB2021112
21. OECD (2017) Silver nanoparticles: summary of the dossier
22. Occupational Safety and Health Administration. Nickel, metal & insoluble compounds (as Ni)†. OSHA
23. Liberda EN, Cuevas AK, Qu Q, Chen LC (2014) The acute exposure effects of inhaled nickel nanoparticles on murine endothelial progenitor cells. Inhal Toxicol 26(10):588–597. https://doi.org/10.3109/08958378.2014.937882
24. Kang GS, Gillespie PA, Gunnison A, Moreira AL, Tchou-Wong KM, Chen LC (2011) Long-term inhalation exposure to nickel nanoparticles exacerbated atherosclerosis in a susceptible mouse model. Environ Health Perspect 119(2):176–181. https://doi.org/10.1289/ehp.1002508
25. Pelin M, Sosa S, Prato M, Tubaro A (2018) Occupational exposure to graphene based nanomaterials: risk assessment. Nanoscale. https://doi.org/10.1039/c8nr04950e

26. Lee YS et al (2019) Derivation of occupational exposure limits for multi-walled carbon nanotubes and graphene using subchronic inhalation toxicity data and a multi-path particle dosimetry model. Toxicol Res (Camb) 8(4):580–586. https://doi.org/10.1039/c9tx00026g
27. Schulte PA, Leso V, Niang M, Iavicoli I (2019) Current state of knowledge on the health effects of engineered nanomaterials in workers: a systematic review of human studies and epidemiological investigations. Scand J Work Environ Health. https://doi.org/10.5271/sjweh.3800
28. Hassan MR, Yasmin F, Noor FK, Rahman MS, Uddin MS, Bhowmik S (2023) Synthesis and applications of nickel nanoparticles (NiNPs)-Comprehensive Review. J Ultra Chem 19(1):9–37. https://doi.org/10.22147/juc/190102
29. Jaji ND, Lee HL, Hussin MH, Akil HM, Zakaria MR, Othman MBH (2020) Advanced nickel nanoparticles technology: from synthesis to applications. Nanotechnol Rev 9(1):1456–1480. https://doi.org/10.1515/ntrev-2020-0109
30. Maynard AD, Kuempel ED (2005) Airborne nanostructured particles and occupational health. J Nanopart Res 7(6):587–614. https://doi.org/10.1007/s11051-005-6770-9
31. Schulte P et al (2008) Sharpening the focus on occupational safety and health in nanotechnology. Scand J Work Environ Health 34(6):471–478. https://doi.org/10.5271/sjweh.1292
32. EC (2013) Guidance on the protection of the health and safety of workers from the potential risks related to nanomaterials at work: Guidance for employers and health and safety practitioners. (Online). Available: http://ec.europa.eu/progress
33. NIOSH (2018) Controlling health hazards when working with nanomaterials: questions to ask before you start. https://doi.org/10.26616/NIOSHPUB2018103
34. NIOSH (2009) Approaches to safe nanotechnology managing the health and safety concerns associated with engineered nanomaterials
35. Dahm MM, Bertke S, Schubauer-Berigan MK (2019) Predicting occupational exposures to carbon nanotubes and nanofibers based on workplace determinants modeling. Ann Work Expo Health 63(2):158–172. https://doi.org/10.1093/annweh/wxy102
36. Qamar W et al. (2024) An insight into impact of nanomaterials toxicity on human health. PeerJ Inc. https://doi.org/10.7717/peerj.17807
37. Hoseini-Alfatemi SM, Fallah F, Armin S, Hafizi M, Karimi A, Kalanaky S (2020) Evaluation of blood and liver cytotoxicity and apoptosis-necrosis induced by nanochelating based silver nanoparticles in mouse model. Iran J Pharm Res 19(2):207–218. https://doi.org/10.22037/IJPR.2020.1101026
38. National Industrial Chemicals Notification and Assessment Scheme (NICNAS) (2016) Titanium dioxide: human health tier II assessment chemicals in this assessment preface. (Online). Available: https://www.nicnas.gov.au/chemical-information/imap-assessments/imap-group-assessment-report?assessment_id=1722
39. American Chemical Society Committee on Chemical Safety. Nanotechnology safety resources. American Chemical Society Committee on Chemical Safety
40. Kovrižnych JA, Sotńiková R, Zeljenková D, Rollerová E, Szabová E, Wimmerova S (2013) Acute toxicity of 31 different nanoparticles to zebrafish (Danio rerio) tested in adulthood and in early life stages—comparative study. Interdisc Toxicol 6(2):67–73. https://doi.org/10.2478/intox-2013-0012
41. NIOSH (2009) Progress toward safe nanotechnology in the workplace. (Online). Available: www.cdc.gov/niosh
42. Occupational health and safety administration in Finland. Risk assessment. https://tyosuojelu.fi/en/safety-and-health-in-workplace/risk-assessment
43. EPA (2007) Nanotechnology white paper. (Online). Available: www.epa.gov/osa
44. Schulte PA et al (2013) Overview of risk management for engineered nanomaterials. In: Nanosafe 2012: international conferences on safe production and use of nanomaterials. Institute of Physics Publishing. https://doi.org/10.1088/1742-6596/429/1/012062
45. The Royal Society & The Royal Academy of Engineering (2004) Nanoscience and nanotechnologies: opportunities and uncertainties. The Royal Society & the Royal Academy of Engineering

46. Schulte P, Geraci C, Zumwalde R, Hoover M, Kuempe E (2008) Occupational risk management of engineered nanoparticles. J Occup Environ Hyg 5(4):239–249. https://doi.org/10.1080/15459620801907840
47. Jamen S (2020) Nanomaterials risk assessment tools 01
48. Hristozov D et al (2014) Risk assessment of engineered nanomaterials. In: Monteiro-Riviere NA, Tran CL (eds) Nanotoxicology, 2nd edn, vol 25. CRC Press, pp. 459–478. https://doi.org/10.1201/b16562-32
49. EC (1988) Council directive 98/24/EC on the protection of the health and safety of workers from the risks related to chemical agents at work
50. OECD (2022) Important issues on risk assessment of manufactured nanomaterials. (Online). Available: www.oecd.org/chemicalsafety/
51. Kuempel ED, Geraci CL, Schulte PA (2007) Risk assessment approaches and research needs for nanomaterials: an examination of data and information from current studies. In: Nanotechnology—toxicological issues and environmental safety, pp 119–145
52. ISO (2024) ISO/TS 12901–1 Nanotechnologies-occupational risk management applied to engineered nanomaterials-part 1: principles and approaches. (Online). Available: www.iso.org
53. OECD (2021) Evaluation of tools and models for assessing occupational and consumer exposure to manufactured nanomaterials-part III: performance testing results of tools/models for consumer exposure. (Online). Available: www.oecd.org/chemicalsafety/
54. OECD (2021) Evaluation of tools and models for assessing occupational and consumer exposure to manufactured nanomaterials-part I: compilation of tools/models and analysis for further evaluation. (Online). Available: www.oecd.org/chemicalsafety/
55. OECD (2009) Preliminary review of OECD test guidelines for their applicability to manufactured nanomaterials
56. OECD (2012) Guidance on sample preparation and dosimetry for the safety testing of manufactured nanomaterials
57. OECD (2024) A tiered approach for reliable bioaccumulation assessment of manufactured nanomaterials in the environment whilst minimising the use of vertebrate testing
58. OECD (2018) Evaluation of in vitro methods for human hazard assessment applied in the OECD testing programme for the safety of manufactured nanomaterials—2
59. OECD (2018) Evaluation of in vitro methods for human hazard assessment applied in the OECD testing programme for the safety of manufactured nanomaterials—1
60. OECD (2021) Evaluation of tools and models for assessing occupational and consumer exposure to manufactured nanomaterials-part II: performance testing results of tools/models for occupational exposure. (Online). Available: www.oecd.org/chemicalsafety/
61. OECD (2023) Early awareness and action system for advanced materials (Early4AdMa): pre-regulatory and anticipatory risk governance tool to Advanced Materials. OECD series on the safety of manufactured nanomaterials No. 108. (Online). Available: https://www.oecd.org/chemicalsafety/safer-and-sustainable
62. OECD (2021) Evaluation of tools and models used for assessing environmental exposure to manufactured nanomaterials functional assessment and statistical analysis of nano-specific environmental exposure tools and models. (Online). Available: www.oecd.org/chemicalsafety/
63. OECD (2016) Categorisation of manufactured nanomaterials workshop report
64. OECD (2024) Advanced materials: case study on nanocarriers—workshop report
65. OECD (2019) Physical-chemical decision framework to inform decisions for risk assessment of manufactured nanomaterials
66. OECD (2019) Guiding principles for measurements and reporting for nanomaterials: physical chemical parameters
67. OECD (2016) Physical-chemical parameters: measurements and methods relevant for the regulation of nanomaterials
68. OECD (2009) Identification, compilation and analysis of guidance information for exposure measurement and exposure mitigation: manufactured nanomaterials
69. OECD (2016) Gold nanoparticle occupational exposure assessment in a pilot scale facility

70. OECD (2009) Emission assessment for identification of sources and release of airborne manufactured nanomaterials in the workplace: compilation of existing guidance
71. OECD (2022) Determination of the volume specific surface area of manufactured nanomaterials. (Online). Available: http://www.oecd.org/termsandconditions/
72. Ray PC, Yu H, Fu PP (2009) Toxicity and environmental risks of nanomaterials: challenges and future needs. https://doi.org/10.1080/10590500802708267
73. German Federal Institute for Risk Assesment. Health risk assessment of nanomaterials
74. NIOSH (2008) Safe nanotechnology in the workplace, an introduction for employers, managers, and safety and health professionals. (Online). Available: www.cdc.gov/niosh/topics/nanotech/
75. NIOSH (2013) Current strategies for engineering controls in nanomaterial production and downstream handling processes. (Online). Available: www.cdc.gov/niosh/eNews
76. NIOSH (2012) General safe practices for working with engineered nanomaterials in research laboratories
77. OECD (2010) Compilation of nanomaterial exposure mitigation guidelines relating to laboratories
78. Dunn KH, Topmiller JL, McCleery T, Whalen J (2018) Workplace design solutions: protecting workers during intermediate and downstream processing of nanomaterials. https://doi.org/10.26616/NIOSHPUB2018122
79. Dunn KH, Topmiller JL, McCleery T, Cincinnati WJ (2018) Protecting workers during the handling of nanomaterials. https://doi.org/10.26616/NIOSHPUB2018121
80. OECD (2016) Single walled carbon nanotubes (SWCNTs): summary of the dossier
81. OECD (2009) Comparison of guidance on selection of skin protective equipment and respirators for use in the workplace: manufactured nanomaterials
82. AIHA (2018) Personal protective equipment for engineered nanoparticles
83. Ostiguy C, Debia M, Roberge B, Dufresne A (2015) Best practices guidance for nanomaterial risk management in the workplace—report R-899. (Online). Available: https://abonnement.lacsst.com/prevention
84. Daniels R et al (2020) Current intelligence bulletin 69: NIOSH practices in occupational risk assessment. https://doi.org/10.26616/NIOSHPUB2020106re
85. NIOSH (2009) Current intelligence bulletin 60: interim guidance for medical screening and hazard surveillance for workers potentially exposed to engineered nanoparticles
86. Hoover MD, Stefaniak AB, Day GA (2007) Exposure assessment considerations for nanoparticles in the workplace. In: Monteiro-Riviere NA, Tran CL (eds) Nanotoxicology: characterization, dosing and health effects, vol 5. CRC Press, New York, pp 71–83. https://doi.org/10.3109/9781420045154-6
87. NIOSH (2022) Occupational exposure sampling for engineered nanomaterials. Technical report by Hodson L, Eastlake A. Cincinnati
88. Isigonis P et al (2020) Risk governance of emerging technologies demonstrated in terms of its applicability to nanomaterials. Small 16(36). https://doi.org/10.1002/smll.202003303
89. EPA. Control of nanoscale materials under the toxic substances control act
90. OECD (2022) Safe(r) and sustainable innovation approach (SSIA)
91. OECD (2020) Moving towards a safe(r) innovation approach (SIA) for more sustainable nanomaterials and nano-enabled products
92. OECD. How is your country addressing the risks associated with industrial and consumer chemicals? OECD

MIX
Papier aus verantwortungsvollen Quellen
Paper from responsible sources
FSC® C105338

If you have any concerns about our products,
you can contact us on
ProductSafety@springernature.com

In case Publisher is established outside the EU,
the EU authorized representative is:
Springer Nature Customer Service Center GmbH
Europaplatz 3, 69115 Heidelberg, Germany

Printed by Libri Plureos GmbH
in Hamburg, Germany